Birkhäuser

Birkhäuser Advanced Texts

Series Editors
Steven G. Krantz, Washington University, St. Louis, MO, USA
Shrawan Kumar, University of North Carolina at Chapel Hill, NC, USA
Jan Nekovář, Université Pierre et Marie Curie, Paris, France

For further volumes:
http://www.springer.com/series/4842

Donald L. Cohn

Measure Theory

Second Edition

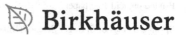 Birkhäuser

Donald L. Cohn
Department of Mathematics
    and Computer Science
Suffolk University
Boston, MA, USA

ISBN 978-1-4899-9762-3   ISBN 978-1-4614-6956-8 (eBook)
DOI 10.1007/978-1-4614-6956-8
Springer New York Heidelberg Dordrecht London

Mathematics Subject Classification (2010): 28-01, 60-01, 28C05, 28C15, 28A05, 26A39, 28A51

Printed on acid-free paper

Springer is part of Springer Science+Business Media (www.birkhauser-science.com)

To Linda, Henry, Edward, and Susan

Preface

In this new edition there are two types of changes: I have made improvements to the text of the first edition and have added some new topics.

In addition to making some corrections and reworking some arguments from the first edition, I have added an introduction before Chap. 1, in which I have said a bit about how the Lebesgue integral arose and indicated something about how the topics covered are related to one another. I hope that this will make it easier for the reader to see the structure of what he or she is studying. I have also improved the layout of the pages a bit, with the examples now easier to find.

There are a number of new topics. These main additions are the Henstock–Kurzweil integral, the Banach–Tarski paradox, and an introduction to measure-theoretic probability theory. These are, of course, supplementary to the main lines of the book, but they should give the reader a better feel for the relationship between measure theory and other parts of mathematics. As minor additions there are introductions to the Daniell integral and to the theory of liftings.

The mathematical level of the book and the background expected of the reader have not changed from the first edition.

There are several people and organizations that I would like to thank. Suffolk University's College of Arts and Sciences, together with its Department of Mathematics and Computer Science, made possible a sabbatical leave to work on this new edition. Richard Dudley and the Department of Mathematics at MIT provided office space and library access during that leave. Henry Cohn, Carl Offner, and Xinxin Jiang read and commented on parts of the manuscript. A number of people, some of whom I can no longer name, sent me useful comments on and corrections for the first edition. Ann Kostant, Tom Grasso, Kate Ghezzi, and Allen Mann, along with the production staff at Birkhäuser, were very helpful. My wife, Linda, typed parts of the manuscript, did a large amount of proofreading, and put up with my schedule as I worked on the book. I thank them all.

The Preface from the First Edition

This book is intended as a straightforward treatment of the parts of measure theory necessary for analysis and probability. The first five or six chapters form an introduction to measure and integration, while the last three chapters should provide the reader with some tools that are necessary for study and research in any of a number of directions. (For instance, one who has studied Chaps. 7 and 9 should be able to go on to interesting topics in harmonic analysis, without having to pause to learn a new theory of integration and to reconcile it with the one he or she already knows.) I hope that the last three chapters will also prove to be a useful reference.

Chapters 1 through 5 deal with abstract measure and integration theory and presuppose only the familiarity with the topology of Euclidean spaces that a student should acquire in an advanced calculus course. Lebesgue measure on $\mathbb{R}$ (and on $\mathbb{R}^d$) is constructed in Chap. 1 and is used as a basic example thereafter.

Chapter 6, on differentiation, begins with a treatment of changes of variables in $\mathbb{R}^d$ and then gives the basic results on the almost everywhere differentiation of functions on $\mathbb{R}$ (and measures on $\mathbb{R}^d$). The first section of this chapter makes use of the derivative (as a linear transformation) of a function from $\mathbb{R}^d$ to $\mathbb{R}^d$; the necessary definitions and facts are recalled, with appropriate references. The rest of the chapter has the same prerequisites as the earlier chapters.

Chapter 7 contains a rather thorough treatment of integration on locally compact Hausdorff spaces. I hope that the beginner can learn the basic facts from Sects. 7.2 and 7.3 without too much trouble. These sections, together with Sect. 7.4 and the first part of Sect. 7.6, cover almost everything the typical analyst needs to know about regular measures. The technical facts needed for dealing with very large locally compact Hausdorff spaces are included in Sects. 7.5 and 7.6.

In Chap. 8 I have tried to collect those parts of the theory of analytic sets that are of everyday use in analysis and probability. I hope it will serve both as an introduction and as a useful reference.

Chapter 9 is devoted to integration on locally compact groups. In addition to a construction and discussion of Haar measure, I have included a brief introduction to convolution on $L^1(G)$ and on the space of finite signed or complex regular Borel measures on G. The details are provided for arbitrary locally compact groups but in such a way that a reader who is interested only in second countable groups should find it easy to make the appropriate omissions.

Chapters 7 through 9 presuppose a little background in general topology. The necessary facts are reviewed, and so some facility with arguments involving topological spaces and metric spaces is actually all that is required. The reader who can work through Sects. 7.1 and 8.1 should have no trouble.

In addition to the main body of the text, there are five appendices. The first four explain the notation used and contain some elementary facts from set theory, calculus, and topology; they should remind the reader of a few things he or she may have forgotten and should thereby make the book quite self-contained. The fifth appendix contains an introduction to the Bochner integral.

Each section ends with some exercises. They are, for the most part, intended to give the reader practice with the concepts presented in the text. Some contain examples, additional results, or alternative proofs and should provide a bit of perspective. Only a few of the exercises are used later in the text itself; these few are provided with hints, as needed, that should make their solution routine.

I believe that no result in this book is new. Hence the lack of a bibliographic citation should never be taken as a claim of originality. The notes at the ends of chapters occasionally tell where a theorem or proof first appeared; most often, however, they point the reader to alternative presentations or to sources of further information.

The system used for cross-references within the book should be almost self-explanatory. For example, Proposition 1.3.5 and Exercise 1.3.7 are to be found in Sect. 1.3 of Chap. 1, while C.1 and Theorem C.8 are to be found in Appendix C.

There are a number of people to whom I am indebted and whom I would like to thank. First there are those from whom I learned integration theory, whether through courses, books, papers, or conversations; I won't try to name them, but I thank them all. I would like to thank R.M. Dudley and W.J. Buckingham, who read the original manuscript, and J.P. Hajj, who helped me with the proofreading. These three read the book with much care and thought and provided many useful suggestions. (I must, of course, accept responsibility for ignoring a few of their suggestions and for whatever mistakes remain.) Finally, I thank my wife, Linda, for typing and providing editorial advice on the manuscript, for helping with the proofreading, and especially for her encouragement and patience during the years it took to write this book.

Boston, MA, USA   Donald L. Cohn

Contents

Introduction

In this introduction we

- briefly review the Riemann integral as studied in calculus and elementary analysis,
- sketch how some difficulties with the Riemann integral led to the Lebesgue integral, and
- outline the main topics in this book and note how they relate to the Riemann and Lebesgue integrals.

The Riemann Integral—Darboux's Definition

Let $[a,b]$ be a closed bounded interval. A *partition* of $[a,b]$ is a finite sequence $\{a_i\}_{i=0}^{k}$ of real numbers such that

$$a = a_0 < a_1 < \cdots < a_k = b.$$

Sometimes we will call the values a_i the *division points* of the partition. We will generally denote a partition by a symbol such as $\mathscr{P}$.

Suppose that f is a bounded real-valued function on $[a,b]$ and that $\mathscr{P}$ is a partition of $[a,b]$, say with division points $\{a_i\}_{i=0}^{k}$. For $i = 1, \ldots, k$ define numbers m_i and M_i by $m_i = \inf\{f(x) : x \in [a_{i-1}, a_i]\}$ and $M_i = \sup\{f(x) : x \in [a_{i-1}, a_i]\}$. Then the *lower sum* $l(f, \mathscr{P})$ corresponding to f and $\mathscr{P}$ is defined to be $\sum_{i=1}^{k} m_i(a_i - a_{i-1})$. Similarly, the *upper sum* $u(f, \mathscr{P})$ corresponding to f and $\mathscr{P}$ is defined to be $\sum_{i=1}^{k} M_i(a_i - a_{i-1})$. See Fig. 1 below.

Since f is bounded, there are real numbers m and M such that $m \le f(x) \le M$ holds for each x in $[a,b]$. Then each lower sum of f satisfies

$$l(f, \mathscr{P}) = \sum_{i=1}^{k} m_i(a_i - a_{i-1}) \le \sum_{i=1}^{k} M(a_i - a_{i-1}) = M(b-a),$$

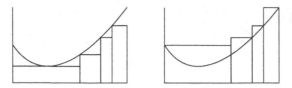

Fig. 1 A lower sum and an upper sum

and so the set of lower sums of f is bounded above, in fact by $M(b-a)$. It follows that the set of lower sums has a supremum (a least upper bound); this supremum is called the lower integral of f over $[a,b]$ and is denoted by $\underline{\int}_a^b f$. A similar argument shows that the set of upper sums of f is bounded below, and so one can define the upper integral of f, written $\overline{\int}_a^b f$, to be the infimum (the greatest lower bound) of the set of upper sums. It is not difficult to show (see Sect. 2.5 for details) that $\underline{\int}_a^b f \leq \overline{\int}_a^b f$. If $\underline{\int}_a^b f = \overline{\int}_a^b f$, then f is said to be *Riemann integrable* on $[a,b]$, and the common value of $\underline{\int}_a^b f$ and $\overline{\int}_a^b f$ is called the *Riemann integral* of f over $[a,b]$ and is denoted by $\int_a^b f$ or $\int_a^b f(x)\,dx$.

The Riemann Integral—Riemann's Definition

It is sometimes useful to view Riemann integrals as limits of what are called Riemann sums. For this we need a couple of definitions. A *tagged partition* of an interval $[a,b]$ is a partition $\{a_i\}_{i=0}^k$ of $[a,b]$, together with a sequence $\{x_i\}_{i=1}^k$ of numbers (called *tags*) such that $a_{i-1} \leq x_i \leq a_i$ holds for $i = 1, \ldots, k$. (In other words, each tag x_i must belong to the corresponding interval $[a_{i-1}, a_i]$.) As with partitions, we will often denote a tagged partition by a symbol such as $\mathscr{P}$.

The *mesh* $\|\mathscr{P}\|$ of a partition (or of a tagged partition) $\mathscr{P}$ is defined by $\|\mathscr{P}\| = \max_i(a_i - a_{i-1})$, where $\{a_i\}$ is the sequence of division points for $\mathscr{P}$. In other words, the mesh of a partition is the length of the longest of its subintervals.

The *Riemann sum* $\mathscr{R}(f, \mathscr{P})$ corresponding to the function f and the tagged partition $\mathscr{P}$ is defined by

$$\mathscr{R}(f, \mathscr{P}) = \sum_{i=1}^k f(x_i)(a_i - a_{i-1}).$$

Then, according to Riemann's definition, the function f is integrable over $[a,b]$ if there is a number L (which will be the value of the integral) such that

$$\lim_{\mathscr{P}} \mathscr{R}(f, \mathscr{P}) = L,$$

where the limit is taken as the mesh of $\mathscr{P}$ approaches 0. If we express this in terms of ε's and δ's, we see that the function f is Riemann integrable, with integral L, if for every positive ε there is a positive δ such that $|\mathscr{R}(f, \mathscr{P}) - L| < \varepsilon$ holds for each tagged partition $\mathscr{P}$ of $[a, b]$ that satisfies $\|\mathscr{P}\| < \delta$.

Darboux's and Riemann's definitions are equivalent:[1] they give exactly the same classes of integrable functions, with the same values for the integrals (see Proposition 2.5.7).

Another standard result is that every continuous function on $[a, b]$ is Riemann integrable; see Example 2.5.2 (or, for a somewhat stronger result, see Theorem 2.5.4).

The final thing to recall is the fundamental theorem of calculus (see Exercise 2.5.6 for a sketch of its proof):

Theorem 1 (The Fundamental Theorem of Calculus). *Suppose that $f: [a, b] \to \mathbb{R}$ is continuous and that $F: [a, b] \to \mathbb{R}$ is defined by $F(x) = \int_a^x f(t)\,dt$. Then F is differentiable at each x in $[a, b]$ and its derivative is given by $F'(x) = f(x)$.*

From Riemann to Lebesgue

In many situations involving integrals (for example, when integrating an infinite series term by term or when differentiating under the integral sign), it is necessary to be able to reverse the order of taking limits and evaluating integrals—that is, to be able to say things like

$$\int_a^b \lim_n f_n(x)\,dx = \lim_n \int_a^b f_n(x)\,dx.$$

Thus one needs to have theorems of the following sort:

Theorem 2. *Suppose that $\{f_n\}$ is a sequence of integrable functions on the interval $[a, b]$ and that f is a function such that $\{f_n\}$ converges to f in a suitable[2] way. Then f is integrable and*

$$\int_a^b f(x)\,dx = \lim_n \int_a^b f_n(x)\,dx.$$

In elementary analysis courses one sees that Theorem 2 is valid for the Riemann integral if by "converges to f in a suitable way," we mean "converges uniformly to f" (see Exercise 2.5.7). On the other hand, if we do not assume uniform

---

[1]The reader may well be asking why people consider two definitions of the Riemann integral. The general answer is that Darboux's definition is simpler and more elegant, while Riemann's is useful for various calculations of limits (see, for example, Exercise 2.5.8). For our purposes, the Darboux approach makes our discussion of the relationship between the Riemann and Lebesgue integrals simpler, while the Riemann approach is more closely related to the Henstock–Kurzweil and McShane integrals (see Appendix H).

[2]The problem is, of course, to figure out what "suitable" might mean and to define the integral in such a way that theorems like this one will be applicable in many situations.

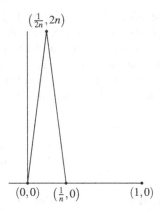

Fig. 2 Function defined in Example 3

convergence of $\{f_n\}$ to f, but only pointwise[3] convergence, then, as we see in the following examples, Theorem 2 may fail.

Example 3. For each positive integer n let f_n be the piecewise linear function on $[0,1]$ whose graph is made up of three line segments, connecting the points $(0,0)$, $(\frac{1}{2n},2n)$, $(\frac{1}{n},0)$, and $(1,0)$. See Fig. 2. Then for each n the triangle formed by the graph of f_n and the x-axis has area 1, and so f_n satisfies $\int_0^1 f_n(x)\,dx = 1$. Furthermore, for each x in $[0,1]$ we have $\lim_n f_n(x) = 0$. Thus $\lim_n \int_0^1 f_n(x)\,dx = 1$ but $\int_0^1 \lim_n f_n(x)\,dx = 0$, and the conclusion of Theorem 2 fails for the sequence $\{f_n\}$.   □

The failure of the conclusion of Theorem 2 in the preceding example comes from the fact that the sequence $\{f_n\}$ is not uniformly bounded—that is, from the fact that there is no constant M such that $|f_n(x)| \le M$ holds for all n and x. Next let us look at an example in which the functions f_n are uniformly bounded, in fact, in which we have $0 \le f_n(x) \le 1$ for all n and all x, and yet the conclusion to Theorem 2 fails.

Example 4. Recall that the set of rational numbers is countable (see A.6). Hence we can choose an enumeration $\{x_n\}$ of the rational numbers in the interval $[0,1]$ (that is, a sequence whose members are the rational numbers in $[0,1]$, with each rational in that interval occurring exactly once in the sequence). For each n define a function $f_n : [0,1] \to \mathbb{R}$ by

$$f_n(x) = \begin{cases} 1 & \text{if } x \in \{x_1, x_2, \ldots, x_n\}, \text{ and} \\ 0 & \text{otherwise.} \end{cases}$$

---

[3]Recall that $\{f_n\}$ converges pointwise to f on $[a,b]$ if $\lim_n f_n(x) = f(x)$ for each x in $[a,b]$.

Thus $f_n(x)$ has value 1 for n values of x (namely for $x_1, \ldots, x_n$) and has value 0 otherwise. It is easy to check that for each n, all the lower sums of f_n are 0 and hence that the lower integral $\underline{\int}_0^1 f_n$ is 0. On the other hand, it is not hard to construct, for each n and each positive δ, a partition $\mathscr{P}$ of $[0, 1]$ in which each of $x_1, x_2, \ldots, x_n$ is in the interior of some subinterval that belongs to $\mathscr{P}$ and has length at most δ/n. It follows that $u(f_n, \mathscr{P}) \leq \delta$. Since this can be done for each positive δ, it follows that the upper integral $\overline{\int}_0^1 f_n$ is also 0. Consequently f_n is Riemann integrable over $[0, 1]$ and $\int_0^1 f_n(x)\,dx = 0$.

For each x let us consider the behavior of the sequence $\{f_n(x)\}$. If x is rational, then $f_n(x) = 1$ for all large n, while if x is irrational, then $f_n(x) = 0$ for all n. Thus $\{f_n(x)\}$ converges pointwise to the function $f\colon [0, 1] \to \mathbb{R}$ defined by

$$f(x) = \begin{cases} 1 & \text{if } x \text{ is rational and belongs to } [0, 1], \text{ and} \\ 0 & \text{if } x \text{ is irrational and belongs to } [0, 1]. \end{cases}$$

Since the rationals are dense in $[0, 1]$, as are the irrationals, it follows that every lower sum for f has value 0 and every upper sum for f has value 1. Thus the lower and upper integrals of f are given by $\underline{\int}_0^1 f = 0$ and $\overline{\int}_0^1 f = 1$, and f is not Riemann integrable. Thus the conclusion of Theorem 2 fails for this example.   □

Example 5. It may seem that the difficulty in the previous example comes from the fact that the functions f_n fail to be continuous. However, one can also produce a sequence $\{f_n\}$ such that

(a) each f_n is continuous,
(b) $0 \leq f_n(x) \leq 1$ holds for each n and each x, and
(c) $\{f_n\}$ converges pointwise to a function that is not Riemann integrable.

(See Exercise 2.5.4.)   □

The questions involved in making Theorem 2 precise were important unresolved issues in the late nineteenth century; they arose, for example, in the study of Fourier series.

In the early twentieth century, Lebesgue defined a new integral, which he used to give very useful answers to questions of the sort discussed above. For example, Lebesgue showed that Theorem 2, when formulated in terms of his new integral, holds for pointwise convergence of the sequence $\{f_n\}$, subject only to some rather natural boundedness conditions on that sequence (see the dominated convergence theorem, Theorem 2.4.5). It is hard to overemphasize the simplicity and ease of application of the limit theorems for the Lebesgue integral.

Let us briefly sketch how the Lebesgue integral is defined. For simplicity, we will for now restrict our attention to functions $f\colon [a, b] \to \mathbb{R}$ that are nonnegative and bounded (those assumptions are in no way necessary). So let c be a positive number such that $0 \leq f(x) < c$ holds for each x in $[a, b]$. As we have seen, the definition of the Riemann integral deals with partitions of the interval $[a, b]$, that is, of the domain of f. One way of defining the Lebesgue integral deals with partitions of the range of f, rather than of the domain. So suppose that $\mathscr{P}$ is a partition of $[0,c]$, say given by a sequence of $\{a_i\}_{i=0}^{k}$ of dividing points. For $i = 1, \ldots, k$ define A_i by

$$A_i = \{x \in [a,b] : f(x) \in [a_{i-1}, a_i)\}. \tag{1}$$

(Note that the sets A_i are not necessarily subintervals of $[a,b]$—they can also be empty, unions of finite collections of subintervals, or even more complicated sets.) Let us consider the sum $s(f, \mathscr{P})$ given by

$$s(f, \mathscr{P}) = \sum_{i=1}^{k} a_{i-1} \text{meas}(A_i), \tag{2}$$

where $\text{meas}(A_i)$ is the size, in a sense still to be defined, of the set A_i. Subject to the condition that the function f must be simple enough that $\text{meas}(A_i)$ makes sense for all sets A_i as defined by (1), the Lebesgue integral of f is defined to be the supremum of the set of all sums of the form (2), where these sums are considered for all partitions $\mathscr{P}$ of the interval $[0,c]$. (One can check that this does not depend on the value of c, as long as it is large enough that $f(x) < c$ holds for all x.)

Now let us survey some of the contents of this book.

The first issue that needs resolving is the meaning of the expression $\text{meas}(A_i)$ that occurs in Eq. (2). That is the goal of Chap. 1, which begins with the question of how to describe and organize the subsets of $\mathbb{R}$ whose size can reasonably be measured (that is, the *measurable* sets) and then continues with the question of how to measure the sizes of those subsets (the study of Lebesgue measure and of more general measures). Since it is useful to consider integration not just for functions defined on $\mathbb{R}$ or on subintervals of $\mathbb{R}$ but also in more general settings, including $\mathbb{R}^d$, some of the discussion in Chap. 1 is rather abstract. This abstractness does not add much to the level of difficulty of the chapter.

Appendix G is in some sense a continuation of Chap. 1. It gives an exposition of the Banach–Tarski paradox, which is a very famous result that quite vividly shows that Lebesgue measure on $\mathbb{R}^3$ cannot be extended in any reasonable way to all the subsets of $\mathbb{R}^3$. (Appendix G is deeper than Chap. 1 and requires more background on the reader's part.)

The main objective of Chap. 2 is the definition of the Lebesgue integral. Section 2.1 deals with measurable functions, those functions that are tame enough that the sets A_i in Eq. (2) are measurable. Section 2.2 introduces properties that hold *almost everywhere* and in particular considers convergence almost everywhere, which can often be used in place of pointwise convergence. The integral is finally defined in Sect. 2.3, and the basic limit theorems for the integral are proved in Sect. 2.4.

Chapter 3 deals more deeply with limits and convergence in integration theory, while Chap. 4 deals with measures that have signed or complex values and with relationships between measures.

In multivariable calculus courses one learns how to calculate integrals over subsets of $\mathbb{R}^d$ by repeatedly calculating one-dimensional integrals. Chapter 5 deals with such matters for the Lebesgue integral. Section 6.1 deals with another aspect of integration on $\mathbb{R}^d$, namely with change of variable in integrals over subsets of $\mathbb{R}^d$.

The fundamental theorem of calculus (Theorem 1 above) relates Riemann integrals to derivatives. Such relationships for the Lebesgue integral are discussed in the last two sections of Chap. 6.

In the discussion above of Chap. 1 we noted that our treatment of measures and measurable sets is fairly general. This generality is useful for a number of applications, such as to cases where integration on locally compact topological spaces is needed (see Chaps. 7 and 9) and to the study of probability theory (see Chap. 10 for a brief introduction to the application of measure theory to probability theory).

Many deeper questions about measurable sets and functions arise naturally. Some useful and classical results along these lines are given in Chap. 8.

Let us return for a moment to the second of our definitions of the Riemann integral, the one expressed in terms of limits of Riemann sums. In the second half of the twentieth century Henstock and Kurzweil gave what may seem to be a small modification of this definition. The resulting integral is known as the Henstock–Kurzweil integral or the generalized Riemann integral. Although their definition seems very simple, their integral (for functions on $\mathbb{R}$) turns out to be more general than the Lebesgue integral and to have what is in some ways a more natural relationship to derivatives. See Appendix H for an introduction to the Henstock–Kurzweil integral.

Chapter 1
Measures

Suppose that X is a set and $f: X \to \mathbb{R}$ is a function that we want to integrate. As we noted in the introduction, we need to deal with the sizes of subsets of X in order to define the integral of f. In this chapter we introduce measures, the basic tool for dealing with such sizes. The first two sections of the chapter are abstract (but elementary). Section 1.1 looks at σ-algebras, the collections of sets whose sizes we measure, while Sect. 1.2 introduces measures themselves. The heart of the chapter is in the following two sections, where we look at some general techniques for constructing measures (Sect. 1.3) and at the basic properties of Lebesgue measure (Sect. 1.4). The chapter ends with Sects. 1.5 and 1.6, which introduce some additional fundamental techniques for handling measures and σ-algebras.

1.1 Algebras and Sigma-Algebras

Let X be an arbitrary set. A collection $\mathscr{A}$ of subsets of X is an *algebra* on X if

(a) $X \in \mathscr{A}$,
(b) for each set A that belongs to $\mathscr{A}$, the set A^c belongs to $\mathscr{A}$,
(c) for each finite sequence $A_1, \ldots, A_n$ of sets that belong to $\mathscr{A}$, the set $\cup_{i=1}^{n} A_i$ belongs to $\mathscr{A}$, and
(d) for each finite sequence $A_1, \ldots, A_n$ of sets that belong to $\mathscr{A}$, the set $\cap_{i=1}^{n} A_i$ belongs to $\mathscr{A}$.

Of course, in conditions (b), (c), and (d), we have required that $\mathscr{A}$ be *closed* under complementation, under the formation of finite unions, and under the formation of finite intersections. It is easy to check that closure under complementation and closure under the formation of finite unions together imply closure under the

D.L. Cohn, *Measure Theory: Second Edition*, Birkhäuser Advanced Texts Basler Lehrbücher, DOI 10.1007/978-1-4614-6956-8_1,
© Springer Science+Business Media, LLC 2013

formation of finite intersections (use that fact that $\cap_{i=1}^n A_i = (\cup_{i=1}^n A_i^c)^c$). Thus we could have defined an algebra using only conditions (a), (b), and (c). A similar argument shows that we could have used only conditions (a), (b), and (d).

Again let X be an arbitrary set. A collection $\mathscr{A}$ of subsets of X is a σ-*algebra*[1] on X if

(a) $X \in \mathscr{A}$,
(b) for each set A that belongs to $\mathscr{A}$, the set A^c belongs to $\mathscr{A}$,
(c) for each infinite sequence $\{A_i\}$ of sets that belong to $\mathscr{A}$, the set $\cup_{i=1}^\infty A_i$ belongs to $\mathscr{A}$, and
(d) for each infinite sequence $\{A_i\}$ of sets that belong to $\mathscr{A}$, the set $\cap_{i=1}^\infty A_i$ belongs to $\mathscr{A}$.

Thus a σ-algebra on X is a family of subsets of X that contains X and is closed under complementation, under the formation of countable unions, and under the formation of countable intersections. Note that, as in the case of algebras, we could have used only conditions (a), (b), and (c), or only conditions (a), (b), and (d), in our definition.

Each σ-algebra on X is an algebra on X since, for example, the union of the finite sequence $A_1, A_2, \ldots, A_n$ is the same as the union of the infinite sequence $A_1, A_2, \ldots,$ $A_n, A_n, A_n, \ldots$.

If X is a set and $\mathscr{A}$ is a family of subsets of X that is closed under complementation, then X belongs to $\mathscr{A}$ if and only if $\varnothing$ belongs to $\mathscr{A}$. Thus in the definitions of algebras and σ-algebras given above, we can replace condition (a) with the requirement that $\varnothing$ be a member of $\mathscr{A}$. Furthermore, if $\mathscr{A}$ is a family of subsets of X that is nonempty, closed under complementation, and closed under the formation of finite or countable unions, then $\mathscr{A}$ must contain X: if the set A belongs to $\mathscr{A}$, then X, since it is the union of A and A^c, must also belong to $\mathscr{A}$. Thus in our definitions of algebras and σ-algebras, we can replace condition (a) with the requirement that $\mathscr{A}$ be nonempty.

If $\mathscr{A}$ is a σ-algebra on the set X, it is sometimes convenient to call a subset of X $\mathscr{A}$-*measurable* if it belongs to $\mathscr{A}$.

Examples 1.1.1 (Some Families of Sets That Are Algebras or σ-algebras, and Some That Are Not).

(a) Let X be a set, and let $\mathscr{A}$ be the collection of all subsets of X. Then $\mathscr{A}$ is a σ-algebra on X.
(b) Let X be a set, and let $\mathscr{A} = \{\varnothing, X\}$. Then $\mathscr{A}$ is a σ-algebra on X.
(c) Let X be an infinite set, and let $\mathscr{A}$ be the collection of all finite subsets of X. Then $\mathscr{A}$ does not contain X and is not closed under complementation; hence it is not an algebra (or a σ-algebra) on X.

---

[1] The terms *field* and *σ-field* are sometimes used in place of algebra and σ-algebra.

(d) Let X be an infinite set, and let $\mathscr{A}$ be the collection of all subsets A of X such that either A or A^c is finite. Then $\mathscr{A}$ is an algebra on X (check this) but is not closed under the formation of countable unions; hence it is not a σ-algebra.

(e) Let X be an uncountable set, and let $\mathscr{A}$ be the collection of all countable (i.e., finite or countably infinite) subsets of X. Then $\mathscr{A}$ does not contain X and is not closed under complementation; hence it is not an algebra.

(f) Let X be a set, and let $\mathscr{A}$ be the collection of all subsets A of X such that either A or A^c is countable. Then $\mathscr{A}$ is a σ-algebra.

(g) Let $\mathscr{A}$ be the collection of all subsets of $\mathbb{R}$ that are unions of finitely many intervals of the form $(a,b]$, $(a,+\infty)$, or $(-\infty,b]$. It is easy to check that each set that belongs to $\mathscr{A}$ is the union of a finite disjoint collection of intervals of the types listed above, and then to check that $\mathscr{A}$ is an algebra on $\mathbb{R}$ (the empty set belongs to $\mathscr{A}$, since it is the union of the empty, and hence finite, collection of intervals). The algebra $\mathscr{A}$ is not a σ-algebra; for example, the bounded open subintervals of $\mathbb{R}$ are unions of sequences of sets in $\mathscr{A}$ but do not themselves belong to $\mathscr{A}$. $\qquad\square$

Next we consider ways of constructing σ-algebras.

Proposition 1.1.2. *Let X be a set. Then the intersection of an arbitrary nonempty collection of σ-algebras on X is a σ-algebra on X.*

Proof. Let $\mathscr{C}$ be a nonempty collection of σ-algebras on X, and let $\mathscr{A}$ be the intersection of the σ-algebras that belong to $\mathscr{C}$. It is enough to check that $\mathscr{A}$ contains X, is closed under complementation, and is closed under the formation of countable unions. The set X belongs to $\mathscr{A}$, since it belongs to each σ-algebra that belongs to $\mathscr{C}$. Now suppose that $A \in \mathscr{A}$. Each σ-algebra that belongs to $\mathscr{C}$ contains A and so contains A^c; thus A^c belongs to the intersection $\mathscr{A}$ of these σ-algebras. Finally, suppose that $\{A_i\}$ is a sequence of sets that belong to $\mathscr{A}$ and hence to each σ-algebra in $\mathscr{C}$. Then $\cup_i A_i$ belongs to each σ-algebra in $\mathscr{C}$ and so to $\mathscr{A}$. $\qquad\square$

The reader should note that the union of a family of σ-algebras can fail to be a σ-algebra (see Exercise 5).

Proposition 1.1.2 implies the following result, which is a basic tool for the construction of σ-algebras.

Corollary 1.1.3. *Let X be a set, and let $\mathscr{F}$ be a family of subsets of X. Then there is a smallest σ-algebra on X that includes $\mathscr{F}$.*

Of course, to say that $\mathscr{A}$ is the smallest σ-algebra on X that includes $\mathscr{F}$ is to say that $\mathscr{A}$ is a σ-algebra on X that includes $\mathscr{F}$ and that every σ-algebra on X that includes $\mathscr{F}$ also includes $\mathscr{A}$. If $\mathscr{A}_1$ and $\mathscr{A}_2$ are both smallest σ-algebras that include $\mathscr{F}$, then $\mathscr{A}_1 \subseteq \mathscr{A}_2$ and $\mathscr{A}_2 \subseteq \mathscr{A}_1$, and so $\mathscr{A}_1 = \mathscr{A}_2$; thus the smallest σ-algebra on X that includes $\mathscr{F}$ is unique. The smallest σ-algebra is called the σ-algebra *generated* by $\mathscr{F}$ and is often denoted by $\sigma(\mathscr{F})$.

Proof. Let $\mathscr{C}$ be the collection of all σ-algebras on X that include $\mathscr{F}$. Then $\mathscr{C}$ is nonempty, since it contains the σ-algebra that consists of all subsets of

X. The intersection of the σ-algebras that belong to $\mathscr{C}$ is, according to Proposition 1.1.2, a σ-algebra; it includes $\mathscr{F}$ and is included in every σ-algebra in $\mathscr{C}$—that is, it is included in every σ-algebra on X that includes $\mathscr{F}$. $\square$

We now use the preceding corollary to define an important family of σ-algebras. The *Borel σ-algebra* on $\mathbb{R}^d$ is the σ-algebra on $\mathbb{R}^d$ generated by the collection of open subsets of $\mathbb{R}^d$; it is denoted by $\mathscr{B}(\mathbb{R}^d)$. The *Borel subsets* of $\mathbb{R}^d$ are those that belong to $\mathscr{B}(\mathbb{R}^d)$. In case $d = 1$, one generally writes $\mathscr{B}(\mathbb{R})$ in place of $\mathscr{B}(\mathbb{R}^1)$.

Proposition 1.1.4. *The σ-algebra $\mathscr{B}(\mathbb{R})$ of Borel subsets of $\mathbb{R}$ is generated by each of the following collections of sets:*

(a) *the collection of all closed subsets of $\mathbb{R}$;*
(b) *the collection of all subintervals of $\mathbb{R}$ of the form $(-\infty, b]$;*
(c) *the collection of all subintervals of $\mathbb{R}$ of the form $(a, b]$.*

Proof. Let $\mathscr{B}_1$, $\mathscr{B}_2$, and $\mathscr{B}_3$ be the σ-algebras generated by the collections of sets in parts (a), (b), and (c) of the proposition. We will show that $\mathscr{B}(\mathbb{R}) \supseteq \mathscr{B}_1 \supseteq \mathscr{B}_2 \supseteq \mathscr{B}_3$ and then that $\mathscr{B}_3 \supseteq \mathscr{B}(\mathbb{R})$; this will establish the proposition. Since $\mathscr{B}(\mathbb{R})$ includes the family of open subsets of $\mathbb{R}$ and is closed under complementation, it includes the family of closed subsets of $\mathbb{R}$; thus it includes the σ-algebra generated by the closed subsets of $\mathbb{R}$, namely $\mathscr{B}_1$. The sets of the form $(-\infty, b]$ are closed and so belong to $\mathscr{B}_1$; consequently $\mathscr{B}_1 \supseteq \mathscr{B}_2$. Since $(a, b] = (-\infty, b] \cap (-\infty, a]^c$, each set of the form $(a, b]$ belongs to $\mathscr{B}_2$; thus $\mathscr{B}_2 \supseteq \mathscr{B}_3$. Finally, note that each open subinterval of $\mathbb{R}$ is the union of a sequence of sets of the form $(a, b]$ and that each open subset of $\mathbb{R}$ is the union of a sequence of open intervals (see Proposition C.4). Thus each open subset of $\mathbb{R}$ belongs to $\mathscr{B}_3$, and so $\mathscr{B}_3 \supseteq \mathscr{B}(\mathbb{R})$. $\square$

As we proceed, the reader should note the following properties of the σ-algebra $\mathscr{B}(\mathbb{R})$:

(a) It contains virtually[2] every subset of $\mathbb{R}$ that is of interest in analysis.
(b) It is small enough that it can be dealt with in a fairly constructive manner.

It is largely these properties that explain the importance of $\mathscr{B}(\mathbb{R})$.

Proposition 1.1.5. *The σ-algebra $\mathscr{B}(\mathbb{R}^d)$ of Borel subsets of $\mathbb{R}^d$ is generated by each of the following collections of sets:*

(a) *the collection of all closed subsets of $\mathbb{R}^d$;*
(b) *the collection of all closed half-spaces in $\mathbb{R}^d$ that have the form $\{(x_1, \ldots, x_d) : x_i \leq b\}$ for some index i and some b in $\mathbb{R}$;*
(c) *the collection of all rectangles in $\mathbb{R}^d$ that have the form*

$$\{(x_1, \ldots, x_d) : a_i < x_i \leq b_i \text{ for } i = 1, \ldots, d\}.$$

---

[2]See Chap. 8 for some interesting and useful sets that are not Borel sets.

Proof. This proposition can be proved with essentially the argument that was used for Proposition 1.1.4, and so most of the proof is omitted. To see that the σ-algebra generated by the rectangles of part (c) is included in the σ-algebra generated by the half-spaces of part (b), note that each strip that has the form

$$\{(x_1, \ldots, x_d) : a < x_i \leq b\}$$

for some i is the difference of two of the half-spaces in part (b) and that each of the rectangles in part (c) is the intersection of d such strips. $\qquad \Box$

Let us look in more detail at some of the sets in $\mathscr{B}(\mathbb{R}^d)$. Let $\mathscr{G}$ be the family of all open subsets of $\mathbb{R}^d$, and let $\mathscr{F}$ be the family of all closed subsets of $\mathbb{R}^d$. (Of course $\mathscr{G}$ and $\mathscr{F}$ depend on the dimension d, and it would have been more precise to write $\mathscr{G}(\mathbb{R}^d)$ and $\mathscr{F}(\mathbb{R}^d)$.) Let $\mathscr{G}_\delta$ be the collection of all intersections of sequences of sets in $\mathscr{G}$, and let $\mathscr{F}_\sigma$ be the collection of all unions of sequences of sets in $\mathscr{F}$. Sets in $\mathscr{G}_\delta$ are often called G_δ's, and sets in $\mathscr{F}_\sigma$ are often called F_σ's. The letters G and F presumably stand for the German word *Gebiet* and the French word *fermé*, and the letters σ and δ for the German words *Summe* and *Durchschnitt*.

Proposition 1.1.6. *Each closed subset of $\mathbb{R}^d$ is a G_δ, and each open subset of $\mathbb{R}^d$ is an F_σ.*

Proof. Suppose that F is a closed subset of $\mathbb{R}^d$. We need to construct a sequence $\{U_n\}$ of open subsets of $\mathbb{R}^d$ such that $F = \cap_n U_n$. For this define U_n by

$$U_n = \{x \in \mathbb{R}^d : \|x - y\| < 1/n \text{ for some } y \text{ in } F\}.$$

(Note that U_n is empty if F is empty.) It is clear that each U_n is open and that $F \subseteq \cap_n U_n$. The reverse inclusion follows from the fact that F is closed (note that each point in $\cap_n U_n$ is the limit of a sequence of points in F). Hence each closed subset of $\mathbb{R}^d$ is a G_δ.

If U is open, then U^c is closed and so is a G_δ. Thus there is a sequence $\{U_n\}$ of open sets such that $U^c = \cap_n U_n$. The sets U_n^c are then closed, and $U = \cup_n U_n^c$; hence U is an F_σ. $\qquad \Box$

For an arbitrary family $\mathscr{S}$ of sets, let $\mathscr{S}_\sigma$ be the collection of all unions of sequences of sets in $\mathscr{S}$, and let $\mathscr{S}_\delta$ be the collection of all intersections of sequences of sets in $\mathscr{S}$. We can iterate the operations represented by σ and δ, obtaining from the class $\mathscr{G}$ the classes $\mathscr{G}_\delta$, $\mathscr{G}_{\delta\sigma}$, $\mathscr{G}_{\delta\sigma\delta}$, ..., and from the class $\mathscr{F}$ the classes $\mathscr{F}_\sigma$, $\mathscr{F}_{\sigma\delta}$, $\mathscr{F}_{\sigma\delta\sigma}$, ... (Note that $\mathscr{G} = \mathscr{G}_\sigma$ and $\mathscr{F} = \mathscr{F}_\delta$. Note also that $\mathscr{G}_{\delta\delta} = \mathscr{G}_\delta$, that $\mathscr{F}_{\sigma\sigma} = \mathscr{F}_\sigma$, and so on.) It now follows (see Proposition 1.1.6) that all the inclusions in Fig. 1.1 below are valid.

It turns out that no two of these classes of sets are equal and that there are Borel sets that belong to none of them (see Exercises 7 and 9 in Sect. 8.2).

A sequence $\{A_i\}$ of sets is called *increasing* if $A_i \subseteq A_{i+1}$ holds for each i and *decreasing* if $A_i \supseteq A_{i+1}$ holds for each i.

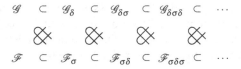

$$\mathscr{G} \quad \subset \quad \mathscr{G}_\delta \quad \subset \quad \mathscr{G}_{\delta\sigma} \quad \subset \quad \mathscr{G}_{\delta\sigma\delta} \quad \subset \quad \cdots$$

$$\mathscr{F} \quad \subset \quad \mathscr{F}_\sigma \quad \subset \quad \mathscr{F}_{\sigma\delta} \quad \subset \quad \mathscr{F}_{\sigma\delta\sigma} \quad \subset \quad \cdots$$

Fig. 1.1

Proposition 1.1.7. *Let X be a set, and let $\mathscr{A}$ be an algebra on X. Then $\mathscr{A}$ is a σ-algebra if either*

(a) *$\mathscr{A}$ is closed under the formation of unions of increasing sequences of sets, or*
(b) *$\mathscr{A}$ is closed under the formation of intersections of decreasing sequences of sets.*

Proof. First suppose that condition (a) holds. Since $\mathscr{A}$ is an algebra, we can check that it is a σ-algebra by verifying that it is closed under the formation of countable unions. Suppose that $\{A_i\}$ is a sequence of sets that belong to $\mathscr{A}$. For each n let $B_n = \cup_{i=1}^{n} A_i$. The sequence $\{B_n\}$ is increasing, and, since $\mathscr{A}$ is an algebra, each B_n belongs to $\mathscr{A}$; thus assumption (a) implies that $\cup_n B_n$ belongs to $\mathscr{A}$. However, $\cup_i A_i$ is equal to $\cup_n B_n$ and so belongs to $\mathscr{A}$. Thus $\mathscr{A}$ is closed under the formation of countable unions and so is a σ-algebra.

Now suppose that condition (b) holds. It is enough to check that condition (a) holds. If $\{A_i\}$ is an increasing sequence of sets that belong to $\mathscr{A}$, then $\{A_i^c\}$ is a decreasing sequence of sets that belong to $\mathscr{A}$, and so condition (b) implies that $\cap_i A_i^c$ belongs to $\mathscr{A}$. Since $\cup_i A_i = (\cap_i A_i^c)^c$, it follows that $\cup_i A_i$ belongs to $\mathscr{A}$. Thus condition (a) follows from condition (b), and the proof is complete. $\qquad\square$

Exercises

1. Find the σ-algebra on $\mathbb{R}$ that is generated by the collection of all one-point subsets of $\mathbb{R}$.
2. Show that $\mathscr{B}(\mathbb{R})$ is generated by the collection of intervals $(-\infty, b]$ for which the endpoint b is a rational number.
3. Show that $\mathscr{B}(\mathbb{R})$ is generated by the collection of all compact subsets of $\mathbb{R}$.
4. Show that if $\mathscr{A}$ is an algebra of sets, and if $\cup_n A_n$ belongs to $\mathscr{A}$ whenever $\{A_n\}$ is a sequence of disjoint sets in $\mathscr{A}$, then $\mathscr{A}$ is a σ-algebra.
5. Show by example that the union of a collection of σ-algebras on a set X can fail to be a σ-algebra on X. (Hint: There are examples in which X is a small finite set.)
6. Find an infinite collection of subsets of $\mathbb{R}$ that contains $\mathbb{R}$, is closed under the formation of countable unions, and is closed under the formation of countable intersections, but is not a σ-algebra.

7. Let $\mathscr{S}$ be a collection of subsets of the set X. Show that for each A in $\sigma(\mathscr{S})$, there is a countable subfamily $\mathscr{C}_0$ of $\mathscr{S}$ such that $A \in \sigma(\mathscr{C}_0)$. (Hint: Let $\mathscr{A}$ be the union of the σ-algebras $\sigma(\mathscr{C})$, where $\mathscr{C}$ ranges over the countable subfamilies of $\mathscr{S}$, and show that $\mathscr{A}$ is a σ-algebra that satisfies $\mathscr{S} \subseteq \mathscr{A} \subseteq \sigma(\mathscr{S})$ and hence is equal to $\sigma(\mathscr{S})$.)
8. Find all σ-algebras on $\mathbb{N}$.
9. (a) Show that $\mathbb{Q}$ is an F_σ, but not a G_δ, in $\mathbb{R}$. (Hint: Use the Baire category theorem, Theorem D.37.)
   (b) Find a subset of $\mathbb{R}$ that is neither an F_σ nor a G_δ.

1.2 Measures

Let X be a set, and let $\mathscr{A}$ be a σ-algebra on X. A function μ whose domain is the σ-algebra $\mathscr{A}$ and whose values belong to the extended half-line $[0, +\infty]$ is said to be *countably additive* if it satisfies

$$\mu(\cup_{i=1}^\infty A_i) = \sum_{i=1}^\infty \mu(A_i)$$

for each infinite sequence $\{A_i\}$ of disjoint sets that belong to $\mathscr{A}$. (Since $\mu(A_i)$ is nonnegative for each i, the sum $\sum_{i=1}^\infty \mu(A_i)$ always exists, either as a real number or as $+\infty$; see Appendix B.) A *measure* (or a *countably additive measure*) on $\mathscr{A}$ is a function $\mu : \mathscr{A} \to [0, +\infty]$ that satisfies $\mu(\varnothing) = 0$ and is countably additive.

We should note a related concept which is sometimes of interest. Let $\mathscr{A}$ be an algebra (not necessarily a σ-algebra) on the set X. A function μ whose domain is $\mathscr{A}$ and whose values belong to $[0, +\infty]$ is *finitely additive* if it satisfies

$$\mu(\cup_{i=1}^n A_i) = \sum_{i=1}^n \mu(A_i)$$

for each finite sequence $A_1, \ldots, A_n$ of disjoint sets that belong to $\mathscr{A}$. A *finitely additive measure* on the algebra $\mathscr{A}$ is a function $\mu : \mathscr{A} \to [0, +\infty]$ that satisfies $\mu(\varnothing) = 0$ and is finitely additive.

It is easy to check that every countably additive measure is finitely additive: simply extend the finite sequence $A_1, \ldots, A_n$ to an infinite sequence $\{A_i\}$ by letting $A_i = \varnothing$ if $i > n$, and then use the fact that $\mu(\varnothing) = 0$. There are, however, finitely additive measures that are not countably additive (see Example 1.2.1(d) and Exercise 8 in Sect. 3.5).

Finite additivity might at first seem to be a more natural property than countable additivity. However, countably additive measures on the one hand seem to be sufficient for almost all applications and, on the other hand, support a much more powerful theory of integration than do finitely additive measures. Thus we will follow the usual practice and devote almost all of our attention to countably additive measures.

We should emphasize that in this book the word "measure" (without modifiers) will always denote a countably additive measure. The expression "finitely additive measure" will always be written out in full.

If X is a set, if $\mathscr{A}$ is a σ-algebra on X, and if μ is a measure on $\mathscr{A}$, then the triplet $(X, \mathscr{A}, \mu)$ is often called a *measure space*. Likewise, if X is a set and if $\mathscr{A}$ is a σ-algebra on X, then the pair $(X, \mathscr{A})$ is often called a *measurable space*. If $(X, \mathscr{A}, \mu)$ is a measure space, then one often says that μ is a *measure on* $(X, \mathscr{A})$, or, if the σ-algebra $\mathscr{A}$ is clear from context, a *measure on* X.

Examples 1.2.1.

(a) Let X be an arbitrary set, and let $\mathscr{A}$ be a σ-algebra on X. Define a function $\mu : \mathscr{A} \to [0, +\infty]$ by letting $\mu(A)$ be n if A is a finite set with n elements and letting $\mu(A)$ be $+\infty$ if A is an infinite set. Then μ is a measure; it is often called *counting measure* on $(X, \mathscr{A})$.

(b) Let X be a nonempty set, and let $\mathscr{A}$ be a σ-algebra on X. Let x be a member of X. Define a function $\delta_x : \mathscr{A} \to [0, +\infty]$ by letting $\delta_x(A)$ be 1 if $x \in A$ and letting $\delta_x(A)$ be 0 if $x \notin A$. Then δ_x is a measure; it is called a *point mass* concentrated at x.

(c) Consider the set $\mathbb{R}$ of all real numbers and the σ-algebra $\mathscr{B}(\mathbb{R})$ of Borel subsets of $\mathbb{R}$. In Sect. 1.3 we will construct a measure on $\mathscr{B}(\mathbb{R})$ that assigns to each subinterval of $\mathbb{R}$ its length; this measure is known as Lebesgue measure and will be denoted by λ in this book.

(d) Let X be the set of all positive integers, and let $\mathscr{A}$ be the collection of all subsets A of X such that either A or A^c is finite. Then $\mathscr{A}$ is an algebra, but not a σ-algebra (see Example 1.1.1(d)). Define a function $\mu : \mathscr{A} \to [0, +\infty]$ by letting $\mu(A)$ be 1 if A is infinite and letting $\mu(A)$ be 0 if A is finite. It is easy to check that μ is a finitely additive measure; however, it is impossible to extend μ to a countably additive measure on the σ-algebra generated by $\mathscr{A}$ (if $A_k = \{k\}$ for each k, then $\mu(\cup_{k=1}^{\infty} A_k) = \mu(X) = 1$, while $\sum_{k=1}^{\infty} \mu(A_k) = 0$).

(e) Let X be an arbitrary set, and let $\mathscr{A}$ be an arbitrary σ-algebra on X. Define a function $\mu : \mathscr{A} \to [0, +\infty]$ by letting $\mu(A)$ be $+\infty$ if $A \neq \varnothing$, and letting $\mu(A)$ be 0 if $A = \varnothing$. Then μ is a measure.

(f) Let X be a set that has at least two members, and let $\mathscr{A}$ be the σ-algebra consisting of all subsets of X. Define a function $\mu : \mathscr{A} \to [0, +\infty]$ by letting $\mu(A)$ be 1 if $A \neq \varnothing$ and letting $\mu(A)$ be 0 if $A = \varnothing$. Then μ is not a measure, nor even a finitely additive measure, for if A_1 and A_2 are disjoint nonempty subsets of X, then $\mu(A_1 \cup A_2) = 1$, while $\mu(A_1) + \mu(A_2) = 2$.   □

Proposition 1.2.2. *Let $(X, \mathscr{A}, \mu)$ be a measure space, and let A and B be subsets of X that belong to $\mathscr{A}$ and satisfy $A \subseteq B$. Then $\mu(A) \leq \mu(B)$. If in addition A satisfies $\mu(A) < +\infty$, then $\mu(B - A) = \mu(B) - \mu(A)$.*

Proof. The sets A and $B - A$ are disjoint and satisfy $B = A \cup (B - A)$; thus the additivity of μ implies that

$$\mu(B) = \mu(A) + \mu(B - A).$$

Since $\mu(B-A) \geq 0$, it follows that $\mu(A) \leq \mu(B)$. In case $\mu(A) < +\infty$, the relation $\mu(B) - \mu(A) = \mu(B-A)$ also follows. $\qquad\square$

Let μ be a measure on a measurable space $(X, \mathscr{A})$. Then μ is a *finite* measure if $\mu(X) < +\infty$ and is a *σ-finite* measure if X is the union of a sequence $A_1, A_2, \ldots$ of sets that belong to $\mathscr{A}$ and satisfy $\mu(A_i) < +\infty$ for each i. More generally, a set in $\mathscr{A}$ is σ-finite under μ if it is the union of a sequence of sets that belong to $\mathscr{A}$ and have finite measure under μ. The measure space $(X, \mathscr{A}, \mu)$ is also called *finite* or *σ-finite* if μ is finite or σ-finite. Most of the constructions and basic properties that we will consider are valid for all measures. For a few important theorems, however, we will need to assume that the measures involved are finite or σ-finite.

If the measure space $(X, \mathscr{A}, \mu)$ is σ-finite, then X is the union of a sequence $\{B_i\}$ of disjoint sets that belong to $\mathscr{A}$ and have finite measure under μ; such a sequence $\{B_i\}$ can be formed by choosing a sequence $\{A_i\}$ as in the definition of σ-finiteness, and then letting $B_1 = A_1$ and $B_i = A_i - (\cup_{j=1}^{i-1} A_j)$ if $i > 1$.

Examples 1.2.3 (Dealing with σ-Finiteness). Note that the measure defined in Example 1.2.1(a) is finite if and only if the set X is finite and is σ-finite if and only if the set X is the union of a sequence of finite sets that belong to $\mathscr{A}$.[3] The measure defined in Example 1.2.1(b) is finite. Lebesgue measure, described in Example 1.2.1(c), is σ-finite, since $\mathbb{R}$ is the union of a sequence of bounded intervals. See also Exercises 2 and 7 below. $\qquad\square$

The following propositions give some elementary but useful properties of measures.

Proposition 1.2.4. *Let* $(X, \mathscr{A}, \mu)$ *be a measure space. If* $\{A_k\}$ *is an arbitrary sequence of sets that belong to* $\mathscr{A}$, *then*

$$\mu(\cup_{k=1}^{\infty} A_k) \leq \sum_{k=1}^{\infty} \mu(A_k).$$

Proof. Define a sequence $\{B_k\}$ of subsets of X by letting $B_1 = A_1$ and letting $B_k = A_k - (\cup_{i=1}^{k-1} A_i)$ if $k > 1$. Then each B_k belongs to $\mathscr{A}$ and is a subset of the corresponding A_k, and so satisfies $\mu(B_k) \leq \mu(A_k)$. Since in addition the sets B_k are disjoint and satisfy $\cup_k B_k = \cup_k A_k$, it follows that

$$\mu(\cup_k A_k) = \mu(\cup_k B_k) = \sum_k \mu(B_k) \leq \sum_k \mu(A_k). \qquad\square$$

In other words, the countable additivity of μ implies the *countable subadditivity* of μ.

---

[3] If in Example 1.2.1(a) the σ-algebra $\mathscr{A}$ contains all the subsets of X, then μ is σ-finite if and only if X is at most countably infinite.

Proposition 1.2.5. *Let* $(X, \mathscr{A}, \mu)$ *be a measure space.*

(a) *If* $\{A_k\}$ *is an increasing sequence of sets that belong to* $\mathscr{A}$, *then* $\mu(\cup_k A_k) = \lim_k \mu(A_k)$.

(b) *If* $\{A_k\}$ *is a decreasing sequence of sets that belong to* $\mathscr{A}$, *and if* $\mu(A_n) < +\infty$ *holds for some* n, *then* $\mu(\cap_k A_k) = \lim_k \mu(A_k)$.

Proof. First suppose that $\{A_k\}$ is an increasing sequence of sets that belong to $\mathscr{A}$, and define a sequence $\{B_i\}$ of sets by letting $B_1 = A_1$ and letting $B_i = A_i - A_{i-1}$ if $i > 1$. The sets just constructed are disjoint, belong to $\mathscr{A}$, and satisfy $A_k = \cup_{i=1}^k B_i$ for each k. It follows that $\cup_k A_k = \cup_i B_i$ and hence that

$$\mu(\cup_k A_k) = \sum_i \mu(B_i) = \lim_k \sum_{i=1}^k \mu(B_i) = \lim_k \mu(\cup_{i=1}^k B_i) = \lim_k \mu(A_k).$$

This completes the proof of (a).

Now suppose that $\{A_k\}$ is a decreasing sequence of sets that belong to $\mathscr{A}$ and that $\mu(A_n) < +\infty$ holds for some n. We can assume that $n = 1$. For each k let $C_k = A_1 - A_k$. Then $\{C_k\}$ is an increasing sequence of sets that belong to $\mathscr{A}$ and satisfy

$$\cup_k C_k = A_1 - (\cap_k A_k).$$

It follows from part (a) that $\mu(\cup_k C_k) = \lim_k \mu(C_k)$ and hence that

$$\mu(A_1 - (\cap_k A_k)) = \mu(\cup_k C_k) = \lim_k \mu(C_k) = \lim_k \mu(A_1 - A_k).$$

In view of Proposition 1.2.2 and the assumption that $\mu(A_1) < +\infty$, this implies that $\mu(\cap_k A_k) = \lim_k \mu(A_k)$. □

The preceding proposition has the following partial converse, which is sometimes useful for checking that a finitely additive measure is in fact countably additive.

Proposition 1.2.6. *Let* $(X, \mathscr{A})$ *be a measurable space, and let* μ *be a finitely additive measure on* $(X, \mathscr{A})$. *Then* μ *is a measure if either*

(a) $\lim_k \mu(A_k) = \mu(\cup_k A_k)$ *holds for each increasing sequence* $\{A_k\}$ *of sets that belong to* $\mathscr{A}$, *or*

(b) $\lim_k \mu(A_k) = 0$ *holds for each decreasing sequence* $\{A_k\}$ *of sets that belong to* $\mathscr{A}$ *and satisfy* $\cap_k A_k = \varnothing$.

Proof. We need to verify the countable additivity of μ. Let $\{B_j\}$ be a sequence of disjoint sets that belong to $\mathscr{A}$; we will prove that $\mu(\cup_j B_j) = \sum_j \mu(B_j)$.

First assume that condition (a) holds, and for each k let $A_k = \cup_{j=1}^k B_j$. Then the finite additivity of μ implies that $\mu(A_k) = \sum_{j=1}^k \mu(B_j)$, while condition (a) implies that $\mu(\cup_{k=1}^\infty A_k) = \lim_k \mu(A_k)$; since $\cup_{j=1}^\infty B_j = \cup_{k=1}^\infty A_k$, it follows that

$$\mu(\cup_{j=1}^\infty B_j) = \mu(\cup_{k=1}^\infty A_k) = \lim_k \mu(A_k) = \sum_{j=1}^\infty \mu(B_j).$$

Now assume that condition (b) holds, and for each k let $A_k = \cup_{j=k}^{\infty} B_j$. Then the finite additivity of μ implies that

$$\mu(\cup_{j=1}^{\infty} B_j) = \sum_{j=1}^{k} \mu(B_j) + \mu(A_{k+1}),$$

while condition (b) implies that $\lim_k \mu(A_{k+1}) = 0$; hence $\mu(\cup_{j=1}^{\infty} B_j) = \sum_{j=1}^{\infty} \mu(B_j)$.

$\square$

Let us close this section by introducing a bit of terminology. A measure on $(\mathbb{R}^d, \mathscr{B}(\mathbb{R}^d))$ is often called a *Borel measure* on $\mathbb{R}^d$. More generally, if X is a Borel subset of $\mathbb{R}^d$ and if $\mathscr{A}$ is the σ-algebra consisting of those Borel subsets of $\mathbb{R}^d$ that are included in X, then a measure on $(X, \mathscr{A})$ is called a *Borel measure* on X.

Now suppose that $(X, \mathscr{A})$ is a measurable space such that for each x in X the set $\{x\}$ belongs to $\mathscr{A}$. A finite or σ-finite measure μ on $(X, \mathscr{A})$ is *continuous* if $\mu(\{x\}) = 0$ holds for each x in X and is *discrete* if there is a countable subset D of X such that $\mu(D^c) = 0$. (More elaborate definitions are needed if $\mathscr{A}$ does not contain each $\{x\}$ or if μ is not σ-finite. We will, however, not need to consider such matters.)

Exercises

1. Suppose that μ is a finite measure on $(X, \mathscr{A})$.
   (a) Show that if A and B belong to $\mathscr{A}$, then

   $$\mu(A \cup B) = \mu(A) + \mu(B) - \mu(A \cap B).$$

   (b) Show that if A, B, and C belong to $\mathscr{A}$, then

   $$\mu(A \cup B \cup C) = \mu(A) + \mu(B) + \mu(C)$$
   $$- \mu(A \cap B) - \mu(A \cap C) - \mu(B \cap C)$$
   $$+ \mu(A \cap B \cap C).$$

   (c) Find and prove a corresponding formula for the measure of the union of n sets.
2. Define μ on $(\mathbb{R}, \mathscr{B}(\mathbb{R}))$ by letting $\mu(A)$ be the number of rational numbers in A (of course $\mu(A) = +\infty$ if there are infinitely many rational numbers in A). Show that μ is a σ-finite measure under which each open subinterval of $\mathbb{R}$ has infinite measure.
3. Let $\mathscr{A}$ be the σ-algebra of all subsets of $\mathbb{N}$, and let μ be counting measure on $(\mathbb{N}, \mathscr{A})$. Give a decreasing sequence $\{A_k\}$ of sets in $\mathscr{A}$ such that $\mu(\cap_k A_k) \neq \lim_k \mu(A_k)$. Hence the finiteness assumption cannot be removed from part (b) of Proposition 1.2.5.

4. Let $(X, \mathscr{A})$ be a measurable space.
   (a) Suppose that μ is a nonnegative countably additive function on $\mathscr{A}$. Show that if $\mu(A)$ is finite for some A in $\mathscr{A}$, then $\mu(\varnothing) = 0$. (Thus μ is a measure.)
   (b) Show by example that in general the condition $\mu(\varnothing) = 0$ does not follow from the remaining parts of the definition of a measure.
5. Let $(X, \mathscr{A})$ be a measurable space, and let x and y belong to X. Show that the point masses δ_x and δ_y are equal if and only if x and y belong to exactly the same sets in $\mathscr{A}$.
6. Let $(X, \mathscr{A})$ be a measurable space.
   (a) Show that if $\{\mu_n\}$ is an increasing sequence of measures on $(X, \mathscr{A})$ (here "increasing" means that $\mu_n(A) \le \mu_{n+1}(A)$ holds for each A and each n), then the formula $\mu(A) = \lim_n \mu_n(A)$ defines a measure on $(X, \mathscr{A})$.
   (b) Show that if $\{\mu_n\}$ is an arbitrary sequence of measures on $(X, \mathscr{A})$, then the formula $\mu(A) = \sum_n \mu_n(A)$ defines a measure on $(X, \mathscr{A})$.
7. Let $\{x_n\}$ be a sequence of real numbers, and define a measure μ on $(\mathbb{R}, \mathscr{B}(\mathbb{R}))$ by $\mu = \sum_n \delta_{x_n}$ (see Exercise 6).
   (a) Show that μ assigns finite values to the bounded subintervals of $\mathbb{R}$ if and only if $\lim_n |x_n| = +\infty$.
   (b) For which sequences $\{x_n\}$ is the measure μ σ-finite?
8. Let $(X, \mathscr{A}, \mu)$ be a measure space, and define $\mu^\bullet : \mathscr{A} \to [0, +\infty]$ by

$$\mu^\bullet(A) = \sup\{\mu(B) : B \subseteq A, B \in \mathscr{A}, \text{ and } \mu(B) < +\infty\}.$$

   (a) Show that $\mu^\bullet$ is a measure on $(X, \mathscr{A})$.
   (b) Show that if μ is σ-finite, then $\mu^\bullet = \mu$.
   (c) Find $\mu^\bullet$ if X is nonempty and μ is the measure defined by

$$\mu(A) = \begin{cases} +\infty & \text{if } A \in \mathscr{A} \text{ and } A \ne \varnothing, \text{ and} \\ 0 & \text{if } A = \varnothing. \end{cases}$$

9. Let μ be a measure on $(X, \mathscr{A})$, and let $\{A_k\}$ be a sequence of sets in $\mathscr{A}$ such that $\sum_k \mu(A_k) < +\infty$. Show that the set of points that belong to A_k for infinitely many values of k has measure zero under μ. (Hint: Consider the set $\cap_{n=1}^\infty \cup_{k=n}^\infty A_k$, and note that $\mu(\cap_{n=1}^\infty \cup_{k=n}^\infty A_k) \le \mu(\cup_{k=p}^\infty A_k)$ holds for each p.)

1.3   Outer Measures

In this section we develop one of the standard techniques for constructing measures; then we use it to construct Lebesgue measure on $\mathbb{R}^d$.

Let X be a set, and let $\mathscr{P}(X)$ be the collection of all subsets of X. An *outer measure* on X is a function $\mu^* : \mathscr{P}(X) \to [0, +\infty]$ such that

(a) $\mu^*(\varnothing) = 0$,
(b) if $A \subseteq B \subseteq X$, then $\mu^*(A) \le \mu^*(B)$, and
(c) if $\{A_n\}$ is an infinite sequence of subsets of X, then $\mu^*(\cup_n A_n) \le \sum_n \mu^*(A_n)$.

Thus an outer measure on X is a *monotone* and *countably subadditive* function from $\mathscr{P}(X)$ to $[0, +\infty]$ whose value at $\varnothing$ is 0.

Note that a measure can fail to be an outer measure; in fact, a measure on X is an outer measure if and only if its domain is $\mathscr{P}(X)$ (see Propositions 1.2.2 and 1.2.4). On the other hand, an outer measure generally fails to be countably additive and so fails to be a measure.

In Theorem 1.3.6, we will prove that for each outer measure μ^* on X there is a relatively natural σ-algebra $\mathscr{M}_{\mu^*}$ on X such that the restriction of μ^* to $\mathscr{M}_{\mu^*}$ is countably additive, and hence a measure. Many important measures can be derived from outer measures in this way.

Examples 1.3.1.

(a) Let X be an arbitrary set, and define μ^* on $\mathscr{P}(X)$ by $\mu^*(A) = 0$ if $A = \varnothing$ and $\mu^*(A) = 1$ otherwise. Then μ^* is an outer measure.
(b) Let X be an arbitrary set, and define μ^* on $\mathscr{P}(X)$ by $\mu^*(A) = 0$ if A is countable, and $\mu^*(A) = 1$ if A is uncountable. Then μ^* is an outer measure.
(c) Let X be an infinite set, and define μ^* on $\mathscr{P}(X)$ by $\mu^*(A) = 0$ if A is finite, and $\mu^*(A) = 1$ if A is infinite. Then μ^* fails to be countably subadditive and so is not an outer measure.
(d) *Lebesgue outer measure* on $\mathbb{R}$, which we will denote by λ^*, is defined as follows. For each subset A of $\mathbb{R}$, let $\mathscr{C}_A$ be the set of all infinite sequences $\{(a_i, b_i)\}$ of bounded open intervals such that $A \subseteq \cup_i (a_i, b_i)$. Then $\lambda^*: \mathscr{P}(\mathbb{R}) \to [0, +\infty]$ is defined by

$$\lambda^*(A) = \inf \left\{ \sum_i (b_i - a_i) : \{(a_i, b_i)\} \in \mathscr{C}_A \right\}.$$

(Note that the set of sums involved here is nonempty and that the infimum of the set consisting of $+\infty$ alone is $+\infty$. We check in the following proposition that λ^* is indeed an outer measure.) □

Proposition 1.3.2. *Lebesgue outer measure on $\mathbb{R}$ is an outer measure, and it assigns to each subinterval of $\mathbb{R}$ its length.*

Proof. We begin by verifying that λ^* is an outer measure. The relation $\lambda^*(\varnothing) = 0$ holds, since for each positive number ε there is a sequence $\{(a_i, b_i)\}$ of open intervals (whose union necessarily includes $\varnothing$) such that $\sum_i (b_i - a_i) < \varepsilon$. For the monotonicity of λ^*, note that if $A \subseteq B$, then each sequence of open intervals that covers B also covers A, and so $\lambda^*(A) \le \lambda^*(B)$. Now consider the countable subadditivity of λ^*. Let $\{A_n\}_{n=1}^{\infty}$ be an arbitrary sequence of subsets of $\mathbb{R}$. If $\sum_n \lambda^*(A_n) = +\infty$, then $\lambda^*(\cup_n A_n) \le \sum_n \lambda^*(A_n)$ certainly holds. So suppose that

$\sum_n \lambda^*(A_n) < +\infty$, and let ε be an arbitrary positive number. For each n choose a sequence $\{(a_{n,i}, b_{n,i})\}_{i=1}^{\infty}$ that covers A_n and satisfies

$$\sum_{i=1}^{\infty} (b_{n,i} - a_{n,i}) < \lambda^*(A_n) + \varepsilon/2^n.$$

If we combine these sequences into one sequence $\{(a_j, b_j)\}$ (see, for example, the construction in the last paragraph of A.6), then the combined sequence satisfies

$$\cup_n A_n \subseteq \cup_j (a_j, b_j)$$

and

$$\sum_j (b_j - a_j) < \sum_n (\lambda^*(A_n) + \varepsilon/2^n) = \sum_n \lambda^*(A_n) + \varepsilon.$$

These relations, together with the fact that ε is arbitrary, imply that $\lambda^*(\cup_n A_n) \le \sum_n \lambda^*(A_n)$. Thus λ^* is an outer measure.

Now we compute the outer measure of the subintervals of $\mathbb{R}$. First consider a closed bounded interval $[a,b]$. It is easy to see that $\lambda^*([a,b]) \le b - a$ (cover $[a,b]$ with sequences of open intervals in which the first interval is barely larger than $[a,b]$, and the sum of the lengths of the other intervals is very small). We turn to the reverse inequality. Let $\{(a_i, b_i)\}$ be a sequence of bounded open intervals whose union includes $[a,b]$. Since $[a,b]$ is compact, there is a positive integer n such that $[a,b] \subseteq \cup_{i=1}^{n} (a_i, b_i)$. It is easy to check that $b - a \le \sum_{i=1}^{n} (b_i - a_i)$ (use induction on n) and hence that $b - a \le \sum_{i=1}^{\infty} (b_i - a_i)$. Since $\{(a_i, b_i)\}$ was an arbitrary sequence whose union includes $[a,b]$, it follows that $b - a \le \lambda^*([a,b])$. Thus $\lambda^*([a,b]) = b - a$.

The outer measure of an arbitrary bounded interval is its length, since such an interval I includes and is included in closed bounded intervals of length arbitrarily close to the length of I. Finally, an unbounded interval has infinite outer measure, since it includes arbitrarily long closed bounded intervals. □

Let us look at another basic example.

Example 1.3.3. *Lebesgue outer measure* on $\mathbb{R}^d$, which we will denote by λ^* (or, if necessary in order to avoid ambiguity, by λ_d^*) is defined as follows. A *d-dimensional interval* is a subset of $\mathbb{R}^d$ of the form $I_1 \times \cdots \times I_d$, where $I_1, \ldots, I_d$ are subintervals of $\mathbb{R}$ and $I_1 \times \cdots \times I_d$ is given by

$$I_1 \times \cdots \times I_d = \{(x_i, \ldots, x_d) : x_i \in I_i \text{ for } i = 1, \ldots, d\}.$$

Note that the intervals $I_1, \ldots, I_d$, and hence the d-dimensional interval $I_1 \times \cdots \times I_d$, can be open, closed, or neither open nor closed. The *volume* of the d-dimensional interval $I_1 \times \cdots \times I_d$ is the product of the lengths of the intervals $I_1, \ldots, I_d$, and will be denoted by $\mathrm{vol}(I_1 \times \cdots \times I_d)$. For each subset A of $\mathbb{R}^d$ let $\mathscr{C}_A$ be the set of all sequences $\{R_i\}$ of bounded and open d-dimensional intervals for which $A \subseteq \cup_{i=1}^{\infty} R_i$. Then $\lambda^*(A)$, the outer measure of A, is the infimum of the set

$$\left\{ \sum_{i=1}^{\infty} \text{vol}(R_i) : \{R_i\} \in \mathscr{C}_A \right\}. \qquad\qquad \square$$

We note the following analogue of Proposition 1.3.2.

Proposition 1.3.4. *Lebesgue outer measure on* $\mathbb{R}^d$ *is an outer measure, and it assigns to each d-dimensional interval its volume.*

Proof. Most of the details are omitted, since they are very similar to those in the proof of Proposition 1.3.2. Note, however, that if K is a compact d-dimensional interval and if $\{R_i\}_{i=1}^{\infty}$ is a sequence of bounded and open d-dimensional intervals for which $K \subseteq \cup_{i=1}^{\infty} R_i$, then there is a positive integer n such that $K \subseteq \cup_{i=1}^{n} R_i$, and K can be decomposed into a finite collection $\{K_j\}$ of d-dimensional intervals that overlap only on their boundaries and are such that for each j the interior of K_j is included in some R_i (where $i \leq n$). From this it follows that

$$\text{vol}(K) = \sum_{j} \text{vol}(K_j) \leq \sum_{i} \text{vol}(R_i)$$

and hence that $\text{vol}(K) \leq \lambda^*(K)$. The remaining modifications needed to convert our proof of Proposition 1.3.2 into a proof of the present result are straightforward. $\square$

Let X be a set, and let μ^* be an outer measure on X. A subset B of X is μ^*-*measurable* (or *measurable with respect to* μ^*) if

$$\mu^*(A) = \mu^*(A \cap B) + \mu^*(A \cap B^c)$$

holds for *every* subset A of X. Thus a μ^*-measurable subset of X is one that divides each subset of X in such a way that the sizes (as measured by μ^*) of the pieces add properly. A *Lebesgue measurable* subset of $\mathbb{R}$ or of $\mathbb{R}^d$ is of course one that is measurable with respect to Lebesgue outer measure.

Note that the subadditivity of the outer measure μ^* implies that

$$\mu^*(A) \leq \mu^*(A \cap B) + \mu^*(A \cap B^c)$$

holds for all subsets A and B of X. Thus to check that a subset B of X is μ^*-measurable, we need only check that

$$\mu^*(A) \geq \mu^*(A \cap B) + \mu^*(A \cap B^c) \qquad\qquad (1)$$

holds for each subset A of X. Note also that inequality (1) certainly holds if $\mu^*(A) = +\infty$. Thus the μ^*-measurability of B can be verified by checking that (1) holds for each A that satisfies $\mu^*(A) < +\infty$.

Proposition 1.3.5. *Let X be a set, and let μ^* be an outer measure on X. Then each subset B of X that satisfies $\mu^*(B) = 0$ or that satisfies $\mu^*(B^c) = 0$ is μ^*-measurable.*

Proof. Assume that $\mu^*(B) = 0$ or that $\mu^*(B^c) = 0$. According to the remarks above, we need only check that each subset A of X satisfies

$$\mu^*(A) \geq \mu^*(A \cap B) + \mu^*(A \cap B^c).$$

However our assumption about B and the monotonicity of μ^* imply that one of the terms on the right-hand side of this inequality vanishes and that the other is at most $\mu^*(A)$; thus the required inequality follows. $\square$

It follows that the sets $\varnothing$ and X are measurable for every outer measure on X.

The following theorem is the fundamental fact about outer measures; it will be the key to many of our constructions of measures.

Theorem 1.3.6. *Let X be a set, let μ^* be an outer measure on X, and let $\mathscr{M}_{\mu^*}$ be the collection of all μ^*-measurable subsets of X. Then*

(a) *$\mathscr{M}_{\mu^*}$ is a σ-algebra, and*
(b) *the restriction of μ^* to $\mathscr{M}_{\mu^*}$ is a measure on $\mathscr{M}_{\mu^*}$.*

Proof. We begin by showing that $\mathscr{M}_{\mu^*}$ is an algebra of sets. First note that Proposition 1.3.5 implies that X belongs to $\mathscr{M}_{\mu^*}$. Note also that the equation

$$\mu^*(A) = \mu^*(A \cap B) + \mu^*(A \cap B^c)$$

is not changed if the sets B and B^c are interchanged; thus the μ^*-measurability of B implies that of B^c, and so $\mathscr{M}_{\mu^*}$ is closed under complementation. Now suppose that B_1 and B_2 are μ^*-measurable subsets of X; we will show that $B_1 \cup B_2$ is μ^*-measurable. For this, let A be an arbitrary subset of X. The μ^*-measurability of B_1 implies

$$\mu^*(A \cap (B_1 \cup B_2)) = \mu^*(A \cap (B_1 \cup B_2) \cap B_1) + \mu^*(A \cap (B_1 \cup B_2) \cap B_1^c)$$
$$= \mu^*(A \cap B_1) + \mu^*(A \cap B_1^c \cap B_2).$$

If we use this identity and the fact that $(B_1 \cup B_2)^c = B_1^c \cap B_2^c$, and then simplify the resulting expression by appealing first to the measurability of B_2 and then to the measurability of B_1, we find

$$\mu^*(A \cap (B_1 \cup B_2)) + \mu^*(A \cap (B_1 \cup B_2)^c)$$
$$= \mu^*(A \cap B_1) + \mu^*(A \cap B_1^c \cap B_2) + \mu^*(A \cap B_1^c \cap B_2^c)$$
$$= \mu^*(A \cap B_1) + \mu^*(A \cap B_1^c)$$
$$= \mu^*(A).$$

Since A was an arbitrary subset of X, the set $B_1 \cup B_2$ must be measurable. Thus $\mathscr{M}_{\mu^*}$ is an algebra.

Next suppose that $\{B_i\}$ is an infinite sequence of disjoint μ^*-measurable sets; we will show by induction that

$$\mu^*(A) = \sum_{i=1}^{n} \mu^*(A \cap B_i) + \mu^*(A \cap (\cap_{i=1}^{n} B_i^c)) \qquad (2)$$

holds for each subset A of X and each positive integer n. Equation (2) is, in the case where $n = 1$, simply a restatement of the measurability of B_1. As to the induction step, note that the μ^*-measurability of B_{n+1} and the disjointness of the sequence $\{B_i\}$ imply that

$$\mu^*(A \cap (\cap_{i=1}^n B_i^c))$$
$$= \mu^*(A \cap (\cap_{i=1}^n B_i^c) \cap B_{n+1}) + \mu^*(A \cap (\cap_{i=1}^n B_i^c) \cap B_{n+1}^c)$$
$$= \mu^*(A \cap B_{n+1}) + \mu^*(A \cap (\cap_{i=1}^{n+1} B_i^c)).$$

With this (2) is proved.

Note that we do not increase the right-hand side of Eq. (2) if we replace $\mu^*(A \cap (\cap_{i=1}^n B_i^c))$ with $\mu^*(A \cap (\cap_{i=1}^\infty B_i^c))$, and thus with $\mu^*(A \cap (\cup_{i=1}^\infty B_i)^c)$; by letting the n in the sum in the resulting inequality approach infinity, we find

$$\mu^*(A) \geq \sum_{i=1}^\infty \mu^*(A \cap B_i) + \mu^*(A \cap (\cup_{i=1}^\infty B_i)^c). \tag{3}$$

This and the countable subadditivity of μ^* imply that

$$\mu^*(A) \geq \sum_{i=1}^\infty \mu^*(A \cap B_i) + \mu^*(A \cap (\cup_{i=1}^\infty B_i)^c)$$
$$\geq \mu^*(A \cap (\cup_{i=1}^\infty B_i)) + \mu^*(A \cap (\cup_{i=1}^\infty B_i)^c)$$
$$\geq \mu^*(A);$$

it follows that each inequality in the preceding calculation must in fact be an equality and hence that $\cup_{i=1}^\infty B_i$ is μ^*-measurable. Thus $\mathscr{M}_{\mu^*}$ is closed under the formation of unions of disjoint sequences of sets. Since the union of an arbitrary sequence $\{B_i\}$ of sets in $\mathscr{M}_{\mu^*}$ is the union of a disjoint sequence of sets in $\mathscr{M}_{\mu^*}$, namely of the sequence

$$B_1, B_1^c \cap B_2, \ldots, B_1^c \cap B_2^c \cap \cdots \cap B_{n-1}^c \cap B_n, \ldots,$$

the algebra $\mathscr{M}_{\mu^*}$ is closed under the formation of countable unions. With this we have proved that $\mathscr{M}_{\mu^*}$ is a σ-algebra.

To show that the restriction of μ^* to $\mathscr{M}_{\mu^*}$ is a measure, we need to verify its countable additivity. If $\{B_i\}$ is a sequence of disjoint sets in $\mathscr{M}_{\mu^*}$, then replacing A with $\cup_{i=1}^\infty B_i$ in inequality (3) yields

$$\mu^*(\cup_{i=1}^\infty B_i) \geq \sum_{i=1}^\infty \mu^*(B_i) + 0;$$

since the reverse inequality is automatic, the countable additivity of the restriction of μ^* to $\mathscr{M}_{\mu^*}$ follows. □

We turn to applications of Theorem 1.3.6 and begin with Lebesgue measure. We will denote the *collection of Lebesgue measurable subsets* of $\mathbb{R}$ by $\mathscr{M}_{\lambda^*}$.

Proposition 1.3.7. *Every Borel subset of* $\mathbb{R}$ *is Lebesgue measurable.*

Proof. We begin by checking that every interval of the form $(-\infty, b]$ is Lebesgue measurable. Let B be such an interval. According to the remarks made just before the statement of Proposition 1.3.5, we need only check that

$$\lambda^*(A) \geq \lambda^*(A \cap B) + \lambda^*(A \cap B^c) \tag{4}$$

holds for each subset A of $\mathbb{R}$ for which $\lambda^*(A) < +\infty$. Let A be such a set, let ε be an arbitrary positive number, and let $\{(a_n, b_n)\}$ be a sequence of open intervals that covers A and satisfies $\sum_{n=1}^{\infty}(b_n - a_n) < \lambda^*(A) + \varepsilon$. Then for each n the sets $(a_n, b_n) \cap B$ and $(a_n, b_n) \cap B^c$ are disjoint intervals (one of which may instead be the empty set) whose union is (a_n, b_n), and so

$$b_n - a_n = \lambda^*((a_n, b_n)) = \lambda^*((a_n, b_n) \cap B) + \lambda^*((a_n, b_n) \cap B^c) \tag{5}$$

(see Proposition 1.3.2). Since the sequence $\{(a_n, b_n) \cap B\}$ covers $A \cap B$ and the sequence $\{(a_n, b_n) \cap B^c\}$ covers $A \cap B^c$, we have from Eq. (5) and the countable subadditivity of λ^* that

$$\lambda^*(A \cap B) + \lambda^*(A \cap B^c) \leq \sum_n \lambda^*((a_n, b_n) \cap B) + \sum_n \lambda^*((a_n, b_n) \cap B^c)$$
$$= \sum_n (b_n - a_n) < \lambda^*(A) + \varepsilon.$$

However, ε was arbitrary, and so inequality (4) and the Lebesgue measurability of B follow.

Thus the collection $\mathscr{M}_{\lambda^*}$ of Lebesgue measurable sets is a σ-algebra on $\mathbb{R}$ (Theorem 1.3.6) that contains each interval of the form $(-\infty, b]$. However $\mathscr{B}(\mathbb{R})$ is the smallest σ-algebra on $\mathbb{R}$ that contains all these intervals (Proposition 1.1.4), and so $\mathscr{B}(\mathbb{R}) \subseteq \mathscr{M}_{\lambda^*}$. $\qquad\square$

We will also use $\mathscr{M}_{\lambda^*}$ to denote the *collection of Lebesgue measurable subsets* of $\mathbb{R}^d$.

Proposition 1.3.8. *Every Borel subset of* $\mathbb{R}^d$ *is Lebesgue measurable.*

Proof. It is easy to give a proof of Proposition 1.3.8 by modifying that of Proposition 1.3.7; the details are left to the reader. $\qquad\square$

The restriction of Lebesgue outer measure on $\mathbb{R}$ (or on $\mathbb{R}^d$) to the collection $\mathscr{M}_{\lambda^*}$ of Lebesgue measurable subsets of $\mathbb{R}$ (or of $\mathbb{R}^d$) is called *Lebesgue measure* and will be denoted by λ or by λ_d. The restriction of Lebesgue outer measure to $\mathscr{B}(\mathbb{R})$ or to $\mathscr{B}(\mathbb{R}^d)$ is also called *Lebesgue measure*, and it too will be denoted by λ or by λ_d. We can specify which version of Lebesgue measure we intend by referring, for example, to Lebesgue measure on $(\mathbb{R}, \mathscr{B}(\mathbb{R}))$ or to Lebesgue measure on $(\mathbb{R}, \mathscr{M}_{\lambda^*})$. We will deal most often with Lebesgue measure on the Borel sets; its relation to the other version of Lebesgue measure is treated in Sect. 1.5.

Two questions arise immediately. Is every subset of $\mathbb{R}$ Lebesgue measurable? Is every Lebesgue measurable set a Borel set? The answer to each of these questions is no; see Sects. 1.4 and 2.1 for details.

We close this section with a technique for constructing and representing all finite measures on $(\mathbb{R}, \mathscr{B}(\mathbb{R}))$. We begin with the following elementary fact.

Proposition 1.3.9. *Let μ be a finite measure on $(\mathbb{R}, \mathscr{B}(\mathbb{R}))$, and let $F_\mu : \mathbb{R} \to \mathbb{R}$ be defined by $F_\mu(x) = \mu((-\infty, x])$. Then F_μ is bounded, nondecreasing, and right-continuous, and satisfies $\lim_{x \to -\infty} F_\mu(x) = 0$.*

Proof. It follows from Proposition 1.2.2 that $0 \leq \mu((-\infty, x]) \leq \mu(\mathbb{R})$ holds for all x in $\mathbb{R}$ and that $\mu((-\infty, x]) \leq \mu((-\infty, y])$ holds for all x and y in $\mathbb{R}$ such that $x \leq y$; hence F_μ is bounded and nondecreasing. Next suppose that $x \in \mathbb{R}$ and that $\{x_n\}$ is the sequence defined by $x_n = x + 1/n$. Then $(-\infty, x] = \cap_{n=1}^{\infty}(-\infty, x_n]$, and so Proposition 1.2.5 implies that $F_\mu(x) = \lim_n F_\mu(x_n)$. The right continuity of F_μ follows (note that if $x < y < x_n$, then, since F_μ is nondecreasing, $|F_\mu(y) - F_\mu(x)| \leq |F_\mu(x_n) - F_\mu(x)|$). A similar argument shows that $\lim_{x \to -\infty} F_\mu(x) = 0$. $\square$

Let μ and F_μ be as in Proposition 1.3.9. The interval $(a, b]$ is the difference of the intervals $(-\infty, b]$ and $(-\infty, a]$, and so Proposition 1.2.2 implies that

$$\mu((a, b]) = F_\mu(b) - F_\mu(a). \tag{6}$$

Since F_μ is bounded and nondecreasing, the limit of $F_\mu(t)$ as t approaches x from the left exists for each x in $\mathbb{R}$; this limit is equal to $\sup\{F_\mu(t) : t < x\}$ and will be denoted by $F_\mu(x-)$. Now let $\{a_n\}$ be a sequence that increases to the real number b; if we apply Eq. (6) to each interval $(a_n, b]$ and then use Proposition 1.2.5, we find that

$$\mu(\{b\}) = F_\mu(b) - F_\mu(b-). \tag{7}$$

Consequently F_μ is continuous at b if $\mu(\{b\}) = 0$, and is discontinuous there, with a jump of size $\mu(\{b\})$ in its graph, if $\mu(\{b\}) \neq 0$. Thus the measure μ is continuous (see Sect. 1.2) if and only if the function F_μ is continuous.

Equations (6) and (7) allow one to use F_μ to recover the measure under μ of certain subsets of $\mathbb{R}$ (see also Exercise 4); however, the following proposition allows us to say more, namely that the measure under μ of *every* Borel subset of $\mathbb{R}$ is in fact determined by F_μ.

Proposition 1.3.10. *For each bounded, nondecreasing, and right-continuous function $F : \mathbb{R} \to \mathbb{R}$ that satisfies $\lim_{x \to -\infty} F(x) = 0$, there is a unique finite measure μ on $(\mathbb{R}, \mathscr{B}(\mathbb{R}))$ such that $F(x) = \mu((-\infty, x])$ holds at each x in $\mathbb{R}$.*

Proof. Let F be as in the statement of the proposition. We begin by constructing the required measure μ. Define a function $\mu^* : \mathscr{P}(\mathbb{R}) \to [0, +\infty]$ by letting $\mu^*(A)$ be the infimum of the set of sums $\sum_{n=1}^{\infty}(F(b_n) - F(a_n))$, where $\{(a_n, b_n]\}$ ranges over the set of sequences of half-open intervals that cover A, in the sense that $A \subseteq \cup_{n=1}^{\infty}(a_n, b_n]$. Then μ^* is an outer measure on $\mathbb{R}$; the reader can check this by modifying some of the arguments used in the proof of Proposition 1.3.2.

Next we verify that $\mu^*((-\infty, x]) = F(x)$ holds for each x in $\mathbb{R}$. The inequality $\mu^*((-\infty, x]) \leq F(x)$ holds, since $(-\infty, x]$ can be covered by the intervals in the sequence $\{(x-n, x-n+1]\}_{n=1}^{\infty}$, for which we have $\sum_{n=1}^{\infty}(F(x-n+1) - F(x-n)) = F(x)$. We turn to the reverse inequality. Let $\{(a_n, b_n]\}$ be a sequence that covers $(-\infty, x]$, and let ε be a positive number. Use the fact that $\lim_{t \to -\infty} F(t) = 0$ to choose a number t such that $t < x$ and $F(t) < \varepsilon$, and for each n use the right continuity of F to choose a positive number δ_n such that $F(b_n + \delta_n) < F(b_n) + \varepsilon/2^n$. Then the interval $[t, x]$ is compact, each interval $(a_n, b_n + \delta_n)$ is open, $[t, x] \subseteq \cup_{n=1}^{\infty}(a_n, b_n + \delta_n)$, and $\sum_n (F(b_n + \delta_n) - F(a_n)) \leq \sum_n (F(b_n) - F(a_n)) + \varepsilon$. The compactness of $[t, x]$ implies that there is a positive integer N such that $[t, x] \subseteq \cup_{n=1}^{N}(a_n, b_n + \delta_n)$. It follows that $(t, x]$ is the union of a finite collection of disjoint intervals $(c_j, d_j]$, each of which is included in some $(a_n, b_n + \delta_n]$. Consequently

$$F(x) - F(t) = \sum_j (F(d_j) - F(c_j)) \leq \sum_{n=1}^{\infty}(F(b_n + \delta_n) - F(a_n)),$$

and so

$$F(x) - \varepsilon \leq \sum_{n=1}^{\infty}(F(b_n) - F(a_n)) + \varepsilon.$$

Since ε and the sequence $\{(a_n, b_n]\}$ are arbitrary, the inequality $F(x) \leq \mu^*((-\infty, x])$ follows. With this we have shown that $F(x) = \mu^*((-\infty, x])$.

The reader should check that the proof of Proposition 1.3.7 can be modified so as to show that each interval $(-\infty, b]$ is μ^*-measurable and then that each Borel subset of $\mathbb{R}$ is μ^*-measurable.

Let μ be the restriction of μ^* to $\mathscr{B}(\mathbb{R})$. The preceding steps of our proof, together with Theorem 1.3.6, show that μ is a measure and that it satisfies $\mu((-\infty, x]) = F(x)$ at each x in $\mathbb{R}$. Since F is bounded, while $\mu(\mathbb{R}) = \lim_{n \to \infty} \mu((-\infty, n]) = \lim_{n \to \infty} F(n)$ (Proposition 1.2.5), the measure μ is finite.

Finally we check the uniqueness of μ. Let μ be as constructed above, and let ν be a possibly different measure such that $\nu((-\infty, x]) = F(x)$ holds for each x in $\mathbb{R}$. We first show that

$$\nu(A) \leq \mu(A) \tag{8}$$

is true for each Borel subset A of $\mathbb{R}$. To see this, note that if A is a Borel set and if $\{(a_n, b_n]\}$ is a sequence such that $A \subseteq \cup_n (a_n, b_n]$, then (according to (6), applied to ν)

$$\nu(A) \leq \sum_n \nu((a_n, b_n]) = \sum_n (F(b_n) - F(a_n)). \tag{9}$$

Since $\mu^*(A)$ was defined to be the infimum of the set of values that can occur as sums on the right side of (9), inequality (8) follows. If we apply inequality (8) to A and to A^c, we find

$$v(\mathbb{R}) = v(A) + v(A^c) \le \mu(A) + \mu(A^c) = \mu(\mathbb{R}).$$

Since $v(\mathbb{R}) = \mu(\mathbb{R}) < +\infty$, it follows that $v(A)$ and $v(A^c)$ are equal to $\mu(A)$ and $\mu(A^c)$, respectively. With this the proof that $v = \mu$ is complete. $\square$

The uniqueness assertion on Proposition 1.3.10 can also be proved by means of other standard techniques; see, for example, the discussion following the proof of Corollary 1.6.3.

Exercises

1. Define functions $\mu_1^*, \ldots, \mu_6^*$ on $\mathscr{P}(\mathbb{R})$ by

$$\mu_1^*(A) = \begin{cases} 0 & \text{if } A \text{ is empty,} \\ 1 & \text{if } A \text{ is nonempty,} \end{cases}$$

$$\mu_2^*(A) = \begin{cases} 0 & \text{if } A \text{ is empty,} \\ +\infty & \text{if } A \text{ is nonempty,} \end{cases}$$

$$\mu_3^*(A) = \begin{cases} 0 & \text{if } A \text{ is bounded,} \\ 1 & \text{if } A \text{ is unbounded,} \end{cases}$$

$$\mu_4^*(A) = \begin{cases} 0 & \text{if } A \text{ is empty,} \\ 1 & \text{if } A \text{ is nonempty and bounded,} \\ +\infty & \text{if } A \text{ is unbounded,} \end{cases}$$

$$\mu_5^*(A) = \begin{cases} 0 & \text{if } A \text{ is countable,} \\ 1 & \text{if } A \text{ is uncountable,} \end{cases}$$

$$\mu_6^*(A) = \begin{cases} 0 & \text{if } A \text{ is countable, and} \\ +\infty & \text{if } A \text{ is uncountable.} \end{cases}$$

   (a) Which of μ_2^*, μ_3^*, μ_4^*, and μ_6^* are outer measures? (We noted in Examples 1.3.1(a) and 1.3.1(b) that μ_1^* and μ_5^* are outer measures.)
   (b) For each i such that μ_i^* is an outer measure determine the μ_i^*-measurable subsets of $\mathbb{R}$.
2. Let C be a countable subset of $\mathbb{R}$. Using only the definition of λ^*, show that $\lambda^*(C) = 0$.

3. Show that for each subset A of $\mathbb{R}$ there is a Borel subset B of $\mathbb{R}$ that includes A and satisfies $\lambda(B) = \lambda^*(A)$.

4. Let $F \colon \mathbb{R} \to \mathbb{R}$ be a bounded, nondecreasing, and right-continuous function that satisfies $\lim_{x \to -\infty} F(x) = 0$, and let μ be the measure on $(\mathbb{R}, \mathscr{B}(\mathbb{R}))$ that is associated to F by Proposition 1.3.10. Show that if a and b belong to $\mathbb{R}$ and satisfy $a < b$, then

$$\mu((-\infty, b)) = F(b-),$$

$$\mu((a, b)) = F(b-) - F(a),$$

$$\mu([a, b]) = F(b) - F(a-), \text{ and}$$

$$\mu([a, b)) = F(b-) - F(a-).$$

5. Let X be a set, let $\mathscr{A}$ be an algebra of subsets of X, and let μ be a finitely additive measure on $\mathscr{A}$. For each subset A of X let $\mu^*(A)$ be the infimum of the set of sums $\sum_{k=1}^{\infty} \mu(A_k)$, where $\{A_k\}$ ranges over the sequences of sets in $\mathscr{A}$ for which $A \subseteq \cup_{k=1}^{\infty} A_k$.
   (a) Show that μ^* is an outer measure on X.
   (b) Show that each set in $\mathscr{A}$ is μ^*-measurable.
   (c) Show that if μ is countably additive (in the sense that $\mu(\cup_k A_k) = \sum_k \mu(A_k)$ holds whenever $\{A_k\}$ is a sequence of disjoint sets in $\mathscr{A}$ for which $\cup_k A_k$ belongs to $\mathscr{A}$), then each A in $\mathscr{A}$ satisfies $\mu(A) = \mu^*(A)$.
   (d) Conclude that if μ is a countably additive measure on the algebra $\mathscr{A}$, then there is a countably additive measure on $\sigma(\mathscr{A})$ that agrees with μ on $\mathscr{A}$.

6. (Continuation.) Let X, $\mathscr{A}$, μ, and μ^* be as in Exercise 5, and assume that μ is countably additive.
   (a) Show that if ν is a countably additive measure on $\sigma(\mathscr{A})$ that agrees with μ on $\mathscr{A}$, then $\nu(A) \leq \mu^*(A)$ holds for each A in $\sigma(A)$.
   (b) Conclude that if μ is finite (or if X is the union of a sequence of sets that belong to $\mathscr{A}$ and have finite measure under μ), then μ can be extended to a countably additive measure on $\sigma(\mathscr{A})$ in only one way.

7. Show that a subset B of $\mathbb{R}$ is Lebesgue measurable if and only if

$$\lambda^*(I) = \lambda^*(I \cap B) + \lambda^*(I \cap B^c)$$

holds for each open subinterval I of $\mathbb{R}$.

8. Let I be a bounded subinterval of $\mathbb{R}$. Show that a subset B of I is Lebesgue measurable if and only if it satisfies $\lambda^*(I) = \lambda^*(B) + \lambda^*(I \cap B^c)$.

9. Let λ^* be Lebesgue outer measure on $\mathbb{R}$, and let π be the projection of $\mathbb{R}^2$ onto $\mathbb{R}$ given by $\pi(x, y) = x$. Define a function $\mu^* \colon \mathscr{P}(\mathbb{R}^2) \to [0, +\infty]$ by $\mu^*(A) = \lambda^*(\pi(A))$.
   (a) Show that μ^* is an outer measure on $\mathbb{R}^2$.
   (b) Show that a subset B of $\mathbb{R}^2$ is measurable for the outer measure μ^* defined in this exercise if and only if there are Lebesgue measurable subsets B_0 and B_1 of $\mathbb{R}$ such that $B_0 \subseteq B_1$, $\lambda^*(B_1 - B_0) = 0$, and $B_0 \times \mathbb{R} \subseteq B \subseteq B_1 \times \mathbb{R}$.

1.4 Lebesgue Measure

This section contains a number of the basic properties of Lebesgue measure on $\mathbb{R}^d$. The reader who wants to move quickly on to Chap. 2 might restrict his or her attention to Proposition 1.4.1, Proposition 1.4.4, and Theorem 1.4.9.

Proposition 1.4.1. *Let A be a Lebesgue measurable subset of* $\mathbb{R}^d$*. Then*

(a) $\lambda(A) = \inf\{\lambda(U) : U$ *is open and* $A \subseteq U\}$*, and*
(b) $\lambda(A) = \sup\{\lambda(K) : K$ *is compact and* $K \subseteq A\}$*.*

Proposition 1.4.1 can be put more briefly, namely as the assertion that Lebesgue measure is *regular*. In the interest of simplicity, however, we will delay the study and even the definition of regularity until Sect. 1.5 and Chap. 7.

Proof. Note that the monotonicity of λ implies that

$$\lambda(A) \leq \inf\{\lambda(U) : U \text{ is open and } A \subseteq U\}$$

and

$$\lambda(A) \geq \sup\{\lambda(K) : K \text{ is compact and } K \subseteq A\}.$$

Hence we need only prove the reverse inequalities.

We begin with part (a). Since the required equality clearly holds if $\lambda(A) = +\infty$, we can assume that $\lambda(A) < +\infty$. Let ε be an arbitrary positive number. Then according to the definition of Lebesgue measure, there is a sequence $\{R_i\}$ of open d-dimensional intervals such that $A \subseteq \cup_i R_i$ and $\sum_i \text{vol}(R_i) < \lambda(A) + \varepsilon$. Let U be the union of these intervals. Then U is open, $A \subseteq U$, and (see Propositions 1.2.4 and 1.3.4)

$$\lambda(U) \leq \sum_i \lambda(R_i) = \sum_i \text{vol}(R_i) < \lambda(A) + \varepsilon.$$

Since ε is arbitrary, part (a) is proved.

We turn to part (b) and deal first with the case where A is bounded. Let C be a closed and bounded set that includes A, and let ε be an arbitrary positive number. Use part (a) to choose an open set U that includes $C - A$ and satisfies

$$\lambda(U) < \lambda(C - A) + \varepsilon. \tag{1}$$

Let $K = C - U$. (Drawing a sketch might help the reader.) Then K is a closed and bounded (and hence compact) subset of A; furthermore, $C \subseteq K \cup U$ and so

$$\lambda(C) \leq \lambda(K) + \lambda(U). \tag{2}$$

Inequalities (1) and (2) (and the fact that $\lambda(C - A) = \lambda(C) - \lambda(A)$) imply that $\lambda(A) - \varepsilon < \lambda(K)$. Since ε was arbitrary, part (b) is proved in the case where A is bounded.

Finally, consider the case where A is not bounded. Suppose that b is a real number less than $\lambda(A)$; we will produce a compact subset K of A such that $b < \lambda(K)$. Let $\{A_j\}$ be an increasing sequence of bounded measurable subsets of A such that $A = \cup_j A_j$ (for example, we might let A_j be the intersection of A with the closed ball of radius j about the origin). Proposition 1.2.5 implies that $\lambda(A) = \lim_j \lambda(A_j)$, and so we can choose j_0 such that $\lambda(A_{j_0}) > b$. Now apply to A_{j_0} the weakened form of part (b) that was proved in the preceding paragraph; this gives a compact subset K of A_{j_0} (and hence of A) such that $\lambda(K) > b$. Since b was an arbitrary number less than $\lambda(A)$, the proof is complete. $\square$

The following lemma will be needed for the proof of Proposition 1.4.3. In this lemma we will be dealing with a certain collection of half-open cubes, namely with those that have the form

$$\{(x_1,\ldots,x_d) : j_i 2^{-k} \le x_i < (j_i+1)2^{-k} \text{ for } i = 1, \ldots, d\} \tag{3}$$

for some integers $j_1, \ldots, j_d$ and some positive integer k.

Lemma 1.4.2. *Each open subset of $\mathbb{R}^d$ is the union of a countable disjoint collection of half-open cubes, each of which is of the form given in expression (3).*

Proof. For each positive integer k let $\mathscr{C}_k$ be the collection of all cubes of the form

$$\{(x_1,\ldots,x_d) : j_i 2^{-k} \le x_i < (j_i+1)2^{-k} \text{ for } i = 1, \ldots, d\},$$

where $j_1, \ldots, j_d$ are arbitrary integers. It is easy to see that
(a) each $\mathscr{C}_k$ is a countable partition of $\mathbb{R}^d$, and
(b) if $k_1 < k_2$, then each cube in $\mathscr{C}_{k_2}$ is included in some cube in $\mathscr{C}_{k_1}$.

The reader should keep these facts about the family $\{\mathscr{C}_k\}$ in mind when checking that the collection $\mathscr{D}$ defined below has the properties claimed for it.

Suppose that U is an open subset of $\mathbb{R}^d$. We construct a collection $\mathscr{D}$ of cubes inductively by letting $\mathscr{D}$ be empty at the start, and then at step k (for $k = 1, 2, \ldots$) adding to $\mathscr{D}$ those cubes in $\mathscr{C}_k$ that are included in U but are disjoint from all the cubes put into $\mathscr{D}$ at earlier steps. It is clear that $\mathscr{D}$ is a countable disjoint collection of cubes whose union is included in U. It remains only to check that its union includes U. Let x be a member of U. Since U is open, the cube in $\mathscr{C}_k$ that contains x is included in U if k is sufficiently large. Let k_0 be the smallest such k. Then the cube in $\mathscr{C}_{k_0}$ that contains x belongs to $\mathscr{D}$, and so x belongs to the union of the cubes in $\mathscr{D}$.
$\square$

Proposition 1.4.3. *Lebesgue measure is the only measure on $(\mathbb{R}^d, \mathscr{B}(\mathbb{R}^d))$ that assigns to each d-dimensional interval, or even to each half-open cube of the form given in expression (3), its volume.*

Proof. That Lebesgue measure does assign to each d-dimensional interval its volume was noted in Sect. 1.3. So we need only assume that μ is a measure on $(\mathbb{R}^d, \mathscr{B}(\mathbb{R}^d))$ that assigns to each cube of the form given in expression (3) its volume and prove that $\mu = \lambda$. First suppose that U is an open subset of $\mathbb{R}^d$. Then according to Lemma 1.4.2 there is a disjoint sequence $\{C_j\}$ of half-open cubes that have the form given in expression (3) and whose union is U, and so

$$\mu(U) = \sum_j \mu(C_j) = \sum_j \lambda(C_j) = \lambda(U);$$

hence μ and λ agree on the open subsets of $\mathbb{R}^d$. Next suppose that A is an arbitrary Borel subset of $\mathbb{R}^d$. If U is an open subset of $\mathbb{R}^d$ that includes A, then $\mu(A) \leq \mu(U) = \lambda(U)$; it follows that $\mu(A) \leq \inf\{\lambda(U) : U \text{ is open and } A \subseteq U\}$. The regularity of λ (Proposition 1.4.1) now implies that

$$\mu(A) \leq \lambda(A). \tag{4}$$

We need to show that this inequality can be replaced with an equality. First suppose that A is a bounded Borel subset of $\mathbb{R}^d$ and that V is a bounded open set that includes A. Then inequality (4), applied to the sets A and $V - A$, implies that

$$\mu(V) = \mu(A) + \mu(V - A) \leq \lambda(A) + \lambda(V - A) = \lambda(V);$$

since the extreme members of this inequality are equal, and since $\mu(A)$ and $\mu(V-A)$ are no larger than $\lambda(A)$ and $\lambda(V-A)$, respectively, it follows that $\mu(A)$ and $\lambda(A)$ are equal. Finally, an arbitrary Borel subset A of $\mathbb{R}^d$ is the union of a sequence of disjoint bounded Borel sets and so must satisfy $\mu(A) = \lambda(A)$. $\square$

For each element x and subset A of $\mathbb{R}^d$ we will denote by $A + x$ the subset of $\mathbb{R}^d$ defined by

$$A + x = \{y \in \mathbb{R}^d : y = a + x \text{ for some } a \text{ in } A\};$$

the set $A + x$ is called the *translate* of A by x. We turn to the invariance of Lebesgue measure under such translations.

Proposition 1.4.4. *Lebesgue outer measure on $\mathbb{R}^d$ is translation invariant, in the sense that if $x \in \mathbb{R}^d$ and $A \subseteq \mathbb{R}^d$, then $\lambda^*(A) = \lambda^*(A + x)$. Furthermore, a subset B of $\mathbb{R}^d$ is Lebesgue measurable if and only if $B + x$ is Lebesgue measurable.*

Proof. The equality of $\lambda^*(A)$ and $\lambda^*(A + x)$ follows from the definition of λ^* and the fact that the volume of a d-dimensional interval is invariant under translation. The second assertion follows from the first, together with the definition of a Lebesgue measurable set—note that a set B satisfies

$$\lambda^*(A - x) = \lambda^*((A - x) \cap B) + \lambda^*((A - x) \cap B^c)$$

for all sets $A - x$ if and only if $B + x$ satisfies

$$\lambda^*(A) = \lambda^*(A \cap (B + x)) + \lambda^*(A \cap (B + x)^c)$$

for all sets A. $\square$

Lebesgue measure on $(\mathbb{R}^d, \mathscr{B}(\mathbb{R}^d))$ is characterized up to constant multiples by the following result; see Chap. 9 for analogous results that hold in more general situations.

Proposition 1.4.5. *Let μ be a nonzero measure on $(\mathbb{R}^d, \mathscr{B}(\mathbb{R}^d))$ that is finite on the bounded Borel subsets of $\mathbb{R}^d$ and is translation invariant, in the sense that $\mu(A) = \mu(A + x)$ holds for each A in $\mathscr{B}(\mathbb{R}^d)$ and each x in $\mathbb{R}^d$. Then there is a positive number c such that $\mu(A) = c\lambda(A)$ holds for each A in $\mathscr{B}(\mathbb{R}^d)$.*

Note that for the concept of translation invariance for measures on $(\mathbb{R}^d, \mathscr{B}(\mathbb{R}^d))$ to make sense, the Borel σ-algebra on $\mathbb{R}^d$ must be translation invariant, in the sense that if $A \in \mathscr{B}(\mathbb{R}^d)$ and $x \in \mathbb{R}^d$, then $A + x \in \mathscr{B}(\mathbb{R}^d)$. To check this translation invariance of $\mathscr{B}(\mathbb{R}^d)$, note that $\{A \subseteq \mathbb{R}^d : A + x \in \mathscr{B}(\mathbb{R}^d)\}$ is a σ-algebra that contains the open sets and hence includes $\mathscr{B}(\mathbb{R}^d)$.

Proof. Let $C = \{(x_1, \dots, x_d) : 0 \le x_i < 1 \text{ for each } i\}$, and let $c = \mu(C)$. Then c is finite (since μ is finite on the bounded Borel sets) and positive (if it were 0, then $\mathbb{R}^d$, as the union of a sequence of translates of C, would have measure zero under μ). Define a measure ν on $\mathscr{B}(\mathbb{R}^d)$ by letting $\nu(A) = (1/c)\mu(A)$ hold for each A in $\mathscr{B}(\mathbb{R}^d)$. Then ν is translation invariant, and it assigns to the set C defined above its Lebesgue measure, namely 1. If D is a half-open cube that has the form given in expression (3) and whose edges have length 2^{-k}, then C is the union of 2^{dk} translates of D, and so

$$2^{dk}\nu(D) = \nu(C) = \lambda(C) = 2^{dk}\lambda(D);$$

thus ν and λ agree on all such cubes. Proposition 1.4.3 now implies that $\nu = \lambda$ and hence that $\mu = c\lambda$. $\qquad\square$

Example 1.4.6 (The Cantor Set). We should note a few facts about the *Cantor set*, a set which turns out to be a useful source of examples. Recall that it is defined as follows. Let K_0 be the interval $[0, 1]$. Form K_1 by removing from K_0 the interval $(1/3, 2/3)$. Thus $K_1 = [0, 1/3] \cup [2/3, 1]$. Continue this procedure, forming K_n by removing from K_{n-1} the open middle third of each of the intervals making up K_{n-1}. Thus K_n is the union of 2^n disjoint closed intervals, each of length $(1/3)^n$. The Cantor set (which we will temporarily denote by K) is the set of points that remain; thus $K = \cap_n K_n$.

Of course K is closed and bounded. Furthermore, K has no interior points, since an open interval included in K would for each n be included in one of the intervals making up K_n and so would have length at most $(1/3)^n$. The cardinality of K is that of the continuum: it is easy to check that the map that assigns to a sequence $\{z_n\}$ of 0's and 1's the number $\sum_{n=1}^{\infty} 2z_n/3^n$ is a bijection of the set of all such sequences onto K; hence the cardinality of K is that of the set of all sequences of 0's and 1's and so that of the continuum (see Appendix A). $\qquad\square$

Proposition 1.4.7. *The Cantor set is a compact set that has the cardinality of the continuum but has Lebesgue measure zero.*

Proof. We have already noted that the Cantor set (again call it K) is compact and has the cardinality of the continuum. To compute the measure of K, note that for each n it is included in the set K_n constructed above and that $\lambda(K_n) = (2/3)^n$. Thus $\lambda(K) \leq (2/3)^n$ holds for each n, and so $\lambda(K)$ must be zero. (For an alternative proof, check that the sum of the measures of the intervals removed from $[0,1]$ during the construction of K is the sum of the geometric series

$$\frac{1}{3} + \frac{2}{3} \cdot \frac{1}{3} + \left(\frac{2}{3}\right)^2 \cdot \frac{1}{3} + \left(\frac{2}{3}\right)^3 \cdot \frac{1}{3} + \cdots,$$

and so is 1.) $\qquad\square$

Example 1.4.8 (A Nonmeasurable Set). We now return to one of the promises made in Sect. 1.3 and prove that there is a subset of $\mathbb{R}$ that is not Lebesgue measurable. Note that our proof of this uses the axiom of choice.[4] Whether the use of this axiom is essential was an open question until the mid-1960s, when R.M. Solovay showed that if a certain consistency assumption holds, then the existence of a subset of $\mathbb{R}$ that is not Lebesgue measurable cannot be proved from the axioms of Zermelo–Frankel set theory without the use of the axiom of choice.[5]

Theorem 1.4.9. *There is a subset of $\mathbb{R}$, and in fact of the interval $(0,1)$, that is not Lebesgue measurable.*

Proof. Define a relation $\sim$ on $\mathbb{R}$ by letting $x \sim y$ hold if and only if $x - y$ is rational. It is easy to check that $\sim$ is an equivalence relation: it is reflexive ($x \sim x$ holds for each x), symmetric ($x \sim y$ implies $y \sim x$), and transitive ($x \sim y$ and $y \sim z$ imply $x \sim z$). Note that each equivalence class under $\sim$ has the form $\mathbb{Q} + x$ for some x and so is dense in $\mathbb{R}$. Since these equivalence classes are disjoint, and since each intersects the interval $(0,1)$, we can use the axiom of choice to form a subset E of $(0,1)$ that contains exactly one element from each equivalence class. We will prove that the set E is not Lebesgue measurable.

Let $\{r_n\}$ be an enumeration of the rational numbers in the interval $(-1,1)$, and for each n let $E_n = E + r_n$. We will check that

(a) the sets E_n are disjoint,

(b) $\cup_n E_n$ is included in the interval $(-1,2)$, and

(c) the interval $(0,1)$ is included in $\cup_n E_n$.

To check (a), note that if $E_m \cap E_n \neq \varnothing$, then there are elements e and e' of E such that $e + r_m = e' + r_n$; it follows that $e \sim e'$ and hence that $e = e'$ and $m = n$. Thus (a) is proved. Assertion (b) follows from the inclusion $E \subseteq (0,1)$ and the fact that each term of the sequence $\{r_n\}$ belongs to $(-1,1)$. Now consider assertion (c). Let x be

---

[4] See items A.12 and A.13 in Appendix A.

[5] For details, see Solovay [110].

an arbitrary member of $(0,1)$, and let e be the member of E that satisfies $x \sim e$. Then $x - e$ is rational and belongs to $(-1, 1)$ (recall that both x and e belong to $(0, 1)$) and so has the form r_n for some n. Hence $x \in E_n$, and assertion (c) is proved.

Suppose that the set E is Lebesgue measurable. Then for each n the set E_n is measurable (Proposition 1.4.4), and so property (a) above implies that

$$\lambda(\cup_n E_n) = \sum_n \lambda(E_n);$$

furthermore, the translation invariance of λ implies that $\lambda(E_n) = \lambda(E)$ holds for each n. Hence if $\lambda(E) = 0$, then $\lambda(\cup_n E_n) = 0$, contradicting assertion (c) above, while if $\lambda(E) \neq 0$, then $\lambda(\cup_n E_n) = +\infty$, contradicting assertion (b). Thus the assumption that E is measurable leads to a contradiction, and the proof is complete.

$\square$

Let A be a subset of $\mathbb{R}$. Then $\mathrm{diff}(A)$ is the subset of $\mathbb{R}$ defined by

$$\mathrm{diff}(A) = \{x - y : x \in A \text{ and } y \in A\}.$$

The following fact about such sets is occasionally useful.

Proposition 1.4.10. *Let A be a Lebesgue measurable subset of $\mathbb{R}$ such that $\lambda(A) > 0$. Then $\mathrm{diff}(A)$ includes an open interval that contains 0.*

Proof. According to Proposition 1.4.1, there is a compact subset K of A such that $\lambda(K) > 0$. Since $\mathrm{diff}(K)$ is then included in $\mathrm{diff}(A)$, it is enough to prove that $\mathrm{diff}(K)$ includes an open interval that contains 0. Note that a real number x belongs to $\mathrm{diff}(K)$ if and only if K intersects $x + K$; thus it suffices to prove that if $|x|$ is sufficiently small, then K intersects $x + K$.

Use Proposition 1.4.1 to choose an open set U such that $K \subseteq U$ and $\lambda(U) < 2\lambda(K)$. The distances between the points in K and the points outside U are bounded away from 0 (since the distance from a point x of U to the complement of U is a continuous strictly positive function of x and so has a positive minimum on the compact set K; see D.27 and D.18). Thus there is a positive number ε such that if $|x| < \varepsilon$, then $x + K$ is included in U. Suppose that $|x| < \varepsilon$. If $x + K$ were disjoint from K, then it would follow from the translation invariance of λ and the relation $x + K \subseteq U$ that

$$2\lambda(K) = \lambda(K) + \lambda(x + K) = \lambda(K \cup (x + K)) \leq \lambda(U).$$

However this contradicts the inequality $\lambda(U) < 2\lambda(K)$, and so K and $x + K$ cannot be disjoint. Therefore, $x \in \mathrm{diff}(K)$. Consequently the interval $(-\varepsilon, \varepsilon)$ is included in $\mathrm{diff}(K)$, and thus in $\mathrm{diff}(A)$. $\square$

We can use Proposition 1.4.10, plus a modification of the proof of Theorem 1.4.9, to prove the following rather strong result (see the remark at the end of this section and the one following the proof of Proposition 1.5.4).

Proposition 1.4.11. *There is a subset A of $\mathbb{R}$ such that each Lebesgue measurable set that is included in A or in A^c has Lebesgue measure zero.*

Proof. Define subsets G, G_0, and G_1 of $\mathbb{R}$ by

$$G = \{x : x = r + n\sqrt{2} \text{ for some } r \text{ in } \mathbb{Q} \text{ and } n \text{ in } \mathbb{Z}\},$$

$$G_0 = \{x : x = r + 2n\sqrt{2} \text{ for some } r \text{ in } \mathbb{Q} \text{ and } n \text{ in } \mathbb{Z}\}, \text{ and}$$

$$G_1 = \{x : x = r + (2n+1)\sqrt{2} \text{ for some } r \text{ in } \mathbb{Q} \text{ and } n \text{ in } \mathbb{Z}\}.$$

It is easy to see that G and G_0 are subgroups of $\mathbb{R}$ (under addition), that G_0 and G_1 are disjoint, that $G_1 = G_0 + \sqrt{2}$, and that $G = G_0 \cup G_1$. Define a relation $\sim$ on $\mathbb{R}$ by letting $x \sim y$ hold when $x - y \in G$; the relation $\sim$ is then an equivalence relation on $\mathbb{R}$. Use the axiom of choice to form a subset E of $\mathbb{R}$ that contains exactly one representative of each equivalence class of $\sim$. Let $A = E + G_0$ (that is, let A consist of the points that have the form $e + g_0$ for some e in E and some g_0 in G_0).

We now show that there does not exist a Lebesgue measurable subset B of A such that $\lambda(B) > 0$. For this let us assume that such a set exists; we will derive a contradiction. Proposition 1.4.10 implies that there is an interval $(-\varepsilon, \varepsilon)$ that is included in diff(B) and hence in diff(A). Since G_1 is dense in $\mathbb{R}$, it meets the interval $(-\varepsilon, \varepsilon)$ and hence meets diff(A). This, however, is impossible, since each element of diff(A) is of the form $e_1 - e_2 + g_0$ (where e_1 and e_2 belong to E and g_0 belongs to G_0) and so cannot belong to G_1 (the relation $e_1 - e_2 + g_0 = g_1$ would imply that $e_1 = e_2$ and $g_0 = g_1$, contradicting the disjointness of G_0 and G_1). This completes our proof that every Lebesgue measurable subset of A must have Lebesgue measure zero.

It is easy to check that $A^c = E + G_1$ and hence that $A^c = A + \sqrt{2}$. It follows that each Lebesgue measurable subset of A^c is of the form $B + \sqrt{2}$ for some Lebesgue measurable subset B of A. Since A has no Lebesgue measurable subsets of positive measure, it follows that A^c also has no such subsets, and with this the proof is complete. □

Note that the set A of Proposition 1.4.11 is not Lebesgue measurable: if it were, then both A and A^c would include (in fact, would be) Lebesgue measurable sets of positive Lebesgue measure. Thus we could have presented Theorem 1.4.9 as a corollary of Proposition 1.4.11. (Of course, the proof of Theorem 1.4.9 presented earlier is simpler than the proofs of Propositions 1.4.10 and 1.4.11 taken together and is in fact a classical and well-known argument; hence it was included.)

Exercises

1. Prove that under Lebesgue measure on $\mathbb{R}^2$
   (a) every straight line has measure zero, and
   (b) every circle has measure zero.

2. Let A be a subset of $\mathbb{R}^d$. Show that the conditions

   (i) A is Lebesgue measurable,
   (ii) A is the union of an F_σ and a set of Lebesgue measure zero, and
   (iii) there is a set B that is an F_σ and satisfies $\lambda^*(A \triangle B) = 0$

   are equivalent.
3. Let T be a rotation of $\mathbb{R}^2$ about the origin (or, more generally, a linear map from $\mathbb{R}^d$ to $\mathbb{R}^d$ that preserves distances).
   (a) Show that a subset A of $\mathbb{R}^2$ (or of $\mathbb{R}^d$) is Borel if and only if $T(A)$ is Borel. (Hint: See the remark following the statement of Proposition 1.4.5.)
   (b) Show that each Borel subset A of $\mathbb{R}^2$ (or of $\mathbb{R}^d$) satisfies $\lambda(A) = \lambda(T(A))$. (Hint: Use Proposition 1.4.5.)
4. Show that for each number α that satisfies $0 < \alpha < 1$ there is a closed subset C of $[0,1]$ that satisfies $\lambda(C) = \alpha$ and includes no nonempty open set. (Hint: Imitate the construction of the Cantor set.)
5. Show that there is a Borel subset A of $\mathbb{R}$ such that $0 < \lambda(I \cap A) < \lambda(I)$ holds whenever I is a bounded open subinterval of $\mathbb{R}$.
6. Show that if B is a subset of $\mathbb{R}$ that satisfies $\lambda^*(B) > 0$, then B includes a set that is not Lebesgue measurable. (Hint: Use Proposition 1.4.11.)
7. Show that there exists a decreasing sequence $\{A_n\}$ of subsets of $[0,1]$ such that $\lambda^*(A_n) = 1$ holds for each n, but for which $\cap_n A_n = \varnothing$. (Hint: Let B be a Hamel basis[6] for $\mathbb{R}$ as a vector space over $\mathbb{Q}$, and let $\{B_n\}$ be a strictly increasing sequence of sets such that $B = \cup_n B_n$. For each n let V_n be the subspace of $\mathbb{R}$ spanned by B_n, and let $A_n = [0,1] \cap V_n^c$. Use Proposition 1.4.10 to show that each Borel subset of V_n has Lebesgue measure zero and hence that $\lambda^*(A_n) = 1$.)

1.5   Completeness and Regularity

Let $(X, \mathscr{A}, \mu)$ be a measure space. The measure μ (or the measure space $(X, \mathscr{A}, \mu)$) is *complete* if the relations $A \in \mathscr{A}$, $\mu(A) = 0$, and $B \subseteq A$ together imply that $B \in \mathscr{A}$. It is sometimes convenient to call a subset B of X μ-*negligible* (or μ-*null*) if there is a subset A of X such that $A \in \mathscr{A}$, $B \subseteq A$, and $\mu(A) = 0$. Thus the measure μ is complete if and only if every μ-negligible subset of X belongs to $\mathscr{A}$.

It follows from Proposition 1.3.5 that if μ^* is an outer measure on the set X and if $\mathscr{M}_{\mu^*}$ is the σ-algebra of all μ^*-measurable subsets of X, then the restriction of μ^* to $\mathscr{M}_{\mu^*}$ is complete. In particular, Lebesgue measure on the σ-algebra of

---

[6]This means that B spans $\mathbb{R}$ (i.e., that $\mathbb{R}$ is the smallest linear subspace of $\mathbb{R}$ that includes B) and that no proper subset of B spans $\mathbb{R}$. The axiom of choice implies that such a set B exists; see, for example, Lang [80, Section 5 of Chapter III].

Lebesgue measurable subsets of $\mathbb{R}^d$ is complete. On the other hand, as we will soon see, the restriction of Lebesgue measure to the σ-algebra of Borel subsets of $\mathbb{R}$ is not complete.

It is sometimes convenient to be able to deal with arbitrary subsets of sets of measure zero, and at such times complete measures are desirable. In many such situations the following construction proves useful.

Let $(X, \mathscr{A})$ be a measurable space, and let μ be a measure on $\mathscr{A}$. The *completion* of $\mathscr{A}$ under μ is the collection $\mathscr{A}_\mu$ of subsets A of X for which there are sets E and F in $\mathscr{A}$ such that

$$E \subseteq A \subseteq F \tag{1}$$

and

$$\mu(F - E) = 0. \tag{2}$$

A set that belongs to $\mathscr{A}_\mu$ is sometimes said to be μ-*measurable*.

Suppose that A, E, and F are as in the preceding paragraph. It follows immediately that $\mu(E) = \mu(F)$. Furthermore, if B is a subset of A that belongs to $\mathscr{A}$, then

$$\mu(B) \leq \mu(F) = \mu(E).$$

Hence

$$\mu(E) = \sup\{\mu(B) : B \in \mathscr{A} \text{ and } B \subseteq A\},$$

and so the common value of $\mu(E)$ and $\mu(F)$ depends only on the set A (and the measure μ), and not on the choice of sets E and F satisfying (1) and (2). Thus we can define a function $\overline{\mu} : \mathscr{A}_\mu \to [0, +\infty]$ by letting $\overline{\mu}(A)$ be the common value of $\mu(E)$ and $\mu(F)$, where E and F belong to $\mathscr{A}$ and satisfy (1) and (2). This function $\overline{\mu}$ is called the *completion* of μ.

Proposition 1.5.1. *Let $(X, \mathscr{A})$ be a measurable space, and let μ be a measure on $\mathscr{A}$. Then $\mathscr{A}_\mu$ is a σ-algebra on X that includes $\mathscr{A}$, and $\overline{\mu}$ is a measure on $\mathscr{A}_\mu$ that is complete and whose restriction to $\mathscr{A}$ is μ.*

Proof. It is clear that $\mathscr{A}_\mu$ includes $\mathscr{A}$ (for A in $\mathscr{A}$ let the sets E and F in (1) and (2) equal A), and in particular that $X \in \mathscr{A}_\mu$. Note that the relations $E \subseteq A \subseteq F$ and $\mu(F - E) = 0$ imply the relations $F^c \subseteq A^c \subseteq E^c$ and $\mu(E^c - F^c) = 0$; thus $\mathscr{A}_\mu$ is closed under complementation. Next suppose that $\{A_n\}$ is a sequence of sets in $\mathscr{A}_\mu$. For each n choose sets E_n and F_n in $\mathscr{A}$ such that $E_n \subseteq A_n \subseteq F_n$ and $\mu(F_n - E_n) = 0$. Then $\cup_n E_n$ and $\cup_n F_n$ belong to $\mathscr{A}$ and satisfy $\cup_n E_n \subseteq \cup_n A_n \subseteq \cup_n F_n$ and

$$\mu(\cup_n F_n - \cup_n E_n) \leq \mu(\cup_n (F_n - E_n)) \leq \sum_n \mu(F_n - E_n) = 0;$$

thus $\cup A_n$ belongs to $\mathscr{A}_\mu$. This completes the proof that $\mathscr{A}_\mu$ is a σ-algebra on X that includes $\mathscr{A}$.

Now consider the function $\overline{\mu}$. It is an extension of μ, since for A in $\mathscr{A}$ we can again let E and F equal A. It is clear that $\overline{\mu}$ has nonnegative values and satisfies

$\overline{\mu}(\varnothing) = 0$, and so we need only check its countable additivity. Let $\{A_n\}$ be a sequence of disjoint sets in $\mathscr{A}_\mu$, and for each n again choose sets E_n and F_n in $\mathscr{A}$ that satisfy $E_n \subseteq A_n \subseteq F_n$ and $\mu(F_n - E_n) = 0$. The disjointness of the sets A_n implies the disjointness of the sets E_n, and so we can conclude that

$$\overline{\mu}(\cup_n A_n) = \mu(\cup_n E_n) = \sum_n \mu(E_n) = \sum_n \overline{\mu}(A_n).$$

Thus $\overline{\mu}$ is a measure. It is easy to check that $\overline{\mu}$ is complete. $\square$

We turn to an example.

Proposition 1.5.2. *Lebesgue measure on* $(\mathbb{R}^d, \mathscr{M}_{\lambda*})$ *is the completion of Lebesgue measure on* $(\mathbb{R}^d, \mathscr{B}(\mathbb{R}^d))$.

We begin with the following lemma.

Lemma 1.5.3. *Let A be a Lebesgue measurable subset of* $\mathbb{R}^d$. *Then there exist Borel subsets E and F of* $\mathbb{R}^d$ *such that $E \subseteq A \subseteq F$ and $\lambda(F - E) = 0$.*

Proof. First suppose that A is a Lebesgue measurable subset of $\mathbb{R}^d$ such that $\lambda(A) < +\infty$. For each positive integer n, use Proposition 1.4.1 to choose a compact set K_n such that $K_n \subseteq A$ and $\lambda(A) - 1/n < \lambda(K_n)$ and an open set U_n such that $A \subseteq U_n$ and $\lambda(U_n) < \lambda(A) + 1/n$. Let $E = \cup_n K_n$ and $F = \cap_n U_n$. Then E and F belong to $\mathscr{B}(\mathbb{R}^d)$ and satisfy $E \subseteq A \subseteq F$. The relation

$$\lambda(F - E) \leq \lambda(U_n - K_n) = \lambda(U_n - A) + \lambda(A - K_n) < 2/n$$

holds for each n, and so $\lambda(F - E) = 0$. Thus the lemma is proved in the case where $\lambda(A) < +\infty$.

If A is an arbitrary Lebesgue measurable subset of $\mathbb{R}^d$, then A is the union of a sequence $\{A_n\}$ of Lebesgue measurable sets of finite Lebesgue measure. For each n we can choose Borel sets E_n and F_n such that $E_n \subseteq A_n \subseteq F_n$ and $\lambda(F_n - E_n) = 0$. The sets E and F defined by $E = \cup_n E_n$ and $F = \cup_n F_n$ then satisfy $E \subseteq A \subseteq F$ and $\lambda(F - E) = 0$ (note that $F - E \subseteq \cup_n(F_n - E_n)$). $\square$

Proof of Proposition 1.5.2. Let λ be Lebesgue measure on $(\mathbb{R}^d, \mathscr{B}(\mathbb{R}^d))$, let $\overline{\lambda}$ be the completion of λ, and let λ_m be Lebesgue measure on $(\mathbb{R}^d, \mathscr{M}_{\lambda*})$. Lemma 1.5.3 implies that $\mathscr{M}_{\lambda*}$ is included in the completion of $\mathscr{B}(\mathbb{R}^d)$ under λ and that λ_m is the restriction of $\overline{\lambda}$ to $\mathscr{M}_{\lambda*}$. Thus we need only check that each set A that belongs to the completion of $\mathscr{B}(\mathbb{R}^d)$ under λ is Lebesgue measurable. For such a set A there exist Borel sets E and F such that $E \subseteq A \subseteq F$ and $\lambda(F - E) = 0$. Since $A - E \subseteq F - E$ and $\lambda_m(F - E) = \lambda(F - E) = 0$, the completeness of Lebesgue measure on $\mathscr{M}_{\lambda*}$ implies that $A - E \in \mathscr{M}_{\lambda*}$. Thus A, since it is the union of $A - E$ and E, must belong to $\mathscr{M}_{\lambda*}$. $\square$

We will see in Sect. 2.1 that

(a) there are Lebesgue measurable subsets of $\mathbb{R}$ that are not Borel sets, and
(b) the restriction of Lebesgue measure to $\mathscr{B}(\mathbb{R})$ is not complete.

It should be noted that although replacing a measure space $(X, \mathscr{A}, \mu)$ with its completion $(X, \mathscr{A}_\mu, \overline{\mu})$ enables one to avoid some difficulties, it introduces others. Some difficulties arise because the completed σ-algebra $\mathscr{A}_\mu$ is often more complicated than the original σ-algebra $\mathscr{A}$. Others are caused by the fact that for measures μ and ν defined on a common σ-algebra $\mathscr{A}$, the completions $\mathscr{A}_\mu$ and $\mathscr{A}_\nu$ of $\mathscr{A}$ under μ and ν may not be equal (see Exercise 3). Because of these complications it seems wise whenever possible to avoid arguments that depend on completeness; it turns out that in the basic parts of measure theory this can almost always be done.

Let $(X, \mathscr{A})$ be a measurable space, let μ be a measure on $\mathscr{A}$, and let A be an arbitrary subset of X. Then $\mu^*(A)$, the *outer measure* of A, is defined by

$$\mu^*(A) = \inf\{\mu(B) : A \subseteq B \text{ and } B \in \mathscr{A}\}, \tag{3}$$

and $\mu_*(A)$, the *inner measure* of A, is defined by

$$\mu_*(A) = \sup\{\mu(B) : B \subseteq A \text{ and } B \in \mathscr{A}\}.$$

It is easy to check that $\mu_*(A) \leq \mu^*(A)$ holds for each subset A of X.

Proposition 1.5.4. *Let $(X, \mathscr{A})$ be a measurable space, and let μ be a measure on $(X, \mathscr{A})$. Then the function $\mu^* : \mathscr{P}(X) \to [0, +\infty]$ defined by Eq. (3) is an outer measure (as defined in Sect. 1.3) on X.*

Proof. Certainly μ^* satisfies $\mu^*(\varnothing) = 0$ and is monotone. We turn to its subadditivity. Let $\{A_n\}$ be a sequence of subsets of X. The inequality $\mu^*(\cup_n A_n) \leq \sum_n \mu^*(A_n)$ is clear if $\sum_n \mu^*(A_n) = +\infty$. So suppose that $\sum_n \mu^*(A_n) < +\infty$. Let ε be an arbitrary positive number, and for each n choose a set B_n that belongs to $\mathscr{A}$, includes A_n, and satisfies $\mu(B_n) \leq \mu^*(A_n) + \varepsilon/2^n$. Then the set B defined by $B = \cup_n B_n$ belongs to $\mathscr{A}$, includes $\cup_n A_n$, and satisfies $\mu(B) \leq \sum_n \mu^*(A_n) + \varepsilon$ (see Proposition 1.2.4); thus $\mu^*(\cup_n A_n) \leq \sum_n \mu^*(A_n) + \varepsilon$. Since ε is arbitrary, the proof is complete. $\square$

Note that Proposition 1.4.11 can now be rephrased: there is a subset A of $\mathbb{R}$ such that $\lambda_*(A) = 0$ and $\lambda_*(A^c) = 0$.

Proposition 1.5.5. *Let $(X, \mathscr{A})$ be a measurable space, let μ be a measure on $\mathscr{A}$, and let A be a subset of X such that $\mu^*(A) < +\infty$. Then A belongs to $\mathscr{A}_\mu$ if and only if $\mu_*(A) = \mu^*(A)$.*

Proof. If A belongs to $\mathscr{A}_\mu$, then there are sets E and F that belong to $\mathscr{A}$ and satisfy $E \subseteq A \subseteq F$ and $\mu(F - E) = 0$. Then

$$\mu(E) \leq \mu_*(A) \leq \mu^*(A) \leq \mu(F),$$

and since $\mu(E) = \mu(F)$, the relation $\mu_*(A) = \mu^*(A)$ follows.

One can obtain a proof that the relation $\mu_*(A) = \mu^*(A) < +\infty$ implies that A belongs to $\mathscr{A}_\mu$ by modifying the first paragraph of the proof of Lemma 1.5.3; the details are left to the reader (replace appeals to Proposition 1.4.1 with appeals to the definitions of μ_* and μ^*). $\square$

In this section we have been dealing with one way of approximating sets from above and from below by measurable sets. We turn to another such approximation.

Let $\mathscr{A}$ be a σ-algebra on $\mathbb{R}^d$ that includes the σ-algebra $\mathscr{B}(\mathbb{R}^d)$ of Borel sets. A measure μ on $(\mathbb{R}^d, \mathscr{A})$ is *regular* if

(a) each compact subset K of $\mathbb{R}^d$ satisfies $\mu(K) < +\infty$,
(b) each set A in $\mathscr{A}$ satisfies

$$\mu(A) = \inf\{\mu(U) : U \text{ is open and } A \subseteq U\}, \text{ and}$$

(c) each open subset U of $\mathbb{R}^d$ satisfies

$$\mu(U) = \sup\{\mu(K) : K \text{ is compact and } K \subseteq U\}.$$

Proposition 1.4.1 implies that Lebesgue measure, whether on $(\mathbb{R}^d, \mathscr{M}_{\lambda^*})$ or on $(\mathbb{R}^d, \mathscr{B}(\mathbb{R}^d))$, is regular. Part (b) of that proposition appears to be stronger than condition (c) in the definition of regularity; however, we will see in Chap. 7 that every regular measure on $(\mathbb{R}^d, \mathscr{B}(\mathbb{R}^d))$ satisfies the analogue of part (b) of Proposition 1.4.1. In Chap. 7 we will also see that on more general spaces, the analogue of condition (c) above, rather than of part (b) of Proposition 1.4.1, is the condition that should be used in the definition of regularity.

Proposition 1.5.6. *Let μ be a finite measure on $(\mathbb{R}^d, \mathscr{B}(\mathbb{R}^d))$. Then μ is regular. Moreover, each Borel subset A of $\mathbb{R}^d$ satisfies*

$$\mu(A) = \sup\{\mu(K) : K \subseteq A \text{ and } K \text{ is compact}\}. \tag{4}$$

Let us first prove the following weakened form of Proposition 1.5.6.

Lemma 1.5.7. *Let μ be a finite measure on $(\mathbb{R}^d, \mathscr{B}(\mathbb{R}^d))$. Then each Borel subset A of $\mathbb{R}^d$ satisfies*

$$\mu(A) = \inf\{\mu(U) : A \subseteq U \text{ and } U \text{ is open}\} \text{ and} \tag{5}$$

$$\mu(A) = \sup\{\mu(C) : C \subseteq A \text{ and } C \text{ is closed}\}. \tag{6}$$

Proof. Let $\mathscr{R}$ be the collection of those Borel subsets A of $\mathbb{R}^d$ that satisfy (5) and (6).

We begin by showing that $\mathscr{R}$ contains the open subsets of $\mathbb{R}^d$. Let V be an open subset of $\mathbb{R}^d$. Of course V satisfies

$$\mu(V) = \inf\{\mu(U) : V \subseteq U \text{ and } U \text{ is open}\}.$$

According to Proposition 1.1.6, there is a sequence $\{C_n\}$ of closed subsets of $\mathbb{R}^d$ such that $V = \cup_n C_n$. We can assume that the sequence $\{C_n\}$ is increasing (replace C_n with $\cup_{i=1}^n C_i$ if necessary). Proposition 1.2.5 implies that $\mu(V) = \lim_n \mu(C_n)$, and so V satisfies

$$\mu(V) = \sup\{\mu(C) : C \subseteq V \text{ and } C \text{ is closed}\}.$$

With this we have proved that $\mathscr{R}$ contains all the open subsets of $\mathbb{R}^d$.

It is easy to check (do so) that $\mathscr{R}$ consists of the Borel sets A that satisfy

for each positive ε there exist an open set U and a closed set C

such that $C \subseteq A \subseteq U$ and $\mu(U - C) < \varepsilon$.   (7)

We now show that $\mathscr{R}$ is a σ-algebra. If contains $\mathbb{R}^d$, since $\mathbb{R}^d$ is open. If $A \in \mathscr{R}$, if ε is a positive number, and if C and U are, respectively, closed and open and satisfy $C \subseteq A \subseteq U$ and $\mu(U - C) < \varepsilon$, then U^c and C^c are respectively closed and open and satisfy $U^c \subseteq A^c \subseteq C^c$ and $\mu(C^c - U^c) < \varepsilon$; thus it follows (from (7)) that $\mathscr{R}$ is closed under complementation. Now let $\{A_k\}$ be a sequence of sets in $\mathscr{R}$ and let ε be a positive number. For each k choose a closed set C_k and an open set U_k such that $C_k \subseteq A_k \subseteq U_k$ and $\mu(U_k - C_k) < \varepsilon/2^k$. Let $U = \cup_k U_k$ and $C = \cup_k C_k$. Then U and C satisfy the relations $C \subseteq \cup_k A_k \subseteq U$ and

$$\mu(U - C) \leq \mu(\cup_k(U_k - C_k)) \leq \sum_k (U_k - C_k) < \varepsilon.   \qquad (8)$$

The set U is open, but the set C can fail to be closed. However, for each n the set $\cup_{k=1}^n C_k$ is closed, and it follows from (8), together with the fact that $\mu(U - C) = \lim_n \mu(U - \cup_{k=1}^n C_k)$ that there is a positive integer n such that $\mu(U - \cup_{k=1}^n C_k) < \varepsilon$. Then U and $\cup_{k=1}^n C_k$ are the sets required in (7), and $\mathscr{R}$ is closed under the formation of countable unions.

We have now shown that $\mathscr{R}$ is a σ-algebra on $\mathbb{R}^d$ that contains the open sets. Since $\mathscr{B}(\mathbb{R}^d)$ is the smallest σ-algebra on $\mathbb{R}^d$ that contains the open sets, it follows that $\mathscr{B}(\mathbb{R}^d) \subseteq \mathscr{R}$. With this Lemma 1.5.7 is proved. $\square$

Proof of Proposition 1.5.6. Condition (a) in the definition of regularity follows from the finiteness of μ, while condition (b) follows from Lemma 1.5.7. We turn to condition (c) and Eq. (4). Let A be a Borel subset of $\mathbb{R}^d$ and let ε be a positive number. Then according to Lemma 1.5.7 there is a closed subset C of A such that $\mu(C) > \mu(A) - \varepsilon$. Choose an increasing sequence $\{C_n\}$ of closed and bounded (hence compact) sets whose union is C (these sets can, for example, be constructed by letting $C_n = C \cap \{x \in \mathbb{R}^d : \|x\| \leq n\}$). Proposition 1.2.5 implies that $\mu(C) = \lim_n \mu(C_n)$, and so if n is large enough, then C_n is a compact subset of A such that $\mu(C_n) > \mu(A) - \varepsilon$. Equation (4) and condition (c) follow. $\square$

Exercises

1. Let $(X, \mathscr{A}, \mu)$ be a measure space. Show that $(\mathscr{A}_\mu)_{\overline{\mu}} = \mathscr{A}_\mu$ and $\overline{\overline{\mu}} = \overline{\mu}$.
2. (a) Find the completion of $\mathscr{B}(\mathbb{R})$ under the point mass concentrated at 0.

(b) Let $\mathscr{A}$ be the σ-algebra on $\mathbb{R}^2$ that consists of all unions of (possibly empty) collections of vertical lines. Find the completion of $\mathscr{A}$ under the point mass concentrated at $(0,0)$.

3. Let μ and ν be finite measures on a measurable space $(X,\mathscr{A})$.
   (a) Show by example that $\mathscr{A}_\mu$ and $\mathscr{A}_\nu$ need not be equal.
   (b) Prove or disprove: $\mathscr{A}_\mu = \mathscr{A}_\nu$ if and only if μ and ν have exactly the same sets of measure zero.

4. Show that there is a Lebesgue measurable subset of $\mathbb{R}^2$ whose projection on $\mathbb{R}$ under the map $(x,y) \mapsto x$ is not Lebesgue measurable.

5. Let μ be a measure on $(X,\mathscr{A})$. Show that for each subset A of X there are sets A_0 and A_1 that belong to $\mathscr{A}$ and satisfy the conditions $A_0 \subseteq A \subseteq A_1$, $\mu(A_0) = \mu_*(A)$, and $\mu(A_1) = \mu^*(A)$.

6. Show by example that half of Proposition 1.5.5 can fail if the assumption that $\mu^*(A) < +\infty$ is omitted.

7. Suppose that μ is a measure on $(X,\mathscr{A})$. Show that each subset A of X satisfies $\mu^*(A) + \mu_*(A^c) = \mu(X)$.

8. Show that there is a subset A of the interval $[0,1]$ that satisfies $\lambda^*(A) = 1$ and $\lambda_*(A) = 0$. (Hint: Use Proposition 1.4.11.)

9. Let μ be a σ-finite measure on $(X,\mathscr{A})$, and let μ^* be the outer measure defined in formula (3). Show that $\mathscr{A}_\mu$ is equal to the σ-algebra of μ^*-measurable sets and that $\overline{\mu}$ is the restriction of μ^* to $\mathscr{A}_\mu$.

10. Show that if A is a Lebesgue measurable subset of $\mathbb{R}$, then $\{(x,y) \in \mathbb{R}^2 : x \in A\}$ is a Lebesgue measurable subset of $\mathbb{R}^2$.

11. Let $(X,\mathscr{A})$ be a measurable space, and let C be a subset of X (it is not assumed that C belongs to $\mathscr{A}$).
   (a) Show that the collection of subsets of C that have the form $A \cap C$ for some A in $\mathscr{A}$ is a σ-algebra on C. This σ-algebra is sometimes called the *trace* of $\mathscr{A}$ on C and is denoted by $\mathscr{A}_C$.
   (b) Now suppose that μ is a finite measure on $(X,\mathscr{A})$. Let C_1 be a set that belongs to $\mathscr{A}$, includes C, and satisfies $\mu(C_1) = \mu^*(C)$ (see Exercise 5). Show that if A_1 and A_2 belong to $\mathscr{A}$ and satisfy $A_1 \cap C = A_2 \cap C$, then $\mu(A_1 \cap C_1) = \mu(A_2 \cap C_1)$. Thus we can use the formula $\mu_C(A \cap C) = \mu(A \cap C_1)$ to define a function $\mu_C : \mathscr{A}_C \to [0,+\infty)$.
   (c) Show that $\mu_C(B) = \mu^*(B)$ holds for each B in $\mathscr{A}_C$. Thus μ_C does not depend on the choice of the set C_1.
   (d) Show that μ_C is a measure on $(C,\mathscr{A}_C)$. The measure μ_C is sometimes called the *trace* of μ on C.

12. Let $(X,\mathscr{A})$ be a measurable space, and let C be a subset of X.
   (a) Show that the sets that belong to $\sigma(\mathscr{A} \cup \{C\})$ are exactly those that have the form $(A_1 \cap C) \cup (A_2 \cap C^c)$ for some A_1 and A_2 in $\mathscr{A}$.
   (b) Now suppose that μ is a finite measure on $(X,\mathscr{A})$. Let C_0 and C_1 be $\mathscr{A}$-measurable subsets of C and C^c that satisfy $\mu(C_0) = \mu_*(C)$ and $\mu(C_1) = \mu_*(C^c)$, and let μ_C and μ_{C^c} be the traces of μ on C and C^c (see Exercises 5 and 11). Show that the formulas

$$\mu_0(A) = \mu(A \cap C_0) + \mu_{C^c}(A \cap C^c)$$

and

$$\mu_1(A) = \mu_C(A \cap C) + \mu(A \cap C_1)$$

define measures μ_0 and μ_1 on $\sigma(\mathscr{A} \cup \{C\})$, that these measures agree with μ on $\mathscr{A}$, and that they satisfy $\mu_0(C) = \mu_*(C)$ and $\mu_1(C) = \mu^*(C)$.

(c) Show that for each α between $\mu_*(C)$ and $\mu^*(C)$ there is a measure ν on $\sigma(\mathscr{A} \cup \{C\})$ that agrees with μ on $\mathscr{A}$ and satisfies $\nu(C) = \alpha$. (Hint: Let $\nu = t\mu_0 + (1-t)\mu_1$ for a suitable t.)

1.6 Dynkin Classes

This section is devoted to a technique that is often useful for verifying the equality of measures and the measurability of functions (measurable functions will be defined in Sect. 2.1). We begin with a basic definition.

Let X be a set. A collection $\mathscr{D}$ of subsets of X is a *d-system* (or a *Dynkin class*) on X if

(a) $X \in \mathscr{D}$,
(b) $A - B \in \mathscr{D}$ whenever $A, B \in \mathscr{D}$ and $A \supseteq B$, and
(c) $\cup_n A_n \in \mathscr{D}$ whenever $\{A_n\}$ is an increasing sequence of sets in $\mathscr{D}$.

A collection of subsets of X is a *π-system* on X if it is closed under the formation of finite intersections.

Example 1.6.1. Suppose that X is a set and that $\mathscr{A}$ is a σ-algebra on X. Then $\mathscr{A}$ is certainly a d-system. Furthermore, if μ and ν are finite measures on $\mathscr{A}$ such that $\mu(X) = \nu(X)$, then the collection $\mathscr{S}$ of all sets A that belong to $\mathscr{A}$ and satisfy $\mu(A) = \nu(A)$ is a d-system; it is easy to show by example that $\mathscr{S}$ is not necessarily a σ-algebra (see Exercise 3). The fact that such families $\mathscr{S}$ are d-systems forms the basis for many of the applications of d-systems.   □

Note that the intersection of a nonempty family of d-systems on a set X is a d-system on X and that an arbitrary collection of subsets of X is included in some d-system on X, namely the collection of all subsets of X. Hence if $\mathscr{C}$ is an arbitrary collection of subsets of X, then the intersection of all the d-systems on X that include $\mathscr{C}$ is a d-system on X that includes $\mathscr{C}$; this intersection is the smallest such d-system and is called the d-system *generated* by $\mathscr{C}$. We will sometimes denote this d-system by $d(\mathscr{C})$.

Theorem 1.6.2. *Let X be a set, and let $\mathscr{C}$ be a π-system on X. Then the σ-algebra generated by $\mathscr{C}$ coincides with the d-system generated by $\mathscr{C}$.*

Proof. Let $\mathscr{D}$ be the d-system generated by $\mathscr{C}$, and, as usual, let $\sigma(\mathscr{C})$ be the σ-algebra generated by $\mathscr{C}$. Since every σ-algebra is a d-system, the σ-algebra $\sigma(\mathscr{C})$

is a d-system that includes $\mathscr{C}$; hence $\mathscr{D} \subseteq \sigma(\mathscr{C})$. We can prove the reverse inclusion by showing that $\mathscr{D}$ is a σ-algebra, for then $\mathscr{D}$, as a σ-algebra that includes $\mathscr{C}$, must include the σ-algebra generated by $\mathscr{C}$, namely $\sigma(\mathscr{C})$.

We begin the proof that $\mathscr{D}$ is a σ-algebra by showing that $\mathscr{D}$ is closed under the formation of finite intersections. Define a family $\mathscr{D}_1$ of subsets of X by letting

$$\mathscr{D}_1 = \{A \in \mathscr{D} : A \cap C \in \mathscr{D} \text{ for each } C \text{ in } \mathscr{C}\}.$$

The fact that $\mathscr{C} \subseteq \mathscr{D}$ implies that $X \in \mathscr{D}_1$; furthermore, the identities

$$(A - B) \cap C = (A \cap C) - (B \cap C)$$

and

$$(\cup_n A_n) \cap C = \cup_n (A_n \cap C),$$

together with the fact that $\mathscr{D}$ is a d-system, imply that $\mathscr{D}_1$ is closed under the formation of proper differences and under the formation of unions of increasing sequences of sets. Thus $\mathscr{D}_1$ is a d-system. Since $\mathscr{C}$ is closed under the formation of finite intersections and is included in $\mathscr{D}$, it is included in $\mathscr{D}_1$. Thus $\mathscr{D}_1$ is a d-system that includes $\mathscr{C}$; hence it must include $\mathscr{D}$. With this we have proved that we get a set in $\mathscr{D}$ whenever we take the intersection of a set in $\mathscr{D}$ and a set in $\mathscr{C}$.

Next define $\mathscr{D}_2$ by letting

$$\mathscr{D}_2 = \{B \in \mathscr{D} : A \cap B \in \mathscr{D} \text{ for each } A \text{ in } \mathscr{D}\}.$$

The previous step of this proof shows that $\mathscr{C} \subseteq \mathscr{D}_2$, and a straightforward modification of the argument in the previous step shows that $\mathscr{D}_2$ is a d-system. It follows that $\mathscr{D} \subseteq \mathscr{D}_2$—in other words, that $\mathscr{D}$ is closed under the formation of finite intersections.

It is now easy to complete the proof. Parts (a) and (b) of the definition of a d-system imply that $X \in \mathscr{D}$ and that $\mathscr{D}$ is closed under complementation. As we have just seen, $\mathscr{D}$ is also closed under the formation of finite intersections, and so it is an algebra. Finally $\mathscr{D}$, as a d-system, is closed under the formation of unions of increasing sequences of sets, and so by Proposition 1.1.7 it must be a σ-algebra; with that the proof is complete. $\square$

We turn to some applications of Theorem 1.6.2.

Corollary 1.6.3. *Let $(X, \mathscr{A})$ be a measurable space, and let $\mathscr{C}$ be a π-system on X such that $\mathscr{A} = \sigma(\mathscr{C})$. If μ and ν are finite measures on $\mathscr{A}$ that satisfy $\mu(X) = \nu(X)$ and that satisfy $\mu(C) = \nu(C)$ for each C in $\mathscr{C}$, then $\mu = \nu$.*

Proof. Let $\mathscr{D} = \{A \in \mathscr{A} : \mu(A) = \nu(A)\}$. As we noted above, $\mathscr{D}$ is a d-system. Since $\mathscr{C}$ is a π-system and is included in $\mathscr{D}$, it follows from Theorem 1.6.2 that $\mathscr{D} \supseteq \sigma(\mathscr{C}) = \mathscr{A}$. Thus $\mu(A) = \nu(A)$ holds for each A in $\mathscr{A}$, and the proof is complete. $\square$

Now suppose that μ and v are finite Borel measures on $\mathbb{R}$ such that $\mu(I) = v(I)$ holds for each interval I of the form $(-\infty, b]$. Note that $\mathbb{R}$ is the union of an increasing sequence of intervals of the form $(-\infty, b]$ and hence that $\mu(\mathbb{R}) = v(\mathbb{R})$. Since the collection of all intervals of the form $(-\infty, b]$ is a π-system that generates $\mathscr{B}(\mathbb{R})$ (see Proposition 1.1.4), it follows from Corollary 1.6.3 that $\mu = v$. With this we have another proof of the uniqueness assertion in Proposition 1.3.10.

The following result is essentially an extension of Corollary 1.6.3 to the case of σ-finite measures. Note that it implies that Lebesgue measure is the only measure on $\mathscr{B}(\mathbb{R}^d)$ that assigns to each d-dimensional interval its volume, and so it provides a second proof of part of Proposition 1.4.3.

Corollary 1.6.4. *Let $(X, \mathscr{A})$ be a measurable space, and let $\mathscr{C}$ be a π-system on X such that $\mathscr{A} = \sigma(\mathscr{C})$. If μ and v are measures on $(X, \mathscr{A})$ that agree on $\mathscr{C}$, and if there is an increasing sequence $\{C_n\}$ of sets that belong to $\mathscr{C}$, have finite measure under μ and v, and satisfy $\cup_n C_n = X$, then $\mu = v$.*

Proof. Choose an increasing sequence $\{C_n\}$ of sets that belong to $\mathscr{C}$, have finite measure under μ and v, and satisfy $\cup_n C_n = X$. For each positive integer n define measures μ_n and v_n on $\mathscr{A}$ by $\mu_n(A) = \mu(A \cap C_n)$ and $v_n(A) = v(A \cap C_n)$. Corollary 1.6.3 implies that for each n we have $\mu_n = v_n$. Since

$$\mu(A) = \lim_n \mu_n(A) = \lim_n v_n(A) = v(A)$$

holds for each A in $\mathscr{A}$, the measures μ and v must be equal. □

Exercises

1. Give at least six π-systems on $\mathbb{R}$, each of which generates $\mathscr{B}(\mathbb{R})$.
2. (b) Check that the rectangles of the form considered in part (c) of Proposition 1.1.5, together with the empty set, form a π-system on $\mathbb{R}^d$.
   (b) What is the smallest π-system on $\mathbb{R}^d$ that contains all the half-spaces of the form considered in part (b) of Proposition 1.1.5?
3. Give a measurable space $(X, \mathscr{A})$ and finite measures μ and v on it that satisfy $\mu(X) = v(X)$ but are such that

$$\{A \in \mathscr{A} : \mu(A) = v(A)\}$$

   is not a σ-algebra. (Hint: Don't work too hard; X can be a fairly small finite set.)
4. Show by example that Corollary 1.6.3 would be false if the hypothesis that μ and v are finite were replaced with the hypothesis that they are σ-finite. (See, however, Corollary 1.6.4.)
5. Use Theorem 1.6.2 to give another proof of Proposition 1.5.6. (Hint: Show that the collection consisting of those Borel subsets of $\mathbb{R}^d$ that can be approximated

from below with compact sets and from above with open sets is a d-system, and
that this d-system contains each rectangle of the form considered in part (c) of
Proposition 1.1.5.)

6. Let X be a set. A collection $\mathscr{C}$ of subsets of X is a *monotone class* on X if it is
closed under monotone limits, in the sense that

> (i) if $\{A_n\}$ is an increasing sequence of sets that belong to $\mathscr{C}$, then $\cup_n A_n$
> belongs to $\mathscr{C}$, and
> (ii) if $\{A_n\}$ is a decreasing sequence of sets that belong to $\mathscr{C}$, then $\cap_n A_n$
> belongs to $\mathscr{C}$.

(a) Show that if $\mathscr{A}$ is a collection of subsets of X, then there is a smallest
monotone class on X that includes $\mathscr{A}$. This smallest monotone class is called
the monotone class *generated* by $\mathscr{A}$; let us denote it by $m(\mathscr{A})$.
(b) Prove the *monotone class theorem*: if $\mathscr{A}$ is an algebra of subsets of X, then
$m(\mathscr{A}) = \sigma(\mathscr{A})$. (Hint: Modify the proof of Theorem 1.6.2.)

Notes

Halmos [54] is a standard reference for the theory of measure and integration.
The books by Bartle [3], Berberian [7], Billingsley [8], Bruckner, Bruckner, and
Thomson [23], Dudley [40], Folland [45], Hewitt and Stromberg [59], Munroe [92],
Royden [102], Rudin [105], and Wheeden and Zygmund [127] are also well known
and useful. The reader should see Billingsley [8] and Dudley [40] for applications
to probability theory, Rudin [105] and Benedetto and Czaja [6] for a great variety
of applications to analysis, and Wheeden and Zygmund [127] for applications to
harmonic analysis. Gelbaum and Olmsted [48] contains an interesting collection
of counterexamples. Bogachev's recent two-volume work [15] and Fremlin's five-
volume work [46] are good references. Pap [95] is a collection of survey papers on
measure theory. Federer [44], Krantz and Parks [75], Morgan [89], and Rogers [100]
treat topics in measure theory that are not touched upon here.

Theorem 1.6.2 is due to Dynkin [43] (see also Blumenthal and Getoor [14]).

See Dudley [40] and Bogachev [15] for very thorough historical notes and
bibliographic citations.

Chapter 2
Functions and Integrals

This chapter is devoted to the definition and basic properties of the Lebesgue integral. We first introduce measurable functions—the functions that are simple enough that the integral can be defined for them if their values are not too large (Sect. 2.1). After a brief look in Sect. 2.2 at properties that hold almost everywhere (that is, that may fail on some set of measure zero, as long as they hold everywhere else), we turn to the definition of the Lebesgue integral and to its basic properties (Sects. 2.3 and 2.4). The chapter ends with a sketch of how the Lebesgue integral relates to the Riemann integral (Sect. 2.5) and then with a few more details about measurable functions (Sect. 2.6).

2.1 Measurable Functions

In this section we introduce measurable functions and study some of their basic properties. We begin with the following elementary result.

Proposition 2.1.1. *Let $(X, \mathscr{A})$ be a measurable space, and let A be a subset of X that belongs to $\mathscr{A}$. For a function $f : A \to [-\infty, +\infty]$ the conditions*

(a) *for each real number t the set $\{x \in A : f(x) \leq t\}$ belongs to $\mathscr{A}$,*
(b) *for each real number t the set $\{x \in A : f(x) < t\}$ belongs to $\mathscr{A}$,*
(c) *for each real number t the set $\{x \in A : f(x) \geq t\}$ belongs to $\mathscr{A}$, and*
(d) *for each real number t the set $\{x \in A : f(x) > t\}$ belongs to $\mathscr{A}$*

are equivalent.

Proof. The identity

$$\{x \in A : f(x) < t\} = \bigcup_n \{x \in A : f(x) \leq t - 1/n\}$$

D.L. Cohn, *Measure Theory: Second Edition*, Birkhäuser Advanced Texts Basler Lehrbücher, DOI 10.1007/978-1-4614-6956-8_2, © Springer Science+Business Media, LLC 2013

implies that each of the sets appearing in condition (b) is the union of a sequence of sets appearing in condition (a); hence condition (a) implies condition (b). The sets appearing in condition (c) can be expressed in terms of those appearing in condition (b) by means of the identity

$$\{x \in A : f(x) \geq t\} = A - \{x \in A : f(x) < t\};$$

thus condition (b) implies condition (c). Similar arguments, the details of which are left to the reader, show that condition (c) implies condition (d) and that condition (d) implies condition (a).   □

Let $(X, \mathscr{A})$ be a measurable space, and let A be a subset of X that belongs to $\mathscr{A}$. A function $f \colon A \to [-\infty, +\infty]$ is *measurable with respect to $\mathscr{A}$* if it satisfies one, and hence all, of the conditions of Proposition 2.1.1. A function that is measurable with respect to $\mathscr{A}$ is sometimes called *$\mathscr{A}$-measurable* or, if the σ-algebra $\mathscr{A}$ is clear from context, simply *measurable*. In case $X = \mathbb{R}^d$, a function that is measurable with respect to $\mathscr{B}(\mathbb{R}^d)$ is called *Borel measurable* or a *Borel function*, and a function that is measurable with respect to $\mathscr{M}_{\lambda^*}$ is called *Lebesgue measurable* (recall that $\mathscr{M}_{\lambda^*}$ is the σ-algebra of Lebesgue measurable subsets of $\mathbb{R}^d$). Of course every Borel measurable function on $\mathbb{R}^d$ is Lebesgue measurable.

We turn to a few examples and then to some of the basic facts about measurable functions.

Examples 2.1.2. (a) Let $f \colon \mathbb{R}^d \to \mathbb{R}$ be continuous. Then for each real number t the set $\{x \in \mathbb{R}^d : f(x) < t\}$ is open and so is a Borel set. Thus f is Borel measurable.

(b) Let I be a subinterval of $\mathbb{R}$, and let $f \colon I \to \mathbb{R}$ be nondecreasing. Then for each real number t the set $\{x \in I : f(x) < t\}$ is a Borel set (it is either an interval, a set consisting of only one point, or the empty set). Thus f is Borel measurable.

(c) Let $(X, \mathscr{A})$ be a measurable space, and let B be a subset of X. Then χ_B, the characteristic function of B, is $\mathscr{A}$-measurable if and only if $B \in \mathscr{A}$.

(d) A function is called *simple* if it has only finitely many values. Let $(X, \mathscr{A})$ be a measurable space, let $f \colon X \to [-\infty, +\infty]$ be simple, and let $\alpha_1, \ldots, \alpha_n$ be the values of f. Then f is $\mathscr{A}$-measurable if and only if $\{x \in X : f(x) = \alpha_i\} \in \mathscr{A}$ for $i = 1, \ldots, n$.   □

Proposition 2.1.3. *Let $(X, \mathscr{A})$ be a measurable space, let A be a subset of X that belongs to $\mathscr{A}$, and let f and g be $[-\infty, +\infty]$-valued measurable functions on A. Then the sets $\{x \in A : f(x) < g(x)\}$, $\{x \in A : f(x) \leq g(x)\}$, and $\{x \in A : f(x) = g(x)\}$ belong to $\mathscr{A}$.*

Proof. Note that the inequality $f(x) < g(x)$ holds if and only if there is a rational number r such that $f(x) < r < g(x)$. Thus

$$\{x \in A : f(x) < g(x)\} = \bigcup_{r \in \mathbb{Q}} (\{x \in A : f(x) < r\} \cap \{x \in A : r < g(x)\}),$$

and so $\{x \in A : f(x) < g(x)\}$, as the union of a countable collection of sets that belong to $\mathscr{A}$, itself belongs to $\mathscr{A}$. The set $\{x \in A : g(x) < f(x)\}$ likewise belongs to $\mathscr{A}$. This and the identity

$$\{x \in A : f(x) \le g(x)\} = A - \{x \in A : g(x) < f(x)\}$$

imply that $\{x \in A : f(x) \le g(x)\}$ belongs to $\mathscr{A}$. Finally $\{x \in A : f(x) = g(x)\}$ is the difference of $\{x \in A : f(x) \le g(x)\}$ and $\{x \in A : f(x) < g(x)\}$ and so belongs to $\mathscr{A}$. $\qquad\square$

Let f and g be $[-\infty, +\infty]$-valued functions having a common domain A. The *maximum* and *minimum* of f and g, written $f \vee g$ and $f \wedge g$, are the functions from A to $[-\infty, +\infty]$ defined by

$$(f \vee g)(x) = \max(f(x), g(x))$$

and

$$(f \wedge g)(x) = \min(f(x), g(x)).$$

Equivalently, we can define $f \vee g$ by

$$(f \vee g)(x) = \begin{cases} f(x) & \text{if } f(x) > g(x) \text{ and,} \\ g(x) & \text{otherwise,} \end{cases}$$

with $f \wedge g$ getting a corresponding definition.

If $\{f_n\}$ is a sequence of $[-\infty, +\infty]$-valued functions on A, then $\sup_n f_n : A \to [-\infty, +\infty]$ is defined by

$$(\sup_n f_n)(x) = \sup\{f_n(x) : n = 1, 2, \dots\}$$

and $\inf_n f_n$, $\limsup_n f_n$, $\liminf_n f_n$, and $\lim_n f_n$ are defined in analogous ways. The domain of $\lim_n f_n$ consists of those points in A at which $\limsup_n f_n$ and $\liminf_n f_n$ agree; the domain of each of the other four functions is A. Each of these functions can have infinite values, even if all the f_n's have only finite values; in particular, $\lim_n f_n(x)$ can be $+\infty$ or $-\infty$.

Proposition 2.1.4. *Let $(X, \mathscr{A})$ be a measurable space, let A be a subset of X that belongs to $\mathscr{A}$, and let f and g be $[-\infty, +\infty]$-valued measurable functions on A. Then $f \vee g$ and $f \wedge g$ are measurable.*

Proof. The measurability of $f \vee g$ follows from the identity

$$\{x \in A : (f \vee g)(x) \le t\} = \{x \in A : f(x) \le t\} \cap \{x \in A : g(x) \le t\},$$

and the measurability of $f \wedge g$ follows from the identity

$$\{x \in A : (f \wedge g)(x) \leq t\} = \{x \in A : f(x) \leq t\} \cup \{x \in A : g(x) \leq t\}. \qquad \square$$

Proposition 2.1.5. *Let* $(X, \mathscr{A})$ *be a measurable space, let A be a subset of X that belongs to* $\mathscr{A}$*, and let* $\{f_n\}$ *be a sequence of* $[-\infty, +\infty]$*-valued measurable functions on A. Then*

(a) *the functions* $\sup_n f_n$ *and* $\inf_n f_n$ *are measurable,*
(b) *the functions* $\limsup_n f_n$ *and* $\liminf_n f_n$ *are measurable, and*
(c) *the function* $\lim_n f_n$ *(whose domain is* $\{x \in A : \limsup_n f_n(x) = \liminf_n f_n(x)\}$*) is measurable.*

Proof. The measurability of $\sup_n f_n$ and $\inf_n f_n$ follows from the identities

$$\{x \in A : \sup_n f_n(x) \leq t\} = \bigcap_n \{x \in A : f_n(x) \leq t\}$$

and

$$\{x \in A : \inf_n f_n(x) < t\} = \bigcup_n \{x \in A : f_n(x) < t\}.$$

For each positive integer k define functions g_k and h_k by $g_k = \sup_{n \geq k} f_n$ and $h_k = \inf_{n \geq k} f_n$. Part (a) of the proposition implies first that each g_k is measurable and that each h_k is measurable and then that $\inf_k g_k$ and $\sup_k h_k$ are measurable. Since $\limsup_n f_n$ and $\liminf_n f_n$ are equal to $\inf_k g_k$ and $\sup_k h_k$, they too are measurable.

Let A_0 be the domain of $\lim_n f_n$. Then A_0 is equal to $\{x \in A : \limsup_n f_n(x) = \liminf_n f_n(x)\}$, which according to Proposition 2.1.3 belongs to $\mathscr{A}$. Since

$$\{x \in A_0 : \lim_n f_n(x) \leq t\} = A_0 \cap \{x \in A : \limsup_n f_n(x) \leq t\},$$

the measurability of $\lim_n f_n$ follows. $\qquad \square$

In the following two propositions we deal with arithmetic operations on $[0, +\infty]$-valued measurable functions (see B.4) and on $\mathbb{R}$-valued measurable functions. Arithmetic operations on $[-\infty, +\infty]$-valued functions are trickier and are seldom needed.

Proposition 2.1.6. *Let* $(X, \mathscr{A})$ *be a measurable space, let A be a subset of X that belongs to* $\mathscr{A}$*, let f and g be* $[0, +\infty]$*-valued measurable functions on A, and let* α *be a nonnegative real number. Then* αf *and* $f + g$ *are measurable.*[1]

Proof. For the measurability of αf, note that if $\alpha = 0$, then αf is identically 0 and so measurable, while if $\alpha > 0$, then for each t the set $\{x \in A : \alpha f(x) < t\}$ is equal to $\{x \in A : f(x) < t/\alpha\}$ and so belongs to $\mathscr{A}$.

---

[1]Recall that $0 \cdot (+\infty) = 0$ and that if $x \neq -\infty$, then $x + (+\infty) = (+\infty) + x = +\infty$. See Appendix B.

We turn to $f+g$. It is easy to check that $(f+g)(x) < t$ holds if and only if there is a rational number r such that $f(x) < r$ and $g(x) < t - r$. Thus

$$\{x \in A : (f+g)(x) < t\}$$
$$= \bigcup_{r \in \mathbb{Q}} (\{x \in A : f(x) < r\} \bigcap \{x \in A : g(x) < t - r\}),$$

and so $\{x \in A : (f+g)(x) < t\}$, as the union of a countable collection of sets that belong to $\mathscr{A}$, itself belongs to $\mathscr{A}$. The measurability of $f+g$ follows.   □

Proposition 2.1.7. *Let $(X, \mathscr{A})$ be a measurable space, let A be a subset of X that belongs to $\mathscr{A}$, let f and g be measurable real-valued functions on A, and let α be a real number. Then αf, $f+g$, $f-g$, fg, and f/g (where the domain of f/g is $\{x \in A : g(x) \neq 0\}$) are measurable.*

Proof. The measurability of αf and $f+g$ can be verified by modifying the proof of Proposition 2.1.6, and so the details are omitted (note that if $\alpha < 0$, then $\{x \in A : \alpha f(x) < t\} = \{x \in A : f(x) > t/\alpha\}$). The measurability of $f-g$ follows from the identity $f - g = f + (-1)g$.

We turn to the product of measurable functions and begin by showing that if $h : A \to \mathbb{R}$ is measurable, then h^2 is measurable. For this note that if $t \leq 0$, then

$$\{x \in A : h^2(x) < t\} = \emptyset,$$

while if $t > 0$, then

$$\{x \in A : h^2(x) < t\} = \{x \in A : -\sqrt{t} < h(x) < \sqrt{t}\};$$

the measurability of h^2 follows. Hence if f and g are measurable, then f^2, g^2, and $(f+g)^2$ are measurable, and the measurability of fg follows from the identity

$$fg = \frac{1}{2}((f+g)^2 - f^2 - g^2).$$

Let $A_0 = \{x \in A : g(x) \neq 0\}$, so that A_0 is the domain of f/g. It is easy to check (do so) that A_0 belongs to $\mathscr{A}$. Since for each t the set $\{x \in A_0 : (f/g)(x) < t\}$ is the union of

$$\{x \in A : g(x) > 0\} \cap \{x \in A : f(x) < tg(x)\}$$

and

$$\{x \in A : g(x) < 0\} \cap \{x \in A : f(x) > tg(x)\},$$

the measurability of f/g follows (see Proposition 2.1.3).   □

Let A be a set, and let f be an extended real-valued function[2] on A. The *positive part* f^+ and the *negative part* f^- of f are the extended real-valued functions defined by

$$f^+(x) = \max(f(x), 0)$$

and

$$f^-(x) = -\min(f(x), 0).$$

Thus $f^+ = f \vee 0$ and $f^- = (-f) \vee 0$. It is easy to check that if $(X, \mathscr{A})$ is a measurable space and if f is a $[-\infty, +\infty]$-valued function defined on a subset of X, then f is measurable if and only if f^+ and f^- are both measurable. It follows from this remark, together with Proposition 2.1.6, that the absolute value $|f|$ of a measurable function f is measurable (note that $|f| = f^+ + f^-$).

Let $(X, \mathscr{A})$ be a measurable space, let A be a subset of X that belongs to $\mathscr{A}$, and let f be a $[-\infty, +\infty]$-valued function on A. The following relationships between the measurability of f and the measurability of restrictions of f to subsets of A are sometimes useful:

(a) If f is $\mathscr{A}$-measurable and if B is a subset of A that belongs to $\mathscr{A}$, then the restriction f_B of f to B is $\mathscr{A}$-measurable; this follows from the identity

$$\{x \in B : f_B(x) < t\} = B \cap \{x \in A : f(x) < t\}.$$

(b) If $\{B_n\}$ is a sequence of sets that belong to $\mathscr{A}$, if $A = \cup_n B_n$, and if for each n the restriction f_{B_n} of f to B_n is $\mathscr{A}$-measurable, then f is $\mathscr{A}$-measurable; this follows from the identity

$$\{x \in A : f(x) < t\} = \bigcup_n \{x \in B_n : f_{B_n}(x) < t\}.$$

We will repeatedly have need for the following basic result.

Proposition 2.1.8. *Let $(X, \mathscr{A})$ be a measurable space, let A be a subset of X that belongs to $\mathscr{A}$, and let f be a $[0, +\infty]$-valued measurable function on A. Then there is a sequence $\{f_n\}$ of simple $[0, +\infty)$-valued measurable functions on A that satisfy*

$$f_1(x) \le f_2(x) \le \dots \tag{1}$$

and

$$f(x) = \lim_n f_n(x) \tag{2}$$

at each x in A.

---

[2] An *extended real-valued function* is, of course, a $[-\infty, +\infty]$-valued function.

Proof. For each positive integer n and for $k = 1, 2, \ldots, n2^n$ let $A_{n,k} = \{x \in A : (k-1)/2^n \leq f(x) < k/2^n\}$. The measurability of f implies that each $A_{n,k}$ belongs to $\mathscr{A}$. Define a sequence $\{f_n\}$ of functions from A to $\mathbb{R}$ by requiring f_n to have value $(k-1)/2^n$ at each point in $A_{n,k}$ (for $k = 1, 2, \ldots, n2^n$) and to have value n at each point in $A - \cup_k A_{n,k}$. The functions so defined are simple and measurable, and it is easy to check that they satisfy (1) and (2) at each x in A. □

Suppose that $(X, \mathscr{A})$ is a measurable space and that f is a $[-\infty, +\infty]$-valued $\mathscr{A}$-measurable function defined on an $\mathscr{A}$-measurable subset A of X. Then by applying Proposition 2.1.8 to the positive and negative parts of f, we can construct a sequence $\{f_n\}$ of simple $\mathscr{A}$-measurable functions from A to $\mathbb{R}$ such that $f(x) = \lim_n f_n(x)$ holds at each x in A.

The following proposition gives some additional ways of viewing measurable functions; part (d) suggests a way to deal with more general situations (see Sect. 2.6).

Proposition 2.1.9. *Let $(X, \mathscr{A})$ be a measurable space, and let A be a subset of X that belongs to $\mathscr{A}$. For a function $f : A \to \mathbb{R}$, the conditions*

(a) *f is measurable with respect to $\mathscr{A}$,*
(b) *for each open subset U of $\mathbb{R}$ the set $f^{-1}(U)$ belongs to $\mathscr{A}$,*
(c) *for each closed subset C of $\mathbb{R}$ the set $f^{-1}(C)$ belongs to $\mathscr{A}$, and*
(d) *for each Borel subset B of $\mathbb{R}$ the set $f^{-1}(B)$ belongs to $\mathscr{A}$*

are equivalent.

Proof. Let $\mathscr{F} = \{B \subseteq \mathbb{R} : f^{-1}(B) \in \mathscr{A}\}$. Then the fact that $f^{-1}(\mathbb{R}) = A$ and the identities

$$f^{-1}(B^c) = A - f^{-1}(B)$$

and

$$f^{-1}\left(\bigcup_n B_n\right) = \bigcup_n f^{-1}(B_n)$$

imply that $\mathscr{F}$ is a σ-algebra on $\mathbb{R}$. To require that f be measurable is to require that $\mathscr{F}$ contain all the intervals of the form $(-\infty, b]$ or equivalently (since $\mathscr{F}$ is a σ-algebra) to require that $\mathscr{F}$ include the σ-algebra on $\mathbb{R}$ generated by these intervals. Since the σ-algebra generated by these intervals is the σ-algebra of Borel subsets of $\mathbb{R}$ (Proposition 1.1.4), conditions (a) and (d) are equivalent. However the σ-algebra of Borel subsets of $\mathbb{R}$ is also generated by the collection of all open subsets of $\mathbb{R}$ and by the collection of all closed subsets of $\mathbb{R}$, and so conditions (b) and (c) are equivalent to the others. □

We close this section by returning to one of the promises made in Sect. 1.3 and proving that there are Lebesgue measurable subsets of $\mathbb{R}$ that are not Borel sets. For this we will use the following example.

Example 2.1.10. Recall the construction of the Cantor set given in Sect. 1.4. There we let K_0 be the interval $[0, 1]$, and for each positive integer n we constructed a compact set K_n by removing from K_{n-1} the open middle third of each of the intervals making up K_{n-1}. The Cantor set K is given by $K = \cap_n K_n$.

The *Cantor function* (also known as the *Cantor singular function*) is the function $f: [0, 1] \rightarrow [0, 1]$ defined as follows (the concept of singularity will be defined and studied in Chap. 4). For each x in the interval $(1/3, 2/3)$ let $f(x) = 1/2$. Thus f is now defined at each point removed from $[0, 1]$ in the construction of K_1. Next define f at each point removed from K_1 in the construction of K_2 by letting $f(x) = 1/4$ if $x \in (1/9, 2/9)$ and letting $f(x) = 3/4$ if $x \in (7/9, 8/9)$. Continue in this way, letting $f(x)$ be $1/2^n$, $3/2^n$, $5/2^n$, ... on the various intervals removed from K_{n-1} in the construction of K_n. After all these steps, f is defined on the open set $[0, 1] - K$, is nondecreasing, and has values in $[0, 1]$. Extend it to all of $[0, 1]$ by letting $f(0) = 0$ and letting

$$f(x) = \sup\{f(t) : t \in [0, 1] - K \text{ and } t < x\}$$

if $x \in K$ and $x \neq 0$. This completes the definition of the Cantor function.

It is easy to check that f is nondecreasing and continuous, and it is clear that $f(0) = 0$ and $f(1) = 1$. The intermediate value theorem (Theorem C.13) thus implies that for each y in $[0, 1]$ there is at least one x in $[0, 1]$ such that $f(x) = y$, and so we can define a function $g: [0, 1] \rightarrow [0, 1]$ by

$$g(y) = \inf\{x \in [0, 1] : f(x) = y\}. \tag{3}$$

The continuity of f implies that $f(g(y)) = y$ holds for each y in $[0, 1]$; hence g is injective. It is easy to check that all the values of g lie in the Cantor set. The fact that f is nondecreasing implies that g is nondecreasing and hence that g is Borel measurable (see Example 2.1.2(b)).   □

Proposition 2.1.11. *There is a Lebesgue measurable subset of $\mathbb{R}$ that is not a Borel set.*

Proof. Let g be the function constructed above, let A be a subset of $[0, 1]$ that is not Lebesgue measurable (see Theorem 1.4.9), and let $B = g(A)$. Then B is a subset of the Cantor set and so is Lebesgue measurable (recall that $\lambda(K) = 0$ and that Lebesgue measure on the σ-algebra of Lebesgue measurable sets is complete). If B were a Borel set, then $g^{-1}(B)$ would also be a Borel set (recall that g is Borel measurable, and see Proposition 2.1.9). However the injectivity of g implies that $g^{-1}(B)$ is the set A, which is not Lebesgue measurable and hence is not a Borel set. Consequently the Lebesgue measurable set B is not a Borel set.   □

Example 2.1.12. The proof of Proposition 2.1.11 gives a Borel set of Lebesgue measure 0 (the Cantor set) that has a subset that is not a Borel set. It follows that Lebesgue measure on $(\mathbb{R}, \mathcal{B}(\mathbb{R}))$ is not complete.   □

Exercises

1. Let X be a set, let $\{A_k\}$ be a sequence of subsets of X, let $B = \cup_{n=1}^{\infty} \cap_{k=n}^{\infty} A_k$, and let $C = \cap_{n=1}^{\infty} \cup_{k=n}^{\infty} A_k$. Show that
   (a) $\liminf_k \chi_{A_k} = \chi_B$, and
   (b) $\limsup_k \chi_{A_k} = \chi_C$.
2. Show that the supremum of an uncountable family of $[-\infty, +\infty]$-valued Borel measurable functions on $\mathbb{R}$ can fail to be Borel measurable.
3. Show that if $f : \mathbb{R} \to \mathbb{R}$ is differentiable everywhere on $\mathbb{R}$, then its derivative f' is Borel measurable.
4. Let $(X, \mathscr{A})$ be a measurable space, and let $\{f_n\}$ be a sequence of $[-\infty, +\infty]$-valued measurable functions on X. Show that

$$\{x \in X : \lim_n f_n(x) \text{ exists and is finite}\}$$

   belongs to $\mathscr{A}$.
5. Let $(X, \mathscr{A})$ be a measurable space.
   (a) Show directly (i.e., without using Proposition 2.1.6 or Proposition 2.1.7) that if $f, g : X \to \mathbb{R}$ are $\mathscr{A}$-measurable *simple* functions, then $f + g$ and fg are $\mathscr{A}$-measurable.
   (b) Now let $f, g : X \to \mathbb{R}$ be arbitrary $\mathscr{A}$-measurable functions. Use Propositions 2.1.4, 2.1.5, and 2.1.8, together with part (a) of this exercise, to show that $f + g$ and fg are $\mathscr{A}$-measurable.
6. Let $(X, \mathscr{A})$ be a measurable space, and let $f, g : X \to \mathbb{R}$ be measurable. Give still another proof of the measurability of $f + g$, this time by checking that for each real t the function $x \mapsto t - f(x)$ is measurable and then using Proposition 2.1.3. (Hint: Consider $\{x : g(x) < t - f(x)\}$.)
7. Let f be the Cantor function, and let μ be the Borel measure on $\mathbb{R}$ associated to f by Proposition 1.3.10 (actually, one should apply Proposition 1.3.10 to the function from $\mathbb{R}$ to $\mathbb{R}$ that agrees with f on $[0,1]$, vanishes on $(-\infty, 0)$, and is identically 1 on $(1, +\infty)$). Show that
   (a) each of the 2^n intervals remaining after the nth step in the construction of the Cantor set has measure $1/2^n$ under μ,
   (b) the Cantor set has measure 1 under μ, and
   (c) each x in $\mathbb{R}$ satisfies $\mu(\{x\}) = 0$.
   Thus all the mass of μ is concentrated on a set of Lebesgue measure zero (the Cantor set), but μ is not a sum of multiples of point masses.
8. Let g be the inverse of the Cantor function (that is, let g be defined by formula (3)). Show that the points x that have the form $x = g(y)$ for some y in $[0,1]$ are exactly those that belong to the Cantor set and are not right-hand endpoints of intervals removed from $[0,1]$ during the construction of the Cantor set.
9. Let $(X, \mathscr{A})$ be a measurable space and let C be a subset of X that does not belong to $\mathscr{A}$. Show that a function $f : X \to \mathbb{R}$ is $\sigma(\mathscr{A} \cup \{C\})$-measurable if and only if there exist $\mathscr{A}$-measurable functions $f_1, f_2 : X \to \mathbb{R}$ such that $f = f_1 \chi_C + f_2 \chi_{C^c}$. (See part (a) of Exercise 1.5.12.)

10. Let $\mathscr{V}_0$ be the collection of all Borel measurable functions from $\mathbb{R}$ to $\mathbb{R}$. Show that $\mathscr{V}_0$ is the smallest of those collections $\mathscr{V}$ of functions from $\mathbb{R}$ to $\mathbb{R}$ for which

   (i) $\mathscr{V}$ is a vector space over $\mathbb{R}$,
   (ii) $\mathscr{V}$ contains each continuous function, and
   (iii) if $\{f_n\}$ is an increasing sequence of nonnegative functions in $\mathscr{V}$ and if $\lim_n f_n(x)$ is finite for each x in $\mathbb{R}$, then $\lim_n f_n$ belongs to $\mathscr{V}$.

   (Hint: Suppose that $\mathscr{V}$ satisfies conditions (a), (b), and (c), and define $S(\mathscr{V})$ by $S(\mathscr{V}) = \{A \subseteq \mathbb{R} : \chi_A \in \mathscr{V}\}$. Show that $S(\mathscr{V})$ contains each interval of the form $(-\infty, a)$, and then use Theorem 1.6.2 to show that $S(\mathscr{V})$ contains each Borel set.)

2.2 Properties That Hold Almost Everywhere

Let $(X, \mathscr{A}, \mu)$ be a measure space. A property of points of X is said to hold μ-*almost everywhere* if the set of points in X at which it fails to hold is μ-negligible. In other words, a property holds μ-almost everywhere if there is a set N that belongs to $\mathscr{A}$, satisfies $\mu(N) = 0$, and contains every point at which the property fails to hold. More generally, if E is a subset of X, then a property is said to hold μ-*almost everywhere on E* if the set of points in E at which it fails to hold is μ-negligible. The expression μ-almost everywhere is often abbreviated to μ-a.e. or to a.e.$[\mu]$. In cases where the measure μ is clear from context, the expressions almost everywhere and a.e. are also used.

Consider a property that holds almost everywhere, and let F be the set of points in X at which it fails. Then it is not necessary that F belong to $\mathscr{A}$; it is only necessary that there be a set N that belongs to $\mathscr{A}$, includes F, and satisfies $\mu(N) = 0$. Of course, if μ is complete, then F will belong to $\mathscr{A}$.

Examples 2.2.1. Suppose that f and g are functions on X. Then $f = g$ almost everywhere if the set of points x at which $f(x) \neq g(x)$ is μ-negligible, and $f \geq g$ almost everywhere if the set of points x at which $f(x) < g(x)$ is μ-negligible. Note that the sets $\{x \in X : f(x) \neq g(x)\}$ and $\{x \in X : f(x) < g(x)\}$ belong to $\mathscr{A}$ if f and g are $\mathscr{A}$-measurable; otherwise these sets may fail to belong to $\mathscr{A}$. If $\{f_n\}$ is a sequence of functions on X and f is a function on X, then $\{f_n\}$ converges to f almost everywhere if the set of points x at which the relation $f(x) = \lim_n f_n(x)$ fails to hold is μ-negligible. In this case one also says that $f = \lim_n f_n$ almost everywhere. □

Proposition 2.2.2. *Let $(X, \mathscr{A}, \mu)$ be a measure space, and let f and g be extended real-valued functions on X that are equal almost everywhere. If μ is complete and if f is $\mathscr{A}$-measurable, then g is $\mathscr{A}$-measurable.*

Proof. Let t be a real number and let N be a set that belongs to $\mathscr{A}$, satisfies $\mu(N) = 0$, and is such that f and g agree everywhere outside N. Then

$$\{x \in X : g(x) \leq t\} = (\{x \in X : f(x) \leq t\} \cap N^c) \cup (\{x \in X : g(x) \leq t\} \cap N). \quad (1)$$

The measurability of f and N implies that $\{x \in X : f(x) \le t\} \cap N^c$ belongs to $\mathscr{A}$, while the completeness of μ implies that $\{x \in X : g(x) \le t\} \cap N$ belongs to $\mathscr{A}$. The measurability of g follows.   □

Corollary 2.2.3. *Let* $(X, \mathscr{A}, \mu)$ *be a measure space, let* $\{f_n\}$ *be a sequence of extended real-valued functions on* X, *and let* f *be an extended real-valued function on* X *such that* $\{f_n\}$ *converges to* f *almost everywhere. If* μ *is complete and if each* f_n *is* $\mathscr{A}$-*measurable, then* f *is* $\mathscr{A}$-*measurable.*

Proof. According to Proposition 2.1.5 the function $\liminf_n f_n$ is $\mathscr{A}$-measurable. Since f and $\liminf_n f_n$ agree almost everywhere, Proposition 2.2.2 implies that f is $\mathscr{A}$-measurable.   □

Example 2.2.4. Suppose that $(X, \mathscr{A}, \mu)$ is a measure space that is not complete, and let N be a μ-negligible subset of X that does not belong to $\mathscr{A}$. Then the characteristic function χ_N and the constant function 0 agree almost everywhere, but 0 is $\mathscr{A}$-measurable while χ_N is not. Thus Proposition 2.2.2 would fail if the hypothesis of completeness were removed. Furthermore, the sequence each term of which is 0 converges almost everywhere to χ_N; consequently Corollary 2.2.3 would also fail if the hypothesis of completeness were removed.   □

Proposition 2.2.5. *Let* $(X, \mathscr{A}, \mu)$ *be a measure space, and let* $\mathscr{A}_\mu$ *be the completion of* $\mathscr{A}$ *under* μ. *Then a function* $f \colon X \to [-\infty, +\infty]$ *is* $\mathscr{A}_\mu$-*measurable if and only if there are* $\mathscr{A}$-*measurable functions* $f_0, f_1 \colon X \to [-\infty, +\infty]$ *such that*

$$f_0 \le f \le f_1 \text{ holds everywhere on } X \tag{2}$$

and

$$f_0 = f_1 \text{ holds } \mu\text{-almost everywhere on } X. \tag{3}$$

In the context of Proposition 2.2.5, it is natural to ask whether it is always possible, given an $\mathscr{A}_\mu$-measurable function f with values in $\mathbb{R}$, rather than in $[-\infty, +\infty]$, to find *real*-valued functions f_0 and f_1 that satisfy (2) and (3). It turns out that the answer is no; see Exercise 8.3.3.

Proof. First suppose that there exist $\mathscr{A}$-measurable functions f_0 and f_1 that satisfy (2) and (3). Then f_0 is $\mathscr{A}_\mu$-measurable and $f = f_0$ holds $\overline{\mu}$-almost everywhere, and so Proposition 2.2.2, applied to the space $(X, \mathscr{A}_\mu, \overline{\mu})$, implies that f is $\mathscr{A}_\mu$-measurable.

Now suppose that $f \colon X \to [-\infty, +\infty]$ is $\mathscr{A}_\mu$-measurable. If f is simple and $[0, +\infty)$-valued, say attaining values $a_1, \dots, a_k$ on the sets $A_1, \dots, A_k$, then there are sets $B_1, \dots, B_k$ and $C_1, \dots, C_k$ that belong to $\mathscr{A}$ and satisfy $C_i \subseteq A_i \subseteq B_i$ and $\mu(B_i - C_i) = 0$ for each i. The functions f_0 and f_1 defined by $f_0 = \sum_i a_i \chi_{C_i}$ and $f_1 = \sum_i a_i \chi_{B_i}$ then satisfy (2) and (3).

We can deal with the case where f is simple and real-valued by applying the preceding argument to the positive and negative parts of f.

Finally, let $f\colon X \to [-\infty, +\infty]$ be an arbitrary $\mathscr{A}_\mu$-measurable function, and choose a sequence $\{g_n\}$ of simple $\mathscr{A}_\mu$-measurable functions from X to $\mathbb{R}$ such that $f(x) = \lim_n g_n(x)$ holds at each x in X (see the remark following the proof of Proposition 2.1.8). If for each n we choose $\mathscr{A}$-measurable functions $g_{0,n}$ and $g_{1,n}$ such that

$$g_{0,n} \leq g_n \leq g_{1,n} \text{ holds everywhere on } X$$

and

$$g_{0,n} = g_{1,n} \text{ holds } \mu\text{-almost everywhere on } X,$$

then the required functions f_0 and f_1 can be constructed by letting f_0 be $\overline{\lim}_n g_{0,n}$ and f_1 be $\underline{\lim}_n g_{1,n}$.   □

Exercises

1. Give Borel functions $f, g\colon \mathbb{R} \to \mathbb{R}$ that agree on some dense subset of $\mathbb{R}$ but are such that $f(x) \neq g(x)$ holds at λ-almost every x in $\mathbb{R}$.
2. Let $\{x_n\}$ be a sequence of real numbers, and define μ on $(\mathbb{R}, \mathscr{B}(\mathbb{R}))$ by $\mu = \sum_n \delta_{x_n}$ (see Exercise 1.2.6). Show that functions $f, g\colon \mathbb{R} \to \mathbb{R}$ agree μ-almost everywhere if and only if $f(x_n) = g(x_n)$ holds for each n.
3. Let f and g be continuous real-valued functions on $\mathbb{R}$. Show that if $f = g$ λ-almost everywhere, then $f = g$ (i.e., $f(x) = g(x)$ for every x in $\mathbb{R}$).
4. Let μ be the finite Borel measure on $\mathbb{R}$ that is associated to the Cantor function by Proposition 1.3.10 (see Exercise 2.1.7). Show that continuous real-valued functions on $\mathbb{R}$ agree μ-almost everywhere if and only if they agree at every point in the Cantor set.
5. Let $(X, \mathscr{A}, \mu)$ be a measure space, and let f and $f_1, f_2, \ldots$ be $[-\infty, +\infty]$-valued $\mathscr{A}$-measurable functions on X. Show that if $\{f_n\}$ converges to f almost everywhere, then there are $\mathscr{A}$-measurable functions $g_1, g_2, \ldots$ that are equal to $f_1, f_2, \ldots$ almost everywhere and satisfy $f = \lim_n g_n$ everywhere.
6. Show that the function $f\colon \mathbb{R} \to \mathbb{R}$ defined by

$$f(x) = \begin{cases} 0 & \text{if } x \text{ is irrational,} \\ 1 & \text{if } x \text{ is rational} \end{cases}$$

is nowhere continuous and that the function $g\colon \mathbb{R} \to \mathbb{R}$ defined by

$$g(x) = \begin{cases} 0 & \text{if } x = 0 \text{ or } x \text{ is irrational,} \\ \frac{1}{q} & \text{if } x = \frac{p}{q}, \text{ where } p \text{ and } q \text{ are relatively prime and } q > 0 \end{cases}$$

is continuous λ-almost everywhere.

2.3   The Integral

In this section we construct the integral and study some of its basic properties. The construction will take place in three stages.

We begin with the simple functions. Let $(X, \mathscr{A})$ be a measurable space. We will denote by $\mathscr{S}$ the collection of all simple real-valued $\mathscr{A}$-measurable functions on X and by $\mathscr{S}_+$ the collection of nonnegative functions in $\mathscr{S}$.

Let μ be a measure on $(X, \mathscr{A})$. If f belongs to $\mathscr{S}_+$ and is given by $f = \sum_{i=1}^{m} a_i \chi_{A_i}$, where $a_1, \ldots, a_m$ are nonnegative real numbers and $A_1, \ldots, A_m$ are disjoint subsets of X that belong to $\mathscr{A}$, then $\int f \, d\mu$, the *integral* of f with respect to μ, is defined to be $\sum_{i=1}^{m} a_i \mu(A_i)$ (note that this sum is either a nonnegative real number or $+\infty$). We need to check that $\int f \, d\mu$ depends only on f and not on $a_1, \ldots, a_m$ and $A_1, \ldots, A_m$. So suppose that f is also given by $\sum_{j=1}^{n} b_j \chi_{B_j}$, where $b_1, \ldots, b_n$ are nonnegative real numbers and $B_1, \ldots, B_n$ are disjoint subsets of X that belong to $\mathscr{A}$. We can assume that $\cup_{i=1}^{m} A_i = \cup_{j=1}^{n} B_j$ (if necessary eliminate those sets A_i for which $a_i = 0$ and those sets B_j for which $b_j = 0$). Then the additivity of μ and the fact that $a_i = b_j$ if $A_i \cap B_j \neq 0$ imply that

$$\sum_{i=1}^{m} a_i \mu(A_i) = \sum_{i=1}^{m} \sum_{j=1}^{n} a_i \mu(A_i \cap B_j)$$

$$= \sum_{i=1}^{m} \sum_{j=1}^{n} b_j \mu(A_i \cap B_j) = \sum_{j=1}^{n} b_j \mu(B_j);$$

hence $\int f \, d\mu$ does not depend on the representation of f used in its definition.

Before proceeding to the next stage of our construction, we verify a few properties of the integral of a nonnegative simple function.

Proposition 2.3.1. *Let $(X, \mathscr{A}, \mu)$ be a measure space, let f and g belong to $\mathscr{S}_+$, and let α be a nonnegative real number. Then*

(a) $\int \alpha f \, d\mu = \alpha \int f \, d\mu$,
(b) $\int (f + g) \, d\mu = \int f \, d\mu + \int g \, d\mu$, and
(c) *if $f(x) \leq g(x)$ holds at each x in X, then $\int f \, d\mu \leq \int g \, d\mu$.*

Proof. Suppose that $f = \sum_{i=1}^{m} a_i \chi_{A_i}$, where $a_1, \ldots, a_m$ are nonnegative real numbers and $A_1, \ldots, A_m$ are disjoint subsets of X that belong to $\mathscr{A}$, and that $g = \sum_{j=1}^{n} b_j \chi_{B_j}$, where $b_1, \ldots, b_n$ are nonnegative real numbers and $B_1, \ldots, B_n$ are disjoint subsets of X that belong to $\mathscr{A}$. We can again assume that $\cup_{i=1}^{m} A_i = \cup_{j=1}^{n} B_j$. Then parts (a) and (b) follow from the calculations

$$\int \alpha f \, d\mu = \sum_{i=1}^{m} \alpha a_i \mu(A_i) = \alpha \sum_{i=1}^{m} a_i \mu(A_i) = \alpha \int f \, d\mu$$

and

$$\int (f+g)\,d\mu = \sum_{i=1}^{m}\sum_{j=1}^{n}(a_i+b_j)\mu(A_i\cap B_j)$$

$$= \sum_{i=1}^{m}\sum_{j=1}^{n}a_i\mu(A_i\cap B_j) + \sum_{i=1}^{m}\sum_{j=1}^{n}b_j\mu(A_i\cap B_j)$$

$$= \sum_{i=1}^{m}a_i\mu(A_i) + \sum_{j=1}^{n}b_j\mu(B_j) = \int f\,d\mu + \int g\,d\mu.$$

Next suppose that $f(x) \leq g(x)$ holds at each x in X. Then $g-f$ belongs to $\mathscr{S}_+$, and so part (c) follows from the calculation

$$\int g\,d\mu = \int (f+(g-f))\,d\mu = \int f\,d\mu + \int (g-f)\,d\mu \geq \int f\,d\mu. \qquad \square$$

Proposition 2.3.2. *Let $(X,\mathscr{A},\mu)$ be a measure space, let f belong to $\mathscr{S}_+$, and let $\{f_n\}$ be a nondecreasing sequence of functions in $\mathscr{S}_+$ such that $f(x) = \lim_n f_n(x)$ holds at each x in X. Then $\int f\,d\mu = \lim_n \int f_n\,d\mu$.*

This proposition is a weak version of one of the fundamental properties of the Lebesgue integral, the monotone convergence theorem (Theorem 2.4.1). We need this weakened version now for use as a tool in completing the definition of the integral.

Proof. It follows from Proposition 2.3.1 that

$$\int f_1\,d\mu \leq \int f_2\,d\mu \leq \cdots \leq \int f\,d\mu;$$

hence $\lim_n \int f_n\,d\mu$ exists and satisfies $\lim_n \int f_n\,d\mu \leq \int f\,d\mu$. We turn to the reverse inequality. Let ε be a number such that $0 < \varepsilon < 1$. We will construct a nondecreasing sequence $\{g_n\}$ of functions in $\mathscr{S}_+$ such that $g_n \leq f_n$ holds for each n and such that $\lim_n \int g_n\,d\mu = (1-\varepsilon)\int f\,d\mu$. Since $\int g_n\,d\mu \leq \int f_n\,d\mu$, this will imply that $(1-\varepsilon)\int f\,d\mu \leq \lim_n \int f_n\,d\mu$ and, since ε is arbitrary, that $\int f\,d\mu \leq \lim_n \int f_n\,d\mu$. Consequently $\int f\,d\mu = \lim_n \int f_n\,d\mu$.

We turn to the construction of the sequence $\{g_n\}$. Suppose that $a_1, \ldots, a_k$ are the nonzero values of f and that $A_1, \ldots, A_k$ are the sets on which these values occur. Thus $f = \sum_{i=1}^{k} a_i\chi_{A_i}$. For each n and i let

$$A(n,i) = \{x \in A_i : f_n(x) \geq (1-\varepsilon)a_i\}.$$

Then each $A(n,i)$ belongs to $\mathscr{A}$, and for each i the sequence $\{A(n,i)\}_{n=1}^{\infty}$ is nondecreasing and satisfies $A_i = \cup_n A(n,i)$. If we let $g_n = \sum_{i=1}^{k}(1-\varepsilon)a_i\chi_{A(n,i)}$, then g_n belongs to $\mathscr{S}_+$ and satisfies $g_n \leq f_n$, and we can use Proposition 1.2.5 to conclude that

$$\lim_n \int g_n \, d\mu = \lim_n \sum_{i=1}^{k} (1-\varepsilon) a_i \mu(A(n,i))$$

$$= \sum_{i=1}^{k} (1-\varepsilon) a_i \mu(A_i) = (1-\varepsilon) \int f \, d\mu. \qquad \square$$

As our next step, we define the integral of an arbitrary $[0,+\infty]$-valued $\mathscr{A}$-measurable function on X. For such a function f, let

$$\int f \, d\mu = \sup \left\{ \int g \, d\mu : g \in \mathscr{S}_+ \text{ and } g \leq f \right\}.$$

It is easy to see that for functions f in $\mathscr{S}_+$, this agrees with the previous definition.

Let us check a few properties of the integral on the class of $[0,+\infty]$-valued measurable functions. The first of these properties is an extension of Proposition 2.3.2 and will itself be generalized in Theorem 2.4.1 (the monotone convergence theorem). It is included here so that it can be used in the proof of Proposition 2.3.4.

Proposition 2.3.3. *Let $(X, \mathscr{A}, \mu)$ be a measure space, let f be a $[0,+\infty]$-valued $\mathscr{A}$-measurable function on X, and let $\{f_n\}$ be a nondecreasing sequence of functions in $\mathscr{S}_+$ such that $f(x) = \lim_n f_n(x)$ holds at each x in X. Then $\int f \, d\mu = \lim_n \int f_n \, d\mu$.*

Proof. It is clear that

$$\int f_1 \, d\mu \leq \int f_2 \, d\mu \leq \cdots \leq \int f \, d\mu;$$

hence $\lim_n \int f_n \, d\mu$ exists and satisfies $\lim_n \int f_n \, d\mu \leq \int f \, d\mu$. We turn to the reverse inequality. Recall that $\int f \, d\mu$ is the supremum of those elements of $[0,+\infty]$ of the form $\int g \, d\mu$, where g ranges over the set of functions that belong to $\mathscr{S}_+$ and satisfy $g \leq f$. Thus to prove that $\int f \, d\mu \leq \lim_n \int f_n \, d\mu$, it is enough to check that if g is a function in $\mathscr{S}_+$ that satisfies $g \leq f$, then $\int g \, d\mu \leq \lim_n \int f_n \, d\mu$. Let g be such a function. Then $\{g \wedge f_n\}$ is a nondecreasing sequence of functions in $\mathscr{S}_+$ for which $g = \lim_n (g \wedge f_n)$, and so Proposition 2.3.2 implies that $\int g \, d\mu = \lim_n \int (g \wedge f_n) \, d\mu$. Since $\int (g \wedge f_n) \, d\mu \leq \int f_n \, d\mu$, it follows that $\int g \, d\mu \leq \lim_n \int f_n \, d\mu$, and the proof is complete. $\qquad \square$

Proposition 2.3.4. *Let $(X, \mathscr{A}, \mu)$ be a measure space, let f and g be $[0,+\infty]$-valued $\mathscr{A}$-measurable functions on X, and let α be a nonnegative real number. Then*

(a) $\int \alpha f \, d\mu = \alpha \int f \, d\mu$,
(b) $\int (f+g) \, d\mu = \int f \, d\mu + \int g \, d\mu$, and
(c) *if $f(x) \leq g(x)$ holds at each x in X, then $\int f \, d\mu \leq \int g \, d\mu$.*

Proof. Choose nondecreasing sequences $\{f_n\}$ and $\{g_n\}$ of functions in $\mathscr{S}_+$ such that $f = \lim_n f_n$ and $g = \lim_n g_n$ (see Proposition 2.1.8). Then $\{\alpha f_n\}$ and $\{f_n + g_n\}$ are nondecreasing sequences of functions in $\mathscr{S}_+$ that satisfy $\alpha f = \lim_n \alpha f_n$ and

$f + g = \lim_n (f_n + g_n)$, and so we can use Proposition 2.3.3, together with the homogeneity and additivity of the integral on $\mathscr{S}_+$, to conclude that

$$\int \alpha f \, d\mu = \lim_n \int \alpha f_n \, d\mu = \lim_n \alpha \int f_n \, d\mu = \alpha \int f \, d\mu$$

and

$$\int (f + g) \, d\mu = \lim_n \int (f_n + g_n) \, d\mu$$

$$= \lim_n \left(\int f_n \, d\mu + \int g_n \, d\mu \right) = \int f \, d\mu + \int g \, d\mu.$$

Thus parts (a) and (b) are proved. For part (c), note that if $f \le g$, then the class of functions h in $\mathscr{S}_+$ that satisfy $h \le f$ is included in the class of functions h in $\mathscr{S}_+$ that satisfy $h \le g$; it follows that $\int f \, d\mu \le \int g \, d\mu$.   □

Finally, let f be an arbitrary $[-\infty, +\infty]$-valued $\mathscr{A}$-measurable function on X. If $\int f^+ \, d\mu$ and $\int f^- \, d\mu$ are both finite, then f is called *integrable* (or μ-*integrable* or *summable*), and its *integral* $\int f \, d\mu$ is defined by

$$\int f \, d\mu = \int f^+ \, d\mu - \int f^- \, d\mu.$$

The integral of f is said to *exist* if at least one of $\int f^+ \, d\mu$ and $\int f^- \, d\mu$ is finite, and again in this case, $\int f \, d\mu$ is defined to be $\int f^+ \, d\mu - \int f^- \, d\mu$. In either case one sometimes writes $\int f(x) \, \mu(dx)$ or $\int f(x) \, d\mu(x)$ in place of $\int f \, d\mu$.

Suppose that $f \colon X \to [-\infty, +\infty]$ is $\mathscr{A}$-measurable and that $A \in \mathscr{A}$. Then f is *integrable over* A if the function $f \chi_A$ is integrable, and in this case $\int_A f \, d\mu$, the *integral of f over* A, is defined to be $\int f \chi_A \, d\mu$. Likewise, if $A \in \mathscr{A}$ and if f is a measurable function whose domain is A (rather than the entire space X), then the integral of f over A is defined to be the integral (if it exists) of the function on X that agrees with f on A and vanishes on A^c. In case $\mu(A^c) = 0$, one often writes $\int f \, d\mu$ in place of $\int_A f \, d\mu$ and calls f integrable, rather than integrable over A.

In case $X = \mathbb{R}^d$ and $\mu = \lambda$, one often refers to *Lebesgue integrability* and the *Lebesgue integral*. The Lebesgue integral of a function f on $\mathbb{R}$ is often written $\int f(x) \, dx$. In case we are integrating over the interval $[a, b]$, we may write $\int_a^b f$ or $\int_a^b f(x) \, dx$ or, if we need to emphasize that we mean the Lebesgue integral, $(L) \int_a^b f$ or $(L) \int_a^b f(x) \, dx$.

We define $\mathscr{L}^1(X, \mathscr{A}, \mu, \mathbb{R})$ (or sometimes simply $\mathscr{L}^1$) to be the set of all real-valued (rather than $[-\infty, +\infty]$-valued) integrable functions on X. According to Proposition 2.3.6 below, $\mathscr{L}^1(X, \mathscr{A}, \mu, \mathbb{R})$ is a vector space and the integral is a linear functional on $\mathscr{L}^1(X, \mathscr{A}, \mu, \mathbb{R})$.

Lemma 2.3.5. *Let $(X, \mathscr{A}, \mu)$ be a measure space, and let f_1, f_2, g_1, and g_2 be nonnegative real-valued integrable functions on X such that $f_1 - f_2 = g_1 - g_2$. Then $\int f_1 \, d\mu - \int f_2 \, d\mu = \int g_1 \, d\mu - \int g_2 \, d\mu$.*

Proof. Since the functions f_1, f_2, g_1, and g_2 satisfy $f_1 - f_2 = g_1 - g_2$, they also satisfy $f_1 + g_2 = g_1 + f_2$ and so satisfy

$$\int f_1 \, d\mu + \int g_2 \, d\mu = \int g_1 \, d\mu + \int f_2 \, d\mu$$

(Proposition 2.3.4); since all the integrals involved are finite, this implies that

$$\int f_1 \, d\mu - \int f_2 \, d\mu = \int g_1 \, d\mu - \int g_2 \, d\mu. \qquad \square$$

Proposition 2.3.6. *Let* $(X, \mathscr{A}, \mu)$ *be a measure space, let f and g be real-valued integrable functions on X, and let α be a real number. Then*

(a) *αf and $f + g$ are integrable,*
(b) *$\int \alpha f \, d\mu = \alpha \int f \, d\mu$,*
(c) *$\int (f + g) \, d\mu = \int f \, d\mu + \int g \, d\mu$, and*
(d) *if $f(x) \leq g(x)$ holds at each x in X, then $\int f \, d\mu \leq \int g \, d\mu$.*

Proof. The integrability of αf and the relation $\int \alpha f \, d\mu = \alpha \int f \, d\mu$ are clear if $\alpha = 0$. If α is positive, then $(\alpha f)^+ = \alpha f^+$ and $(\alpha f)^- = \alpha f^-$; thus $(\alpha f)^+$ and $(\alpha f)^-$, and hence αf, are integrable, and

$$\int \alpha f \, d\mu = \int (\alpha f)^+ \, d\mu - \int (\alpha f)^- \, d\mu$$

$$= \alpha \int f^+ \, d\mu - \alpha \int f^- \, d\mu = \alpha \int f \, d\mu.$$

If α is negative, then $(\alpha f)^+ = -\alpha f^-$ and $(\alpha f)^- = -\alpha f^+$, and we can modify the preceding argument so as to show that αf is integrable and that $\int \alpha f \, d\mu = \alpha \int f \, d\mu$.

Now consider the sum of f and g. Note that $(f + g)^+ \leq f^+ + g^+$ and $(f + g)^- \leq f^- + g^-$; thus (Proposition 2.3.4)

$$\int (f + g)^+ \, d\mu \leq \int f^+ \, d\mu + \int g^+ \, d\mu < +\infty$$

and

$$\int (f + g)^- \, d\mu \leq \int f^- \, d\mu + \int g^- \, d\mu < +\infty,$$

and so $f + g$ is integrable. Since $f + g$ is equal to $(f + g)^+ - (f + g)^-$ and to $f^+ + g^+ - (f^- + g^-)$, it follows from Lemma 2.3.5 that

$$\int (f + g) \, d\mu = \int (f^+ + g^+) \, d\mu - \int (f^- + g^-) \, d\mu,$$

and hence that $\int (f + g) \, d\mu = \int f \, d\mu + \int g \, d\mu$.

If $f(x) \leq g(x)$ holds at each x in X, then $g - f$ is a nonnegative integrable function; hence $\int (g - f) \, d\mu \geq 0$, and so $\int g \, d\mu - \int f \, d\mu = \int (g - f) \, d\mu \geq 0$. $\qquad \square$

Examples 2.3.7.

(a) If μ is a finite measure, then every bounded measurable function on $(X,\mathscr{A},\mu)$ is integrable.

(b) In particular, every bounded Borel function, and hence every continuous function, on $[a,b]$ is Lebesgue integrable. (We'll see in Sect. 2.5 that the Lebesgue integral of a continuous function on $[a,b]$ can be found by calculating its Riemann integral.)

(c) Suppose that $\mathscr{A}$ is the σ-algebra on $\mathbb{N}$ containing all subsets of $\mathbb{N}$ and that μ is counting measure on $\mathscr{A}$. It follows from Proposition 2.3.3 that a nonnegative function f on $\mathbb{N}$ is μ-integrable if and only if the infinite series $\sum_n f(n)$ is convergent, and that in that case the integral and the sum of the series agree. Since a not necessarily nonnegative function f is integrable if and only if f^+ and f^- are integrable, it follows that f is integrable if and only if the infinite series $\sum_n f(n)$ is absolutely convergent. Once again, the integral and the sum of the series have the same value.

(d) Note that a simple measurable function that vanishes almost everywhere is integrable, with integral 0. We can reach the same conclusion for arbitrary measurable functions that vanish almost everywhere by first using Proposition 2.3.3 to deal with nonnegative functions and then using the decomposition $f = f^+ - f^-$. For a converse, see Corollary 2.3.12.   □

We now consider a few elementary properties of the integral; the basic limit theorems for the integral will be presented in the next section.

Proposition 2.3.8. *Let $(X,\mathscr{A},\mu)$ be a measure space, and let f be a $[-\infty,+\infty]$-valued $\mathscr{A}$-measurable function on X. Then f is integrable if and only if $|f|$ is integrable. If these functions are integrable, then $|\int f\,d\mu| \leq \int |f|\,d\mu$.*

Proof. Recall that by definition f is integrable if and only if f^+ and f^- are integrable. On the other hand, since $|f| = f^+ + f^-$, part (b) of Proposition 2.3.4 implies that $|f|$ is integrable if and only if f^+ and f^- are integrable. Thus the integrability of f is equivalent to the integrability of $|f|$. In case f and $|f|$ are integrable, the inequality $|\int f\,d\mu| \leq \int |f|\,d\mu$ follows from the calculation

$$\left| \int f\,d\mu \right| = \left| \int f^+\,d\mu - \int f^-\,d\mu \right| \leq \int f^+\,d\mu + \int f^-\,d\mu = \int |f|\,d\mu. \qquad \square$$

The reader should note that there are functions that are not measurable, and hence not integrable, but that have an integrable absolute value (see Exercise 3). Hence we needed to include the measurability of f among the hypotheses of Proposition 2.3.8.

Proposition 2.3.9. *Let $(X,\mathscr{A},\mu)$ be a measure space, and let f and g be $[-\infty,+\infty]$-valued $\mathscr{A}$-measurable functions on X that agree almost everywhere. If either $\int f\,d\mu$ or $\int g\,d\mu$ exists, then both exist, and $\int f\,d\mu = \int g\,d\mu$.*

Proof. First consider the case where f and g are nonnegative. Let $A = \{x \in X : f(x) \neq g(x)\}$, and let h be the function defined by

$$h(x) = \begin{cases} +\infty & \text{if } x \in A, \\ 0 & \text{if } x \notin A. \end{cases}$$

Then $\int h \, d\mu = 0$ (apply Proposition 2.3.3 to the sequence $\{h_n\}$ defined by $h_n = n\chi_A$). In view of Proposition 2.3.4 and the inequality $f \leq g + h$, this implies that $\int f \, d\mu \leq \int g \, d\mu + \int h \, d\mu = \int g \, d\mu$. A similar argument shows that $\int g \, d\mu \leq \int f \, d\mu$. Thus $\int f \, d\mu = \int g \, d\mu$.

The case where f and g are not necessarily nonnegative can be reduced to the case just treated through the decompositions $f = f^+ - f^-$ and $g = g^+ - g^-$. $\square$

Proposition 2.3.10. *Let* $(X, \mathscr{A}, \mu)$ *be a measure space, and let* f *be a* $[0, +\infty]$-*valued* $\mathscr{A}$-*measurable function on* X. *If* t *is a positive real number and if* A_t *is defined by* $A_t = \{x \in X : f(x) \geq t\}$, *then*

$$\mu(A_t) \leq \frac{1}{t} \int_{A_t} f \, d\mu \leq \frac{1}{t} \int f \, d\mu.$$

Proof. The relation $0 \leq t\chi_{A_t} \leq f\chi_{A_t} \leq f$ and part (c) of Proposition 2.3.4 imply that

$$\int t\chi_{A_t} \, d\mu \leq \int_{A_t} f \, d\mu \leq \int f \, d\mu.$$

Since $\int t\chi_{A_t} \, d\mu = t\mu(A_t)$, the proposition follows. $\square$

Corollary 2.3.11. *Let* $(X, \mathscr{A}, \mu)$ *be a measure space, and let* f *be a* $[-\infty, +\infty]$-*valued integrable function on* X. *Then* $\{x \in X : f(x) \neq 0\}$ *is* σ-*finite under* μ.

Proof. Proposition 2.3.10, applied to the function $|f|$, implies that the sets A_1, A_2, ... defined by

$$A_n = \left\{ x \in X : |f(x)| \geq \frac{1}{n} \right\}$$

have finite measure under μ. Thus $\{x \in X : f(x) \neq 0\}$, since it is equal to $\cup_n A_n$, is σ-finite under μ. $\square$

Corollary 2.3.12. *Let* $(X, \mathscr{A}, \mu)$ *be a measure space, and let* f *be a* $[-\infty, +\infty]$-*valued* $\mathscr{A}$-*measurable function on* X *that satisfies* $\int |f| \, d\mu = 0$. *Then* f *vanishes* μ-*almost everywhere.*

Proof. Proposition 2.3.10, applied to the function $|f|$, implies that

$$\mu\left(\left\{ x \in X : |f(x)| \geq \frac{1}{n} \right\} \right) \leq n \int |f| \, d\mu = 0$$

holds for each positive integer n. Since

$$\{x \in X : f(x) \neq 0\} = \bigcup_n \left\{x \in X : |f(x)| \geq \frac{1}{n}\right\},$$

the countable subadditivity of μ implies that $\mu(\{x \in X : f(x) \neq 0\}) = 0$. Thus f vanishes almost everywhere. □

Corollary 2.3.13. *Let $(X, \mathscr{A}, \mu)$ be a measure space, and let f be a $[-\infty, +\infty]$- valued integrable function on X such that $\int_A f \, d\mu \geq 0$ holds for all A in $\mathscr{A}$ (or even just for all A in the smallest σ-algebra on X that makes f measurable). Then $f \geq 0$ holds μ-almost everywhere.*

Proof. Let $A = \{x \in X : f(x) < 0\}$. Then $\int f\chi_A \, d\mu = \int_A f \, d\mu = 0$ (since $f < 0$ on A, yet we are assuming that $\int_A f \, d\mu \geq 0$). It follows from Corollary 2.3.12 that $f\chi_A$ vanishes almost everywhere and hence that $f \geq 0$ holds almost everywhere. □

Corollary 2.3.14. *Let $(X, \mathscr{A}, \mu)$ be a measure space, and let f be a $[-\infty, +\infty]$- valued integrable function on X. Then $|f(x)| < +\infty$ holds at μ-almost every x in X.*

Proof. Proposition 2.3.10, applied to the function $|f|$, implies that

$$\mu(\{x \in X : |f(x)| \geq n\}) \leq \frac{1}{n} \int |f| \, d\mu$$

holds for each positive integer n. Thus

$$\mu(\{x \in X : |f(x)| = +\infty\}) \leq \mu(\{x \in X : |f(x)| \geq n\}) \leq \frac{1}{n} \int |f| \, d\mu$$

holds for each n, and so $\mu(\{x \in X : |f(x)| = +\infty\}) = 0$ □

Corollary 2.3.15. *Let $(X, \mathscr{A}, \mu)$ be a measure space, and let f be a $[-\infty, +\infty]$- valued $\mathscr{A}$-measurable function on X. Then f is integrable if and only if there is a function in $\mathscr{L}^1(X, \mathscr{A}, \mu, \mathbb{R})$ that is equal to f almost everywhere.*

In other words, a measurable $[-\infty, +\infty]$-valued function f is integrable if and only if there is an $\mathbb{R}$-valued function that is integrable and equal to f μ-almost everywhere.

Proof. If there is a function in $\mathscr{L}^1(X, \mathscr{A}, \mu, \mathbb{R})$ that is equal to f almost everywhere, then the integrability of f follows from Proposition 2.3.9. Next suppose that f is integrable, and let $A = \{x \in X : |f(x)| = +\infty\}$. Then $A \in \mathscr{A}$, and Corollary 2.3.14 implies that $\mu(A) = 0$. It follows that the function f_0 defined by $f_0 = f\chi_{A^c}$ is $\mathscr{A}$- measurable and agrees with f almost everywhere. Proposition 2.3.9 now implies that f_0 is integrable and hence a member of $\mathscr{L}^1(X, \mathscr{A}, \mu, \mathbb{R})$. □

Exercises

1. Let $(X, \mathscr{A}, \mu)$ be a measure space, and let f and g belong to $\mathscr{L}^1(X, \mathscr{A}, \mu, \mathbb{R})$. Show that $f \vee g$ and $f \wedge g$ belong to $\mathscr{L}^1(X, \mathscr{A}, \mu, \mathbb{R})$.
2. Give Borel functions $f, g \colon \mathbb{R} \to \mathbb{R}$ that are Lebesgue integrable but are such that fg is not Lebesgue integrable.
3. Show that there is a function $f \colon \mathbb{R} \to \mathbb{R}$ that is not Lebesgue integrable, but is such that $|f|$ is Lebesgue integrable. (Hint: Let $f = \chi_A - \chi_B$, where A and B are suitable subsets of $\mathbb{R}$.)
4. Let $(X, \mathscr{A}, \mu)$ be a measure space, let $f, g \colon X \to [-\infty, +\infty]$ be integrable, and let $h \colon X \to [-\infty, +\infty]$ be an $\mathscr{A}$-measurable function that satisfies $h(x) = f(x) + g(x)$ at μ-almost every x in X. Show that h is integrable and that $\int h \, d\mu = \int f \, d\mu + \int g \, d\mu$.
5. Let $(X, \mathscr{A}, \mu)$ be a measure space, and let $f \colon X \to [-\infty, +\infty]$ be an $\mathscr{A}$-measurable function whose integral exists and is not equal to $-\infty$. Show that if $g \colon X \to [-\infty, +\infty]$ is an $\mathscr{A}$-measurable function that satisfies $f \leq g$ μ-almost everywhere, then the integral of g exists and satisfies $\int f \, d\mu \leq \int g \, d\mu$.
6. Let $(X, \mathscr{A}, \mu)$ be a measure space, let $\{f_n\}$ be a nondecreasing sequence of $[0, +\infty]$-valued $\mathscr{A}$-measurable functions on X, and let f be the function on X that satisfies $f(x) = \lim_n f_n(x)$ at each x in X.
   (a) Show that if g belongs to $\mathscr{S}_+$ and satisfies $g \leq f$, then for each ε in the interval $(0, 1)$, there is a sequence $\{g_n\}$ in $\mathscr{S}_+$ such that $g_n \leq f_n$ holds for each n and such that $\lim_n \int g_n \, d\mu = (1 - \varepsilon) \int g \, d\mu$. (Hint: See the proof of Proposition 2.3.2).
   (b) Use part (a) to prove that $\lim_n \int f_n \, d\mu = \int f \, d\mu$. Thus we have another proof of Proposition 2.3.3 and, at the same time, of Theorem 2.4.1 below (see, however, the last paragraph of the proof of Theorem 2.4.1).

2.4 Limit Theorems

In this section we prove the basic limit theorems of integration theory. These results are extremely important and account for much of the power of the Lebesgue integral. We will use them often in the rest of the book.

Theorem 2.4.1 (The Monotone Convergence Theorem). *Let $(X, \mathscr{A}, \mu)$ be a measure space, and let f and $f_1, f_2, \ldots$ be $[0, +\infty]$-valued $\mathscr{A}$-measurable functions on X. Suppose that*

$$f_1(x) \leq f_2(x) \leq \ldots \tag{1}$$

and

$$f(x) = \lim_n f_n(x) \tag{2}$$

hold at μ-almost every x in X. Then $\int f \, d\mu = \lim_n \int f_n \, d\mu$.

In this theorem the functions f and $f_1, f_2, \ldots$ are only assumed to be nonnegative and measurable; there are no assumptions about whether they are integrable.

Proof. First suppose that relations (1) and (2) hold at each x in X. The monotonicity of the integral (part (c) of Proposition 2.3.4) implies that

$$\int f_1 \, d\mu \leq \int f_2 \, d\mu \leq \cdots \leq \int f \, d\mu;$$

hence the sequence $\{\int f_n \, d\mu\}$ converges (perhaps to $+\infty$), and its limit satisfies $\lim_n \int f_n \, d\mu \leq \int f \, d\mu$. We turn to the reverse inequality. For each n choose a nondecreasing sequence $\{g_{n,k}\}_{k=1}^{\infty}$ of simple $[0, +\infty)$-valued measurable functions such that $f_n = \lim_k g_{n,k}$ (Proposition 2.1.8). For each n define a function h_n by

$$h_n = \max(g_{1,n}, g_{2,n}, \ldots, g_{n,n}).$$

Then $\{h_n\}$ is a nondecreasing sequence of simple $[0, +\infty)$-valued measurable functions that satisfy $h_n \leq f_n$ and $f = \lim_n h_n$. It follows from these remarks, Proposition 2.3.3, and the monotonicity of the integral that

$$\int f \, d\mu = \lim_n \int h_n \, d\mu \leq \lim_n \int f_n \, d\mu.$$

Hence $\int f \, d\mu = \lim_n \int f_n \, d\mu$.

Now suppose that we only require that relations (1) and (2) hold for almost every x in X. Let N be a set that belongs to $\mathscr{A}$, has measure zero under μ, and contains all points at which one or more of these relations fails. The function $f\chi_{N^c}$ and the sequence $\{f_n\chi_{N^c}\}$ satisfy the hypotheses made in the first part of the proof, and so

$$\int f\chi_{N^c} \, d\mu = \lim_n \int f_n\chi_{N^c} \, d\mu. \tag{3}$$

Since $f_n\chi_{N^c}$ agrees with f_n almost everywhere and $f\chi_{N^c}$ agrees with f almost everywhere, Eq. (3) and Proposition 2.3.9 imply that

$$\int f \, d\mu = \lim_n \int f_n \, d\mu. \qquad \square$$

Corollary 2.4.2 (Beppo Levi's Theorem). *Let $(X, \mathscr{A}, \mu)$ be a measure space, and let $\sum_{k=1}^{\infty} f_k$ be an infinite series whose terms are $[0, +\infty]$-valued $\mathscr{A}$-measurable functions on X. Then*

$$\int \sum_{k=1}^{\infty} f_k \, d\mu = \sum_{k=1}^{\infty} \int f_k \, d\mu.$$

Proof. Use the linearity of the integral, and apply Theorem 2.4.1 to the sequence $\{\sum_{k=1}^{n} f_k\}_{n=1}^{\infty}$ of partial sums of the series $\sum_{k=1}^{\infty} f_k$. $\qquad \square$

Example 2.4.3. Corollary 2.4.2 can be applied as follows to construct a large class of measures. Suppose that $(X, \mathscr{A}, \mu)$ is a measure space and that $f: X \to [0, +\infty]$ is $\mathscr{A}$-measurable. Define a function $v: \mathscr{A} \to [0, +\infty]$ by $v(A) = \int_A f \, d\mu$. Then $v(\varnothing) = 0$, and Corollary 2.4.2, applied to the series $\sum_n f \chi_{A_n}$, implies that if $\{A_n\}$ is a sequence of disjoint sets in $\mathscr{A}$, then $v(\cup_n A_n) = \sum_n v(A_n)$. Thus v is a measure on $(X, \mathscr{A})$. Moreover v is a finite measure if and only if f is μ-integrable. $\qquad \square$

The next result is often used to show that a function is integrable or to provide an upper bound for the value of an integral.

Theorem 2.4.4 (Fatou's Lemma). *Let $(X, \mathscr{A}, \mu)$ be a measure space, and let $\{f_n\}$ be a sequence of $[0, +\infty]$-valued $\mathscr{A}$-measurable functions on X. Then*

$$\int \varliminf_n f_n \, d\mu \leq \varliminf_n \int f_n \, d\mu.$$

Proof. For each positive integer n let $g_n = \inf_{k \geq n} f_k$. Each g_n is $\mathscr{A}$-measurable (Proposition 2.1.5), and the relations

$$g_1(x) \leq g_2(x) \leq \cdots$$

and

$$\varliminf_n f_n(x) = \lim_n g_n(x)$$

hold at each x in X. It follows from the monotone convergence theorem (Theorem 2.4.1) and the inequality $g_n \leq f_n$ that

$$\int \varliminf_n f_n \, d\mu = \int \lim_n g_n \, d\mu = \lim_n \int g_n \, d\mu \leq \varliminf_n \int f_n \, d\mu. \qquad \square$$

Theorem 2.4.5 (Lebesgue's Dominated Convergence Theorem). *Let $(X, \mathscr{A}, \mu)$ be a measure space, let g be a $[0, +\infty]$-valued integrable function on X, and let f and $f_1, f_2, \dots$ be $[-\infty, +\infty]$-valued $\mathscr{A}$-measurable functions on X such that*

$$f(x) = \lim_n f_n(x) \tag{4}$$

and

$$|f_n(x)| \leq g(x), \ n = 1, 2, \dots \tag{5}$$

hold at μ-almost every x in X. Then f and $f_1, f_2, \dots$ are integrable, and $\int f \, d\mu = \lim_n \int f_n \, d\mu$.

Proof. The integrability of f and $f_1, f_2, \dots$ follows from that of g; see Proposition 2.3.8, Proposition 2.3.9, and part (c) of Proposition 2.3.4.

Let us begin our proof that $\int f \, d\mu = \lim_n \int f_n \, d\mu$ by supposing that relations (4), (5), and

$$g(x) < +\infty \tag{6}$$

hold at *every* x in X. Then $\{g + f_n\}$ is a sequence of nonnegative $\mathscr{A}$-measurable functions such that $(g + f)(x) = \lim_n (g + f_n)(x)$ holds at each x in X, and so Fatou's lemma (Theorem 2.4.4) implies that

$$\int (g + f) \, d\mu \leq \varliminf_n \int (g + f_n) \, d\mu$$

and hence that

$$\int f \, d\mu \leq \varliminf_n \int f_n \, d\mu.$$

A similar argument, applied to the sequence $\{g - f_n\}$, shows that

$$\int (g - f) \, d\mu \leq \varliminf_n \int (g - f_n) \, d\mu$$

and hence that

$$\varlimsup_n \int f_n \, d\mu \leq \int f \, d\mu.$$

Consequently $\int f \, d\mu = \lim_n \int f_n \, d\mu$.

Next suppose that we only require that relations (4), (5), and (6) hold at almost every x in X (note that, according to Corollary 2.3.14, the hypothesis $\int g \, d\mu < +\infty$ implies that relation (6) holds at almost every x in X). We can reduce the present case to the one we have just dealt with by using a modified version of the final part of the proof of Theorem 2.4.1; the details are left to the reader. □

Example 2.4.6. Let us note how Theorem 2.4.5 can be used to justify "differentiation under the integral sign." Let $(X, \mathscr{A}, \mu)$ be a measure space, let $g : X \to [0, +\infty]$ be an integrable function, let I be an open subinterval of $\mathbb{R}$, and let $f : X \times I \to \mathbb{R}$ be such that

(a) for each t in I the function $x \mapsto f(x, t)$ is integrable,
(b) for each x in X the function $t \mapsto f(x, t)$ is differentiable on I, and
(c) the inequality

$$\left| \frac{f(x, t) - f(x, t_0)}{t - t_0} \right| \leq g(x) \tag{7}$$

holds for all t, t_0 in I and all x in X.

Define $g : I \to \mathbb{R}$ by $g(t) = \int_X f(x, t) \, \mu(dx)$. Let us use the dominated convergence theorem to show that g is differentiable on I, with g' given by $g'(t) = \int_X f_t(x, t) \, \mu(dx)$ at each t in I (here $f_t(x, t)$ denotes the partial derivative with respect to t). Suppose that $\{t_n\}$ is a sequence of elements of I, all different from t_0, such that

$\lim_n t_n = t_0$. Then, in view of inequality (7), the dominated convergence theorem implies that $x \mapsto f_t(x, t_0)$ is integrable and that

$$\lim_n \frac{g(t_n) - g(t_0)}{t_n - t_0} = \int_X f_t(x, t_0) \mu(dx).$$

Combining this with item C.7 in Appendix C finishes the argument.   □

Exercises

1. Give sequences $\{f_n\}$, $\{g_n\}$, and $\{h_n\}$ of functions in $\mathscr{L}^1(\mathbb{R}, \mathscr{B}(\mathbb{R}), \lambda, \mathbb{R})$ that converge to zero almost everywhere, but satisfy
   (a) $\lim_n \int f_n \, d\lambda = +\infty$,
   (b) $\lim_n \int g_n \, d\lambda = 1$, and
   (c) $\limsup_n \int h_n \, d\lambda = 1$ and $\liminf_n \int h_n \, d\lambda = -1$.

2. Prove that the monotone convergence theorem still holds if the assumption that the functions $f_1, f_2, \ldots$ are nonnegative is dropped, and the assumption that f_1 is integrable is added (note that in this case the integrals of the functions f and $f_2, f_3, \ldots$ exist, but may be $+\infty$).

3. Let $(X, \mathscr{A}, \mu)$ be a measure space. Use Exercise 2 to show that if $\{f_n\}$ is a decreasing sequence of measurable functions and if f_1 is integrable, then $\int \lim_n f_n \, d\mu = \lim_n \int f_n \, d\mu$ (as in Exercise 2 the integrals involved exist, but may be infinite).

4. Let f, g, and $f_1, f_2, \ldots$ be as in the dominated convergence theorem, and define sequences $\{p_n\}$ and $\{q_n\}$ by $p_n = \inf_{k \geq n} f_k$ and $q_n = \sup_{k \geq n} f_k$. Use Exercises 2 and 3, together with the inequality $p_n \leq f_n \leq q_n$, to give another proof of the dominated convergence theorem.

5. Use Exercise 3, applied to the sequence $\{h_n\}$ defined by $h_n = \sup_{k \geq n} |f_k - f|$, to give still another proof of the dominated convergence theorem. (Of course the functions f and $f_1, f_2, \ldots$ can be modified so that they are real valued and hence so that $f_k - f$ makes sense.)

6. Let $(X, \mathscr{A}, \mu)$ be a measure space, and let $f: X \to [0, +\infty]$ be $\mathscr{A}$-measurable.
   (a) Show that if each value of f is a nonnegative integer or $+\infty$, then $\int f \, d\mu = \sum_{n=1}^{\infty} \mu(\{x : f(x) \geq n\})$.
   (b) Now suppose that the values of f are arbitrary elements of $[0, +\infty]$ and that μ is finite. Show that the integrability of f is equivalent to the convergence of the series $\sum_{n=1}^{\infty} \mu(\{x : f(x) \geq n\})$.

7. Let $(X, \mathscr{A})$ and $(Y, \mathscr{B})$ be measurable spaces. A function $K: X \times \mathscr{B} \to [0, +\infty]$ is called a *kernel* from $(X, \mathscr{A})$ to $(Y, \mathscr{B})$ if

   (i) for each x in X the function $B \mapsto K(x, B)$ is a measure on $(Y, \mathscr{B})$, and
   (ii) for each B in $\mathscr{B}$ the function $x \mapsto K(x, B)$ is $\mathscr{A}$-measurable.

Suppose that K is a kernel from $(X, \mathscr{A})$ to $(Y, \mathscr{B})$, that μ is a measure on $(X, \mathscr{A})$, and that f is a $[0, +\infty]$-valued $\mathscr{B}$-measurable function on Y. Show that
(a) $B \mapsto \int K(x, B) \mu(dx)$ is a measure on $(Y, \mathscr{B})$,
(b) $x \mapsto \int f(y) K(x, dy)$ is an $\mathscr{A}$-measurable function on X, and
(c) if ν is the measure on $(Y, \mathscr{B})$ defined in part (a), then $\int f(y) \nu(dy) = \int (\int f(y) K(x, dy)) \mu(dx)$. (Hint: Begin with the case where f is a characteristic function.)

8. (Continuation.) Now suppose that μ is finite, that $\sup\{K(x, Y) : x \in X\}$ is finite, and that the measurable function f is bounded but not necessarily nonnegative. Show that
(a) $x \mapsto \int f(y) K(x, dy)$ is a bounded $\mathscr{A}$-measurable function on X, and
(b) $\int f(y) \nu(dy) = \int (\int f(y) K(x, dy)) \mu(dx)$. (Here again ν is the measure defined in part (a) of Exercise 7.)

9. Let $(X, \mathscr{A}, \mu)$ be a measure space, let g be a $[0, +\infty]$-valued integrable function on X, and let f and f_t (for t in $[0, +\infty)$) be real-valued $\mathscr{A}$-measurable functions on X such that

$$f(x) = \lim_{t \to +\infty} f_t(x)$$

and

$$|f_t(x)| \le g(x) \text{ for } t \text{ in } [0, +\infty)$$

hold at almost every x in X. Show that $\int f \, d\mu = \lim_{t \to +\infty} \int f_t \, d\mu$. (Hint: Give a simplified version of the argument in Example 2.4.6.)

10. Let I be an open subinterval of $\mathbb{R}$, and let $f : \mathbb{R} \to \mathbb{R}$ be a Borel measurable function such that $x \mapsto e^{tx} f(x)$ is Lebesgue integrable for each t in I. Define $h : I \to \mathbb{R}$ by $h(t) = \int_{\mathbb{R}} e^{tx} f(x) \lambda(dx)$. Show that h is differentiable, with derivative given by $h'(t) = \int_{\mathbb{R}} x e^{tx} f(x) \lambda(dx)$, at each t in I. Of course, it is part of your task to show that $x \mapsto x e^{tx} f(x)$ is integrable for each t in I. (Hint: Use the Maclaurin expansion of e^u to show that $|e^u - 1| \le |u| e^{|u|}$ holds for each u in $\mathbb{R}$, and use the argument from Example 2.4.6.)

11. Let $(X, \mathscr{A}, \mu)$ be a measure space, and let f and $f_1, f_2, \ldots$ be nonnegative functions that belong to $\mathscr{L}^1(X, \mathscr{A}, \mu, \mathbb{R})$ and satisfy

(i) $\{f_n\}$ converges to f almost everywhere, and
(ii) $\int f \, d\mu = \lim_n \int f_n \, d\mu$.

Show that $\lim_n \int |f_n - f| \, d\mu = 0$.

2.5   The Riemann Integral

This section contains the standard facts that relate the Lebesgue integral to the Riemann integral. We begin by recalling Darboux's definition of the Riemann integral, as given in the Introduction (we use it as our basic definition), and then we give a number of details that we omitted earlier. We also give the standard characterization of the Riemann integrable functions on a closed bounded interval as the bounded functions on that interval that are almost everywhere continuous.

Let $[a,b]$ be a closed bounded interval. A *partition* of $[a,b]$ is a finite sequence $\{a_i\}_{i=0}^k$ of real numbers such that

$$a = a_0 < a_1 < \cdots < a_k = b.$$

We will generally denote a partition by a symbol such as $\mathscr{P}$ or $\mathscr{P}_n$.

If $\{a_i\}_{i=0}^k$ and $\{b_i\}_{i=0}^j$ are partitions of $[a,b]$ and if each term of $\{a_i\}_{i=0}^k$ appears among the terms of $\{b_i\}_{i=0}^j$, then $\{b_i\}_{i=0}^j$ is a *refinement of* or is *finer than* $\{a_i\}_{i=0}^k$.

Let f be a bounded real-valued function on $[a,b]$. If $\mathscr{P}$ is the partition $\{a_i\}_{i=0}^k$ of $[a,b]$ and if $m_i = \inf\{f(x) : x \in [a_{i-1}, a_i]\}$ and $M_i = \sup\{f(x) : x \in [a_{i-1}, a_i]\}$ for $i = 1, \ldots, k$, then the *lower sum* $l(f, \mathscr{P})$ corresponding to f and $\mathscr{P}$ is defined to be $\sum_{i=1}^k m_i(a_i - a_{i-1})$, and the *upper sum* $u(f, \mathscr{P})$ corresponding to f and $\mathscr{P}$ is defined to be $\sum_{i=1}^k M_i(a_i - a_{i-1})$.

It is easy to check that if $\mathscr{P}$ is an arbitrary partition of $[a,b]$, then

$$l(f, \mathscr{P}) \leq u(f, \mathscr{P})$$

and that if $\mathscr{P}_1$ and $\mathscr{P}_2$ are partitions of $[a,b]$ such that $\mathscr{P}_2$ is a refinement of $\mathscr{P}_1$, then

$$l(f, \mathscr{P}_1) \leq l(f, \mathscr{P}_2)$$

and

$$u(f, \mathscr{P}_2) \leq u(f, \mathscr{P}_1)$$

(first consider the case where $\mathscr{P}_2$ contains exactly one more point than $\mathscr{P}_1$, and then use induction on the difference between the number of points in $\mathscr{P}_2$ and the number of points in $\mathscr{P}_1$). It follows that if $\mathscr{P}_1$ and $\mathscr{P}_2$ are arbitrary partitions of $[a,b]$, then

$$l(f, \mathscr{P}_1) \leq u(f, \mathscr{P}_2)$$

(let $\mathscr{P}_3$ be a partition of $[a,b]$ that is a refinement of both $\mathscr{P}_1$ and $\mathscr{P}_2$ and note that

$$l(f, \mathscr{P}_1) \leq l(f, \mathscr{P}_3) \leq u(f, \mathscr{P}_3) \leq u(f, \mathscr{P}_2)).$$

Hence the set of all lower sums for f is bounded above by each of the upper sums for f. The supremum of this set of lower sums is the *lower integral* of f over $[a,b]$ and is denoted by $\underline{\int}_a^b f$. The lower integral satisfies $\underline{\int}_a^b f \leq u(f, \mathscr{P})$ for each upper sum $u(f, \mathscr{P})$ and so is a lower bound for the set of all upper sums for f. The infimum of this set of upper sums is the *upper integral* of f over $[a,b]$ and is denoted by $\overline{\int}_a^b f$. It follows immediately that $\underline{\int}_a^b f \leq \overline{\int}_a^b f$. If $\underline{\int}_a^b f = \overline{\int}_a^b f$, then f is *Riemann integrable* on $[a,b]$, and the common value of $\underline{\int}_a^b f$ and $\overline{\int}_a^b f$ is called the *Riemann integral* of f over $[a,b]$ and is denoted by $\int_a^b f$ or $\int_a^b f(x)\,dx$ (we'll occasionally write $(R)\int_a^b f$ or $(R)\int_a^b f(x)\,dx$ when we need to make clear which integral we mean).

The following reformulation of the definition of Riemann integrability is often useful.

Lemma 2.5.1. *A bounded function $f\colon [a,b] \to \mathbb{R}$ is Riemann integrable if and only if for every positive ε there is a partition $\mathscr{P}$ of $[a,b]$ such that $u(f,\mathscr{P})-l(f,\mathscr{P})<\varepsilon$.*

Proof. This is an immediate consequence of the fact that f is Riemann integrable if and only if

$$\sup_{\mathscr{P}} l(f,\mathscr{P}) = \inf_{\mathscr{P}} u(f,\mathscr{P}),$$

together with the fact that if $\mathscr{P}_1$ and $\mathscr{P}_2$ are partitions such that

$$u(f,\mathscr{P}_1) - l(f,\mathscr{P}_2) < \varepsilon,$$

then taking a common refinement $\mathscr{P}$ of $\mathscr{P}_1$ and $\mathscr{P}_2$ gives a partition $\mathscr{P}$ such that $u(f,\mathscr{P}) - l(f,\mathscr{P}) < \varepsilon$. $\square$

Example 2.5.2. Suppose that f is a continuous, and hence bounded, function on $[a,b]$. Then f is uniformly continuous (Theorem C.12), and so for each positive number ε there is a positive number δ such that if x and y are elements of $[a,b]$ that satisfy $|x-y|<\delta$, then $|f(x)-f(y)|<\varepsilon$. If ε and δ are related in this way and if $\mathscr{P}$ is a partition of $[a,b]$ into intervals each of which has length less than δ, then $u(f,\mathscr{P})-l(f,\mathscr{P}) \le \varepsilon(b-a)$. It follows that every continuous function on $[a,b]$ is Riemann integrable. $\square$

Example 2.5.3. Let $f\colon [0,1] \to \mathbb{R}$ be the characteristic function of the set of rational numbers in $[0,1]$. Then f is Lebesgue integrable, and $\int_{[0,1]} f\,d\lambda = 0$. However, as we noted in the Introduction, every lower sum of f is equal to 0 and every upper sum of f is equal to 1; thus f is not Riemann integrable. $\square$

Theorem 2.5.4. *Let $[a,b]$ be a closed bounded interval, and let f be a bounded real-valued function on $[a,b]$. Then*

(a) *f is Riemann integrable if and only if it is continuous at almost every point of $[a,b]$, and*

(b) *if f is Riemann integrable, then f is Lebesgue integrable and the Riemann and Lebesgue integrals of f coincide.*

Proof. Suppose that f is Riemann integrable. Then for each positive integer n we can choose a partition $\mathscr{P}_n$ of $[a,b]$ such that $u(f,\mathscr{P}_n) - l(f,\mathscr{P}_n) < 1/n$. By replacing the $\mathscr{P}_n$'s with finer partitions if necessary, we can assume that for each n the partition $\mathscr{P}_{n+1}$ is a refinement of the partition $\mathscr{P}_n$. Define sequences $\{g_n\}$ and $\{h_n\}$ of functions on $[a,b]$ by letting g_n and h_n agree with f at the point a and letting them be constant on each interval of the form $(a_{i-1},a_i]$ determined by $\mathscr{P}_n$, there having the values $\inf\{f(x) : a_{i-1} \le x \le a_i\}$ and $\sup\{f(x) : a_{i-1} \le x \le a_i\}$, respectively. Then $\{g_n\}$ is an increasing sequence of simple Borel functions that satisfy $g_n \le f$ and $\int_{[a,b]} g_n \, d\lambda = l(f, \mathscr{P}_n)$ for each n, and $\{h_n\}$ is a decreasing sequence of simple Borel functions that satisfy $h_n \ge f$ and $\int_{[a,b]} h_n \, d\lambda = u(f, \mathscr{P}_n)$ for each n. Since f is bounded, the sequences $\{g_n\}$ and $\{h_n\}$ are bounded. Define functions g and h by $g = \lim_n g_n$ and $h = \lim_n h_n$. Then g and h are Borel measurable, and the dominated convergence theorem (Theorem 2.4.5) implies that g and h are Lebesgue integrable, with $\int_{[a,b]} g \, d\lambda$ and $\int_{[a,b]} h \, d\lambda$ equal to $\lim_n l(f, \mathscr{P}_n)$ and $\lim_n u(f, \mathscr{P}_n)$, respectively, and so to the Riemann integral of f. Thus $\int_{[a,b]} (h - g) \, d\lambda = 0$. Since in addition $h - g \ge 0$, Corollary 2.3.12 implies that

$$g(x) = h(x) \text{ for almost every } x \text{ in } [a,b]. \tag{1}$$

This relation has two consequences. For the first, note that if $g(x) = h(x)$ and if x is a point in $[a,b]$ that is not a division point in any of the partitions $\mathscr{P}_n$, then f is continuous at x. Thus (1) implies that f is continuous almost everywhere, and so half of part (a) is proved. Note also that $g \le f \le h$, and so (1) implies that f is equal to g almost everywhere. It follows that f is Lebesgue measurable and Lebesgue integrable (Propositions 2.2.2 and 2.3.9) and that the Riemann and Lebesgue integrals of f are the same; thus part (b) is proved.

We turn to the remaining half of part (a). For this suppose that f is continuous almost everywhere. For each n let $\mathscr{P}_n$ be the partition of $[a,b]$ that divides $[a,b]$ into 2^n subintervals of equal length. Use these partitions $\mathscr{P}_n$ to construct functions g_n and h_n as in the first part of this proof. The relations $f(x) = \lim_n g_n(x)$ and $f(x) = \lim_n h_n(x)$ clearly hold at each x at which f is continuous and so at almost every x in $[a,b]$. Thus $\lim_n (h_n - g_n) = 0$ holds almost everywhere, and so, since $\int_{[a,b]} g_n \, d\lambda = l(f, \mathscr{P}_n)$ and $\int_{[a,b]} h_n \, d\lambda = u(f, \mathscr{P}_n)$, the dominated convergence theorem implies that

$$\lim_n (u(f, \mathscr{P}_n) - l(f, \mathscr{P}_n)) = 0.$$

Thus for each ε there is a partition $\mathscr{P}$ of $[a,b]$ such that $u(f, \mathscr{P}) - l(f, \mathscr{P}) < \varepsilon$, and the Riemann integrability of f follows.   □

Example 2.5.5. Since the characteristic function of the set of rational numbers in $[0,1]$ is not continuous anywhere in $[0,1]$, part (a) of Theorem 2.5.4 gives another proof that this characteristic function is not Riemann integrable.   □

Example 2.5.6. We saw in the Introduction that the pointwise limit of a bounded sequence of Riemann integrable functions may fail to be Riemann integrable. Thus a simple rewriting of the dominated convergence theorem so as to apply to the Riemann integral will fail. However, in view of Theorem 2.5.4 and the dominated convergence theorem for the Lebesgue integral, we can repair this difficulty by adding the hypothesis that the limit function be Riemann integrable. The repaired assertion is still not as powerful as the dominated convergence theorem for the Lebesgue integral, since it can only be applied when we can prove the Riemann integrability of the limit function.   □

It is sometimes useful to view Riemann integrals as the limits of what are called Riemann sums. For this we need a couple of definitions. A *tagged partition* of an interval $[a,b]$ is a partition $\{a_i\}_{i=0}^k$ of $[a,b]$, together with a sequence $\{x_i\}_{i=1}^k$ of numbers (called *tags*) such that $a_{i-1} \leq x_i \leq a_i$ holds for $i = 1, \ldots, k$. (In other words, each tag x_i must belong to the corresponding interval $[a_{i-1}, a_i]$.) As with partitions, we will often denote a tagged partition by a symbol such as $\mathscr{P}$.

The *mesh* or *norm* $\|\mathscr{P}\|$ of a partition (or a tagged partition) $\mathscr{P}$ is defined by $\|\mathscr{P}\| = \max_i(a_i - a_{i-1})$, where $\{a_i\}$ is the sequence of division points for $\mathscr{P}$. In other words, the mesh of a partition is the length of the longest of its subintervals.

The Riemann sum $\mathscr{R}(f, \mathscr{P})$ corresponding to the function f and the tagged partition $\mathscr{P}$ is defined by

$$\mathscr{R}(f, \mathscr{P}) = \sum_{i=1}^k f(x_i)(a_i - a_{i-1}).$$

Since for each i the value $f(x_i)$ lies between the infimum m_i and the supremum M_i of the values of f on the interval $[a_{i-1}, a_i]$, we have

$$l(f, \mathscr{P}) \leq \mathscr{R}(f, \mathscr{P}) \leq u(f, \mathscr{P})$$

for each tagged partition $\mathscr{P}$.

Proposition 2.5.7. *A function* $f\colon [a,b] \to \mathbb{R}$ *is Riemann integrable if and only if there is a real number L such that*

$$\lim_{\mathscr{P}} \mathscr{R}(f, \mathscr{P}) = L, \tag{2}$$

where the limit is taken as the mesh of the tagged partition $\mathscr{P}$ *approaches* 0. *If this limit exists, then it is equal to the Riemann integral* $\int_a^b f$.

We can make this more precise if we note that saying $\lim_{\mathscr{P}} \mathscr{R}(f, \mathscr{P}) = L$ is the same as saying that for every positive ε there is a positive δ such that $|\mathscr{R}(f, \mathscr{P}) - L| < \varepsilon$ holds whenever $\mathscr{P}$ is a tagged partition whose mesh is less than δ.

Proof. Suppose there exists a number L such that $\lim_{\mathscr{P}} \mathscr{R}(f, \mathscr{P}) = L$. Let ε be a positive number, choose a corresponding δ, and then choose a partition $\mathscr{P}_0$ whose mesh is less than δ. Consider the collection of all tagged partitions $\mathscr{P}$ that have the same division points as $\mathscr{P}_0$. Each of these tagged partitions has mesh less than δ and so satisfies $|\mathscr{R}(f, \mathscr{P}) - L| < \varepsilon$. By choosing the tags appropriately, we can find tagged partitions $\mathscr{P}_1$ and $\mathscr{P}_2$ in this collection that make $\mathscr{R}(f, \mathscr{P}_1)$ and $\mathscr{R}(f, \mathscr{P}_2)$ arbitrarily close to $l(f, \mathscr{P}_0)$ and $u(f, \mathscr{P}_0)$, which gives us $|l(f, \mathscr{P}_0) - L| \leq \varepsilon$ and $|u(f, \mathscr{P}_0) - L| \leq \varepsilon$. It then follows from Lemma 2.5.1 that f is Riemann integrable. It is easy to check that $L = \int_a^b f$ (note that $\int_a^b f$ lies between $l(f, \mathscr{P}_0)$ and $u(f, \mathscr{P}_0)$).

Now suppose that f is Riemann integrable. Let ε be a positive number, and choose a partition $\mathscr{P}_0$ such that $u(f,\mathscr{P}_0) - l(f,\mathscr{P}_0) < \varepsilon$ (see Lemma 2.5.1). Let N be the number of subintervals in $\mathscr{P}_0$. We will produce a positive number δ such that each tagged partition $\mathscr{P}$ with mesh less than δ satisfies $|\mathscr{R}(f,\mathscr{P}) - \int_a^b f| < 2\varepsilon$. We begin by assuming that δ is smaller than the mesh of $\mathscr{P}_0$; we will presently see how much smaller we should make it. So let $\mathscr{P}$ be a tagged partition with mesh less than δ. If it happens that $\mathscr{P}$ is a refinement of $\mathscr{P}_0$ (i.e., every subinterval of $\mathscr{P}$ is a subset of some subinterval of $\mathscr{P}_0$), then $\mathscr{R}(f,\mathscr{P})$ satisfies

$$l(f,\mathscr{P}_0) \le \mathscr{R}(f,\mathscr{P}) \le u(f,\mathscr{P}_0)$$

and so belongs to the interval $[l(f,\mathscr{P}_0), u(f,\mathscr{P}_0)]$. Since $\int_a^b f$ also belongs to this interval, it follows that

$$\left| \mathscr{R}(f,\mathscr{P}) - \int_a^b f \right| \le u(f,\mathscr{P}_0) - l(f,\mathscr{P}_0) \le \varepsilon.$$

We turn to the general case, where $\mathscr{P}$ might not be a refinement of $\mathscr{P}_0$. Some of the intervals $[a_{i-1}, a_i]$ in $\mathscr{P}$ might contain a division point of $\mathscr{P}_0$ as an interior point. Since there are only N subintervals in $\mathscr{P}_0$, at most $N-1$ subintervals of $\mathscr{P}$ can have a division point of $\mathscr{P}_0$ as an interior point. Build a new tagged partition $\mathscr{P}'$ of $[a,b]$ by taking the subintervals and tags from $\mathscr{P}$ but splitting each subinterval whose interior contains a division point into two subintervals (dividing it at the corresponding division point) and choosing arbitrary tags in the new intervals. The differences between $\mathscr{R}(f,\mathscr{P}')$ and $\mathscr{R}(f,\mathscr{P})$ arise only from the split intervals, and it is easy to check that $|\mathscr{R}(f,\mathscr{P}) - \mathscr{R}(f,\mathscr{P}')| \le 2M(N-1)\delta$, where M is an upper bound for the values of $|f|$. If we require that δ be so small that $2M(N-1)\delta < \varepsilon$ and note that $|\mathscr{R}(f,\mathscr{P}') - \int_a^b f| \le \varepsilon$ (since $\mathscr{P}'$ is a refinement of $\mathscr{P}_0$), then we have

$$\left| \mathscr{R}(f,\mathscr{P}) - \int_a^b f \right| \le \left| \mathscr{R}(f,\mathscr{P}) - \mathscr{R}(f,\mathscr{P}') \right| + \left| \mathscr{R}(f,\mathscr{P}') - \int_a^b f \right|$$

$$\le 2M(N-1)\delta + \varepsilon < 2\varepsilon,$$

and the proof is complete.   □

Note that although in the Riemann theory integrals over all of $\mathbb{R}$ are defined as improper integrals, in the Lebesgue theory they can be[3] defined directly. If f is a Lebesgue integrable function on $\mathbb{R}$, then the relation

---

[3]There are also cases of functions defined on $\mathbb{R}$ that are not Lebesgue integrable over $\mathbb{R}$ but for which the corresponding improper integral exists. For instance, define $f \colon \mathbb{R} \to \mathbb{R}$ by $f(x) = 0$ if $x < 1$ and $f(x) = (-1)^n/n$ if $n \le x < n+1$, where $n = 1, 2, \ldots$.

$$\int_{\mathbb{R}} f \, d\lambda = \lim_{\substack{a \to -\infty \\ b \to +\infty}} \int_{[a,b]} f \, d\lambda$$

holds, but as a consequence of the dominated convergence theorem (see Exercise 5), and not as a definition.

Exercises

1. Suppose that $a < b < c$ and that f is a real-valued function on $[a,c]$. Show directly (i.e., without using Theorem 2.5.4) that f is Riemann integrable on $[a,c]$ if and only if it is Riemann integrable on $[a,b]$ and $[b,c]$. Also show that

$$\int_a^c f = \int_a^b f + \int_b^c f$$

   if f is Riemann integrable on these intervals.
2. Let $\mathscr{R}_{[a,b]}$ be the set of all Riemann integrable functions on the interval $[a,b]$. Show directly (i.e., without using Theorem 2.5.4) that
   (a) $\mathscr{R}_{[a,b]}$ is a vector space over $\mathbb{R}$, and
   (b) $f \mapsto \int_a^b f$ is a linear functional on $\mathscr{R}_{[a,b]}$.
3. Show that a Riemann integrable function is not necessarily Borel measurable. (Hint: Consider χ_B, where B is the set constructed in the proof of Proposition 2.1.11.)
4. Show that there is an increasing sequence $\{f_n\}$ of continuous functions on $[0,1]$ such that

   (i) $0 \le f_n(x) \le 1$ holds for each n and x, and
   (ii) $\lim_n f_n$ is not Riemann integrable.

   (Hint: Let C be one of the closed sets constructed in Exercise 1.4.4, let $U = [0,1] - C$, and choose $\{f_n\}$ so that $\lim_n f_n = \chi_U$.)
5. Show that if $f \in \mathscr{L}^1(\mathbb{R}, \mathscr{B}(\mathbb{R}), \lambda, \mathbb{R})$, then

$$\int_{\mathbb{R}} f \, d\lambda = \lim_{\substack{a \to -\infty \\ b \to +\infty}} \int_{[a,b]} f \, d\lambda.$$

   (Hint: Use the dominated convergence theorem and a modification of the hint given for Exercise 2.4.9.)
6.(a) Show that if $f \colon [a,b] \to \mathbb{R}$ is Riemann integrable and if $m \le f(t) \le M$ holds for all t in the subinterval $[c,d]$ of $[a,b]$, then

$$m(d-c) \le \int_c^d f(t) \, dt \le M(d-c).$$

(b) Prove the fundamental theorem of calculus, in the form given in the Introduction to this book. (Hint: Use part (a) to estimate $\frac{F(x)-F(x_0)}{x-x_0}$.)

7. (a) Suppose that f and g are bounded functions on the interval $[a,b]$ and that ε is positive. Show that if $|f(x) - g(x)| \leq \varepsilon$ holds for all x in $[a,b]$, then $|\underline{\int}_a^b f - \underline{\int}_a^b g| \leq \varepsilon(b-a)$ and $|\overline{\int}_a^b f - \overline{\int}_a^b g| \leq \varepsilon(b-a)$.

(b) Suppose that $\{f_n\}$ is a sequence of Riemann integrable functions on the interval $[a,b]$ and that $\{f_n\}$ converges uniformly to a function f. Show that f is Riemann integrable and that $\int_a^b f(x)\,dx = \lim_n \int_a^b f_n(x)\,dx$. (Hint: Use part (a).)

8. Show that as n approaches infinity, the mean of the n values $n/(n+1)$, $n/(n+2)$, ..., $n/(n+n)$ approaches $\ln(2)$. (Hint: Write the mean of those values as a Riemann sum for the integral $\int_0^1 \frac{1}{1+x}\,dx$.)

2.6 Measurable Functions Again, Complex-Valued Functions, and Image Measures

In this section we give a general definition of measurable functions, and then we discuss some related concepts and some examples.

Let $(X,\mathscr{A})$ and $(Y,\mathscr{B})$ be measurable spaces. A function $f\colon X \to Y$ is *measurable with respect to $\mathscr{A}$ and $\mathscr{B}$* if for each B in $\mathscr{B}$ the set $f^{-1}(B)$ belongs to $\mathscr{A}$. Instead of saying that f is measurable with respect to $\mathscr{A}$ and $\mathscr{B}$, we will sometimes say that f is a *measurable function* from $(X,\mathscr{A})$ to $(Y,\mathscr{B})$ or simply that $f\colon (X,\mathscr{A}) \to (Y,\mathscr{B})$ is *measurable*. Likewise, if A belongs to $\mathscr{A}$, a function $f\colon A \to Y$ is *measurable* if $f^{-1}(B) \in \mathscr{A}$ holds whenever B belongs to $\mathscr{B}$.

Proposition 2.6.1. *Let $(X,\mathscr{A})$, $(Y,\mathscr{B})$, and $(Z,\mathscr{C})$ be measurable spaces, and let $f\colon (Y,\mathscr{B}) \to (Z,\mathscr{C})$ and $g\colon (X,\mathscr{A}) \to (Y,\mathscr{B})$ be measurable. Then $f \circ g\colon (X,\mathscr{A}) \to (Z,\mathscr{C})$ is measurable.*

Proof. Suppose that $C \in \mathscr{C}$. Then $f^{-1}(C) \in \mathscr{B}$, and so $g^{-1}(f^{-1}(C)) \in \mathscr{A}$. Since $(f \circ g)^{-1}(C) = g^{-1}(f^{-1}(C))$, the measurability of $f \circ g$ follows. $\qquad\square$

See Exercises 1 and 2 for some applications of the preceding proposition.

The following result is often useful for verifying the measurability of a function.

Proposition 2.6.2. *Let $(X,\mathscr{A})$ and $(Y,\mathscr{B})$ be measurable spaces, and let $\mathscr{B}_0$ be a collection of subsets of Y such that $\sigma(\mathscr{B}_0) = \mathscr{B}$. Then a function $f\colon X \to Y$ is measurable with respect to $\mathscr{A}$ and $\mathscr{B}$ if and only if $f^{-1}(B) \in \mathscr{A}$ holds for each B in $\mathscr{B}_0$.*

Proof. Of course, every function f that is measurable with respect to $\mathscr{A}$ and $\mathscr{B}$ satisfies $f^{-1}(B) \in \mathscr{A}$ for each B in $\mathscr{B}_0$. We turn to the converse, and assume that $f^{-1}(B) \in \mathscr{A}$ holds for each B in $\mathscr{B}_0$. Let $\mathscr{F}$ be the collection of all subsets B of Y such that $f^{-1}(B) \in \mathscr{A}$. The identities $f^{-1}(Y) = X$, $f^{-1}(B^c) = (f^{-1}(B))^c$, and

$f^{-1}(\cup_n B_n) = \cup_n f^{-1}(B_n)$ imply that $\mathscr{F}$ is a σ-algebra on Y. Since $\mathscr{F}$ includes $\mathscr{B}_0$, it must include the σ-algebra generated by $\mathscr{B}_0$, namely $\mathscr{B}$. Thus f is measurable with respect to $\mathscr{A}$ and $\mathscr{B}$. $\square$

Example 2.6.3. Suppose that $(X, \mathscr{A})$ is a measurable space and that f is a real-valued function on X. Proposition 2.1.9 implies that f is $\mathscr{A}$-measurable (in the sense of Sect. 2.1) if and only if it is measurable with respect to $\mathscr{A}$ and $\mathscr{B}(\mathbb{R})$. This conclusion can also be derived from Proposition 2.6.2 (let the collection $\mathscr{B}_0$ in Proposition 2.6.2 consist of all intervals of the form $(-\infty, t]$; see Proposition 1.1.4).
                                                                                              $\square$

Next we consider extended real-valued functions. Let $\mathscr{B}(\overline{\mathbb{R}})$ be the collection of all subsets of $\overline{\mathbb{R}}$ of the form $B \cup C$, where $B \in \mathscr{B}(\mathbb{R})$ and $C \subseteq \{-\infty, +\infty\}$. It is easy to check that $\mathscr{B}(\overline{\mathbb{R}})$ is a σ-algebra on $\overline{\mathbb{R}}$.

Proposition 2.6.4. *Let $(X, \mathscr{A})$ be a measurable space, and let f be an extended real-valued function on X. Then f is $\mathscr{A}$-measurable (in the sense of Sect. 2.1) if and only if it is measurable with respect to $\mathscr{A}$ and $\mathscr{B}(\overline{\mathbb{R}})$.*

Proof. If f is measurable with respect to $\mathscr{A}$ and $\mathscr{B}(\overline{\mathbb{R}})$, then for each t in $\mathbb{R}$ the set $\{x \in X : f(x) \leq t\}$, as the inverse image under f of the set $\{-\infty\} \cup (-\infty, t]$, belongs to $\mathscr{A}$; hence f must be $\mathscr{A}$-measurable.

Now assume that f is $\mathscr{A}$-measurable. Then $f^{-1}(\{+\infty\})$ and $f^{-1}(\{-\infty\})$ are equal to $\cap_{n=1}^{\infty}\{x \in X : f(x) > n\}$ and $\cap_{n=1}^{\infty}\{x \in X : f(x) < -n\}$, respectively, and so the inverse image under f of each subset of $\{-\infty, +\infty\}$ belongs to $\mathscr{A}$. In addition $\{x \in X : -\infty < f(x) < +\infty\}$ belongs to $\mathscr{A}$, and Proposition 2.1.9 (applied to the restriction of f to $\{x \in X : -\infty < f(x) < +\infty\}$) implies that $f^{-1}(B)$ belongs to $\mathscr{A}$ whenever B is a Borel subset of $\mathbb{R}$. Thus $f^{-1}(B \cup C) \in \mathscr{A}$ if $B \in \mathscr{B}(\mathbb{R})$ and $C \subseteq \{-\infty, +\infty\}$, and so f is measurable with respect to $\mathscr{A}$ and $\mathscr{B}(\overline{\mathbb{R}})$. $\square$

See Exercise 4 for another proof of Proposition 2.6.4.

Example 2.6.5. Let $(X, \mathscr{A})$ be a measurable space, and let f be an $\mathbb{R}^d$-valued function on X. Let f_1, ..., f_d be the components of f, i.e., the real-valued functions on X that satisfy $f(x) = (f_1(x), f_2(x), \ldots, f_d(x))$ at each x in X. Then Proposition 2.6.2 and part (b) of Proposition 1.1.5 imply that f is measurable with respect to $\mathscr{A}$ and $\mathscr{B}(\mathbb{R}^d)$ if and only if $f_1, \ldots, f_d$ are $\mathscr{A}$-measurable. It follows from this remark and Propositions 2.1.5 and 2.1.7 that the class of measurable functions from $(X, \mathscr{A})$ to $(\mathbb{R}^d, \mathscr{B}(\mathbb{R}^d))$ is closed under the formation of sums, scalar multiples, and limits. $\square$

Example 2.6.6. Now consider the space $\mathbb{R}^2$, and identify it with the set $\mathbb{C}$ of complex numbers. The remarks just above imply that a complex-valued function on $(X, \mathscr{A})$ is measurable with respect to $\mathscr{A}$ and $\mathscr{B}(\mathbb{C})$, that is, with respect to $\mathscr{A}$ and $\mathscr{B}(\mathbb{R}^2)$, if and only if its real and imaginary parts are $\mathscr{A}$-measurable, and that the collection of measurable functions from $(X, \mathscr{A})$ to $(\mathbb{C}, \mathscr{B}(\mathbb{C}))$ is closed under the formation of sums and limits and under multiplication by real constants. Similar arguments show that the product of two measurable complex-valued functions on X is measurable; in particular, the product of a complex number and a complex-valued measurable function is measurable.   □

Let $(X, \mathscr{A}, \mu)$ be a measure space. A complex-valued function f on X is *integrable* if its real and imaginary parts $\Re(f)$ and $\Im(f)$ are integrable; if f is integrable, then its *integral* is defined by

$$\int f \, d\mu = \int \Re(f) \, d\mu + i \int \Im(f) \, d\mu.$$

It is easy to check that if f and g are integrable complex-valued functions on X and if α is a complex number, then

(a) $f + g$ and αf are integrable,
(b) $\int (f + g) \, d\mu = \int f \, d\mu + \int g \, d\mu$, and
(c) $\int (\alpha f) \, d\mu = \alpha \int f \, d\mu$.

The dominated convergence theorem (Theorem 2.4.5) is valid if the functions f and $f_1, f_2, \ldots$ appearing in it are complex-valued (consider the real and imaginary parts of these functions separately).

Proposition 2.6.7. *Let $(X, \mathscr{A}, \mu)$ be a measure space, and let f be a complex-valued function on X that is measurable with respect to $\mathscr{A}$ and $\mathscr{B}(\mathbb{C})$. Then f is integrable if and only if $|f|$ is integrable. If these functions are integrable, then $|\int f \, d\mu| \leq \int |f| \, d\mu$.*

Proof. The measurability of $|f|$ is easy to check (see Exercise 2). Let $\Re(f)$ and $\Im(f)$ be the real and imaginary parts of f. If f is integrable, then the integrability of $|f|$ follows from the inequality $|f| \leq |\Re(f)| + |\Im(f)|$, while if $|f|$ is integrable, then the integrability of f follows from the inequalities $|\Re(f)| \leq |f|$ and $|\Im(f)| \leq |f|$ (see Proposition 2.3.8). Now suppose that f is integrable. Write the complex number $\int f \, d\mu$ in its polar form, letting w be a complex number of absolute value 1 such that

$$\int f \, d\mu = w \left| \int f \, d\mu \right|.$$

If we divide by w and use that fact that $|w^{-1}| = 1$, we find

$$\left| \int f \, d\mu \right| = w^{-1} \int f \, d\mu = \int (w^{-1} f) \, d\mu = \int \Re(w^{-1} f) \, d\mu \leq \int |f| \, d\mu,$$

and the proof is complete.   □

Let $(X, \mathscr{A}, \mu)$ be a measure space, let $(Y, \mathscr{B})$ be a measurable space, and let $f : (X, \mathscr{A}) \to (Y, \mathscr{B})$ be measurable. Define a $[0, +\infty]$-valued function μf^{-1} on $\mathscr{B}$ by letting $\mu f^{-1}(B) = \mu(f^{-1}(B))$ for each B in $\mathscr{B}$. Clearly $\mu f^{-1}(\varnothing) = 0$. Note that if $\{B_n\}$ is a sequence of disjoint sets that belong to $\mathscr{B}$, then $\{f^{-1}(B_n)\}$ is a sequence of disjoint sets that belong to $\mathscr{A}$ and satisfy $f^{-1}(\cup_n B_n) = \cup_n f^{-1}(B_n)$;

it follows that $\mu f^{-1}(\cup_n B_n) = \sum_n \mu f^{-1}(B_n)$ and hence that μf^{-1} is a measure on $(Y, \mathscr{B})$. The measure[4] μf^{-1} is sometimes called the *image of μ under f*.

Proposition 2.6.8. *Let $(X, \mathscr{A}, \mu)$ be a measure space, let $(Y, \mathscr{B})$ be a measurable space, and let $f: (X, \mathscr{A}) \to (Y, \mathscr{B})$ be measurable. Let g be an extended real-valued $\mathscr{B}$-measurable function on Y. Then g is μf^{-1}-integrable if and only if $g \circ f$ is μ-integrable. If these functions are integrable, then*

$$\int_Y g \, d(\mu f^{-1}) = \int_X (g \circ f) \, d\mu.$$

Proof. The measurability of $g \circ f$ follows from Propositions 2.6.1 and 2.6.4. We turn to the integrability of g and $g \circ f$. First suppose that g is the characteristic function of a set B in $\mathscr{B}$. Then $g \circ f$ is the characteristic function of $f^{-1}(B)$, and $\int_Y g \, d(\mu f^{-1})$ and $\int_X (g \circ f) \, d\mu$ are both equal to $\mu(f^{-1}(B))$. Thus the identity

$$\int_Y g \, d(\mu f^{-1}) = \int_X (g \circ f) \, d\mu$$

holds for characteristic functions. The additivity and homogeneity of the integral (Proposition 2.3.4) imply that this identity holds for nonnegative simple $\mathscr{B}$-measurable functions, and an approximation argument (use Proposition 2.1.8 and Theorem 2.4.1) shows that it holds for all $[0, +\infty]$-valued $\mathscr{B}$-measurable functions. Since an arbitrary $\mathscr{B}$-measurable function can be separated into its positive and negative parts, the proposition follows. ☐

We derive two elementary consequences of Proposition 2.6.8. First suppose that $f: \mathbb{R} \to \mathbb{R}$ is defined by $f(x) = -x$. Then $\lambda f^{-1} = \lambda$, and so a Borel function g on $\mathbb{R}$ is Lebesgue integrable if and only if the function $x \mapsto g(-x)$ is Lebesgue integrable. If these functions are integrable, then

$$\int g(x) \lambda(dx) = \int g(-x) \lambda(dx).$$

A similar argument shows that if $y \in \mathbb{R}$, then a Borel function g is Lebesgue integrable if and only if the function $x \mapsto g(x+y)$ is Lebesgue integrable. If these functions are integrable, then

$$\int g(x) \lambda(dx) = \int g(x+y) \lambda(dx).$$

---

[4]Another notation for μf^{-1} is $\mu \circ f^{-1}$.

Exercises

1. Let $(X, \mathscr{A})$ be a measurable space. Use Proposition 2.6.1 and Example 2.1.2(a) to give another proof that if $f, g \colon X \to \mathbb{R}$ are measurable, then $f + g$ and fg are measurable. (Hint: Consider the function $H \colon X \to \mathbb{R}^2$ defined by $H(x) = (f(x), g(x))$.)

2. Show that if f is a measurable complex-valued function on $(X, \mathscr{A})$, then $|f|$ is also measurable.

3. Let $(X, \mathscr{A})$ be a measurable space, and let $f, g \colon X \to \mathbb{C}$ be measurable. Show that if g does not vanish, then f/g is measurable.

4. (a) Show that $\mathscr{B}(\overline{\mathbb{R}})$ is the σ-algebra on $\overline{\mathbb{R}}$ generated by the intervals of the form $[-\infty, t]$.

   (b) Use part (a) of this exercise, together with Proposition 2.6.2, to give another proof of Proposition 2.6.4.

5. Let X and Y be sets, and let f be a function from X to Y. Show that

   (a) if $\mathscr{A}$ is a σ-algebra on X, then $\{B \subseteq Y : f^{-1}(B) \in \mathscr{A}\}$ is a σ-algebra on Y,

   (b) if $\mathscr{B}$ is a σ-algebra on Y, then $\{f^{-1}(B) : B \in \mathscr{B}\}$ is a σ-algebra on X, and

   (c) if $\mathscr{C}$ is a collection of subsets of Y, then

$$\sigma(\{f^{-1}(C) : C \in \mathscr{C}\}) = \{f^{-1}(B) : B \in \sigma(\mathscr{C})\}.$$

6. Let μ be a nonzero finite Borel measure on $\mathbb{R}$, and let $F \colon \mathbb{R} \to \mathbb{R}$ be the function defined by $F(x) = \mu((-\infty, x])$. Define a function g on the interval $(0, \lim_{x \to +\infty} F(x))$ by

$$g(x) = \inf\{t \in \mathbb{R} : F(t) \geq x\}.$$

   (a) Show that g is nondecreasing, finite valued, and Borel measurable.

   (b) Show that $\mu = \lambda g^{-1}$. (Hint: Start by showing that $\mu(B) = \lambda(g^{-1}(B))$ when B has the form $(-\infty, b]$.)

7. Show that a convex subset of $\mathbb{R}^2$ need not be a Borel set. (Hint: Consider an open ball, together with part of its boundary.)

Notes

See the notes for Chap. 1 for some alternative expositions of basic integration theory. At some point the reader should work through the constructions of the integral given in some of those references. The construction given by Halmos [54] is useful for the study of vector-valued functions (see also Appendix E).

There is an approach to integration theory, due to Daniell [32] and Stone [114], in which the integral is developed before measures are introduced. For an outline of this approach, see Sect. 7.7, and see the notes at the end of Chap. 7.

Chapter 3
Convergence

In this chapter we look in some detail at the convergence of sequences of functions. In Sect. 3.1 we define convergence in measure and convergence in mean, and we compare those modes of convergence with pointwise and almost everywhere convergence. In Sect. 3.2 we recall the definitions of norms and seminorms on vector spaces, and in Sects. 3.3 and 3.4 we apply these concepts to the study of the L^p spaces and to the convergence of functions in certain (semi-)norms, the p-norms. Finally, in Sect. 3.5 we begin to look at dual spaces (the spaces of continuous linear functionals on normed vector spaces). We will continue the study of dual spaces in Sects. 4.5, 7.3, and 7.5, by which time we will have developed enough tools to analyze and characterize a number of standard dual spaces.

3.1 Modes of Convergence

In this section we define and study a few modes of convergence for sequences of measurable functions. For simplicity we will discuss only real-valued functions. It is easy to check that everything can be extended so as to apply to complex-valued functions and to $[-\infty, +\infty]$-valued functions that are finite almost everywhere.[1]

Let $(X, \mathscr{A}, \mu)$ be a measure space, and let f and $f_1, f_2, \dots$ be real-valued $\mathscr{A}$-measurable functions on X. The sequence $\{f_n\}$ converges to f *in measure* if

$$\lim_n \mu(\{x \in X : |f_n(x) - f(x)| > \varepsilon\}) = 0$$

---

[1] We can verify our results in the case of $[-\infty, +\infty]$-valued functions that are finite almost everywhere by choosing a μ-null set N such that the functions f and $f_1, f_2, \dots$ are all finite outside N and then replacing f and $f_1, f_2, \dots$ with the functions g and $g_1, g_2, \dots$ defined by $g = f \chi_{N^c}$ and $g_n = f_n \chi_{N^c}$. This enables us to avoid the complications caused by expressions like $f_n(x) - f(x)$ when $f_n(x)$ or $f(x)$ is infinite.

D.L. Cohn, *Measure Theory: Second Edition*, Birkhäuser Advanced
Texts Basler Lehrbücher, DOI 10.1007/978-1-4614-6956-8_3,
© Springer Science+Business Media, LLC 2013

holds for each positive ε. As we noted in Sect. 2.2, the sequence $\{f_n\}$ converges to *f almost everywhere* if $f(x) = \lim_n f_n(x)$ holds at μ-almost every point x in X.

Examples 3.1.1. We should note that in general convergence in measure neither implies nor is implied by convergence almost everywhere.

(a) To see that convergence almost everywhere does not imply convergence in measure, consider the space $(\mathbb{R}, \mathscr{B}(\mathbb{R}), \lambda)$ and the sequence whose nth term is the characteristic function of the interval $[n, +\infty)$. This sequence clearly converges to the zero function almost everywhere (in fact, everywhere) but not in measure.

(b) Next consider the interval $[0, 1)$, together with the σ-algebra of Borel subsets of $[0, 1)$ and Lebesgue measure. Let $\{f_n\}$ be the sequence whose first term is the characteristic function of $[0, 1)$, whose next two terms are the characteristic functions of $[0, 1/2)$ and $[1/2, 1)$, whose next four terms are the characteristic functions of $[0, 1/4)$, $[1/4, 1/2)$, $[1/2, 3/4)$, and $[3/4, 1)$, and so on. Then $\{f_n\}$ converges to the zero function in measure, but for each x in $[0, 1)$ the sequence $\{f_n(x)\}$ contains infinitely many ones and infinitely many zeros and so is not convergent. $\square$

Nevertheless there are some useful relations, given by the following two propositions, between convergence in measure and convergence almost everywhere (see also Exercise 6).

Proposition 3.1.2. *Let $(X, \mathscr{A}, \mu)$ be a measure space, and let f and f_1, f_2, ... be real-valued $\mathscr{A}$-measurable functions on X. If μ is finite and if $\{f_n\}$ converges to f almost everywhere, then $\{f_n\}$ converges to f in measure.*

Proof. We must show that

$$\lim_n \mu(\{x \in X : |f_n(x) - f(x)| > \varepsilon\}) = 0$$

holds for each positive ε. So let ε be a positive number, and define sets $A_1, A_2, \ldots$ and $B_1, B_2, \ldots$ by

$$A_n = \{x \in X : |f_n(x) - f(x)| > \varepsilon\}$$

and $B_n = \cup_{k=n}^{\infty} A_k$. The sequence $\{B_n\}$ is decreasing, and its intersection is included in

$$\{x \in X : \{f_n(x)\} \text{ does not converge to } f(x)\}.$$

Thus $\mu(\cap_n B_n) = 0$, and so (Proposition 1.2.5) $\lim_n \mu(B_n) = 0$. Since $A_n \subseteq B_n$, it follows that

$$\lim_n \mu(\{x \in X : |f_n(x) - f(x)| > \varepsilon\}) = \lim_n \mu(A_n) = 0.$$

Thus $\{f_n\}$ converges to f in measure. $\square$

Proposition 3.1.3. *Let $(X, \mathscr{A}, \mu)$ be a measure space, and let f and f_1, f_2, ... be real-valued $\mathscr{A}$-measurable functions on X. If $\{f_n\}$ converges to f in measure, then there is a subsequence of $\{f_n\}$ that converges to f almost everywhere.*

Proof. The hypothesis that $\{f_n\}$ converges to f in measure means that

$$\lim_n \mu(\{x \in X : |f_n(x) - f(x)| > \varepsilon\}) = 0$$

holds for each positive ε. We use this relation to construct a sequence $\{n_k\}$ of positive integers, choosing n_1 so that

$$\mu(\{x \in X : |f_{n_1}(x) - f(x)| > 1\}) \leq \frac{1}{2},$$

and then choosing the remaining terms of $\{n_k\}$ inductively so that the relations $n_k > n_{k-1}$ and

$$\mu\left(\left\{x \in X : |f_{n_k}(x) - f(x)| > \frac{1}{k}\right\}\right) \leq \frac{1}{2^k}$$

hold for $k = 2, 3, \dots$. Define sets A_k, $k = 1, 2, \dots$, by

$$A_k = \left\{x \in X : |f_{n_k}(x) - f(x)| > \frac{1}{k}\right\}.$$

If $x \notin \cap_{j=1}^{\infty} \cup_{k=j}^{\infty} A_k$, then there is a positive integer j such that $x \notin \cup_{k=j}^{\infty} A_k$ and hence such that $|f_{n_k}(x) - f(x)| \leq 1/k$ holds for $k = j, j+1, \dots$. Thus $\{f_{n_k}\}$ converges to f at each x outside $\cap_{j=1}^{\infty} \cup_{k=j}^{\infty} A_k$. Since

$$\mu\left(\bigcup_{k=j}^{\infty} A_k\right) \leq \sum_{k=j}^{\infty} \mu(A_k) \leq \sum_{k=j}^{\infty} \frac{1}{2^k} = \frac{1}{2^{j-1}}$$

holds for each j, it follows that $\mu(\cap_{j=1}^{\infty} \cup_{k=j}^{\infty} A_k) = 0$, and the proof is complete. $\square$

Proposition 3.1.4 (Egoroff's Theorem).   *Let $(X, \mathscr{A}, \mu)$ be a measure space, and let f and f_1, f_2, ... be real-valued $\mathscr{A}$-measurable functions on X. If μ is finite and if $\{f_n\}$ converges to f almost everywhere, then for each positive number ε there is a subset B of X that belongs to $\mathscr{A}$, satisfies $\mu(B^c) < \varepsilon$, and is such that $\{f_n\}$ converges to f uniformly on B.*

Proof. Let ε be a positive number, and for each n let $g_n = \sup_{j \geq n} |f_j - f|$. It is easy to check that each g_n is finite almost everywhere. The sequence $\{g_n\}$ converges to 0 almost everywhere, and so in measure (see Proposition 3.1.2 and the footnote at the beginning of this section). Hence for each positive integer k we can choose a positive integer n_k such that

$$\mu\left(\left\{x \in X : g_{n_k}(x) > \frac{1}{k}\right\}\right) < \frac{\varepsilon}{2^k}.$$

Define sets $B_1, B_2, \ldots$ by $B_k = \{x \in X : g_{n_k}(x) \leq 1/k\}$, and let $B = \cap_k B_k$. The set B satisfies

$$\mu(B^c) = \mu\left(\bigcup_k B_k^c\right) \leq \sum_k \mu(B_k^c) < \sum_k \frac{\varepsilon}{2^k} = \varepsilon.$$

If δ is a positive number and if k is a positive integer such that $1/k < \delta$, then, since $B \subseteq B_k$,

$$|f_n(x) - f(x)| \leq g_{n_k}(x) \leq \frac{1}{k} < \delta$$

holds for all x in B and all positive integers n such that $n \geq n_k$; thus $\{f_n\}$ converges to f uniformly on B.   □

Egoroff's theorem provides motivation for the following definition. Let $(X, \mathscr{A}, \mu)$ be a measure space, and let f and $f_1, f_2, \ldots$ be real-valued $\mathscr{A}$-measurable functions on X. Then $\{f_n\}$ converges to f *almost uniformly* if for each positive number ε there is a subset B of X that belongs to $\mathscr{A}$, satisfies $\mu(B^c) < \varepsilon$, and is such that $\{f_n\}$ converges to f uniformly on B. It is clear that if $\{f_n\}$ converges to f almost uniformly, then $\{f_n\}$ converges to f almost everywhere. It follows from this remark and Egoroff's theorem that on a finite measure space almost everywhere convergence is equivalent to almost uniform convergence.

Suppose that $(X, \mathscr{A}, \mu)$ is a measure space and that f and $f_1, f_2, \ldots$ belong to $\mathscr{L}^1(X, \mathscr{A}, \mu, \mathbb{R})$. Then $\{f_n\}$ converges to f *in mean* if

$$\lim_n \int |f_n - f|\, d\mu = 0.$$

Proposition 3.1.5. *Let $(X, \mathscr{A}, \mu)$ be a measure space, and let f and $f_1, f_2, \ldots$ belong to $\mathscr{L}^1(X, \mathscr{A}, \mu, \mathbb{R})$. If $\{f_n\}$ converges to f in mean, then $\{f_n\}$ converges to f in measure.*

Proof. This is an immediate consequence of the inequality

$$\mu(\{x \in X : |f_n(x) - f(x)| > \varepsilon\}) \leq \frac{1}{\varepsilon} \int |f_n - f|\, d\mu$$

(see Proposition 2.3.10).   □

Convergence in mean does not, however, imply convergence almost everywhere (see the example given above of a sequence that converges in measure but not almost everywhere). On the other hand, if $\{f_n\}$ converges to f in mean, then $\{f_n\}$ does have a subsequence that converges to f almost everywhere; this follows from Propositions 3.1.3 and 3.1.5 (or, alternatively, from Exercise 4).

Neither convergence almost everywhere nor convergence in measure implies convergence in mean. To see this, consider the space $(\mathbb{R}, \mathscr{B}(\mathbb{R}), \lambda)$, and define a sequence $\{f_n\}$ by letting f_n have value n on the interval $[0, 1/n]$ and value 0 elsewhere. Then $\{f_n\}$ converges to 0 almost everywhere and in measure, but not in mean (note that $\int |f_n - 0|\, d\lambda = 1$). There are, however, supplementary hypotheses under which convergence almost everywhere or in measure does imply convergence in mean; such hypotheses are given in the following proposition and in Exercise 4.2.16.

Proposition 3.1.6. *Let $(X, \mathscr{A}, \mu)$ be a measure space, and let f and f_1, f_2, ... belong to $\mathscr{L}^1(X, \mathscr{A}, \mu, \mathbb{R})$. If $\{f_n\}$ converges to f almost everywhere or in measure, and if there is a nonnegative extended real-valued integrable function g such that*

$$|f_n| \le g \ (\text{for } n = 1, 2, \ldots) \text{ and } |f| \le g \tag{1}$$

hold almost everywhere, then $\{f_n\}$ converges to f in mean.

Proof. First suppose that $\{f_n\}$ converges to f almost everywhere and hence that $\{|f_n - f|\}$ converges to 0 almost everywhere. Relation (1) implies that

$$|f_n - f| \le |f_n| + |f| \le 2g$$

holds almost everywhere. Thus we can use the dominated convergence theorem (Theorem 2.4.5) to conclude that $\lim_n \int |f_n - f| \, d\mu = 0$.

Now suppose that $\{f_n\}$ converges to f in measure and satisfies condition (1). Then every subsequence of $\{f_n\}$ has a subsequence that converges to f almost everywhere (Proposition 3.1.3), and so by what we have just proved, in mean. If the original sequence $\{f_n\}$ did not converge to f in mean, then there would be a positive number ε and a subsequence $\{f_{n_k}\}$ of $\{f_n\}$ such that $\int |f_{n_k} - f| \, d\mu \ge \varepsilon$ holds for each k. Since this subsequence could have no subsequence converging to f in mean, we have a contradiction. Thus $\{f_n\}$ must converge to f in mean.   □

Exercises

1. Let $(X, \mathscr{A}, \mu)$ be a measure space, and let A and $A_1, A_2, \ldots$ belong to $\mathscr{A}$. Show that
   (a) $\{\chi_{A_n}\}$ converges to 0 in measure if and only if $\lim_n \mu(A_n) = 0$,
   (b) $\{\chi_{A_n}\}$ converges to 0 almost everywhere if and only if $\mu(\cap_{n=1}^{\infty} \cup_{k=n}^{\infty} A_k) = 0$, and
   (c) $\{\chi_{A_n}\}$ converges to χ_A almost everywhere if and only if the three sets A, $\cap_{n=1}^{\infty} \cup_{k=n}^{\infty} A_k$, and $\cup_{n=1}^{\infty} \cap_{k=n}^{\infty} A_k$ differ only by μ-null sets. (Hint: See Exercise 2.1.1.)
2. Let μ be counting measure on the σ-algebra of all subsets of $\mathbb{N}$, and let f and f_1, $f_2, \ldots$ be real-valued functions on $\mathbb{N}$. Show that $\{f_n\}$ converges to f in measure if and only if it converges uniformly to f.
3. Let $(X, \mathscr{A}, \mu)$ be a measure space, let f and f_1, f_2, ... be real-valued $\mathscr{A}$-measurable functions on X, and let $g: \mathbb{R} \to \mathbb{R}$ be Borel measurable. Show that if $\{f_n\}$ converges to f almost everywhere and if g is continuous at $f(x)$ for almost every x, then $\{g \circ f_n\}$ converges to $g \circ f$ almost everywhere.

4. Suppose that $(X, \mathscr{A}, \mu)$ is a measure space and that f and $f_1, f_2, \ldots$ belong to $\mathscr{L}^1(X, \mathscr{A}, \mu, \mathbb{R})$. Show that if $\{f_n\}$ converges to f in mean so fast that

$$\sum_n \int |f_n - f| \, d\mu < +\infty,$$

   then $\{f_n\}$ converges to f almost everywhere.
5. Let μ be a measure on $(X, \mathscr{A})$, and let $f, f_1, f_2, \ldots$ and $g, g_1, g_2, \ldots$ be real-valued $\mathscr{A}$-measurable functions on X.
   (a) Show that if μ is finite, if $\{f_n\}$ converges to f in measure, and if $\{g_n\}$ converges to g in measure, then $\{f_n g_n\}$ converges to fg in measure.
   (b) Can the assumption that μ is finite be omitted in part (a)?
6. Let μ be a finite measure on $(X, \mathscr{A})$ and f and $f_1, f_2, \ldots$ be real-valued $\mathscr{A}$-measurable functions on X. Show that $\{f_n\}$ converges to f in measure if and only if each subsequence of $\{f_n\}$ has a subsequence that converges to f almost everywhere.
7. Egoroff's theorem applies to *sequences* of measurable functions on a finite measure space. One can ask about the situation where one has a family $\{f_t\}_{t \in T}$ on a finite measure space $(X, \mathscr{A}, \mu)$, where T is a subinterval of $\mathbb{R}$ of the form $[t_0, +\infty)$. (The following results are due to Walter [125].)
   (a) For each n in $\mathbb{N}$ define g_n by $g_n(x) = \sup\{|f_t(x) - f(x)| : t \in [n, +\infty)\}$. Show that if each g_n is measurable, then the conclusion of Egoroff's theorem holds for the family $\{f_t\}_{t \in T}$.
   (b) Let $\{A_n\}$ be a sequence of disjoint subsets of $[0, 1]$ that are not Lebesgue measurable and are such that all the A_n's have the same (strictly positive) Lebesgue outer measure. (See the discussion of nonmeasurable sets in Sect. 1.4.) Define a subset B of $[0, 1] \times [1, +\infty)$ by

$$B = \{(x, t) : x \in A_n \text{ and } t = x + n \text{ for some } n\},$$

   and for each t let f_t be the characteristic function of the set $\{x \in [0, 1] : (x, t) \in B\}$. Show that each f_t is Borel measurable but that the conclusion of Egoroff's theorem fails for the family $\{f_t\}_{t \in [1, +\infty)}$

3.2 Normed Spaces

Let V be a vector space over $\mathbb{R}$ (or over $\mathbb{C}$). A *norm* on V is a function $\|\cdot\| : V \to \mathbb{R}$ that satisfies

(a) $\|v\| \geq 0$,
(b) $\|v\| = 0$ if and only if $v = 0$,
(c) $\|\alpha v\| = |\alpha| \|v\|$, and
(d) $\|u + v\| \leq \|u\| + \|v\|$

for each u and v in V and each α in $\mathbb{R}$ (or in $\mathbb{C}$). Condition (c) says that $\|\cdot\|$ is *homogeneous*, and condition (d) says that it satisfies the *triangle inequality*. If in condition (b) the words "if and only if" are replaced with the word "if," but conditions (a), (c), and (d) remain unchanged, then $\|\cdot\|$ is called a *seminorm*. Thus a norm is a seminorm for which 0 is the only vector that satisfies $\|v\| = 0$. A *normed vector space* (or a *normed linear space*) is a vector space together with a norm.

Examples 3.2.1. Let us consider a few examples.

(a) The function that assigns to each number its absolute value is a norm on $\mathbb{R}$ (or on $\mathbb{C}$). This is the norm that will be assumed whenever we deal with $\mathbb{R}$ or $\mathbb{C}$ as a normed space.

(b) The formula $\|(x_1,\ldots,x_d)\|_2 = (\sum_{i=1}^{d} |x_i|^2)^{1/2}$ defines a norm on $\mathbb{R}^d$ and on $\mathbb{C}^d$ (the triangle inequality follows from Exercise 9 or from Minkowski's inequality (Proposition 3.3.3)).

(c) Let $(X, \mathscr{A}, \mu)$ be a measure space and let $\mathscr{L}^1(X, \mathscr{A}, \mu, \mathbb{R})$ be the set of all real-valued integrable functions on X. Then $\mathscr{L}^1(X, \mathscr{A}, \mu, \mathbb{R})$ is a vector space over $\mathbb{R}$, and the formula

$$\|f\|_1 = \int |f| d\mu$$

defines a seminorm on $\mathscr{L}^1(X, \mathscr{A}, \mu, \mathbb{R})$. If f is an $\mathscr{A}$-measurable function on X such that $f = 0$ holds almost everywhere but not everywhere, then f satisfies $\|f\|_1 = 0$ but not $f = 0$. Thus for many choices of $(X, \mathscr{A}, \mu)$ the seminorm $\|\cdot\|_1$ is not a norm.

(d) Let $[a, b]$ be a closed bounded interval, and let $C[a, b]$ be the vector space of all continuous real-valued functions on $[a, b]$. The function $\|\cdot\|_1 : C[a, b] \to \mathbb{R}$ defined by

$$\|f\|_1 = \int_a^b |f| d\lambda$$

is a norm (note that a continuous function on $[a, b]$ that vanishes almost everywhere must vanish everywhere).

(e) The function $\|\cdot\|_\infty : C[a, b] \to \mathbb{R}$ defined by the formula

$$\|f\|_\infty = \sup\{|f(x)| : x \in [a, b]\}$$

is a norm (the continuity of f and the compactness of $[a, b]$ imply that $\|f\|_\infty$ is finite; see Theorem C.12). It is called the *uniform norm* or the *sup* (for supremum) *norm* on $C[a, b]$.

(f) More generally, let X be an arbitrary nonempty set, and let V be a vector space of bounded real-valued (or complex-valued) functions on X. Then the formula

$$\|f\|_\infty = \sup\{|f(x)| : x \in X\}$$

defines a norm on V. □

Recall that a *metric* on a set S is a function $d \colon S \times S \to \mathbb{R}$ that satisfies

(a) $d(s,t) \geq 0$,
(b) $d(s,t) = 0$ if and only if $s = t$,
(c) $d(s,t) = d(t,s)$, and
(d) $d(r,t) \leq d(r,s) + d(s,t)$

for all r, s, and t in S. Condition (d) says that d satisfies the *triangle inequality*. If in condition (b) the words "if and only if" are replaced with the word "if," but conditions (a), (c), and (d) remain unchanged, then d is called a *semimetric*. A *metric space* is a set S together with a metric on S.

It is easy to check that if V is a vector space and if $\| \cdot \|$ is a norm (or a seminorm) on V, then the formula

$$d(u,v) = \|u - v\|$$

defines a metric (or a semimetric) on V.

Recall that if S is a metric space and if s and $s_1, s_2, \ldots$ are elements of S, then the sequence $\{s_n\}$ *converges* to s if $\lim_n d(s_n,s) = 0$; the point s is then called the *limit* of $\{s_n\}$ and is denoted by $\lim_n s_n$ (see Exercise 1). In particular, if V is a normed linear space and if v and $v_1, v_2, \ldots$ are elements of V, then the sequence $\{v_n\}$ converges to v (with respect to the metric induced by the norm on V) if and only if $\lim_n \|v_n - v\| = 0$.

Examples 3.2.2. Let us return to some of the examples above. The metric induced on $\mathbb{R}^d$ by the norm defined in Example 3.2.1(b) is the usual one, stemming from the Pythagorean theorem. If $(X, \mathscr{A}, \mu)$ is a measure space and if f and $f_1, f_2, \ldots$ belong to $\mathscr{L}^1(X, \mathscr{A}, \mu, \mathbb{R})$, then $\{f_n\}$ converges to f with respect to the seminorm[2] defined in Example 3.2.1(c) if and only if it converges to f in mean (see Sect. 3.1). Finally, if f and $f_1, f_2, \ldots$ are continuous functions on $[a,b]$, then $\{f_n\}$ converges to f with respect to the norm defined in Example 3.2.1(e) if and only if it converges uniformly to f. $\square$

Let d be a metric (or a semimetric) on a set S. Then a subset A of S is *dense* in S if for each s in S and each positive ε there is an element a of A that satisfies $d(s,a) < \varepsilon$. It is clear that A is dense in S if and only if for each s in S there is a sequence $\{a_n\}$ of elements of A such that $\lim_n d(a_n,s) = 0$. A metric (or semimetric) space is *separable* if it has a countable dense subset. For example, the rational numbers form a countable dense subset of $\mathbb{R}$, and so $\mathbb{R}$ is separable.

Now let S be an arbitrary metric space. A sequence $\{s_n\}$ of elements of S is a *Cauchy sequence* if for each positive number ε there is a positive integer N such that $d(s_m,s_n) < \varepsilon$ holds whenever $m \geq N$ and $n \geq N$. Of course, every convergent sequence is a Cauchy sequence (let s be the limit of $\{s_n\}$, and note that $d(s_m,s_n) \leq$

---

[2]Convergence with respect to a semimetric or a seminorm is defined in the same way as convergence with respect to a metric or a norm. Note, however, that a sequence that is convergent with respect to a semimetric or a seminorm might have several limits.

$d(s_m,s) + d(s,s_n))$. On the other hand, if every Cauchy sequence in S converges to a point in S, then S is called *complete*. A normed linear space that is complete (with respect to the metric induced by its norm) is called a *Banach space*.

It is a basic consequence of the axioms for the real number system that $\mathbb{R}$ is complete under the metric defined by $(x,y) \mapsto |x-y|$.[3] The proofs of completeness that we give for other spaces will depend ultimately on this fact.

Example 3.2.3. Let us show that $C[a,b]$ is complete under the uniform norm. Let $\{f_n\}$ be a Cauchy sequence in $C[a,b]$. For each x in $[a,b]$ the sequence $\{f_n(x)\}$ satisfies $|f_m(x) - f_n(x)| \le \|f_m - f_n\|_\infty$ and so is a Cauchy sequence of real numbers; thus it is convergent. Define a function $f: [a,b] \to \mathbb{R}$ by letting $f(x) = \lim_n f_n(x)$ hold at each x in $[a,b]$. We need to show that $\{f_n\}$ converges uniformly to f and that f is continuous. Let us begin by showing that the convergence of $\{f_n\}$ to f is uniform. Let ε be a positive number, and use the fact that $\{f_n\}$ is a Cauchy sequence to choose a positive integer N such that $\|f_m - f_n\|_\infty < \varepsilon$ holds whenever m and n satisfy $m \ge N$ and $n \ge N$. Then

$$|f_m(x) - f_n(x)| < \varepsilon$$

holds for all x in $[a,b]$ and all m and n satisfying $m \ge N$ and $n \ge N$, and so (take limits as m approaches infinity)

$$|f(x) - f_n(x)| \le \varepsilon$$

holds for all x in $[a,b]$ and all n satisfying $n \ge N$. Thus $\|f_n - f\|_\infty \le \varepsilon$ holds[4] when $n \ge N$. Since ε was arbitrary, we have shown that $\{f_n\}$ converges uniformly to f.

We turn to the continuity of f. Let x_0 belong to $[a,b]$, and let ε be an arbitrary positive number. Choose a positive integer N such that $\|f_n - f\|_\infty < \varepsilon/3$ holds whenever n satisfies $n \ge N$, and then use the continuity of f_N to choose a positive number δ such that $|f_N(x) - f_N(x_0)| < \varepsilon/3$ holds if x belongs to $[a,b]$ and satisfies $|x - x_0| < \delta$. It follows that if $x \in [a,b]$ and $|x - x_0| < \delta$, then

$$|f(x) - f(x_0)| \le |f(x) - f_N(x)| + |f_N(x) - f_N(x_0)| + |f_N(x_0) - f(x_0)|$$

$$< \frac{\varepsilon}{3} + \frac{\varepsilon}{3} + \frac{\varepsilon}{3} = \varepsilon.$$

Since ε and x_0 were arbitrary, the continuity of f follows. This finishes our proof of the completeness of $C[a,b]$ under $\|\cdot\|_\infty$. $\square$

---

[3] See, for instance, Gleason [49], Hoffman [60], Rudin [104], or Thomson, Bruckner, and Bruckner [117].

[4] Actually, the norm here and in the following paragraph is the norm from Example 3.2.1(f). We can't say that it is the norm from $C[a,b]$ until we show that f is continuous.

Example 3.2.4. Let us also note an example of a normed linear space that is not complete. Consider the space $C[-1,1]$, together with the norm defined by $\|f\|_1 = \int_{-1}^1 |f| \, d\lambda$. For each n define a function $f_n \colon [-1,1] \to \mathbb{R}$ by

$$
f_n(x) = \begin{cases} 0 & \text{if } -1 \le x \le 0, \\ nx & \text{if } 0 < x \le \frac{1}{n}, \\ 1 & \text{if } \frac{1}{n} < x \le 1. \end{cases}
$$

It is easy to check that $\{f_n\}$ is a Cauchy sequence in $C[-1,1]$, but that there is no continuous function f such that $\lim_n \|f_n - f\|_1 = 0$. Hence $C[a,b]$ is not complete under $\| \cdot \|_1$.   □

We close this section with a sometimes useful criterion for the completeness of a normed linear space. Let V be a normed linear space, and let $\sum_{k=1}^\infty v_k$ be an infinite series with terms in V. The series $\sum_{k=1}^\infty v_k$ is *convergent* if $\lim_n \sum_{k=1}^n v_k$ exists, and is *absolutely convergent* if the series $\sum_{k=1}^\infty \|v_k\|$ of real numbers is convergent. Recall that every absolutely convergent series of real numbers is convergent; for more general normed linear spaces we have the following result.

Proposition 3.2.5. *Let V be a normed linear space. Then V is complete if and only if every absolutely convergent series with terms in V is convergent.*

Proof. First suppose that V is complete, and let $\sum_{k=1}^\infty v_k$ be an absolutely convergent series in V. Let $\{s_n\}$ be the sequence of partial sums of the series $\sum_{k=1}^\infty v_k$, and let $\{t_n\}$ be the sequence of partial sums of the series $\sum_{k=1}^\infty \|v_k\|$; thus $s_n = \sum_{k=1}^n v_k$ and $t_n = \sum_{k=1}^n \|v_k\|$. Note that if $m < n$, then

$$
\|s_n - s_m\| = \left\| \sum_{k=m+1}^n v_k \right\| \le \sum_{k=m+1}^n \|v_k\| = t_n - t_m. \tag{1}
$$

The convergence of $\sum_{k=1}^\infty \|v_k\|$ implies that $\{t_n\}$ is a Cauchy sequence and, in view of (1), that $\{s_n\}$ is a Cauchy sequence. Since V is complete, the sequence $\{s_n\}$, and hence the series $\sum_{k=1}^\infty v_k$, must converge.

Next suppose that every absolutely convergent series in V is convergent, and let $\{u_n\}$ be a Cauchy sequence in V. Since $\{u_n\}$ is a Cauchy sequence, we can choose (how?) a subsequence $\{u_{n_k}\}$ of $\{u_n\}$ such that $\|u_{n_{k+1}} - u_{n_k}\| \le 1/2^{k+1}$ holds for each k. Define a series $\sum_{k=1}^\infty v_k$ by letting $v_1 = u_{n_1}$ and letting $v_k = u_{n_k} - u_{n_{k-1}}$ if $k > 1$; thus $\{u_{n_k}\}$ is the sequence of partial sums of the series $\sum_{k=1}^\infty v_k$. Since $\|v_k\| \le 1/2^k$ holds if $k > 1$, the series $\sum_{k=1}^\infty v_k$ is absolutely convergent and hence convergent. Thus the sequence $\{u_{n_k}\}$ converges, say to u. The inequality

$$
\|u - u_n\| \le \|u - u_{n_k}\| + \|u_{n_k} - u_n\|
$$

implies that $\|u - u_n\|$ can be made small by making n (and k) large, and so the original sequence $\{u_n\}$ also converges to u. The completeness of V follows.   □

Exercises

1. Let S be a metric space, and let $\{s_n\}$ be a sequence of elements of S. Show that $\{s_n\}$ converges to at most one point in S. (Thus the expression "$\lim_n s_n$" makes sense.)

2. Let $C^1[0,1]$ consist of those functions $f: [0,1] \to \mathbb{R}$ such that f' is defined and continuous at each point in $[0,1]$ (of course $f'(0)$ and $f'(1)$ are to be interpreted as one-sided derivatives). Show that
   (a) the formula $\|f\| = \int_0^1 |f'(x)| dx$ defines a seminorm, but not a norm, on $C^1[0,1]$, and
   (b) the formula $\|f\| = |f(0)| + \int_0^1 |f'(x)| dx$ defines a norm on $C^1[0,1]$.

3. Let ℓ^∞ be the set of all bounded sequences of real numbers (of course ℓ^∞ is a vector space over $\mathbb{R}$.) Show that ℓ^∞ is complete under the norm defined in Example 3.2.1(f).

4. Let c_0 be the set of all sequences $\{x_n\}$ of real numbers for which $\lim_n x_n = 0$. Show that c_0 is a closed linear subspace of ℓ^∞ (see Exercise 3) and hence that c_0 is complete under the norm $\|\cdot\|_\infty$ defined by $\|\{x_n\}\|_\infty = \sup_n |x_n|$.

5. Let μ be a finite measure on $(X, \mathscr{A})$. Show that
   (a) the formula

   $$d(f,g) = \int \frac{|f-g|}{1+|f-g|} d\mu$$

   defines a semimetric on the collection of all real-valued $\mathscr{A}$-measurable functions on X, and
   (b) $\lim_n d(f_n, f) = 0$ holds if and only if $\{f_n\}$ converges to f in measure.

6. Now let us consider an analogous result for the space $(\mathbb{R}, \mathscr{B}(\mathbb{R}), \lambda)$. Suppose that $h: \mathbb{R} \to \mathbb{R}$ is defined by $h(t) = 1/(1+t^2)$. Show that
   (a) the formula

   $$d(f,g) = \int \frac{|f-g|}{1+|f-g|} h \, d\lambda$$

   defines a semimetric on the collection of all real-valued Borel measurable functions on $\mathbb{R}$, and
   (b) $\lim_n d(f_n, f) = 0$ holds if and only if $\{f_n\}$ converges to f in measure on each bounded subinterval of $\mathbb{R}$.

7. Let V be a vector space over $\mathbb{R}$. A function $(\cdot, \cdot): V \times V \to \mathbb{R}$ is an *inner product* on V if
   (i) $(x,x) \geq 0$,
   (ii) $(x,x) = 0$ if and only if $x = 0$,
   (iii) $(x,y) = (y,x)$, and
   (iv) $(\alpha x + \beta y, z) = \alpha(x,z) + \beta(y,z)$

hold for all x, y, z in V and all α, β in $\mathbb{R}$.[5] An *inner product space* is a vector space, together with an inner product on it. The *norm* $\|\cdot\|$ associated to the inner product $(\cdot,\cdot)$ is defined by $\|x\| = \sqrt{(x,x)}$.

(a) Prove that an inner product satisfies the *Cauchy–Schwarz inequality*: if $x,y \in V$, then $|(x,y)| \le \|x\|\|y\|$. (Hint: Define a function $p\colon \mathbb{R} \to \mathbb{R}$ by $p(t) = \|x\|^2 + 2t(x,y) + t^2\|y\|^2$, and note that $p(t) = \|x+ty\|^2 \ge 0$ holds for each real t; then recall that a quadratic polynomial $at^2 + bt + c$ is nonnegative for each t only if $b^2 - 4ac \le 0$.)

(b) Verify that the norm associated to $(\cdot,\cdot)$ is indeed a norm. (Hint: Use the Cauchy–Schwarz inequality when checking the triangle inequality.)

8. Let $(\cdot,\cdot)$ be an inner product on the real vector space V, and let $\|\cdot\|$ be the associated norm. Show that

(a) $\|x+y\|^2 + \|x-y\|^2 = 2\|x\|^2 + 2\|y\|^2$ and

(b) $\|x+y\|^2 - \|x-y\|^2 = 4(x,y)$

hold for all x,y in V. (The identity in part (a) is called the *parallelogram law*.)

9. (a) Check that the formula $(x,y) = \sum_{i=1}^{d} x_i y_i$ defines an inner product on $\mathbb{R}^d$ (here, x and y are the vectors $(x_1,\ldots,x_d)$ and $(y_1,\ldots,y_d)$).

(b) Conclude that the function $\|\cdot\|_2\colon \mathbb{R}^d \to \mathbb{R}$ defined by $\|x\|_2 = (\sum_{i=1}^{d} x_i^2)^{1/2}$ is indeed a norm. (See part (b) of Exercise 7.)

10. Let ℓ^2 be the set of all infinite sequences $\{x_n\}$ of real numbers for which $\sum_n x_n^2 < +\infty$.

(a) Show that ℓ^2 is a vector space over $\mathbb{R}$. (Hint: Note that $(x+y)^2 \le 2x^2 + 2y^2$ holds for all real x and y.)

(b) Show that the formula $(\{x_n\},\{y_n\}) = \sum_n x_n y_n$ defines an inner product on ℓ^2 and hence (see part (b) of Exercise 7) that the formula $\|\{x_n\}\| = (\sum_n x_n^2)^{1/2}$ defines a norm on ℓ^2. (The issue is the convergence of $\sum_n x_n y_n$.)

(c) Show that ℓ^2 is complete under the norm defined in part (b) of this exercise.

11. A *Hilbert space* is an inner product space that is complete under the norm defined by $\|x\| = \sqrt{(x,x)}$. Show that if H is a Hilbert space and if C is a nonempty closed convex subset of H, then there is a unique point y in C that satisfies

$$\|y\| = \inf\{\|z\| : z \in C\}.$$

(Hint: Let $d = \inf\{\|z\| : z \in C\}$, and choose a sequence $\{z_n\}$ in C such that $\lim_n \|z_n\| = d$. Note that the convexity of C implies that $\frac{1}{2}(z_m + z_n) \in C$ and hence that $\|\frac{1}{2}(z_m + z_n)\| \ge d$. Use this inequality, together with part (a) of Exercise 8, to show that $\{z_n\}$ is a Cauchy sequence. Check that $\lim_n z_n$ is the required point y. To check the uniqueness of y, suppose that y' and y'' belong to C and satisfy $\|y'\| = \|y''\| = d$, and apply the preceding argument to the sequence y', y'', y', y'', ...)

---

[5] An inner product on a complex vector space V is a complex-valued function $(\cdot,\cdot)$ on $V \times V$ that satisfies (i), (ii), (iv), and $(x,y) = \overline{(y,x)}$ for all x, y, z in V and all α, β in $\mathbb{C}$. In this book we will deal with inner products only on real vector spaces.

12. Let H be a Hilbert space, and let H_0 be a closed linear subspace of H.

    (a) Show that if $x \in H$, then there is a unique point y in H_0 such that

    $$\|x - y\| = \inf\{\|x - z\| : z \in H_0\}.$$

    (Hint: Apply Exercise 11 to the set $\{x - z : z \in H_0\}$.)

    (b) Show that if x and y are related as in part (a), then $x - y$ is *orthogonal* to H_0, in the sense that $(x - y, z) = 0$ holds for each z in H_0. (Hint: Let $f(t) = \|x - y - tz\|^2 = \|x - y\|^2 - 2t(x - y, z) + t^2\|z\|^2$. Then $f(t)$ is a quadratic polynomial in t, which, by our choice of y, is minimized when $t = 0$. Conclude that $(x - y, z) = 0$.)

13. Let V be a Banach space, and let v and $v_1, v_2, \ldots$ belong to V. The series $\sum_{k=1}^{\infty} v_k$ is said to *converge unconditionally* to v if for each positive ε there is a finite subset F_ε of $\mathbb{N}$ such that $\|\sum_{k \in F} v_k - v\| < \varepsilon$ holds whenever F is a finite subset of $\mathbb{N}$ that includes F_ε.

    (a) Show that if $\sum_{k=1}^{\infty} v_k$ converges absolutely, then it converges unconditionally to some point in V.

    (b) Show that the converse of part (a) holds if $V = \mathbb{R}$.

    (c) Show that the converse of part (a) is not true in general. (Hint: Let V be c_0, ℓ^2, or ℓ^∞.)

14. Let V be the vector space of all real-valued Borel measurable functions on $[0, 1]$. Show that convergence in measure (with respect to Lebesgue measure) is not given by a seminorm on V. That is, show that there is no seminorm $\|\cdot\|$ on V such that elements $f, f_1, f_2, \ldots$ of V satisfy $\lim_n \|f_n - f\| = 0$ if and only if $\{f_n\}$ converges to f in measure. (Hint: Show that if such a seminorm exists, then for each positive ε there are functions $g_1, \ldots, g_n$ in V such that $\|g_i\| \le \varepsilon$ holds for each i but $\frac{1}{n}\sum_{i=1}^{n} g_i$ is equal to the constant function 1. Derive a contradiction.)

15. Again, let V be the vector space of all real-valued Borel measurable functions on $[0, 1]$. Show that convergence almost everywhere (with respect to Lebesgue measure) is not given by a semimetric on V. (Hint: Use Proposition 3.1.3 to show that if such a semimetric existed, then convergence in measure would imply convergence almost everywhere.)

3.3 Definition of $\mathscr{L}^p$ and L^p

Let $(X, \mathscr{A}, \mu)$ be a measure space, and let p satisfy $1 \le p < +\infty$ (p need not be an integer). Then $\mathscr{L}^p(X, \mathscr{A}, \mu, \mathbb{R})$ is the set of all $\mathscr{A}$-measurable functions $f : X \to \mathbb{R}$ such that $|f|^p$ is integrable, and $\mathscr{L}^p(X, \mathscr{A}, \mu, \mathbb{C})$ is the set of all $\mathscr{A}$-measurable functions $f : X \to \mathbb{C}$ such that $|f|^p$ is integrable (see Exercise 2.6.2).

Note that if $a \in \mathbb{R}$ and $f \in \mathscr{L}^p(X, \mathscr{A}, \mu, \mathbb{R})$, then $\alpha f \in \mathscr{L}^p(X, \mathscr{A}, \mu, \mathbb{R})$, and that if $a \in \mathbb{C}$ and $f \in \mathscr{L}^p(X, \mathscr{A}, \mu, \mathbb{C})$, then $\alpha f \in \mathscr{L}^p(X, \mathscr{A}, \mu, \mathbb{C})$. Furthermore, if f and g belong to $\mathscr{L}^p(X, \mathscr{A}, \mu, \mathbb{R})$ or to $\mathscr{L}^p(X, \mathscr{A}, \mu, \mathbb{C})$, then

$$|f(x) + g(x)|^p \le (|f(x)| + |g(x)|)^p \le (2\max\{|f(x)|, |g(x)|\})^p$$
$$\le 2^p |f(x)|^p + 2^p |g(x)|^p$$

holds for each x in X, and so $f + g$ belongs to $\mathscr{L}^p(X, \mathscr{A}, \mu, \mathbb{R})$ or to $\mathscr{L}^p(X, \mathscr{A}, \mu, \mathbb{C})$. Thus $\mathscr{L}^p(X, \mathscr{A}, \mu, \mathbb{R})$ is a vector space over $\mathbb{R}$, and $\mathscr{L}^p(X, \mathscr{A}, \mu, \mathbb{C})$ is a vector space over $\mathbb{C}$.

We turn to the definition of $\mathscr{L}^p(X, \mathscr{A}, \mu, \mathbb{R})$ and $\mathscr{L}^p(X, \mathscr{A}, \mu, \mathbb{C})$ in the case where $p = +\infty$. Let $\mathscr{L}^\infty(X, \mathscr{A}, \mu, \mathbb{R})$ be the set of all bounded real-valued $\mathscr{A}$-measurable functions on X, and let $\mathscr{L}^\infty(X, \mathscr{A}, \mu, \mathbb{C})$ be the set of all bounded[6] complex-valued $\mathscr{A}$-measurable functions on X. It is easy to see that $\mathscr{L}^\infty(X, \mathscr{A}, \mu, \mathbb{R})$ and $\mathscr{L}^\infty(X, \mathscr{A}, \mu, \mathbb{C})$ are vector spaces.

In discussions that are valid for both real- and complex-valued functions we will often use $\mathscr{L}^p(X, \mathscr{A}, \mu)$ to represent either $\mathscr{L}^p(X, \mathscr{A}, \mu, \mathbb{R})$ or $\mathscr{L}^p(X, \mathscr{A}, \mu, \mathbb{C})$.

Let us define, for each p, a function (in fact, a seminorm) $\| \cdot \|_p$ on $\mathscr{L}^p(X, \mathscr{A}, \mu)$. If $1 \le p < +\infty$, we define $\| \cdot \|_p$ by

$$\|f\|_p = \left(\int |f|^p \, d\mu \right)^{1/p}.$$

For the case where $p = +\infty$ we need a few preliminaries. A subset N of X is *locally μ-null* (or simply *locally null*) if for each set A that belongs to $\mathscr{A}$ and satisfies $\mu(A) < +\infty$ the set $A \cap N$ is μ-null. A property of points of X is said to hold *locally almost everywhere* if the set of points at which it fails to hold is locally null. It is easy to check that

(a) every μ-null subset of X is locally μ-null,
(b) if $(X, \mathscr{A}, \mu)$ is σ-finite, then every locally μ-null subset of X is μ-null, and
(c) the union of a sequence of locally μ-null sets is locally μ-null.

See Exercises 5 and 6 for some examples of locally μ-null sets that are not μ-null.

We can define $\| \cdot \|_p$ in the case where $p = +\infty$ by letting $\|f\|_\infty$ be the infimum of those nonnegative numbers M such that $\{x \in X : |f(x)| > M\}$ is locally μ-null.[7] Note that if $f \in \mathscr{L}^\infty(X, \mathscr{A}, \mu)$, then $\{x \in X : |f(x)| > \|f\|_\infty\}$ is locally μ-null, for if $\{M_n\}$ is a nonincreasing sequence of real numbers such that $\|f\|_\infty = \lim_n M_n$ and

---

[6]Some authors define $\mathscr{L}^\infty(X, \mathscr{A}, \mu, \mathbb{R})$ and $\mathscr{L}^\infty(X, \mathscr{A}, \mu, \mathbb{C})$ to consist of functions f that are *essentially bounded*, which means that there is a nonnegative number M such that $\{x \in X : |f(x)| > M\}$ is locally μ-null (locally null sets are defined a bit later in this section). For most purposes, it does not matter which definition of $\mathscr{L}^\infty$ one uses. However, for the study of liftings (see Appendix F), the definition given here is the more convenient one.

[7]We use locally null sets, rather than null sets, here and in the construction of the L^∞ spaces given below in order to make Proposition 3.5.5, Theorem 7.5.4, and Theorem 9.4.8 true.

such that for each n the set $\{x \in X : |f(x)| > M_n\}$ is locally μ-null, then the set $\{x \in X : |f(x)| > \|f\|_\infty\}$ is the union of the sets $\{x \in X : |f(x)| > M_n\}$ and so is locally μ-null. Thus $\|f\|_\infty$ is not only the infimum of the set of numbers M such that $\{x \in X : |f(x)| > M\}$ is locally μ-null but is itself one of those numbers.

We need to derive some standard and important inequalities in order to prove that the functions $\|\cdot\|_p$ are seminorms. Let us begin by introducing some notation. Suppose that p satisfies $1 < p < +\infty$. Then $0 < 1/p < 1$, and so there is a real number q that satisfies $1/p + 1/q = 1$; q satisfies $1 < 1/q < +\infty$. The numbers p and q are sometimes called *conjugate exponents* (see the remarks following the proof of Proposition 3.5.5). The relation $1/p + 1/q = 1$ still holds if when $p = 1$ we let $q = +\infty$, and if when $p = +\infty$ we let $q = 1$; thus we can deal with all p that satisfy $1 \le p \le +\infty$. Note that the relation $1/p + 1/q = 1$ implies that $p + q = pq$, and for finite p and q implies that $p = q(p-1)$ and $q = p(q-1)$.

We turn to the necessary inequalities.

Lemma 3.3.1. *Let p satisfy* $1 < p < +\infty$, *let q be defined by* $1/p + 1/q = 1$, *and let x and y be nonnegative real numbers. Then*

$$xy \le \frac{x^p}{p} + \frac{y^q}{q}.$$

Proof. Since it is clear that the required inequality holds if either $x = 0$ or $y = 0$, we can assume that both x and y are positive. It is enough to prove that

$$u^{1/p}v^{1/q} \le \frac{u}{p} + \frac{v}{q}$$

holds for all positive u and v (let $u = x^p$ and $v = y^q$), and for this it is enough to prove that

$$t^{1/p} \le \frac{t}{p} + \frac{1}{q}$$

holds for all positive t (let $t = u/v$, and then multiply by v). However, this last inequality is easy, since according to elementary calculus the function defined for positive t by

$$t \mapsto \frac{t}{p} + \frac{1}{q} - t^{1/p}$$

has a minimum value of 0. $\square$

Proposition 3.3.2 (Hölder's Inequality).   *Let* $(X, \mathscr{A}, \mu)$ *be a measure space, and let p and q satisfy* $1 \le p \le +\infty$, $1 \le q \le +\infty$, *and* $1/p + 1/q = 1$. *If* $f \in \mathscr{L}^p(X, \mathscr{A}, \mu)$ *and* $g \in \mathscr{L}^q(X, \mathscr{A}, \mu)$, *then fg belongs to* $\mathscr{L}^1(X, \mathscr{A}, \mu)$ *and satisfies* $\int |fg| d\mu \le \|f\|_p \|g\|_q$.

Proof. First suppose that $p = 1$ and $q = +\infty$. If f and g belong to $\mathscr{L}^1(X, \mathscr{A}, \mu)$ and $\mathscr{L}^\infty(X, \mathscr{A}, \mu)$, respectively, then the set $\{x \in X : f(x) \ne 0\}$ is σ-finite under

μ (see Corollary 2.3.11) and the set $\{x \in X : |g(x)| > \|g\|_\infty\}$ is locally μ-null. The intersection of these two sets in thus a μ-null set, and so the inequality

$$|f(x)g(x)| \le \|g\|_\infty |f(x)|$$

holds at almost every x in X. It follows that $fg \in \mathscr{L}^1(X,\mathscr{A},\mu)$ and that

$$\int |fg| \, d\mu \le \int \|g\|_\infty |f| \, d\mu = \|g\|_\infty \|f\|_1.$$

A similar argument applies in case $p = +\infty$ and $q = 1$.

Now suppose that $1 < p < +\infty$ and hence that $1 < q < +\infty$. Let f belong to $\mathscr{L}^p(X,\mathscr{A},\mu)$ and g belong to $\mathscr{L}^q(X,\mathscr{A},\mu)$. Lemma 3.3.1 implies that

$$|f(x)g(x)| \le \frac{1}{p}|f(x)|^p + \frac{1}{q}|g(x)|^q$$

holds for each x in X; hence if $\|f\|_p = 1$ and $\|g\|_q = 1$, then fg belongs to $\mathscr{L}^1(X,\mathscr{A},\mu)$ and satisfies

$$\int |fg| \, d\mu \le \frac{1}{p} \int |f|^p \, d\mu + \frac{1}{q} \int |g|^q \, d\mu = \frac{1}{p} + \frac{1}{q} = 1.$$

In case neither $\|f\|_p$ nor $\|g\|_q$ is 0, we can use this inequality, with f and g replaced by $f/\|f\|_p$ and $g/\|g\|_q$, to conclude that

$$\frac{1}{\|f\|_p \|g\|_q} \int |fg| \, d\mu \le 1$$

and hence that

$$\int |fg| \, d\mu = \|f\|_p \|g\|_q. \tag{1}$$

Since inequality (1) is clear if $\|f\|_p = 0$ or $\|g\|_q = 0$ (for then fg vanishes almost everywhere), the proof is complete. $\qquad\square$

Proposition 3.3.3 (Minkowski's Inequality). *Let $(X,\mathscr{A},\mu)$ be a measure space, and let p satisfy $1 \le p \le +\infty$. If f and g belong to $\mathscr{L}^p(X,\mathscr{A},\mu)$, then $f+g$ belongs to $\mathscr{L}^p(X,\mathscr{A},\mu)$ and $\|f+g\|_p \le \|f\|_p + \|g\|_p$.*

Proof. First suppose that $p = +\infty$. Define N_1 and N_2 by $N_1 = \{x \in X : |f(x)| > \|f\|_\infty\}$ and $N_2 = \{x \in X : |g(x)| > \|g\|_\infty\}$. Then N_1 and N_2 are locally μ-null, and the inequality

$$|f(x) + g(x)| \le |f(x)| + |g(x)| \le \|f\|_\infty + \|g\|_\infty$$

holds at each x outside the locally μ-null set $N_1 \cup N_2$. Thus $f + g \in \mathscr{L}^\infty(X,\mathscr{A},\mu)$ and $\|f+g\|_\infty \le \|f\|_\infty + \|g\|_\infty$.

Next suppose that $p = 1$. Then the inequality $|f(x) + g(x)| \le |f(x)| + |g(x)|$ holds at each x in X, and so $f + g \in \mathscr{L}^1(X, \mathscr{A}, \mu)$ and

$$\|f + g\|_1 = \int |f + g| \, d\mu \le \int |f| \, d\mu + \int |g| \, d\mu = \|f\|_1 + \|g\|_1.$$

Now consider the case where $1 < p < +\infty$. We checked that $f + g \in \mathscr{L}^p(X, \mathscr{A}, \mu)$ earlier in this section. Define q by $1/p + 1/q = 1$. Since $p + q = pq$, it follows that $(|f + g|^{p-1})^q = |f + g|^p$ and hence that $|f + g|^{p-1} \in \mathscr{L}^q(X, \mathscr{A}, \mu)$. Thus if we use the fact that $|f + g|^p \le (|f| + |g|)|f + g|^{p-1}$ and then apply Hölder's inequality (Proposition 3.3.2) to the functions f and $|f + g|^{p-1}$ and to the functions g and $|f + g|^{p-1}$, we can conclude that

$$\int |f + g|^p \, d\mu \le \int (|f| + |g|)|f + g|^{p-1} \, d\mu$$

$$= \int |f| \, |f + g|^{p-1} \, d\mu + \int |g| \, |f + g|^{p-1} \, d\mu$$

$$\le \|f\|_p \| |f + g|^{p-1} \|_q + \|g\|_p \| |f + g|^{p-1} \|_q$$

$$= (\|f\|_p + \|g\|_p) \left(\int |f + g|^p \, d\mu \right)^{1/q}.$$

If $\int |f + g|^p \, d\mu \ne 0$, we can divide the terms of this inequality by $(\int |f + g|^p \, d\mu)^{1/q}$, obtaining

$$\|f + g\|_p \le \|f\|_p + \|g\|_p. \tag{2}$$

Since inequality (2) is clear if $\int |f + g|^p \, d\mu = 0$, the proof is complete. $\qquad \square$

Corollary 3.3.4. *Let $(X, \mathscr{A}, \mu)$ be a measure space, and let p satisfy $1 \le p \le +\infty$. Then $\mathscr{L}^p(X, \mathscr{A}, \mu)$ is a vector space, and $\| \cdot \|_p$ is a seminorm on $\mathscr{L}^p(X, \mathscr{A}, \mu)$.*

Proof. We have already verified that $\mathscr{L}^p(X, \mathscr{A}, \mu)$ is a vector space. The triangle inequality for $\| \cdot \|_p$ is the only other nontrivial thing to check, and it is given by Proposition 3.3.3. $\qquad \square$

Example 3.3.5. Suppose that $1 \le p_1 < p_2 < +\infty$. Then each sequence $\{a_n\}$ that satisfies $\sum |a_n|^{p_1} < +\infty$ also satisfies $\sum |a_n|^{p_2} < +\infty$. Thus if μ is counting measure on the σ-algebra $\mathscr{A}$ of all subsets of $\mathbb{N}$, then $\mathscr{L}^{p_1}(\mathbb{N}, \mathscr{A}, \mu) \subseteq \mathscr{L}^{p_2}(\mathbb{N}, \mathscr{A}, \mu)$. The inclusion is reversed for finite measures: if μ is a finite measure on a measurable space $(X, \mathscr{A})$, then $\mathscr{L}^{p_2}(X, \mathscr{A}, \mu) \subseteq \mathscr{L}^{p_1}(X, \mathscr{A}, \mu)$. See Exercise 9. $\qquad \square$

Note that if there are nonempty subsets A of X that belong to $\mathscr{A}$ and satisfy $\mu(A) = 0$, then there are nonzero functions f that belong to $\mathscr{L}^p(X, \mathscr{A}, \mu)$ and satisfy $\|f\|_p = 0$. Thus for many common measure spaces, the seminorms $\| \cdot \|_p$ are not norms. Since norms and metrics are often easier to deal with than are seminorms and semimetrics, the following construction of normed spaces $L^p(X, \mathscr{A}, \mu)$ from the spaces $\mathscr{L}^p(X, \mathscr{A}, \mu)$ proves useful.

Let $(X, \mathscr{A}, \mu)$ be a measure space, and let $\mathscr{N}^p(X, \mathscr{A}, \mu)$ be the subset of $\mathscr{L}^p(X, \mathscr{A}, \mu)$ that consists of those functions f that belong to $\mathscr{L}^p(X, \mathscr{A}, \mu)$ and satisfy $\|f\|_p = 0$. Thus if $1 \leq p < +\infty$, then $\mathscr{N}^p(X, \mathscr{A}, \mu)$ consists of the $\mathscr{A}$-measurable functions on X that satisfy $\int |f|^p d\mu = 0$ (that is, that vanish almost everywhere), and if $p = +\infty$, then $\mathscr{N}^p(X, \mathscr{A}, \mu)$ consists of the bounded $\mathscr{A}$-measurable functions on X that vanish locally almost everywhere. It is clear that $\mathscr{N}^p(X, \mathscr{A}, \mu)$ is a linear subspace of the vector space $\mathscr{L}^p(X, \mathscr{A}, \mu)$. The space $L^p(X, \mathscr{A}, \mu)$ is defined to be the quotient $\mathscr{L}^p(X, \mathscr{A}, \mu)/\mathscr{N}^p(X, \mathscr{A}, \mu)$. Recall that this means that $L^p(X, \mathscr{A}, \mu)$ is the collection of cosets of $\mathscr{N}^p(X, \mathscr{A}, \mu)$ in $\mathscr{L}^p(X, \mathscr{A}, \mu)$; these cosets[8] are by definition the equivalence classes induced by the equivalence relation $\sim$, where $f \sim g$ holds if and only if $f - g$ belongs to $\mathscr{N}^p(X, \mathscr{A}, \mu)$. Note that if $1 \leq p < +\infty$, then $f \sim g$ holds if and only if f and g are equal almost everywhere, and so $L^p(X, \mathscr{A}, \mu)$ is formed by identifying functions in $\mathscr{L}^p(X, \mathscr{A}, \mu)$ that agree almost everywhere. Likewise, $L^\infty(X, \mathscr{A}, \mu)$ is formed by identifying functions in $\mathscr{L}^\infty(X, \mathscr{A}, \mu)$ that agree locally almost everywhere.

For f in $\mathscr{L}^p(X, \mathscr{A}, \mu)$ let $\langle f \rangle$ be the coset of $\mathscr{N}^p(X, \mathscr{A}, \mu)$ to which f belongs. It is easy to check that the formulas $\langle f \rangle + \langle g \rangle = \langle f + g \rangle$ and $\alpha \langle f \rangle = \langle \alpha f \rangle$ define operations that make $L^p(X, \mathscr{A}, \mu)$ into a vector space. Furthermore, if f and g are functions that belong to $\mathscr{L}^p(X, \mathscr{A}, \mu)$ and satisfy $f \sim g$, then $\|f\|_p = \|g\|_p$ (check this). Thus we can define a function, again called $\|\cdot\|_p$, on $L^p(X, \mathscr{A}, \mu)$ by means of the formula $\|\langle f \rangle\|_p = \|f\|_p$. It is easy to check that $\|\cdot\|_p$ is a norm on $L^p(X, \mathscr{A}, \mu)$ (see Corollary 3.3.4).

We will, of course, write $L^p(X, \mathscr{A}, \mu, \mathbb{R})$ or $L^p(X, \mathscr{A}, \mu, \mathbb{C})$ when the real and complex cases must be distinguished from one another.

It is often convenient to act as though the elements of $L^p(X, \mathscr{A}, \mu)$ were functions, rather than equivalence classes of functions. In fact, some authors use the same symbol for $\mathscr{L}^p(X, \mathscr{A}, \mu)$ and $L^p(X, \mathscr{A}, \mu)$. We will try to avoid this identification of functions and classes of functions, since it can lead to confusion (especially in the study of stochastic processes). However to simplify notation we will often deal with $\mathscr{L}^p(X, \mathscr{A}, \mu)$ when proving theorems about $L^p(X, \mathscr{A}, \mu)$. For example, in the next section we will prove that $L^p(X, \mathscr{A}, \mu)$ is complete by showing that if $\sum_k f_k$ is a series in $\mathscr{L}^p(X, \mathscr{A}, \mu)$ such that $\sum_k \|f_k\|_p < +\infty$, then there is a function f in $\mathscr{L}^p(X, \mathscr{A}, \mu)$ such that $\lim_n \|f - \sum_{k=1}^n f_k\|_p = 0$ (recall Proposition 3.2.5). This will imply the completeness of $L^p(X, \mathscr{A}, \mu)$ and yet avoid the cumbersome notation associated with equivalence classes.

We close this section with a definition. Let $(X, \mathscr{A}, \mu)$ be a measure space, let p satisfy $1 \leq p < +\infty$, and let f and $f_1, f_2, \dots$ belong to $\mathscr{L}^p(X, \mathscr{A}, \mu)$. Then $\{f_n\}$ converges to f in pth mean (or in L^p-norm) if $\lim_n \int |f_n - f|^p d\mu = 0$, or, equivalently, if $\lim_n \|f_n - f\|_p = 0$. There are a number of results relating

---

[8]Equivalently, for each f in $\mathscr{L}^p(X, \mathscr{A}, \mu)$ the coset to which f belongs is the set $f + \mathscr{N}^p(X, \mathscr{A}, \mu)$ and hence the set $\{f + g : g \in \mathscr{N}^p(X, \mathscr{A}, \mu)\}$.

convergence in pth mean to convergence in measure and convergence almost everywhere; the reader would do well to formulate and prove some of them, using the corresponding results in Sect. 3.1 as models (see also Exercise 9).

Exercises

1. Use the inequality $(x-y)^2 \geq 0$ to give an alternate proof of Lemma 3.3.1 in the case where $p = q = 2$.
2. Give an alternate proof of Lemma 3.3.1 by noting that x^p/p and y^q/q are the areas of the shaded regions in Fig. 3.1. (The curve in Fig. 3.1 represents the graph of the equation $t = s^{p-1}$, or, equivalently, of the equation $s = t^{q-1}$.)
3. Let $(X, \mathscr{A}, \mu)$ be a measure space. Check that the formula

$$(\langle f \rangle, \langle g \rangle) = \int f g \, d\mu$$

   defines an inner product on $L^2(X, \mathscr{A}, \mu, \mathbb{R})$ and that the norm associated with this inner product is the usual norm on $L^2(X, \mathscr{A}, \mu, \mathbb{R})$.
4. Let $\mathscr{B}$ be the σ-algebra of Borel subsets of $[0,1]$ and let λ be the restriction of Lebesgue measure to $\mathscr{B}$. Show that if $1 \leq p < 2$ or $2 < p \leq +\infty$ then there is no inner product on $L^p([0,1], \mathscr{B}, \lambda, \mathbb{R})$ that induces the usual norm on $L^p([0,1], \mathscr{B}, \lambda, \mathbb{R})$. (Hint: A norm that comes from an inner product must satisfy the identity in part (a) of Exercise 3.2.8.)

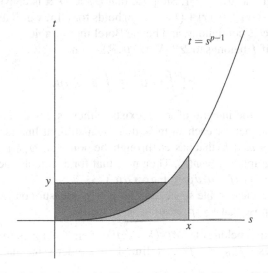

Fig. 3.1 Region used in Exercise 2 for proof of Lemma 3.3.1

5. Let X be a nonempty set, let $\mathscr{A} = \{\varnothing, X\}$, and let $\mu : \mathscr{A} \to [0, +\infty]$ be defined by

$$\mu(A) = \begin{cases} 0 & \text{if } A = \varnothing, \\ +\infty & \text{if } A = X. \end{cases}$$

Show that X is locally μ-null but not μ-null.

6. Suppose that for each subset A of $\mathbb{R}^2$ and each real number x we denote the set $\{y \in \mathbb{R} : (x,y) \in A\}$ by A_x. Let $\mathscr{A}$ consist of those subsets A of $\mathbb{R}^2$ that satisfy $A_x \in \mathscr{B}(\mathbb{R})$ for each x in $\mathbb{R}$, and define $\mu : \mathscr{A} \to [0, +\infty]$ by

$$\mu(A) = \begin{cases} \sum_x \lambda(A_x) & \text{if } A_x \neq \varnothing \text{ for only countably many } x, \\ +\infty & \text{otherwise.} \end{cases}$$

(a) Show that $\mathscr{A}$ is a σ-algebra on $\mathbb{R}^2$ and that μ is a measure on $(\mathbb{R}^2, \mathscr{A})$.
(b) Show that $\{(x,y) \in \mathbb{R}^2 : y = 0\}$ is locally μ-null but not μ-null.

7. Let $(X, \mathscr{A}, \mu)$ be a finite measure space, and let f be an $\mathscr{A}$-measurable real- or complex-valued function on X.

(a) Show that f belongs to $\mathscr{L}^\infty(X, \mathscr{A}, \mu)$ if and only if

(i) f belongs to $\mathscr{L}^p(X, \mathscr{A}, \mu)$ for each p in $[1, +\infty)$, and
(ii) $\sup\{\|f\|_p : 1 \leq p < +\infty\}$ is finite.

(b) Show that if these conditions hold, then $\|f\|_\infty = \lim_{p \to +\infty} \|f\|_p$.

8. (Jensen's inequality.) Let $(X, \mathscr{A})$ be a measurable space, and let μ be a measure on $(X, \mathscr{A})$ such that $\mu(X) = 1$. Suppose that $\varphi : \mathbb{R} \to \mathbb{R}$ is convex, in the sense that $\varphi(tx + (1-t)y) \leq t\varphi(x) + (1-t)\varphi(y)$ holds for all x, y in $\mathbb{R}$ and all t in $[0,1]$.
(a) Show that φ is continuous, and hence Borel measurable.
(b) Show that if f belongs to $\mathscr{L}^1(X, \mathscr{A}, \mu, \mathbb{R})$, then

$$\varphi\left(\int f \, d\mu\right) \leq \int \varphi \circ f \, d\mu.$$

In particular, the integral of $\varphi \circ f$ exists, either as a real number or as $+\infty$. (Hint: Show that for each x_0 in $\mathbb{R}$ there is a straight line (say given by the equation $y = ax + b$) that passes through the point $(x_0, \varphi(x_0))$ and never goes above the graph of $y = \varphi(x)$. Then note that for a suitable such line we have $\varphi(\int f \, d\mu) = \int (af + b) \, d\mu \leq \int \varphi \circ f \, d\mu$.)

9. Let $(X, \mathscr{A})$ be a measurable space, and let μ be a measure on $(X, \mathscr{A})$ such that $\mu(X) = 1$. Suppose that $1 \leq p_1 < p_2 < +\infty$.

(a) Show that if f belongs to $\mathscr{L}^{p_2}(X, \mathscr{A}, \mu)$, then f belongs to $\mathscr{L}^{p_1}(X, \mathscr{A}, \mu)$ and satisfies $\|f\|_{p_1} \leq \|f\|_{p_2}$. (Hint: Use Hölder's inequality or Jensen's inequality.)
(b) Show that if f and $f_1, f_2, \ldots$ belong to $\mathscr{L}^{p_2}(X, \mathscr{A}, \mu)$ and if $\{f_n\}$ converges to f in p_2th mean, then $\{f_n\}$ converges to f in p_1th mean.

3.4   Properties of $\mathscr{L}^p$ and L^p

This section is devoted to some basic properties of the L^p spaces.

Theorem 3.4.1. *Let $(X, \mathscr{A}, \mu)$ be a measure space, and let p satisfy $1 \leq p \leq +\infty$. Then $L^p(X, \mathscr{A}, \mu)$ is complete under the norm $\| \cdot \|_p$.*

Proof. According to Proposition 3.2.5, we need only show that each absolutely convergent series in $L^p(X, \mathscr{A}, \mu)$ is convergent. We do this by considering functions (not equivalence classes) in $\mathscr{L}^p(X, \mathscr{A}, \mu)$, as outlined near the end of Sect. 3.3.

First suppose that $p = +\infty$ and that $\{f_k\}$ is a sequence of functions that belong to $\mathscr{L}^\infty(X, \mathscr{A}, \mu)$ and satisfy $\sum_k \|f_k\|_\infty < +\infty$. For each positive integer k let $N_k = \{x \in X : |f_k(x)| > \|f_k\|_\infty\}$. Then the series $\sum_k f_k(x)$ converges at each x outside $\cup_k N_k$, and the function f defined by

$$f(x) = \begin{cases} \sum_k f_k(x) & \text{if } x \notin \bigcup_k N_k, \\ 0 & \text{if } x \in \bigcup_k N_k \end{cases}$$

is bounded and $\mathscr{A}$-measurable. Since $\cup_k N_k$ is locally null, the inequality

$$\left\| f - \sum_{k=1}^n f_k \right\|_\infty \leq \sum_{k=n+1}^\infty \|f_k\|_\infty$$

holds for each n, and so

$$\lim_n \left\| f - \sum_{k=1}^n f_k \right\|_\infty \leq \lim_n \sum_{k=n+1}^\infty \|f_k\|_\infty = 0.$$

Thus $L^\infty(X, \mathscr{A}, \mu)$ is complete.

Now suppose that $1 \leq p < +\infty$ and that $\{f_k\}$ is a sequence of functions that belong to $\mathscr{L}^p(X, \mathscr{A}, \mu)$ and satisfy $\sum_k \|f_k\|_p < +\infty$. Define $g: X \to [0, +\infty]$ by

$$g(x) = \left(\sum_{k=1}^\infty |f_k(x)| \right)^p$$

(of course $(+\infty)^p = +\infty$). Minkowski's inequality (Proposition 3.3.3), applied to the functions $|f_k|$, implies that

$$\left(\int \left(\sum_{k=1}^n |f_k| \right)^p d\mu \right)^{1/p} = \left\| \sum_{k=1}^n |f_k| \right\|_p \leq \sum_{k=1}^n \|f_k\|_p$$

holds for each n, and so it follows from the monotone convergence theorem that

$$\int g \, d\mu = \lim_n \int \left(\sum_{k=1}^n |f_k| \right)^p d\mu \leq \left(\sum_{k=1}^\infty \|f_k\|_p \right)^p ;$$

thus g is integrable. Consequently $g(x)$ is finite for almost every x (Corollary 2.3.14), and the series $\sum_k f_k(x)$ is absolutely convergent, and hence convergent, for almost every x. Define a function f on X by

$$f(x) = \begin{cases} \sum_{k=1}^{\infty} f_k(x) & \text{if } g(x) < +\infty, \\ 0 & \text{otherwise.} \end{cases}$$

Then f is measurable and satisfies $|f|^p \leq g$, and so it belongs to $\mathscr{L}^p(X,\mathscr{A},\mu)$. Since $\lim_n |\sum_{k=1}^n f_k(x) - f(x)| = 0$ and $|\sum_{k=1}^n f_k(x) - f(x)|^p \leq g(x)$ hold for almost every x, the dominated convergence theorem implies that $\lim_n \|\sum_{k=1}^n f_k - f\|_p = 0$. The completeness of $L^p(X,\mathscr{A},\mu)$ follows.   □

Let $(X,\mathscr{A},\mu)$ be a measure space. We will say that a function f in $\mathscr{L}^p(X,\mathscr{A},\mu)$ *determines* the equivalence class $\langle f \rangle$ in $L^p(X,\mathscr{A},\mu)$ to which it belongs. Likewise, if S is a subset of $\mathscr{L}^p(X,\mathscr{A},\mu)$ and if T is a subset of $L^p(X,\mathscr{A},\mu)$, then we will say that S *determines* T if T consists of the equivalence classes in $L^p(X,\mathscr{A},\mu)$ determined by the elements of S. This terminology will allow us to avoid a fair amount of pedantic notation. (See also the next-to-the-last paragraph in Sect. 3.3.)

Proposition 3.4.2. *Let $(X,\mathscr{A},\mu)$ be a measure space, and let p satisfy $1 \leq p \leq +\infty$. Then the simple functions in $\mathscr{L}^p(X,\mathscr{A},\mu)$ form a dense subspace of $\mathscr{L}^p(X,\mathscr{A},\mu)$ and so determine a dense subspace of $L^p(X,\mathscr{A},\mu)$.*

Proof. We will consider only real-valued functions. The corresponding results for $L^p(X,\mathscr{A},\mu,\mathbb{C})$ can be proved by separating a complex-valued function into its real and imaginary parts.

Let us first consider the case where $1 \leq p < +\infty$. Let f belong to $\mathscr{L}^p(X,\mathscr{A},\mu,\mathbb{R})$. Choose nondecreasing sequences $\{g_k\}$ and $\{h_k\}$ of nonnegative simple $\mathscr{A}$-measurable functions such that $f^+ = \lim_k g_k$ and $f^- = \lim_k h_k$ (Proposition 2.1.8), and define a sequence $\{f_k\}$ by $f_k = g_k - h_k$. Then each f_k is a simple $\mathscr{A}$-measurable function that satisfies $|f_k| \leq |f|$ and so belongs to $\mathscr{L}^p(X,\mathscr{A},\mu,\mathbb{R})$. Since these functions satisfy $|f_k(x) - f(x)| \leq |f(x)|$ and $\lim_k |f_k(x) - f(x)| = 0$ at each x in X, the dominated convergence theorem, applied to the pth powers of the functions $|f_k - f|$, implies that $\lim_k \|f_k - f\|_p = 0$. With this the proof is complete in the case where $1 \leq p < +\infty$.

Now suppose that $p = +\infty$. Let f belong to $\mathscr{L}^\infty(X,\mathscr{A},\mu,\mathbb{R})$, and let ε be a positive number. Choose real numbers $a_0, a_1, \ldots, a_n$ such that

$$a_0 < a_1 < \ldots < a_n$$

and such that the intervals $(a_{i-1}, a_i]$ cover the interval $[-\|f\|_\infty, \|f\|_\infty]$ and have lengths at most ε. Let $A_i = f^{-1}((a_{i-1}, a_i])$ for $i = 1, \ldots, n$, and let $f_\varepsilon = \sum_{i=1}^n a_i \chi_{A_i}$. Then f_ε is a simple $\mathscr{A}$-measurable function that satisfies $\|f - f_\varepsilon\|_\infty \leq \varepsilon$. Since f and ε are arbitrary, the proof is complete.   □

We now turn to some results concerning Lebesgue measure on $\mathbb{R}$. Let $[a,b]$ be a closed bounded interval. A real- or complex-valued function f on $[a,b]$ is a *step function* if there are real numbers $a_0, \ldots, a_n$ such that

(a) $a = a_0 < a_1 < \ldots < a_n = b$, and
(b) f is constant on each interval (a_{i-1}, a_i).

We will use $\mathscr{L}^p([a,b])$ and $L^p([a,b])$ as abbreviations for $\mathscr{L}^p([a,b], \mathscr{B}([a,b]), \lambda)$ and $L^p([a,b], \mathscr{B}([a,b]), \lambda)$, where $\mathscr{B}([a,b])$ is the σ-algebra of Borel subsets of $[a,b]$ and λ is the restriction of Lebesgue measure to $\mathscr{A}$.

The following two propositions are often useful, since step functions and continuous functions are usually easier to deal with than are more general functions.

Proposition 3.4.3. *Suppose that $[a,b]$ is a closed bounded interval and that p satisfies $1 \leq p < +\infty$. Then the subspace of $L^p([a,b])$ determined by the step functions on $[a,b]$ is dense in $L^p([a,b])$.*

Proof. Of course, each step function on $[a,b]$ belongs to $\mathscr{L}^p([a,b])$. Since the Borel measurable simple functions on $[a,b]$ determine a dense subspace of $L^p([a,b])$ (Proposition 3.4.2), it is enough to show that if f is a Borel measurable simple function and if ε is a positive number, then there is a step function g such that $\|f - g\|_p < \varepsilon$, and for this it is enough to check that if χ_A is the characteristic function of a Borel subset A of $[a,b]$, then there are step functions g that make $\|\chi_A - g\|_p$ arbitrarily small. So let A be a Borel subset of $[a,b]$ and let δ be a positive number. Use the construction of Lebesgue outer measure (or Proposition C.4 and the regularity of Lebesgue measure) to choose a sequence $\{(a_n, b_n)\}$ of open intervals such that $A \subseteq \cup_{n=1}^{\infty}(a_n, b_n)$ and $\sum_{n=1}^{\infty}(b_n - a_n) < \lambda(A) + \delta$, and then choose a positive integer N such that $\sum_{n=N+1}^{\infty}(b_n - a_n) < \delta$. Let g be the characteristic function of $[a,b] \cap (\cup_1^N(a_n, b_n))$ and let h be the characteristic function of $[a,b] \cap (\cup_1^{\infty}(a_n, b_n))$. Then g is a step function, and

$$\|\chi_A - g\|_p \leq \|\chi_A - h\|_p + \|h - g\|_p$$

$$\leq \left(\lambda(\bigcup_1^{\infty}(a_n, b_n)) - A\right)^{1/p} + \left(\lambda\left(\bigcup_{N+1}^{\infty}(a_n, b_n)\right)\right)^{1/p}$$

$$< \delta^{1/p} + \delta^{1/p} = 2\delta^{1/p}.$$

Since δ is arbitrary, the proof is complete. $\qquad\square$

Proposition 3.4.4. *Suppose that $[a,b]$ is a closed bounded interval and that p satisfies $1 \leq p < +\infty$. Then the subspace of $L^p([a,b])$ determined by the continuous functions on $[a,b]$ is dense in $L^p([a,b])$.*

Proof. Of course, each continuous function on $[a,b]$ belongs to $\mathscr{L}^p([a,b])$. Since the step functions on $[a,b]$ determine a dense subspace of $L^p([a,b])$ (Proposition 3.4.3), it is enough to prove that for each step function f and each positive number ε there is a continuous function g that satisfies $\|f - g\|_p < \varepsilon$. So let f be a step function on $[a,b]$, let $M = \sup\{|f(x)| : x \in [a,b]\}$, and let δ be a positive number. It is easy to construct a continuous function g on $[a,b]$ such that $\sup\{|g(x)| : x \in [a,b]\} \le M$ and $\lambda(\{x \in [a,b] : f(x) \neq g(x)\}) < \delta$ (take a suitable piecewise linear function). Then

$$\int_a^b |f - g|^p \, d\lambda \le (2M)^p \lambda(\{x : f(x) \neq g(x)\}) \le (2M)^p \delta,$$

and so $\|f - g\|_p \le 2M\delta^{1/p}$. Since δ is arbitrary and M depends only on f, the proof is complete. $\square$

The reader should note that Propositions 3.4.3 and 3.4.4 would fail if p were allowed to be infinite (see Exercises 3 and 4).

Let us call a function on $\mathbb{R}$ a *step function* if for each interval $[a,b]$ its restriction to $[a,b]$ is a step function. Analogues of Propositions 3.4.3 and 3.4.4 hold for $L^p(\mathbb{R}, \mathscr{B}(\mathbb{R}), \lambda)$ if we replace the set of step functions on $[a,b]$ with the set of step functions on $\mathbb{R}$ that vanish outside some bounded interval and if we replace the set of continuous functions on $[a,b]$ with the set of continuous functions on $\mathbb{R}$ that vanish outside some bounded interval. The details are left to the reader. (See also Proposition 7.4.3.)

Let $\mathscr{A}$ be a σ-algebra on the set X. Then $\mathscr{A}$ is *countably generated* if there is a countable subfamily $\mathscr{C}$ of $\mathscr{A}$ such that $\mathscr{A} = \sigma(\mathscr{C})$. For example, the σ-algebra $\mathscr{B}(\mathbb{R})$ of Borel subsets of $\mathbb{R}$ is countably generated (see Exercise 1.1.2).

Proposition 3.4.5. *Let $(X, \mathscr{A}, \mu)$ be a measure space, and let p satisfy $1 \le p < +\infty$. If μ is σ-finite and $\mathscr{A}$ is countably generated, then $L^p(X, \mathscr{A}, \mu)$ is separable.*

The proof will depend on the following two lemmas.

Lemma 3.4.6. *Let $(X, \mathscr{A}, \mu)$ be a finite measure space, and let $\mathscr{A}_0$ be an algebra of subsets of X such that $\mathscr{A} = \sigma(\mathscr{A}_0)$. Then $\mathscr{A}_0$ is dense in $\mathscr{A}$, in the sense that for each A in $\mathscr{A}$ and each positive number ε there is a set A_0 that belongs to $\mathscr{A}_0$ and satisfies $\mu(A \triangle A_0) < \varepsilon$.*

Proof. Let $\mathscr{F}$ be the collection consisting of those sets A in $\mathscr{A}$ such that for each positive ε there is a set A_0 that belongs to $\mathscr{A}_0$ and satisfies $\mu(A \triangle A_0) < \varepsilon$. Of course $\mathscr{A}_0 \subseteq \mathscr{F}$, and so $X \in \mathscr{F}$. The identity $A^c \triangle A_0^c = A \triangle A_0$ implies that if $A \in \mathscr{F}$, then $A^c \in \mathscr{F}$; hence $\mathscr{F}$ is closed under complementation. Now let $\{A_n\}$ be an infinite sequence of sets in $\mathscr{F}$, let $A = \cup_n A_n$, and let ε be a positive number. Choose a positive integer N such that $\mu(A - \cup_1^N A_n) < \varepsilon/2$ (see Proposition 1.2.5), and for $n = 1, 2, \ldots, N$ choose a set B_n that belongs to $\mathscr{A}_0$ and satisfies $\mu(A_n \triangle B_n) < \varepsilon/2N$. The set B defined by $B = \cup_1^N B_n$ then belongs to $\mathscr{A}_0$ and satisfies

$$\mu(A \triangle B) \le \mu\left(A \triangle \left(\bigcup_1^N A_n\right)\right) + \mu\left(\left(\bigcup_1^N A_n\right) \triangle B\right)$$

$$\le \mu\left(A \triangle \left(\bigcup_1^N A_n\right)\right) + \sum_1^N \mu(A_n \triangle B_n)$$

$$< \frac{\varepsilon}{2} + \sum_1^N \frac{\varepsilon}{2N} = \varepsilon.$$

Since we can produce such a set B for each positive ε, it follows that $A \in \mathscr{F}$. Consequently $\mathscr{F}$ is closed under the formation of countable unions and so is a σ-algebra. Since in addition $\mathscr{A}_0 \subseteq \mathscr{F} \subseteq \mathscr{A} = \sigma(\mathscr{A}_0)$, $\mathscr{F}$ must be equal to $\mathscr{A}$. Thus every set in $\mathscr{A}$ can be approximated with sets in $\mathscr{A}_0$. □

Lemma 3.4.7. *Let $(X, \mathscr{A}, \mu)$ be a measure space. Suppose that $\mathscr{A}_0$ is an algebra of subsets of X such that*

(a) $\sigma(\mathscr{A}_0) = \mathscr{A}$, *and*
(b) *X is the union of a sequence of sets that belong to $\mathscr{A}_0$ and have finite measure under μ.*

Then for each positive ε and each set A that belongs to $\mathscr{A}$ and satisfies $\mu(A) < +\infty$ there is a set A_0 that belongs to $\mathscr{A}_0$ and satisfies $\mu(A \triangle A_0) < \varepsilon$.

Proof. Choose a sequence $\{B_n\}$ of sets that belong to $\mathscr{A}_0$, have finite measure under μ, and satisfy $X = \cup_n B_n$. By replacing B_n with $\cup_{k=1}^n B_k$, we can assume that the sequence $\{B_n\}$ is increasing.

Let A belong to $\mathscr{A}$ and satisfy $\mu(A) < +\infty$. Proposition 1.2.5, applied to the sequence $\{A \cap B_n\}$, implies that there is a positive integer N such that $\mu(A \cap B_N) > \mu(A) - \varepsilon/2$. Since the measure $C \mapsto \mu(C \cap B_N)$ is finite, we can use Lemma 3.4.6 to obtain a set A_0 that belongs to $\mathscr{A}_0$ and satisfies $\mu((A \triangle A_0) \cap B_N) < \varepsilon/2$. Then $A_0 \cap B_N$ belongs to $\mathscr{A}_0$ and satisfies

$$\mu(A \triangle (A_0 \cap B_N)) \le \mu(A \triangle (A \cap B_N)) + \mu((A \cap B_N) \triangle (A_0 \cap B_N))$$

$$= \mu(A - (A \cap B_N)) + \mu((A \triangle A_0) \cap B_N)$$

$$< \frac{\varepsilon}{2} + \frac{\varepsilon}{2} = \varepsilon,$$

and the proof of the lemma is complete. □

Proof of Proposition 3.4.5. We can choose a countable subfamily $\mathscr{C}$ of $\mathscr{A}$ that generates $\mathscr{A}$ and contains sets B_n, $n = 1, 2, \ldots$, that have finite measure under μ and satisfy $X = \cup_n B_n$. Let $\mathscr{C}^+$ consist of the sets in $\mathscr{C}$, together with their complements, and let $\mathscr{A}_0$ be the algebra (not the σ-algebra) generated by $\mathscr{C}$. Then $\mathscr{A}_0$ is the set of finite unions of sets that have the form

$$C_1 \cap C_2 \cap \cdots \cap C_N$$

for some choice of N and some choice of sets $C_1, \ldots, C_N$ in $\mathscr{C}^+$. Clearly $\mathscr{A}_0$ is countable and satisfies the hypotheses of Lemma 3.4.7.

Let $\mathscr{S}$ be the collection of all finite sums

$$\sum_j d_j \chi_{D_j},$$

where each d_j is a rational number[9] and each D_j belongs to $\mathscr{A}_0$ and satisfies $\mu(D_j) < +\infty$. The set $\mathscr{S}$ is countable and is included in $\mathscr{L}^p(X, \mathscr{A}, \mu)$; we will show that it determines a dense subset of $L^p(X, \mathscr{A}, \mu)$.

Let f belong to $\mathscr{L}^p(X, \mathscr{A}, \mu)$, and let ε be a positive number. Then there is a simple function g that belongs to $\mathscr{L}^p(X, \mathscr{A}, \mu)$ and satisfies $\|f - g\|_p < \varepsilon$ (Proposition 3.4.2). Suppose that the simple function g has the form $\sum_j a_j \chi_{A_j}$, where each A_j belongs to $\mathscr{A}$ and satisfies $\mu(A_j) < +\infty$. We can choose rational numbers d_j such that

$$\left\| \sum_j a_j \chi_{A_j} - \sum_j d_j \chi_{A_j} \right\|_p \leq \sum_j |a_j - d_j| \|\chi_{A_j}\|_p < \varepsilon,$$

and then we can produce sets D_j in $\mathscr{A}_0$ such that $\| \sum_j d_j \chi_{A_j} - \sum_j d_j \chi_{D_j} \|_p < \varepsilon$ (use Lemma 3.4.7). Since f and ε are arbitrary, and since $\sum_j d_j \chi_{D_j}$ belongs to $\mathscr{S}$ and satisfies

$$\left\| f - \sum_j d_j \chi_{D_j} \right\|_p \leq \|f - g\|_p + \left\| g - \sum_j d_j \chi_{A_j} \right\|_p$$

$$+ \left\| \sum_j d_j \chi_{A_j} - \sum_j d_j \chi_{D_j} \right\|_p < 3\varepsilon,$$

the proof is complete. $\square$

Exercises

1. Use Proposition 3.4.3 to show that if $1 \leq p < +\infty$, then $L^p([a,b])$ is separable.
2. Show that $L^\infty([a,b])$ is not separable. (Hint: Consider the elements of $L^\infty([a,b])$ determined by the characteristic functions of the sets $[a,c]$, where $a < c < b$.)

---

[9] When dealing with the complex L^p spaces, let each d_j be a complex number whose real and imaginary parts are rational.

3. Show that Proposition 3.4.3 would be false if p were allowed to be infinite. (Hint: Construct a Borel subset A of $[a,b]$ such that $\|\chi_A - f\|_\infty \geq 1/2$ holds whenever f is a step function.)
4. Show that Proposition 3.4.4 would be false if p were allowed to be infinite. (Hint: Let $A = [a,c]$, where $a < c < b$. How small can $\|\chi_A - f\|_\infty$ be if f is continuous?)
5. Suppose that for each function $f : \mathbb{R} \to \mathbb{R}$ and each x in $\mathbb{R}$ we define a function $f_x : \mathbb{R} \to \mathbb{R}$ by $f_x(t) = f(t - x)$. (A similar definition applies to complex-valued functions on $\mathbb{R}$.) Show that if $1 \leq p < +\infty$ and if f belongs to $\mathscr{L}^p(\mathbb{R}, \mathscr{B}(\mathbb{R}), \lambda)$, then

$$\lim_{x \to x_0} \|f_x - f_{x_0}\|_p = 0$$

holds for each x_0 in $\mathbb{R}$. (Hint: First, consider the case where f is a step function that vanishes outside some bounded interval. Then use Proposition 3.4.3 (see the remarks following the proof of Proposition 3.4.4).)
6. Show that the hypothesis of σ-finiteness cannot be omitted in Proposition 3.4.5. (Hint: Consider counting measure on $(\mathbb{R}, \mathscr{B}(\mathbb{R}))$.)
7. Show that in Lemma 3.4.7 condition (b) cannot be replaced with the assumption that μ is σ-finite. (Hint: Let $\mathscr{A}_0$ be the algebra on $\mathbb{R}$ defined in Example 1.1.1(g), let $\{r_n\}$ be an enumeration of $\mathbb{Q}$, and let μ be the Borel measure on $\mathbb{R}$ defined by $\mu = \sum_n \delta_{r_n}$.)

3.5 Dual Spaces

Recall that if V_1 and V_2 are vector spaces over $\mathbb{R}$ (or over $\mathbb{C}$), then a function $T : V_1 \to V_2$ is a *linear operator* (or *linear transformation*) if for each v and w in V_1 and each α in $\mathbb{R}$ (or in $\mathbb{C}$) it satisfies $T(v + w) = T(v) + T(w)$ and $T(\alpha v) = \alpha T(v)$. Recall also that if S_1 and S_2 are metric spaces, say with metrics d_1 and d_2, then a function $f : S_1 \to S_2$ is *continuous* if for each point a in S_1 and each positive number ε there is a positive number δ such that $d_2(f(s), f(a)) < \varepsilon$ holds whenever s belongs to S_1 and satisfies $d_1(s, a) < \delta$. Thus if V_1 and V_2 are normed linear spaces, say with norms $\| \cdot \|_1$ and $\| \cdot \|_2$, then a function $f : V_1 \to V_2$ is continuous if and only if for each a in V_1 and each positive number ε there is a positive number δ such that $\|f(v) - f(a)\|_2 < \varepsilon$ holds whenever v belongs to V_1 and satisfies $\|v - a\|_1 < \delta$.

When dealing with several normed spaces, we will often use the symbol $\| \cdot \|$ to denote each of the norms involved. This will of course be done only when there seems to be little chance of confusion.

Proposition 3.5.1. *Let V_1 and V_2 be normed linear spaces, and let $T : V_1 \to V_2$ be a linear operator. Then T is continuous if and only if there is a nonnegative number M such that*

$$\|T(v)\| \leq M\|v\| \tag{1}$$

holds for each v in V_1.

Proof. First suppose that there is a nonnegative number M such that inequality (1) holds for each v in V_1. Then for each v and a in V_1 we have

$$\|T(v) - T(a)\| = \|T(v-a)\| \leq M\|v-a\|;$$

hence if ε is a positive number and if we define δ by $\delta = \varepsilon/M$ (let δ be an arbitrary positive number if $M = 0$), then $\|T(v) - T(a)\| < \varepsilon$ holds whenever $\|v-a\| < \delta$. Thus T is continuous.

Now suppose that T is continuous, and choose a positive number δ such that $\|T(v)\| = \|T(v) - T(0)\| < 1$ if $\|v\| = \|v-0\| < \delta$. Note that (1) holds if $v = 0$, whatever value we use for M. Now suppose that $v \neq 0$ and let $w = v/\|v\|$. It follows that if $0 < t < \delta$, then we have $\|tw\| < \delta$ and $t\|T(w)\| < 1$, from which we get

$$\|T(v)\| < \frac{1}{t}\|v\|.$$

Since t can be chosen arbitrarily close to $1/\delta$, it follows that $\|T(v)\| \leq \frac{1}{\delta}\|v\|$. Thus inequality (1) holds, with M equal to $1/\delta$.   □

Let V_1 and V_2 be normed linear spaces, and let $T: V_1 \to V_2$ be linear. A nonnegative number A such that $\|T(v)\| \leq A\|v\|$ holds for each v in V_1 is called a *bound* for T, and the operator T is called *bounded* if there is a bound for it (see also Exercises 3 and 4). Thus Proposition 3.5.1 says that a linear operator is continuous if and only if it is bounded. It is easy to check that if the operator T is bounded, then the infimum of the set of bounds for T is a bound for T. This smallest bound for T is called the *norm* of T and is written $\|T\|$. It is not hard to check that $\|\cdot\|$ is a norm on the vector space of all bounded linear operators from V_1 to V_2.

We turn to a few special cases. Suppose that V_1 and V_2 are normed linear spaces and that $T: V_1 \to V_2$ is linear. Then T is an *isometry* if $\|T(v)\| = \|v\|$ holds for each v in V_1. Note that if T is an isometry and if v and w belong to V_1, then

$$\|T(v) - T(w)\| = \|T(v-w)\| = \|v-w\|,$$

and so T preserves distances. The linear operator T is an *isometric isomorphism* if it is an isometry that is surjective (note that an isometry is necessarily injective and so is bijective if and only if it is surjective). Thus an isometric isomorphism is a bijection that preserves both linear and metric structure.

Let V be a normed linear space. Recall that a *linear functional* on V is a linear operator on V whose values lie in $\mathbb{R}$ (if V is a vector space over $\mathbb{R}$) or in $\mathbb{C}$ (if V is a vector space over $\mathbb{C}$). We will be particularly concerned with the bounded, that is, continuous, linear functionals on V. It is easy to check that the set of all continuous linear functionals on V is a subspace of the vector space of all linear functionals on V; this subspace is called the *dual space* (or *conjugate space*) of V and is denoted by V^*. The space V^* is sometimes called the *topological dual space* of V in order to distinguish it from the space of *all* linear functionals on V (which is then called the *algebraic dual space* of V).

Note that the function $\| \cdot \| : V^* \to \mathbb{R}$ that assigns to each functional in V^* its norm (as defined above) is in fact a norm on the vector space V^*; for instance, the calculation

$$|(F+G)(v)| \le |F(v)| + |G(v)| \le \|F\| \|v\| + \|G\| \|v\| = (\|F\| + \|G\|)\|v\|$$

shows that $\|F\| + \|G\|$ is a bound for $F+G$ and so implies that $\|F+G\| \le \|F\| + \|G\|$.

Example 3.5.2. Let $[a,b]$ be a closed bounded subinterval of $\mathbb{R}$, and let μ be a finite Borel measure on $[a,b]$. Define $F : C[a,b] \to \mathbb{R}$ by letting

$$F(f) = \int f \, d\mu \tag{2}$$

hold for each f in $C[a,b]$. It is clear that F is a linear functional and that F is *positive*, in the sense that each nonnegative[10] f in $C[a,b]$ satisfies $F(f) \ge 0$. We will see that every positive linear functional on $C[a,b]$ arises in this way (Theorem 7.2.8).   □

Example 3.5.3. Now suppose that $C[a,b]$ is given the norm $\| \cdot \|_\infty$ defined by

$$\|f\|_\infty = \sup\{|f(x)| : x \in [a,b]\}$$

(see Example 3.2.1(e) above). Then the functional F defined by (2) satisfies

$$|F(f)| = \left| \int f \, d\mu \right| \le \int |f| \, d\mu \le \|f\|_\infty \mu([a,b]),$$

and so is bounded and hence continuous. Likewise, if μ_1 and μ_2 are finite Borel measures on $[a,b]$, then the linear functional G defined by

$$G(f) = \int f \, d\mu_1 - \int f \, d\mu_2$$

is continuous. We will see that every continuous linear functional on $C[a,b]$ arises in this way (Theorem 7.3.6). These facts and their generalizations form the basis for many of the applications of measure theory.[11]   □

Example 3.5.4. Suppose that $(X, \mathscr{A}, \mu)$ is an arbitrary measure space, that p satisfies $1 \le p < +\infty$, and that q is defined by $1/p + 1/q = 1$. Let g belong

---

[10] The function f is called nonnegative if $f(x) \ge 0$ holds at each x in $[a,b]$.

[11] The usefulness of these results seems to be attributable to two facts:

(a) If a linear functional on a space of functions can be represented as an integral, then the limit theorems of Sect. 2.4 are applicable.

(b) The methods available for decomposing and analyzing measures are often easier to visualize than those that apply directly to linear functionals.

to $\mathscr{L}^q(X,\mathscr{A},\mu)$. Then fg is integrable whenever f belongs to $\mathscr{L}^p(X,\mathscr{A},\mu)$ (Proposition 3.3.2), and so the formula

$$T_g(f) = \int fg\,d\mu$$

defines a linear functional T_g on $\mathscr{L}^p(X,\mathscr{A},\mu)$. It is clear that if f_1 and f_2 belong to $\mathscr{L}^p(X,\mathscr{A},\mu)$ and agree almost everywhere, then $T_g(f_1) = T_g(f_2)$; thus we can use the formula $T_g(\langle f\rangle) = T_g(f)$ to define a functional, also called T_g, on $L^p(X,\mathscr{A},\mu)$. Hölder's inequality (Proposition 3.3.2) implies that $|T_g(f)| \le \|g\|_q\|f\|_p$ holds for each f in $\mathscr{L}^p(X,\mathscr{A},\mu)$. Thus T_g is continuous on $L^p(X,\mathscr{A},\mu)$, and $\|T_g\| \le \|g\|_q$. We'll see in the following proposition that $\|T_g\| = \|g\|_q$. $\square$

We will denote by T the map from $\mathscr{L}^q(X,\mathscr{A},\mu)$ to $(L^p(X,\mathscr{A},\mu))^*$ that takes the function g to the functional T_g defined above.

Proposition 3.5.5. *Let $(X,\mathscr{A},\mu)$ be a measure space, let p satisfy $1 \le p < +\infty$, and let q be defined by $1/p+1/q = 1$. Then the map $T : \mathscr{L}^q(X,\mathscr{A},\mu) \to (L^p(X,\mathscr{A},\mu))^*$ defined above induces an isometry of $L^q(X,\mathscr{A},\mu)$ into $(L^p(X,\mathscr{A},\mu))^*$.*

Note that Proposition 3.5.5 says that T is an isometry *into* $(L^p(X,\mathscr{A},\mu))^*$; it does not say that T is surjective. Example 4.5.2 in the next chapter gives a case in which T is not surjective. Later we will see that the map T is a surjection, and hence an isometric isomorphism, if

(a) $1 < p < +\infty$ and $(X,\mathscr{A},\mu)$ is arbitrary,
(b) $p = 1$ and μ is σ-finite, or
(c) $p = 1$ and $(X,\mathscr{A},\mu)$ arises through certain topological constructions

(see Theorems 4.5.1, 7.5.4, and 9.4.8). It is because of this relationship between $L^q(X,\mathscr{A},\mu)$ and $(L^p(X,\mathscr{A},\mu))^*$ that numbers p and q satisfying $1/p+1/q = 1$ are called conjugate exponents.

We need a bit of notation for the proof of Proposition 3.5.5. Recall that if z is a complex number, say $z = x+iy$, then $\bar{z}$ (the *complex conjugate* of z) and $\mathrm{sgn}(z)$ are defined by $\bar{z} = x - iy$ and

$$\mathrm{sgn}(z) = \begin{cases} \frac{z}{|z|} & \text{if } z \neq 0 \\ 0 & \text{if } z = 0. \end{cases}$$

It is easy to check that $z\bar{z} = |z|^2$ and $z\,\overline{\mathrm{sgn}(z)} = |z|$ hold for each z and that $|\mathrm{sgn}(z)| = 1$ holds for each nonzero z. If f is a complex-valued function on a set S, then $\bar{f}$ and $\mathrm{sgn}(f)$ are the functions whose values at the point s are $\overline{f(s)}$ and $\mathrm{sgn}(f(s))$.

In the following proof we will assume that the functions involved are complex-valued. The details are essentially the same for real-valued functions (then $\bar{z} = z$ and $\mathrm{sgn}(z)$ is 1, 0, or -1).

Proof of Proposition 3.5.5. It is clear that if g_1 and g_2 are equal almost everywhere (or, in case $q = +\infty$, locally almost everywhere), then $T_{g_1} = T_{g_2}$. Thus T_g depends on g only through the equivalence class $\langle g \rangle$ to which g belongs, and we can define a map, again called T, from $L^q(X, \mathscr{A}, \mu)$ to $(L^p(X, \mathscr{A}, \mu))^*$ by means of the formula $T_{\langle g \rangle} = T_g$. It is clear that T is linear. Since we have already seen that $\|T_g\| \leq \|g\|_q$ holds for each g in $\mathscr{L}^q(X, \mathscr{A}, \mu)$, we need only verify the reverse inequality. Let us consider two cases.

First suppose that $p = 1$ and hence that $q = +\infty$. Let g be an element of $\mathscr{L}^\infty(X, \mathscr{A}, \mu)$ such that $\|g\|_\infty \neq 0$, and let ε be a positive number. Since $\{x \in X : |g(x)| > \|g\|_\infty - \varepsilon\}$ is not locally μ-null,[12] there is a set A that belongs to $\mathscr{A}$, has finite measure under μ, and is such that the set B defined by

$$B = A \cap \{x \in X : |g(x)| > \|g\|_\infty - \varepsilon\}$$

has nonzero measure. Let $f = \overline{\operatorname{sgn}(g)}\chi_B$. Then $f \in \mathscr{L}^1(X, \mathscr{A}, \mu)$,

$$\|f\|_1 = \int |\overline{\operatorname{sgn}(g)}\chi_B|\, d\mu \leq \int \chi_B\, d\mu = \mu(B),$$

and

$$T_g(f) = \int g\,\overline{\operatorname{sgn}(g)}\chi_B\, d\mu = \int |g|\chi_B\, d\mu \geq (\|g\|_\infty - \varepsilon)\mu(B).$$

It is clear that $|T_g(f)| = T_g(f)$, and so the preceding calculations, together with the inequality $|T_g(f)| \leq \|T_g\|\|f\|_1$, imply that $\|g\|_\infty - \varepsilon \leq \|T_g\|$. Since ε can be made arbitrarily close to 0, it follows that $\|g\|_\infty \leq \|T_g\|$. Thus $\|T_g\| = \|g\|_\infty$.

Now suppose that $1 < p < +\infty$ and hence that $1 < q < +\infty$. Let g belong to $\mathscr{L}^q(X, \mathscr{A}, \mu)$, and define a function f by $f = \overline{\operatorname{sgn}(g)}|g|^{q-1}$. The relation $q = p(q-1)$ implies that $|f|^p = |g|^q$; thus f belongs to $\mathscr{L}^p(X, \mathscr{A}, \mu)$ and satisfies $\|f\|_p = (\int |g|^q\, d\mu)^{1/p}$. Furthermore

$$T_g(f) = \int \overline{\operatorname{sgn}(g)}|g|^{q-1}g\, d\mu = \int |g|^q\, d\mu.$$

Consequently it follows from the relation $|T_g(f)| \leq \|T_g\|\|f\|_p$ that

$$\int |g|^q\, d\mu \leq \|T_g\| \left(\int |g|^q\, d\mu \right)^{1/p} \tag{3}$$

and hence that $\|g\|_q \leq \|T_g\|$ (this is clear if $\|g\|_q = 0$; otherwise divide both sides of (3) by $(\int |g|^q\, d\mu)^{1/p}$ and recall that $1 - 1/p = 1/q$.) Thus $\|T_g\| = \|g\|_q$, and the proof is complete. □

---

[12] We are here assuming that the space X is not locally null. If X is locally null, then $L^1(X, \mathscr{A}, \mu)$ and $L^\infty(X, \mathscr{A}, \mu)$ contain only 0, and the proposition is true (but uninteresting).

Exercises

1. Let V_1, V_2, and V_3 be normed linear spaces, and let $S\colon V_1 \to V_2$ and $T\colon V_2 \to V_3$ be bounded linear operators. Show that $T \circ S\colon V_1 \to V_3$ is bounded and that $\|T \circ S\| \le \|T\|\|S\|$.

2. Suppose that V_1 and V_2 are normed linear spaces and that $T\colon V_1 \to V_2$ is an invertible linear operator such that T and T^{-1} are both bounded.
   (a) Show that $1 \le \|T\|\|T^{-1}\|$. (Hint: See Exercise 1.)
   (b) Show by example that equality need not hold in part (a).

3. Let V_1 and V_2 be normed linear spaces, and let $T\colon V_1 \to V_2$ be a linear operator. Show that the subset $T(V_1)$ of V_2 is bounded if and only if T is the zero operator. Thus to say that a linear operator is bounded is *not* to say that its values form a bounded set.

4. Let V_1 and V_2 be normed linear spaces, and let $T\colon V_1 \to V_2$ be a linear operator.
   (a) Show that T is bounded if and only if the set

   $$\{\|T(v)\| : v \in V_1 \text{ and } \|v\| \le 1\}$$

   is bounded above.
   (b) Show that if T is bounded, then

   $$\|T\| = \sup\{\|T(v)\| : v \in V_1 \text{ and } \|v\| \le 1\}.$$

5. Suppose that V_1 and V_2 are normed linear spaces and that $T\colon V_1 \to V_2$ is a linear operator. Show that if T is bounded, then T is uniformly continuous.

6. Let V be a normed linear space. Show that the dual V^* of V is complete under the norm $\|\cdot\|$ defined above. (Hint: Let $\{F_n\}$ be a Cauchy sequence in V^*. Show that for each v in V the sequence $\{F_n(v)\}$ is a Cauchy sequence in $\mathbb{R}$ (or in $\mathbb{C}$) and so is convergent. Then show that the formula $F(v) = \lim_n F_n(v)$ defines a bounded linear functional on V and that $\lim_n \|F_n - F\| = 0$.)

7. Let V be an inner product space, and for each y in V define $F_y\colon V \to \mathbb{R}$ by $F_y(x) = (x, y)$.
   (a) Show that F_y belongs to V^* and satisfies $\|F_y\| = \|y\|$. (Hint: Use the Cauchy–Schwarz inequality; see Exercise 3.2.7. To check that $\|F_y\|$ is equal to (rather than less than) $\|y\|$, consider $F_y(y)$.)
   (b) Show that if $y \ne y'$, then $F_y \ne F_{y'}$.
   (c) Show that if the inner product space V is a Hilbert space and if F belongs to V^*, then there is an element y of V such that $F = F_y$. (Hint: Let $y = 0$ if $F = 0$. Otherwise choose a nonzero element v of V such that $(u, v) = 0$ holds whenever $F(u) = 0$ (see Exercise 3.2.12), and check that a suitable multiple of v works.)

8. (This exercise depends on the Hahn–Banach theorem, which is stated without proof in Appendix E.) Let V be the subspace of ℓ^∞ consisting of those sequences $\{x_n\}$ for which $\lim_n x_n$ exists, and let $F_0\colon V \to \mathbb{R}$ be defined by $F_0(\{x_n\}) = \lim_n x_n$.

(a) Show that F_0 is a bounded linear functional on V and that $\|F_0\| = 1$.
(b) Let F be a bounded linear functional on ℓ^∞ that satisfies $\|F\| = 1$ and agrees with F_0 on V (see Theorem E.7). Show that if $\{x_n\}$ is a nonnegative element of ℓ^∞ (that is, if $\{x_n\}$ belongs to ℓ^∞ and satisfies $x_n \geq 0$ for each n), then $F(\{x_n\}) \geq 0$. (Hint: Consider the sequence $\{x_n'\}$ defined by $x_n' = x_n - c$, where c is a suitably chosen constant.)
(c) For each subset A of $\mathbb{N}$ let $\{\chi_{A,n}\}_{n=1}^\infty$ be the sequence defined by

$$\chi_{A,n} = \begin{cases} 1 & \text{if } n \in A, \\ 0 & \text{if } n \notin A. \end{cases}$$

Show that the function $\mu \colon \mathscr{P}(\mathbb{N}) \to \mathbb{R}$ defined by $\mu(A) = F(\{\chi_{A,n}\})$ is a finitely additive measure, but is not countably additive.

Notes

Kolmogorov and Fomin [73] and Simmons [109] are useful elementary sources of information on metric spaces and normed linear spaces. The basic properties of the L^p-spaces can be found in virtually every book on integration theory.

Chapter 4
Signed and Complex Measures

In this chapter we study signed and complex measures, which are defined to be the countably additive functions from a σ-algebra to $[-\infty, +\infty]$ or to $\mathbb{C}$ that have value 0 on the empty set. We begin in Sect. 4.1 with some basic definitions and facts. Section 4.2 is devoted to the main result of this chapter, the Radon–Nikodym theorem. Let μ be a σ-finite positive measure on a measurable space $(X, \mathscr{A})$. The Radon–Nikodym theorem characterizes those finite positive, signed, or complex measures ν whose values can be computed by integrating some μ-integrable function—in other words, it characterizes those ν for which there is a μ-integrable f such that $\nu(A) = \int_A f \, d\mu$ holds for all A in $\mathscr{A}$. The last part of the chapter is devoted to the relation of the material in the early parts of the chapter to the classical concepts of bounded variation and absolute continuity (Sect. 4.4) and to the use of the Radon–Nikodym theorem to compute the dual spaces of a number of the L^p spaces (Sect. 4.5).

4.1 Signed and Complex Measures

Let $(X, \mathscr{A})$ be a measurable space, and let μ be a function on $\mathscr{A}$ with values in $[-\infty, +\infty]$. The function μ is *finitely additive* if the identity

$$\mu\left(\bigcup_{i=1}^{n} A_i\right) = \sum_{i=1}^{n} \mu(A_i)$$

holds for each finite sequence $\{A_i\}_{i=1}^{n}$ of disjoint sets in $\mathscr{A}$ and is *countably additive* if the identity

$$\mu\left(\bigcup_{i=1}^{\infty} A_i\right) = \sum_{i=1}^{\infty} \mu(A_i)$$

D.L. Cohn, *Measure Theory: Second Edition*, Birkhäuser Advanced Texts Basler Lehrbücher, DOI 10.1007/978-1-4614-6956-8_4,
© Springer Science+Business Media, LLC 2013

holds for each infinite sequence $\{A_i\}$ of disjoint sets in $\mathscr{A}$. If μ is countably additive and satisfies $\mu(\varnothing) = 0$, then it is a *signed measure*. Thus signed measures are the functions that result if in the definition of measures the requirement of nonnegativity is removed. This section is devoted to signed measures and complex measures (to be defined below) and to their relationship to measures.[1]

A signed measure is *finite* if neither $+\infty$ nor $-\infty$ occurs among its values.

Suppose that μ is a signed measure on the measurable space $(X, \mathscr{A})$. Then for each A in $\mathscr{A}$ the sum $\mu(A) + \mu(A^c)$ must be defined (that is, must not be of the form $(+\infty) + (-\infty)$ or $(-\infty) + (+\infty)$) and must equal $\mu(X)$. Hence if there is a set A in $\mathscr{A}$ for which $\mu(A) = +\infty$, then $\mu(X) = +\infty$, and if there is a set A in $\mathscr{A}$ for which $\mu(A) = -\infty$, then $\mu(X) = -\infty$. Consequently a signed measure can attain at most one of the values $+\infty$ and $-\infty$. A similar argument shows that if B is a set in $\mathscr{A}$ for which $\mu(B)$ is finite, then $\mu(A)$ is finite for each $\mathscr{A}$-measurable subset A of B.

Examples 4.1.1.

(a) Let $(X, \mathscr{A}, \mu)$ be a measure space, let f belong to $\mathscr{L}^1(X, \mathscr{A}, \mu, \mathbb{R})$, and define a function ν on $\mathscr{A}$ by $\nu(A) = \int_A f \, d\mu$. Then the linearity of the integral and the dominated convergence theorem imply that ν is a signed measure on $(X, \mathscr{A})$. Note that such a signed measure is the difference of the positive measures ν_1 and ν_2 defined by $\nu_1(A) = \int_A f^+ \, d\mu$ and $\nu_2(A) = \int_A f^- \, d\mu$.

(b) More generally, if ν_1 and ν_2 are positive measures on the measurable space $(X, \mathscr{A})$ and if at least one of them is finite, then $\nu_1 - \nu_2$ is a signed measure on $(X, \mathscr{A})$. We will soon see that every signed measure arises in this way. □

Lemma 4.1.2. *Let $(X, \mathscr{A})$ be a measurable space, and let μ be a signed measure on $(X, \mathscr{A})$. If $\{A_k\}$ is an increasing sequence of sets in $\mathscr{A}$, then*

$$\mu\left(\bigcup_{k=1}^{\infty} A_k\right) = \lim_k \mu(A_k),$$

and if $\{A_k\}$ is a decreasing sequence of sets in $\mathscr{A}$ such that $\mu(A_n)$ is finite for some n, then

$$\mu\left(\bigcap_{k=1}^{\infty} A_k\right) = \lim_k \mu(A_k).$$

Lemma 4.1.3. *Suppose that $(X, \mathscr{A})$ is a measurable space and that μ is an extended real-valued function on $\mathscr{A}$ that is finitely additive and satisfies $\mu(\varnothing) = 0$.*

---

[1] We will try not to abbreviate the phrases "signed measure" and "complex measure" with the word "measure"; thus the word "measure" by itself will continue to mean a nonnegative countably additive function whose value at $\varnothing$ is 0. However, for clarity and emphasis, we will sometimes refer to a measure as a positive measure.

If $\mu(\cup_{k=1}^{\infty} A_k) = \lim_k \mu(A_k)$ *holds for each increasing sequence* $\{A_k\}$ *of sets in* $\mathscr{A}$ *or if* $\lim_k \mu(A_k) = 0$ *holds for each decreasing sequence* $\{A_k\}$ *of sets in* $\mathscr{A}$ *for which* $\cap_{k=1}^{\infty} A_k = \varnothing$, *then* μ *is a signed measure.*

The proofs of these lemmas are very similar to those of Propositions 1.2.5 and 1.2.6 and so are omitted.

Let μ be a signed measure on the measurable space $(X, \mathscr{A})$. A subset A of X is a *positive set* for μ if $A \in \mathscr{A}$ and each $\mathscr{A}$-measurable subset E of A satisfies $\mu(E) \geq 0$. Likewise A is a *negative set* for μ if $A \in \mathscr{A}$ and each $\mathscr{A}$-measurable subset E of A satisfies $\mu(E) \leq 0$.

The role of positive and negative sets is explained by Theorem 4.1.5 and Corollary 4.1.6 below. For the proofs of these results, we will need the following construction.

Lemma 4.1.4. *Let* μ *be a signed measure on the measurable space* $(X, \mathscr{A})$, *and let* A *be a subset of* X *that belongs to* $\mathscr{A}$ *and satisfies* $-\infty < \mu(A) < 0$. *Then there is a negative set* B *that is included in* A *and satisfies*

$$\mu(B) \leq \mu(A). \tag{1}$$

Proof. We will remove a suitable sequence of subsets from A and then let B consist of the points of A that remain. To begin, let

$$\delta_1 = \sup\{\mu(E) : E \in \mathscr{A} \text{ and } E \subseteq A\}, \tag{2}$$

and choose an $\mathscr{A}$-measurable subset A_1 of A that satisfies[2]

$$\mu(A_1) \geq \min\left(\frac{1}{2}\delta_1, 1\right).$$

Then δ_1 and $\mu(A_1)$ are nonnegative (note that (2) implies that $\delta_1 \geq \mu(\varnothing) = 0$). We proceed by induction, constructing sequences $\{\delta_n\}$ and $\{A_n\}$ by letting

$$\delta_n = \sup\left\{\mu(E) : E \in \mathscr{A} \text{ and } E \subseteq \left(A - \bigcup_{i=1}^{n-1} A_i\right)\right\},$$

and then choosing an $\mathscr{A}$-measurable subset A_n of $A - \bigcup_{i=1}^{n-1} A_i$ that satisfies

$$\mu(A_n) \geq \min\left(\frac{1}{2}\delta_n, 1\right).$$

Now define A_∞ and B by $A_\infty = \cup_{n=1}^{\infty} A_n$ and $B = A - A_\infty$.

---

[2] We require that $\mu(A_1)$ be at least $\min(\delta_1/2, 1)$, rather than at least $\delta_1/2$, because we have not yet proved that δ_1 is finite (see Exercise 4).

Let us check that B has the required properties. Since the sets A_n are disjoint and satisfy $\mu(A_n) \geq 0$, it follows that $\mu(A_\infty) \geq 0$ and hence that

$$\mu(A) = \mu(A_\infty) + \mu(B) \geq \mu(B).$$

Thus B satisfies (1).

We turn to the negativity of B. The finiteness of $\mu(A)$ implies the finiteness of $\mu(A_\infty)$ and so, since $\mu(A_\infty) = \sum_n \mu(A_n)$, implies that $\lim_n \mu(A_n) = 0$. Consequently $\lim_n \delta_n = 0$. Since an arbitrary $\mathscr{A}$-measurable subset E of B satisfies $\mu(E) \leq \delta_n$ for each n and so satisfies $\mu(E) \leq 0$, B must be a negative set for μ. $\square$

The following theorem and its corollary give the standard decomposition of signed measures.

Theorem 4.1.5 (Hahn Decomposition Theorem). *Let $(X, \mathscr{A})$ be a measurable space, and let μ be a signed measure on $(X, \mathscr{A})$. Then there are disjoint subsets P and N of X such that P is a positive set for μ, N is a negative set for μ, and $X = P \cup N$.*

Proof. Since the signed measure μ cannot include both $+\infty$ and $-\infty$ among its values, we can for definiteness assume that $-\infty$ is not included. Let

$$L = \inf\{\mu(A) : A \text{ is a negative set for } \mu\} \tag{3}$$

(the set on the right side of (3) is nonempty, since $\varnothing$ is a negative set for μ). Choose a sequence $\{A_n\}$ of negative sets for μ for which $L = \lim_n \mu(A_n)$, and let $N = \cup_{n=1}^{\infty} A_n$. It is easy to check that N is a negative set for μ (each $\mathscr{A}$-measurable subset of N is the union of a sequence of disjoint $\mathscr{A}$-measurable sets, each of which is included in some A_n). Hence $L \leq \mu(N) \leq \mu(A_n)$ holds for each n, and so $L = \mu(N)$. Furthermore, since μ does not attain the value $-\infty$, $\mu(N)$ must be finite.

Let $P = N^c$. Our only remaining task is to check that P is a positive set for μ. If P included an $\mathscr{A}$-measurable set A such that $\mu(A) < 0$, then it would include a negative set B such that $\mu(B) < 0$ (Lemma 4.1.4), and $N \cup B$ would be a negative set such that

$$\mu(N \cup B) = \mu(N) + \mu(B) < \mu(N) = L$$

(recall that $\mu(N)$ is finite). However this contradicts (3), and so P must be a positive set for μ. $\square$

A *Hahn decomposition* of a signed measure μ is a pair (P, N) of disjoint subsets of X such that P is a positive set for μ, N is a negative set for μ, and $X = P \cup N$. Note that a signed measure can have several Hahn decompositions. For example, if X is the interval $[-1, 1]$, if $\mathscr{A}$ is the σ-algebra of Borel subsets of $[-1, 1]$, and if μ is defined by $\mu(A) = \int_A x\lambda(dx)$, then $([0, 1], [-1, 0))$ and $((0, 1], [-1, 0])$ are both Hahn decompositions of μ. On the other hand, if μ is an arbitrary signed measure on a measurable space $(X, \mathscr{A})$ and if (P_1, N_1) and (P_2, N_2) are Hahn decompositions of μ, then $P_1 \cap N_2$ is both a positive set and a negative set for μ, and so each $\mathscr{A}$-measurable subset of $P_1 \cap N_2$ has measure zero under μ. Likewise, each $\mathscr{A}$-measurable subset of $P_2 \cap N_1$ has measure zero under μ. Thus the Hahn decomposition of μ is essentially unique.

Corollary 4.1.6 (Jordan Decomposition Theorem). *Every signed measure is the difference of two positive measures, at least one of which is finite.*

Proof. Let μ be a signed measure on $(X, \mathscr{A})$. Choose a Hahn decomposition (P, N) for μ (see Theorem 4.1.5), and then define functions μ^+ and μ^- on $\mathscr{A}$ by

$$\mu^+(A) = \mu(A \cap P)$$

and

$$\mu^-(A) = -\mu(A \cap N).$$

It is clear that μ^+ and μ^- are positive measures such that $\mu = \mu^+ - \mu^-$. Since $+\infty$ and $-\infty$ cannot both occur among the values of μ, at least one of the values $\mu(P)$ and $\mu(N)$, and hence at least one of the measures μ^+ and μ^-, must be finite. □

Let (P, N) be a Hahn decomposition of the signed measure μ, let μ^+ and μ^- be the measures constructed from (P, N) in the proof of Corollary 4.1.6, and suppose that A belongs to $\mathscr{A}$. Then each $\mathscr{A}$-measurable subset B of A satisfies

$$\mu(B) = \mu^+(B) - \mu^-(B) \le \mu^+(B) \le \mu^+(A).$$

Since in addition $\mu^+(A) = \mu(A \cap P)$, it follows that

$$\mu^+(A) = \sup\{\mu(B) : B \in \mathscr{A} \text{ and } B \subseteq A\}.$$

Likewise the measure μ^- satisfies

$$\mu^-(A) = \sup\{-\mu(B) : B \in \mathscr{A} \text{ and } B \subseteq A\}.$$

Thus μ^+ and μ^- do not depend on the particular Hahn decomposition used in their construction. The measures μ^+ and μ^- are called the *positive part* and the *negative part* of μ, and the representation $\mu = \mu^+ - \mu^-$ is called the *Jordan decomposition* of μ.

The *variation* of the signed measure μ is the positive measure $|\mu|$ defined by $|\mu| = \mu^+ + \mu^-$. It is easy to check that

$$|\mu(A)| \le |\mu|(A)$$

holds for each A in $\mathscr{A}$ and in fact that $|\mu|$ is the smallest of those positive measures ν that satisfy $|\mu(A)| \le \nu(A)$ for each A in $\mathscr{A}$ (see Exercise 5). The *total variation* $\|\mu\|$ of the signed measure μ is defined by $\|\mu\| = |\mu|(X)$.

Let $(X, \mathscr{A})$ be a measurable space. A *complex measure* on $(X, \mathscr{A})$ is a function μ from $\mathscr{A}$ to $\mathbb{C}$ that satisfies $\mu(\varnothing) = 0$ and is *countably additive*, in the sense that

$$\mu\left(\bigcup_{n=1}^{\infty} A_n\right) = \sum_{n=1}^{\infty} \mu(A_n)$$

holds for each infinite sequence $\{A_n\}$ of disjoint sets in $\mathscr{A}$. Note that by definition a complex measure has only complex values and so has no infinite values.

Each complex measure μ on $(X, \mathscr{A})$ can of course be written in the form $\mu = \mu' + i\mu''$, where μ' and μ'' are finite signed measures on $(X, \mathscr{A})$. Hence the Jordan decomposition theorem implies that each complex measure μ can be written in the form

$$\mu = \mu_1 - \mu_2 + i\mu_3 - i\mu_4, \tag{4}$$

where μ_1, μ_2, μ_3, and μ_4 are finite positive measures on $(X, \mathscr{A})$. Such a representation is called the *Jordan decomposition* of μ if $\mu' = \mu_1 - \mu_2$ and $\mu'' = \mu_3 - \mu_4$ are the Jordan decompositions of the real and imaginary parts of μ.

We turn to the *variation* $|\mu|$ of the complex measure μ. For each A in $\mathscr{A}$ let $|\mu|(A)$ be the supremum of the numbers $\sum_{j=1}^{n} |\mu(A_j)|$, where $\{A_j\}_{j=1}^{n}$ ranges over all finite partitions of A into $\mathscr{A}$-measurable sets.

Proposition 4.1.7. *Let $(X, \mathscr{A})$ be a measurable space, and let μ be a complex measure on $(X, \mathscr{A})$. Then the variation $|\mu|$ of μ is a finite measure on $(X, \mathscr{A})$.*

Proof. The relation $|\mu|(\varnothing) = 0$ is immediate.

We can check the finite additivity of $|\mu|$ by showing that if B_1 and B_2 are disjoint sets that belong to $\mathscr{A}$, then $|\mu|(B_1 \cup B_2) = |\mu|(B_1) + |\mu|(B_2)$. For this, note that if $\{A_j\}_{j=1}^{n}$ is a finite partition of $B_1 \cup B_2$ into $\mathscr{A}$-measurable sets, then

$$\sum_{j} |\mu(A_j)| \leq \sum_{j} |\mu(A_j \cap B_1)| + \sum_{j} |\mu(A_j \cap B_2)|$$

$$\leq |\mu|(B_1) + |\mu|(B_2).$$

Since $|\mu|(B_1 \cup B_2)$ is the supremum of the numbers that can appear on the left side of the inequality, it follows that

$$|\mu|(B_1 \cup B_2) \leq |\mu|(B_1) + |\mu|(B_2).$$

A similar argument, based on partitioning B_1 and B_2, shows that

$$|\mu|(B_1) + |\mu|(B_2) \leq |\mu|(B_1 \cup B_2).$$

Thus $|\mu|(B_1 \cup B_2) = |\mu|(B_1) + |\mu|(B_2)$, and the finite additivity of $|\mu|$ is proved.

If $\mu = \mu_1 - \mu_2 + i\mu_3 - i\mu_4$ is the Jordan decomposition of μ, then

$$|\mu|(A) \leq \mu_1(A) + \mu_2(A) + \mu_3(A) + \mu_4(A) \tag{5}$$

holds for each A in $\mathscr{A}$. Since the measures μ_1, μ_2, μ_3, and μ_4 are finite, the finiteness of $|\mu|$ follows. Furthermore, if $\{A_n\}$ is a decreasing sequence of $\mathscr{A}$-measurable sets such that $\cap_n A_n = \varnothing$, then $\lim_n \mu_k(A_n) = 0$ holds for $k = 1$, ..., 4, and so (5) implies that $\lim_n |\mu|(A_n) = 0$. Thus $|\mu|$ is countably additive (Proposition 1.2.6). □

It is easy to check that if μ is a complex measure on $(X, \mathscr{A})$, then $|\mu|$ is the smallest of the positive measures ν that satisfy $|\mu(A)| \leq \nu(A)$ for all A in $\mathscr{A}$ (see Exercise 5). Note that if μ is a finite signed measure, then μ is also a complex measure; it is easy to check that in this case the variation of μ as a signed measure is the same as its variation as a complex measure (Exercise 6).

The *total variation* $\|\mu\|$ of the complex measure μ is defined by $\|\mu\| = |\mu|(X)$.

Suppose that $(X, \mathscr{A})$ is a measurable space. Let $M(X, \mathscr{A}, \mathbb{R})$ be the collection of all finite signed measures on $(X, \mathscr{A})$, and let $M(X, \mathscr{A}, \mathbb{C})$ be the collection of all complex measures on $(X, \mathscr{A})$. It is easy to check that $M(X, \mathscr{A}, \mathbb{R})$ and $M(X, \mathscr{A}, \mathbb{C})$ are vector spaces over $\mathbb{R}$ and $\mathbb{C}$, respectively, and that the total variation gives a norm on each of them.

Proposition 4.1.8. *Let $(X, \mathscr{A})$ be a measurable space. Then the spaces $M(X, \mathscr{A}, \mathbb{R})$ and $M(X, \mathscr{A}, \mathbb{C})$ are complete under the total variation norm.*

Proof. Let $\{\mu_n\}$ be a Cauchy sequence in $M(X, \mathscr{A}, \mathbb{R})$ or in $M(X, \mathscr{A}, \mathbb{C})$. The inequality $|\mu_m(A) - \mu_n(A)| \leq \|\mu_m - \mu_n\|$ implies that for each A in $\mathscr{A}$ the sequence $\{\mu_n(A)\}$ is a Cauchy sequence of real or complex numbers and hence is convergent. Define a real- or complex-valued function μ on $\mathscr{A}$ by letting $\mu(A) = \lim_n \mu_n(A)$ hold at each A in $\mathscr{A}$. We need to check that μ is a signed or complex measure and that $\lim_n \|\mu_n - \mu\| = 0$.

It is clear that $\mu(\varnothing) = 0$ and that μ is at least finitely additive.

As preparation for checking the countable additivity of μ, we will show that the convergence of $\mu_n(A)$ to $\mu(A)$ is uniform in A. If ε is a positive number and if N is a positive integer such that $\|\mu_m - \mu_n\| < \varepsilon$ holds whenever $m \geq N$ and $n \geq N$, then

$$|\mu_m(A) - \mu_n(A)| < \varepsilon \tag{6}$$

holds whenever $A \in \mathscr{A}$, $m \geq N$, and $n \geq N$, and so

$$|\mu(A) - \mu_n(A)| \leq \varepsilon$$

holds whenever $A \in \mathscr{A}$ and $n \geq N$ (let m approach infinity in (6)). Since ε is arbitrary, the uniformity of the convergence of $\mu_n(A)$ to $\mu(A)$ follows.

We now use Lemmas 4.1.2 and 4.1.3 (and their extensions to complex measures) to prove the countable additivity of μ. Let $\{A_k\}$ be a decreasing sequence of sets in $\mathscr{A}$ such that $\cap_k A_k = \varnothing$, and let ε be a positive number. Use the uniformity of the convergence of $\mu_n(A)$ to $\mu(A)$ to choose N so that $|\mu(A) - \mu_n(A)| < \varepsilon/2$ holds whenever $A \in \mathscr{A}$ and $n \geq N$, and then use Lemma 4.1.2 to choose K such that $|\mu_N(A_k)| < \varepsilon/2$ holds whenever $k \geq K$. It follows that if $k \geq K$ then

$$|\mu(A_k)| \leq |\mu(A_k) - \mu_N(A_k)| + |\mu_N(A_k)| < \frac{\varepsilon}{2} + \frac{\varepsilon}{2} = \varepsilon.$$

Thus $\lim_k \mu(A_k) = 0$, and the countable additivity of μ follows.

We turn to the relation $\lim_n \|\mu - \mu_n\| = 0$. Let ε be a positive number, and use the fact that $\{\mu_n\}$ is a Cauchy sequence to choose N so that $\|\mu_m - \mu_n\| < \varepsilon$ holds whenever $m \geq N$ and $n \geq N$. Note that if $m \geq N$ and $n \geq N$, then each partition of X into $\mathscr{A}$-measurable sets A_j, $j = 1, \ldots, k$, satisfies

$$\sum_{j=1}^{k} |\mu_m(A_j) - \mu_n(A_j)| \leq \|\mu_m - \mu_n\| < \varepsilon,$$

and hence satisfies

$$\sum_{j=1}^{k} |\mu(A_j) - \mu_n(A_j)| = \lim_m \sum_{j=1}^{k} |\mu_m(A_j) - \mu_n(A_j)| \leq \varepsilon.$$

Since $\|\mu - \mu_n\|$ is the supremum of the numbers that can appear on the left side of this inequality, it follows that $\|\mu - \mu_n\| \leq \varepsilon$ holds whenever $n \geq N$. Consequently $\lim_n \|\mu - \mu_n\| = 0$. Thus $M(X, \mathscr{A}, \mathbb{R})$ and $M(X, \mathscr{A}, \mathbb{C})$ are complete. $\square$

Let us deal briefly with integration with respect to a finite signed or complex measure.

Suppose that $(X, \mathscr{A})$ is a measurable space. We will denote by $B(X, \mathscr{A}, \mathbb{R})$ the vector space of bounded real-valued $\mathscr{A}$-measurable functions on X and by $B(X, \mathscr{A}, \mathbb{C})$ the vector space of bounded complex-valued $\mathscr{A}$-measurable functions on X. If μ is a finite signed measure on $(X, \mathscr{A})$, if $\mu = \mu^+ - \mu^-$ is the Jordan decomposition of μ, and if f belongs to $B(X, \mathscr{A}, \mathbb{R})$, then the *integral* of f with respect to μ is defined by

$$\int f \, d\mu = \int f \, d\mu^+ - \int f \, d\mu^-.$$

It is clear that $f \mapsto \int f \, d\mu$ defines a linear functional on $B(X, \mathscr{A}, \mathbb{R})$.

If $A \in \mathscr{A}$, then $\int \chi_A \, d\mu = \mu(A)$ holds for each μ in $M(X, \mathscr{A}, \mathbb{R})$. Thus the formula

$$\mu \mapsto \int f \, d\mu$$

defines a linear functional on $M(X, \mathscr{A}, \mathbb{R})$ if f is an $\mathscr{A}$-measurable characteristic function and hence if f is an arbitrary function in $B(X, \mathscr{A}, \mathbb{R})$ (use the linearity of the integral and the dominated convergence theorem).

Similarly, if μ is a complex measure on $(X, \mathscr{A})$, then we can use the Jordan decomposition of μ to define the integral with respect to μ of a function in $B(X, \mathscr{A}, \mathbb{C})$. The expressions $f \mapsto \int f \, d\mu$ and $\mu \mapsto \int f \, d\mu$ define linear functionals on $B(X, \mathscr{A}, \mathbb{C})$ and on $M(X, \mathscr{A}, \mathbb{C})$, respectively.

Now use the formula

$$\|f\|_\infty = \sup\{|f(x)| : x \in X\}$$

to define norms on $B(X, \mathscr{A}, \mathbb{R})$ and $B(X, \mathscr{A}, \mathbb{C})$ (see Example 3.2.1(f)). If μ is a finite signed or complex measure on $(X, \mathscr{A})$ and if f is a simple $\mathscr{A}$-measurable function on X, say with values $a_1, \ldots, a_k$, attained on the sets $A_1, \ldots, A_k$, then

$$\left| \int f \, d\mu \right| = \left| \sum_{j=1}^{k} a_j \mu(A_j) \right| \le \sum_{j=1}^{k} |a_j| |\mu(A_j)| \le \sum_{j=1}^{k} \|f\|_\infty |\mu(A_j)|,$$

and so

$$\left| \int f \, d\mu \right| \le \|f\|_\infty \|\mu\|. \tag{7}$$

Since each function in $B(X, \mathscr{A}, \mathbb{R})$ or in $B(X, \mathscr{A}, \mathbb{C})$ is the uniform limit of a sequence of simple $\mathscr{A}$-measurable functions, it follows that (7) holds whenever f belongs to $B(X, \mathscr{A}, \mathbb{R})$ or $B(X, \mathscr{A}, \mathbb{C})$.

Exercises

1. Let μ be a signed or complex measure on $(X, \mathscr{A})$, and let A belong to $\mathscr{A}$.

   (a) Show that $|\mu|(A) = 0$ holds if and only if each $\mathscr{A}$-measurable subset B of A satisfies $\mu(B) = 0$.
   (b) Show that in general the relation $\mu(A) = 0$ does not imply the relation $|\mu|(A) = 0$.

2. Let μ be a signed measure on $(X, \mathscr{A})$, and let ν_1 and ν_2 be positive measures on $(X, \mathscr{A})$ such that $\mu = \nu_1 - \nu_2$. Show that $\nu_1(A) \ge \mu^+(A)$ and $\nu_2(A) \ge \mu^-(A)$ hold for each A in $\mathscr{A}$.

3. Let μ_1 and μ_2 be finite signed measures on the measurable space $(X, \mathscr{A})$. Define signed measures $\mu_1 \vee \mu_2$ and $\mu_1 \wedge \mu_2$ on $(X, \mathscr{A})$ by $\mu_1 \vee \mu_2 = \mu_1 + (\mu_2 - \mu_1)^+$ and $\mu_1 \wedge \mu_2 = \mu_1 - (\mu_1 - \mu_2)^+$.

   (a) Show that $\mu_1 \vee \mu_2$ is the smallest of those finite signed measures ν that satisfy $\nu(A) \ge \mu_1(A)$ and $\nu(A) \ge \mu_2(A)$ for all A in $\mathscr{A}$.
   (b) Find and prove an analogous characterization of $\mu_1 \wedge \mu_2$.

4. Show that the quantities $\delta_1, \delta_2, \ldots$ defined in the proof of Lemma 4.1.4 are finite. (Hint: Use Theorem 4.1.5; this is legitimate, since Lemma 4.1.4 and Theorem 4.1.5 were proved without using the finiteness of the δ_n's.)

5. Let μ be a signed or complex measure on $(X, \mathscr{A})$, and let ν be a positive measure on $(X, \mathscr{A})$ such that $|\mu(A)| \leq \nu(A)$ holds for each A in $\mathscr{A}$. Show that $|\mu|(A) \leq \nu(A)$ holds for each A in $\mathscr{A}$.

6. Note that if μ is a finite signed measure, then μ is both a signed measure and a complex measure. Show that in this case the two definitions of $|\mu|$ yield the same result.

7. Let μ_1 and μ_2 be finite signed measures, and let ν be the complex measure defined by $\nu = \mu_1 + i\mu_2$. Show that $|\mu_1| \leq |\nu|$, $|\mu_2| \leq |\nu|$ and $|\nu| \leq |\mu_1| + |\mu_2|$. Is it necessarily true that $\|\nu\| \leq \sqrt{\|\mu_1\|^2 + \|\mu_2\|^2}$?

8. Let μ and $\mu_1, \mu_2, \ldots$ be finite signed or complex measures on $(X, \mathscr{A})$. Show that $\lim_n \|\mu_n - \mu\| = 0$ holds if and only if $\mu_n(A)$ converges to $\mu(A)$ uniformly in A as n approaches infinity.

9. Use Proposition 3.2.5, Exercise 1.2.6, and the Jordan decomposition to give another proof of Proposition 4.1.8.

10. Check that the spaces $B(X, \mathscr{A}, \mathbb{R})$ and $B(X, \mathscr{A}, \mathbb{C})$ are complete under the norm $\| \cdot \|_\infty$.

11. Let μ be a finite signed or complex measure on $(X, \mathscr{A})$, and let $\{f_n\}$ be a uniformly bounded sequence of real- or complex-valued $\mathscr{A}$-measurable functions on X (thus there is a positive number B such that $|f_n(x)| \leq B$ holds for each x and n). Show that if $f(x) = \lim_n f_n(x)$ holds at each x in X, then $\int f \, d\mu = \lim_n \int f_n \, d\mu$.

4.2 Absolute Continuity

Let $(X, \mathscr{A})$ be a measurable space, and let μ and ν be positive measures on $(X, \mathscr{A})$. Then ν is *absolutely continuous with respect to* μ if each set A that belongs to $\mathscr{A}$ and satisfies $\mu(A) = 0$ also satisfies $\nu(A) = 0$. One sometimes writes $\nu \ll \mu$ to indicate that ν is absolutely continuous with respect to μ. A measure on $(\mathbb{R}^d, \mathscr{B}(\mathbb{R}^d))$ is simply called *absolutely continuous* if it is absolutely continuous with respect to d-dimensional Lebesgue measure.

Suppose that $(X, \mathscr{A}, \mu)$ is a measure space and that f is a nonnegative function in $\mathscr{L}^1(X, \mathscr{A}, \mu, \mathbb{R})$. We have seen (in Sect. 2.4) that the formula $\nu(A) = \int_A f \, d\mu$ defines a finite positive measure ν on $\mathscr{A}$. If $\mu(A) = 0$, then $f\chi_A$ vanishes μ-almost everywhere, and so $\nu(A) = 0$. Thus ν is absolutely continuous with respect to μ. We will see that if μ is σ-finite, then every finite measure on $(X, \mathscr{A})$ that is absolutely continuous with respect to μ arises in this way.

The following lemma characterizes those finite positive measures that are absolutely continuous with respect to an arbitrary positive measure; this characterization is useful in the classical theory of functions of a real variable (see Sect. 4.4).

Lemma 4.2.1. *Let $(X, \mathscr{A})$ be a measurable space, let μ be a positive measure on $(X, \mathscr{A})$, and let ν be a finite positive measure on $(X, \mathscr{A})$. Then $\nu \ll \mu$ if and only if for each positive ε there is a positive δ such that each $\mathscr{A}$-measurable set A that satisfies $\mu(A) < \delta$ also satisfies $\nu(A) < \varepsilon$.*

Proof. First suppose that for each positive ε there is a corresponding δ. Let A be an $\mathscr{A}$-measurable set that satisfies $\mu(A) = 0$. Then $\mu(A) < \delta$ holds for each δ, and so $\nu(A) < \varepsilon$ holds for each ε; hence A satisfies $\nu(A) = 0$. Thus ν is absolutely continuous with respect to μ.

Next suppose that there is a positive number ε (which we will hold fixed) for which there is no suitable δ. Then for each positive integer k we can (and do) choose an $\mathscr{A}$-measurable set A_k that satisfies $\mu(A_k) < 1/2^k$ and $\nu(A_k) \geq \varepsilon$. Then the inequalities $\mu(\cup_{k=n}^{\infty} A_k) \leq \sum_{k=n}^{\infty} \mu(A_k) < 1/2^{n-1}$ and $\nu(\cup_{k=n}^{\infty} A_k) \geq \nu(A_n) \geq \varepsilon$ hold for each n, and so the set A defined by $A = \cap_{n=1}^{\infty} \cup_{k=n}^{\infty} A_k$ satisfies $\mu(A) = 0$ and $\nu(A) \geq \varepsilon$ (see Proposition 1.2.5). Thus A satisfies $\mu(A) = 0$ but not $\nu(A) = 0$, and so ν is not absolutely continuous with respect to μ. $\square$

We turn to the main result of this section.

Theorem 4.2.2 (Radon–Nikodym Theorem). *Let $(X, \mathscr{A})$ be a measurable space, and let μ and ν be σ-finite positive measures on $(X, \mathscr{A})$. If ν is absolutely continuous with respect to μ, then there is an $\mathscr{A}$-measurable function $g \colon X \to [0, +\infty)$ such that $\nu(A) = \int_A g \, d\mu$ holds for each A in $\mathscr{A}$. The function g is unique up to μ-almost everywhere equality.*

Proof. First consider the case where μ and ν are both finite. Let $\mathscr{F}$ be the set consisting of those $\mathscr{A}$-measurable functions $f \colon X \to [0, +\infty]$ that satisfy $\int_A f \, d\mu \leq \nu(A)$ for each A in $\mathscr{A}$. We will show first that $\mathscr{F}$ contains a function g such that

$$\int g \, d\mu = \sup\left\{ \int f \, d\mu : f \in \mathscr{F} \right\} \tag{1}$$

and then that this function g satisfies $\nu(A) = \int_A g \, d\mu$ for each A in $\mathscr{A}$. Finally, we will show that g can be modified so as to have only finite values.

We begin by checking that if f_1 and f_2 belong to $\mathscr{F}$, then $f_1 \vee f_2$ belongs to $\mathscr{F}$; to see this note that if A is an arbitrary set in $\mathscr{A}$, if $A_1 = \{x \in A : f_1(x) > f_2(x)\}$, and if $A_2 = \{x \in A : f_2(x) \geq f_1(x)\}$, then

$$\int_A (f_1 \vee f_2) \, d\mu = \int_{A_1} f_1 \, d\mu + \int_{A_2} f_2 \, d\mu \leq \nu(A_1) + \nu(A_2) = \nu(A).$$

Furthermore, $\mathscr{F}$ is not empty (the constant 0 belongs to it). Now choose a sequence $\{f_n\}$ of functions in $\mathscr{F}$ for which

$$\lim_n \int f_n \, d\mu = \sup\left\{ \int f \, d\mu : f \in \mathscr{F} \right\}.$$

By replacing f_n with $f_1 \vee \cdots \vee f_n$, we can assume that the sequence $\{f_n\}$ is increasing. Let $g = \lim_n f_n$. The monotone convergence theorem implies that the relation

$$\int_A g \, d\mu = \lim_n \int_A f_n \, d\mu \leq \nu(A)$$

holds for each A and hence that g belongs to $\mathscr{F}$. It also implies that $\int g\,d\mu = \sup\{\int f\,d\mu : f \in \mathscr{F}\}$. Thus g has the first of the properties claimed for it.

We turn to the proof that $\nu(A) = \int_A g\,d\mu$ holds for each A in $\mathscr{A}$. Since g belongs to $\mathscr{F}$, the formula $\nu_0(A) = \nu(A) - \int_A g\,d\mu$ defines a positive measure on $\mathscr{A}$. We need only show that $\nu_0 = 0$. Assume the contrary. Then, since μ is finite, there is a positive number ε such that

$$\nu_0(X) > \varepsilon\mu(X). \tag{2}$$

Let (P,N) be a Hahn decomposition (see Sect. 4.1) for the signed measure $\nu_0 - \varepsilon\mu$. Note that for each A in $\mathscr{A}$ we have $\nu_0(A \cap P) \geq \varepsilon\mu(A \cap P)$, and hence we have

$$\nu(A) = \int_A g\,d\mu + \nu_0(A) \geq \int_A g\,d\mu + \nu_0(A \cap P) \tag{3}$$

$$\geq \int_A g\,d\mu + \varepsilon\mu(A \cap P) = \int_A (g + \varepsilon\chi_P)\,d\mu.$$

Note also that $\mu(P) > 0$, since if $\mu(P) = 0$, then[3] $\nu_0(P) = 0$, and so

$$\nu_0(X) - \varepsilon\mu(X) = (\nu_0 - \varepsilon\mu)(N) \leq 0,$$

contradicting (2). It follows from this, the relation $\int g\,d\mu \leq \nu(X) < +\infty$, and (3) that $g + \varepsilon\chi_P$ belongs to $\mathscr{F}$ and satisfies $\int (g + \varepsilon\chi_P)\,d\mu > \int g\,d\mu$. This, however, contradicts (1) and so implies that $\nu_0 = 0$. Hence $\nu(A) = \int_A g\,d\mu$ holds for each A in $\mathscr{A}$. Since g can have an infinite value only on a μ-null set (Corollary 2.3.14), it can be redefined so as to have only finite values. With this we have constructed the required function in the case where μ and ν are finite.

Now suppose that μ and ν are σ-finite. Then X is the union of a sequence $\{B_n\}$ of disjoint $\mathscr{A}$-measurable sets, each of which has finite measure under μ and under ν. For each n the first part of this proof provides an $\mathscr{A}$-measurable function $g_n : B_n \to [0, +\infty)$ such that $\nu(A) = \int_A g_n\,d\mu$ holds for each $\mathscr{A}$-measurable subset A of B_n. The function $g : X \to [0, +\infty)$ that agrees on each B_n with g_n is then the required function.

We turn to the uniqueness of g. Let $g,h : X \to [0, +\infty)$ be $\mathscr{A}$-measurable functions that satisfy

$$\nu(A) = \int_A g\,d\mu = \int_A h\,d\mu$$

for each A in $\mathscr{A}$. First consider the case where ν is finite. Then $g - h$ is integrable and

$$\int_A (g - h)\,d\mu = 0$$

---

[3]This is where we use the absolute continuity of ν.

holds for each A in $\mathscr{A}$; since in this equation A can be the set where $g > h$ or the set where $g < h$, it follows that $\int (g - h)^+ d\mu = 0$ and $\int (g - h)^- d\mu = 0$ and hence that $(g - h)^+$ and $(g - h)^-$ vanish μ-almost everywhere (Corollary 2.3.12). Thus g and h agree μ-almost everywhere. If ν is σ-finite and if $\{B_n\}$ is a sequence of $\mathscr{A}$-measurable sets that have finite measure under ν and satisfy $X = \cup_n B_n$, then the preceding argument shows that g and h agree μ-almost everywhere on each B_n and hence μ-almost everywhere on X. $\qquad\square$

Example 4.2.3. The assumption that μ is σ-finite cannot simply be omitted from Theorem 4.2.2. To see that, let X be the interval $[0,1]$, let $\mathscr{A}$ be the σ-algebra of Borel subsets of $[0,1]$, let μ be counting measure on $(X, \mathscr{A})$, and let ν be Lebesgue measure on $(X, \mathscr{A})$. Then $\nu \ll \mu$, but there is no measurable function f such that $\nu(A) = \int_A f \, d\mu$ holds for all A. (Concerning the possibility of not requiring that ν be σ-finite, see Exercise 6.) $\qquad\square$

Now suppose that $(X, \mathscr{A})$ is a measurable space, that μ is a positive measure on $(X, \mathscr{A})$, and that ν is a signed or complex measure on $(X, \mathscr{A})$. Then ν is *absolutely continuous with respect to* μ, written $\nu \ll \mu$, if its variation $|\nu|$ is absolutely continuous with respect to μ. It is easy to check that a signed measure ν is absolutely continuous with respect to μ if and only if ν^+ and ν^- are absolutely continuous with respect to μ and that a complex measure ν is absolutely continuous with respect to μ if and only if the measures ν_1, ν_2, ν_3, and ν_4 appearing in its Jordan decomposition $\nu = \nu_1 - \nu_2 + i\nu_3 - i\nu_4$ are absolutely continuous with respect to μ. It is also easy to check that a signed or complex measure ν is absolutely continuous with respect to μ if and only if each A in $\mathscr{A}$ that satisfies $\mu(A) = 0$ also satisfies $\nu(A) = 0$ (be careful: $\nu(A) = 0$ is not equivalent to $|\nu|(A) = 0$; see Exercise 4.1.1).

The Radon–Nikodym theorem can be formulated for signed and complex measures as follows.

Theorem 4.2.4 (Radon–Nikodym Theorem). *Let $(X, \mathscr{A})$ be a measurable space, let μ be a σ-finite positive measure on $(X, \mathscr{A})$, and let ν be a finite signed or complex measure on $(X, \mathscr{A})$. If ν is absolutely continuous with respect to μ, then there is a function g that belongs to $\mathscr{L}^1(X, \mathscr{A}, \mu, \mathbb{R})$ or to $\mathscr{L}^1(X, \mathscr{A}, \mu, \mathbb{C})$ and satisfies $\nu(A) = \int_A g \, d\mu$ for each A in $\mathscr{A}$. The function g is unique up to μ-almost everywhere equality.*

Proof. If ν is a complex measure that is absolutely continuous with respect to μ, then it can be written in the form $\nu = \nu_1 - \nu_2 + i\nu_3 - i\nu_4$, where ν_1, ν_2, ν_3, and ν_4 are finite positive measures that are absolutely continuous with respect to μ. Then Theorem 4.2.2 yields functions g_j, $j = 1, \ldots, 4$, that satisfy $\nu_j(A) = \int_A g_j \, d\mu$ for each A in $\mathscr{A}$. The required function g is now given by $g = g_1 - g_2 + ig_3 - ig_4$. The case of a finite signed measure is similar.

The uniqueness of g can be proved with the method used in the proof of Theorem 4.2.2; in case ν is a complex measure, the real and imaginary parts of g should be considered separately. $\qquad\square$

Let $(X, \mathscr{A})$ be a measurable space, let μ be a σ-finite positive measure on $(X, \mathscr{A})$, and let ν be a finite signed, complex, or σ-finite positive measure on $(X, \mathscr{A})$. Suppose that ν is absolutely continuous with respect to μ. An $\mathscr{A}$-measurable function g on X that satisfies $\nu(A) = \int_A g \, d\mu$ for each A in $\mathscr{A}$ is called a *Radon–Nikodym derivative* of ν with respect to μ or, in view of its uniqueness up to μ-null sets, *the* Radon–Nikodym derivative of ν with respect to μ. A Radon–Nikodym derivative of ν with respect to μ is sometimes denoted by $\frac{d\nu}{d\mu}$.

We close this section with a few facts about the relationship of a finite signed or complex measure to its variation.

Proposition 4.2.5. *Suppose that $(X, \mathscr{A}, \mu)$ is a measure space, that f belongs to $\mathscr{L}^1(X, \mathscr{A}, \mu, \mathbb{R})$ or to $\mathscr{L}^1(X, \mathscr{A}, \mu, \mathbb{C})$, and that ν is the finite signed or complex measure defined by $\nu(A) = \int_A f \, d\mu$. Then*

$$|\nu|(A) = \int_A |f| \, d\mu$$

holds for each A in $\mathscr{A}$.

Proof. Let A belong to $\mathscr{A}$ and let $\{A_j\}_{j=1}^k$ be a finite sequence of disjoint $\mathscr{A}$-measurable sets whose union is A. Then

$$\sum_j |\nu(A_j)| = \sum_j \left| \int_{A_j} f \, d\mu \right| \leq \sum_j \int_{A_j} |f| \, d\mu = \int_A |f| \, d\mu.$$

Since $|\nu|(A)$ is the supremum of the sums that can appear on the left side of this inequality, it follows that $|\nu|(A) \leq \int_A |f| \, d\mu$.

Next construct a sequence $\{g_n\}$ of $\mathscr{A}$-measurable simple functions for which the relations $|g_n(x)| = 1$ and $\lim_n g_n(x) f(x) = |f(x)|$ hold at each x in X (the details of the construction are left to the reader). Suppose that $a_{n,j}$, $j = 1, \ldots, k_n$, are the values of g_n and that these values are attained on the sets $A_{n,j}$, $j = 1, \ldots, k_n$. Then for an arbitrary set A in $\mathscr{A}$ we have

$$\left| \int_A g_n f \, d\mu \right| = \left| \sum_j a_{n,j} \int_{A \cap A_{n,j}} f \, d\mu \right|$$

$$= \left| \sum_j a_{n,j} \nu(A \cap A_{n,j}) \right| \leq \sum_j |\nu(A \cap A_{n,j})| \leq |\nu|(A).$$

Since the dominated convergence theorem implies that $\lim_n \int_A g_n f \, d\mu = \int_A |f| \, d\mu$, it follows that $\int_A |f| \, d\mu \leq |\nu|(A)$. Thus $|\nu|(A) = \int_A |f| \, d\mu$, and the proof is complete. $\square$

Corollary 4.2.6. *Let ν be a finite signed or complex measure on the measurable space $(X, \mathscr{A})$. Then the Radon–Nikodym derivative of ν with respect to $|\nu|$ has absolute value 1 at $|\nu|$-almost every point in X.*

Proof. Proposition 4.2.5, applied in the case where $f = \frac{dv}{d|v|}$ and $\mu = |v|$, implies that

$$|v|(A) = \int_A \left| \frac{dv}{d|v|} \right| d|v|$$

holds for each A in $\mathscr{A}$. Thus $|\frac{dv}{d|v|}|$ is a Radon–Nikodym derivative of $|v|$ with respect to $|v|$. Since the constant 1 is another such Radon–Nikodym derivative, it follows that $|\frac{dv}{d|v|}| = 1$ almost everywhere. $\square$

Recall that in Sect. 4.1 we used the formulas

$$\int f\, dv = \int f\, dv^+ - \int f\, dv^-$$

and

$$\int f\, dv = \int f\, dv_1 - \int f\, dv_2 + i \int f\, dv_3 - i \int f\, dv_4$$

to define the integral of a bounded $\mathscr{A}$-measurable function f with respect to a finite signed or complex measure v. Let $\frac{dv}{d|v|}$ be a Radon–Nikodym derivative of v with respect to $|v|$. Then the relation

$$\int f\, dv = \int f \frac{dv}{d|v|} d|v| \tag{4}$$

holds for each bounded $\mathscr{A}$-measurable function f on X; this is clear in case f is the characteristic function of an $\mathscr{A}$-measurable set and then follows in the general case from the linearity of the integral and the dominated convergence theorem.

Exercises

1. Define a measure v on $(\mathbb{R}, \mathscr{B}(\mathbb{R}))$ by $v(A) = \int_A |x|\, \lambda(dx)$. Show that $v \ll \lambda$, but that for no positive ε does there exist a positive δ such that $v(A) < \varepsilon$ holds whenever A is a Borel set for which $\lambda(A) < \delta$. Thus the assumption that v is finite is essential in Lemma 4.2.1.

2. Let $\{r_n\}$ be an enumeration of the rational numbers, and for each positive integer n let $f_n \colon \mathbb{R} \to \mathbb{R}$ be a nonnegative Borel function that satisfies $\int f_n\, d\lambda = 1$ and vanishes outside the closed interval of length $1/2^n$ centered at r_n. Define μ on $\mathscr{B}(\mathbb{R})$ by $\mu(A) = \int_A \sum_n f_n\, d\lambda$.

    (a) Show that $\sum_n f_n(x) < +\infty$ holds at λ-almost every x in $\mathbb{R}$. (Hint: See Exercise 1.2.9.)
    (b) Show that μ is σ-finite, that $\mu \ll \lambda$, and that each nonempty open subset of $\mathbb{R}$ has infinite measure under μ.

3. Suppose that μ and ν are σ-finite positive measures on $(X,\mathscr{A})$, that $\nu \ll \mu$, and that g is a Radon–Nikodym derivative of ν with respect to μ. Show that

   (a) an $\mathscr{A}$-measurable function $f\colon X \to \mathbb{R}$ is ν-integrable if and only if fg is μ-integrable, and
   (b) if those functions are integrable, then $\int f\,d\nu = \int fg\,d\mu$.

4. Suppose that ν_1, ν_2, and ν_3 are σ-finite positive measures on $(X,\mathscr{A})$, that $\nu_1 \ll \nu_2$, and that $\nu_2 \ll \nu_3$.
   (a) Show that $\nu_1 \ll \nu_3$.
   (b) Make precise and prove the assertion that

$$\frac{d\nu_1}{d\nu_3} = \frac{d\nu_1}{d\nu_2}\frac{d\nu_2}{d\nu_3}.$$

5. Let $(X,\mathscr{A})$ be a measurable space, let μ be a σ-finite positive measure on $(X,\mathscr{A})$, and let ν_1 and ν_2 be finite signed measures on $(X,\mathscr{A})$ that are absolutely continuous with respect to μ.

   (a) Show that $(\nu_1 \vee \nu_2) \ll \mu$ and $(\nu_1 \wedge \nu_2) \ll \mu$ (see Exercise 4.1.3).
   (b) Express the Radon–Nikodym derivatives (with respect to μ) of $\nu_1 \vee \nu_2$ and $\nu_1 \wedge \nu_2$ in terms of those of ν_1 and ν_2.

6. Show that the assumption that ν is σ-finite can be removed from Theorem 4.2.2 if g is allowed to have values in $[0,+\infty]$. (Hint: Reduce the general case to the case where μ is finite. For each positive integer n choose a Hahn decomposition (P_n, N_n) for $\nu - n\mu$; then consider the measures $A \mapsto \nu(A \cap (\cap_n P_n))$ and $A \mapsto \nu(A \cap (\cap_n P_n)^c)$.)

7. Let μ be a σ-finite positive measure on $(X,\mathscr{A})$.

   (a) Show that

$$\{\nu \in M(X,\mathscr{A},\mathbb{R}) : \nu \ll \mu\}$$

   is a closed linear subspace of the normed linear space $M(X,\mathscr{A},\mathbb{R})$.
   (b) Find an isometric isomorphism of $L^1(X,\mathscr{A},\mu,\mathbb{R})$ onto the subspace of $M(X,\mathscr{A},\mathbb{R})$ considered in part (a).

8. Let $(X,\mathscr{A})$ be a measurable space, let μ be a finite signed or complex measure on $(X,\mathscr{A})$, and let f be a bounded real- or complex-valued $\mathscr{A}$-measurable function on X. Show that $|\int f\,d\mu| \le \int |f|\,d|\mu|$.

9. Let μ and ν be σ-finite positive measures on $(X,\mathscr{A})$. Show that the conditions

   (i) $\nu \ll \mu$ and $\mu \ll \nu$,
   (ii) μ and ν have exactly the same sets of measure zero, and
   (iii) there is an $\mathscr{A}$-measurable function g that satisfies $0 < g(x) < +\infty$ at each x in X and is such that $\nu(A) = \int_A g\,d\mu$ holds for each A in $\mathscr{A}$

   are equivalent.

10. Show that if μ is a σ-finite measure on $(X, \mathscr{A})$, then there is a finite measure ν on $(X, \mathscr{A})$ such that $\nu \ll \mu$ and $\mu \ll \nu$. (Hint: See Exercise 9.)

11. Supply the missing details in the following proof of the Radon–Nikodym theorem for finite positive measures. Let $(X, \mathscr{A})$ be a measurable space, and let μ and ν be finite positive measures on $(X, \mathscr{A})$.

   (a) Show that the formula $F(\langle f \rangle) = \int f \, d\nu$ defines a bounded linear functional on $L^2(X, \mathscr{A}, \mu + \nu, \mathbb{R})$.
   (b) Use Exercises 3.3.3 and 3.5.7 to obtain a function g in $\mathscr{L}^2(X, \mathscr{A}, \mu + \nu, \mathbb{R})$ such that $F(\langle f \rangle) = \int f g \, d(\mu + \nu)$ holds for each f in $\mathscr{L}^2(X, \mathscr{A}, \mu + \nu, \mathbb{R})$.
   (c) Show that if $\nu \ll \mu$, then the function g satisfies $0 \leq g(x) < 1$ at $(\mu + \nu)$-almost every x in X and hence can be redefined so that $0 \leq g(x) < 1$ holds at every x in X.
   (d) Show that if $\nu \ll \mu$ and if g has been redefined as in part (c), then $\nu(A) = \int_A g/(1-g) \, d\mu$ holds for each A in $\mathscr{A}$.

12. Let $(X, \mathscr{A}, \mu)$ be a finite measure space, and let $\mathscr{F}$ be a subset of $\mathscr{L}^1(X, \mathscr{A}, \mu)$. Then $\mathscr{F}$ is called L^1-bounded if the set $\{\|f\|_1 : f \in \mathscr{F}\}$ is bounded above, is called uniformly absolutely continuous if for each positive ε there is a positive δ such that $\int_A |f| \, d\mu < \varepsilon$ holds whenever $f \in \mathscr{F}$, $A \in \mathscr{A}$, and $\mu(A) < \delta$, and is called uniformly integrable if it is L^1-bounded and uniformly absolutely continuous. Show that $\mathscr{F}$ is uniformly integrable if and only if it satisfies

$$\lim_{a \to +\infty} \sup \left\{ \int_{\{|f| > a\}} |f| \, d\mu : f \in \mathscr{F} \right\} = 0.$$

(Hint: Recall Proposition 2.3.10.)

13. Show that if $(X, \mathscr{A}, \mu)$ is a finite measure space, then every finite subset of $\mathscr{L}^1(X, \mathscr{A}, \mu)$ is uniformly integrable.

14. Let $(X, \mathscr{A}, \mu)$ be a finite measure space, and let g be a nonnegative function that belongs to $\mathscr{L}^1(X, \mathscr{A}, \mu)$. Show that if $\mathscr{F}$ is a collection of measurable functions such that $|f(x)| \leq g(x)$ holds for each f in $\mathscr{F}$ and each x in X, then $\mathscr{F}$ is uniformly integrable.

15. Construct a finite measure space $(X, \mathscr{A}, \mu)$ and a sequence $\{f_n\}$ of $\mathscr{A}$-measurable functions on X such that $\{f_n : n = 1, 2, \ldots\}$ is uniformly integrable, but $\sup_n |f_n|$ is not integrable. (Compare this with Exercise 14.)

16. Let $(X, \mathscr{A}, \mu)$ be a finite measure space, let $\{f_n\}$ be a sequence of functions in $\mathscr{L}^1(X, \mathscr{A}, \mu)$, and let f be an $\mathscr{A}$-measurable real- or complex-valued function on X.

   (a) Show that if $\{f_n\}$ is uniformly integrable and if $\{f_n\}$ converges to f in measure, then f is integrable and $\int f \, d\mu = \lim_n \int f_n \, d\mu$. (Hint: Use Proposition 3.1.3, Theorem 2.4.4, and the inequality

$$\int |f_n - f| \, d\mu \leq \int_A |f_n - f| \, d\mu + \int_{A^c} |f_n| \, d\mu + \int_{A^c} |f| \, d\mu.)$$

(b) Now suppose that f belongs to $\mathscr{L}^1(X,\mathscr{A},\mu)$. Show that $\{f_n\}$ converges to f in mean if and only if $\{f_n\}$ is uniformly integrable and converges to f in measure.

(c) Use part (a) to give another proof of the dominated convergence theorem in the case where μ is finite. (See Exercise 14.)

4.3 Singularity

Let $(X,\mathscr{A})$ be a measurable space. A positive measure μ on $(X,\mathscr{A})$ is *concentrated* on the $\mathscr{A}$-measurable set E if $\mu(E^c) = 0$. A signed or complex measure μ on $(X,\mathscr{A})$ is *concentrated* on the $\mathscr{A}$-measurable set E if the variation $|\mu|$ of μ is concentrated on E, or equivalently, if each $\mathscr{A}$-measurable subset A of E^c satisfies $\mu(A) = 0$ (see Exercise 4.1.1). Now suppose that μ and ν are positive, signed, or complex measures on $(X,\mathscr{A})$. Then μ and ν are *mutually singular* if there is an $\mathscr{A}$-measurable set E such that μ is concentrated on E and ν is concentrated on E^c. One sometimes writes $\mu \perp \nu$ to indicate that μ and ν are mutually singular. Instead of saying that μ and ν are mutually singular, one sometimes says that μ and ν are singular, that ν is singular with respect to μ, or that μ is singular with respect to ν. A positive, signed, or complex measure on $(\mathbb{R}^d, \mathscr{B}(\mathbb{R}^d))$ is simply called *singular* if it is singular with respect to d-dimensional Lebesgue measure.

Examples 4.3.1.

(a) Let μ be a signed measure on the measurable set $(X,\mathscr{A})$. Then the positive and negative parts μ^+ and μ^- of μ are mutually singular; they are concentrated on the pair of disjoint sets appearing in a Hahn decomposition of μ.

(b) Next let us consider some measures on $(\mathbb{R}, \mathscr{B}(\mathbb{R}))$ that are singular with respect to Lebesgue measure. If μ is a finite discrete measure on $(\mathbb{R}, \mathscr{B}(\mathbb{R}))$, then there is a countable subset C of $\mathbb{R}$ on which μ is concentrated; since Lebesgue measure is concentrated on the complement of C, μ is singular with respect to Lebesgue measure. However not every finite measure on $(\mathbb{R}, \mathscr{B}(\mathbb{R}))$ that is singular with respect to Lebesgue measure is discrete; for example, the measure induced by the Cantor function (defined in Sect. 2.1) is singular with respect to Lebesgue measure but assigns measure zero to each point in $\mathbb{R}$ (see Exercise 2.1.7).   □

Theorem 4.3.2 (Lebesgue Decomposition Theorem). *Let $(X,\mathscr{A})$ be a measurable space, let μ be a positive measure on $(X,\mathscr{A})$, and let ν be a finite signed, complex, or σ-finite positive measure on $(X,\mathscr{A})$. Then there are unique finite signed, complex, or positive measures ν_a and ν_s on $(X,\mathscr{A})$ such that*

(a) *ν_a is absolutely continuous with respect to μ,*
(b) *ν_s is singular with respect to μ, and*
(c) *$\nu = \nu_a + \nu_s$.*

The decomposition $\nu = \nu_a + \nu_s$ is called the *Lebesgue decomposition* of ν, while ν_a and ν_s are called the *absolutely continuous* and *singular parts* of ν.

Proof. We begin with the case in which v is a finite positive measure. Define $\mathcal{N}_\mu$ by

$$\mathcal{N}_\mu = \{B \in \mathscr{A} : \mu(B) = 0\},$$

and choose a sequence $\{B_j\}$ of sets in $\mathcal{N}_\mu$ such that

$$\lim_j v(B_j) = \sup\{v(B) : B \in \mathcal{N}_\mu\}.$$

Let $N = \cup_j B_j$, and define measures v_a and v_s on $(X, \mathscr{A})$ by $v_a(A) = v(A \cap N^c)$ and $v_s(A) = v(A \cap N)$. Of course $v = v_a + v_s$. The countable subadditivity of μ implies that $\mu(N) = 0$ and hence that v_s is singular with respect to μ. Since

$$v(N) = \sup\{v(B) : B \in \mathcal{N}_\mu\},$$

each $\mathscr{A}$-measurable subset B of N^c that satisfies $\mu(B) = 0$ also satisfies $v(B) = 0$ (otherwise $N \cup B$ would belong to $\mathcal{N}_\mu$ and satisfy $v(N \cup B) > v(N)$). The absolute continuity of v_a follows.

In case v is a finite signed or complex measure, we can apply the preceding construction to the finite positive measure $|v|$, obtaining a μ-null set N such that the Lebesgue decomposition of $|v|$ is given by $|v|_a(A) = |v|(A \cap N^c)$ and $|v|_s(A) = |v|(A \cap N)$. It is easy to check that the signed or complex measures v_a and v_s defined by $v_a(A) = v(A \cap N^c)$ and $v_s(A) = v(A \cap N)$ form a Lebesgue decomposition of v.

Now suppose that v is a σ-finite positive measure, and let $\{D_k\}$ be a partition of X into $\mathscr{A}$-measurable sets that have finite measure under v. For each k let $\mathscr{A}_k$ be the σ-algebra on D_k that consists of the $\mathscr{A}$-measurable subsets of D_k, and apply the construction above to the restrictions of the measures μ and v to the spaces $(D_k, \mathscr{A}_k)$. Let $N_1, N_2, \ldots$ be the μ-null subsets of $D_1, D_2, \ldots$ thus constructed, and let $N = \cup_k N_k$. Then the measures v_a and v_s defined by $v_a(A) = v(A \cap N^c)$ and $v_s(A) = v(A \cap N)$ form a Lebesgue decomposition of v.

We turn to the uniqueness of the Lebesgue decomposition. Let $v = v_a + v_s$ and $v = v_a' + v_s'$ be Lebesgue decompositions of v. First suppose that v is a finite signed, complex, or finite positive measure. Then

$$v_a - v_a' = v_s' - v_s,$$

and since $(v_a - v_a') \ll \mu$ and $(v_s' - v_s) \perp \mu$, it follows that

$$v_a - v_a' = v_s' - v_s = 0$$

(see Exercise 1). Thus $v_a = v_a'$ and $v_s = v_s'$. The case where v is a σ-finite positive measure can be dealt with by choosing a partition $\{D_k\}$ of X into $\mathscr{A}$-measurable subsets that have finite measure under v, and applying the preceding argument to the restrictions of v_a, v_s, v_a', and v_s' to the $\mathscr{A}$-measurable subsets of the sets D_k.   □

See Exercise 6 for another proof of the uniqueness of the Lebesgue decomposition.

One sometimes goes a step further for a finite measure v on $(\mathbb{R}, \mathscr{B}(\mathbb{R}))$. Let $C = \{x \in \mathbb{R} : v(\{x\}) \neq 0\}$, and note that C is countable (for each positive integer n, there are only finitely many points x such that $v(\{x\}) \geq 1/n$). Let v_1 be the measure on $\mathscr{B}(\mathbb{R})$ defined by $v_1(A) = v(A \cap C)$, and let v_2 and v_3 be the singular and absolutely continuous (with respect to Lebesgue measure) parts of the measure $A \mapsto v(A \cap C^c)$. Then $v = v_1 + v_2 + v_3$ is a decomposition of v into the sum of a discrete measure, a continuous but singular measure, and an absolutely continuous measure. It is easy to check that the measures appearing in this decomposition are unique.

Exercises

1. Let μ be a positive measure on $(X, \mathscr{A})$, and let v be a positive, signed, or complex measure on $(X, \mathscr{A})$. Show that if $v \ll \mu$ and $v \perp \mu$, then $v = 0$. (Hint: Use the definitions of absolute continuity and of singularity.)
2. Let μ be a positive measure on $(X, \mathscr{A})$. Show that

$$\{v \in M(X, \mathscr{A}, \mathbb{R}) : v \perp \mu\}$$

   is a closed linear subspace of the normed linear space $M(X, \mathscr{A}, \mathbb{R})$.
3. Let μ be a positive measure on $(X, \mathscr{A})$, let v be a finite signed or complex measure on $(X, \mathscr{A})$, and let $v = v_a + v_s$ be the Lebesgue decomposition of v. Show that $\|v\| = \|v_a\| + \|v_s\|$.
4. Let μ and v be positive measures on $(X, \mathscr{A})$ such that for each positive ε there is a set A in $\mathscr{A}$ that satisfies $\mu(A) < \varepsilon$ and $v(A^c) < \varepsilon$. Show that $\mu \perp v$. (Hint: Choose sets $A_1, A_2, \dots$ in such a way that the set A defined by $A = \bigcap_{n=1}^{\infty} \bigcup_{k=n}^{\infty} A_k$ satisfies $\mu(A) = 0$ and $v(A^c) = 0$.)
5. Show by example that in the Lebesgue decomposition theorem, we cannot allow v to be an arbitrary positive measure. (Hint: Let $(X, \mathscr{A}) = (\mathbb{R}, \mathscr{B}(\mathbb{R}))$, let μ be Lebesgue measure on $(X, \mathscr{A})$, and let v be counting measure on $(X, \mathscr{A})$.)
6. (a) Let μ and v be as in Theorem 4.3.2, let $v = v_a + v_s$ be a Lebesgue decomposition of v, and suppose that v_s is concentrated on the μ-null set N. Show that each A in $\mathscr{A}$ satisfies $v_s(A) = v(A \cap N)$ and $v_a(A) = v(A \cap N^c)$.
   (b) Use part (a) to give another proof of the uniqueness assertion in Theorem 4.3.2.
7. (Continuation of Exercise 4.1.3.) Let μ and v be finite positive measures on $(X, \mathscr{A})$. Show that the conditions

   (i) $\mu \perp v$,
   (ii) $\mu \wedge v = 0$, and
   (iii) $\mu \vee v = \mu + v$

   are equivalent.

4.4   Functions of Finite Variation

In Sect. 1.3 we constructed a bijection between the set of all finite positive measures on $(\mathbb{R}, \mathscr{B}(\mathbb{R}))$ and the set of all bounded nondecreasing right-continuous functions $F: \mathbb{R} \to \mathbb{R}$ that vanish at $-\infty$.[4] In this section we will extend this correspondence to a bijection between the set of all finite signed measures on $(\mathbb{R}, \mathscr{B}(\mathbb{R}))$ and a certain set of real-valued functions on $\mathbb{R}$, and we will use this bijection to give a classical characterization of those finite signed measures on $(\mathbb{R}, \mathscr{B}(\mathbb{R}))$ that are absolutely continuous with respect to Lebesgue measure.

Suppose that F is a real-valued function whose domain includes the interval $[a,b]$. Let $\mathscr{S}$ be the collection of finite sequences $\{t_i\}_{i=0}^n$ such that

$$a \leq t_0 < t_1 < \cdots < t_n \leq b.$$

Then $V_F[a,b]$, the *variation of F over* $[a,b]$, is defined by

$$V_F[a,b] = \sup\left\{\sum_i |F(t_i) - F(t_{i-1})| : \{t_i\} \in \mathscr{S}\right\}.$$

The function F is *of finite variation* (or *of bounded variation*) *on* $[a,b]$ if $V_F[a,b]$ is finite.

The *variation of F over* the interval $(-\infty, b]$ and the *variation of F over* $\mathbb{R}$, written $V_F(-\infty, b]$ and $V_F(-\infty, +\infty)$, respectively, are defined in a similar way, now using finite sequences whose members belong to $(-\infty, b]$ or to $(-\infty, +\infty)$. Of course, F is said to be *of finite variation on* $(-\infty, b]$ if $V_F(-\infty, b]$ is finite, and to be *of finite variation* if $V_F(-\infty, +\infty)$ is finite. If $F: \mathbb{R} \to \mathbb{R}$ is of finite variation, then the *variation* of F is the function $V_F: \mathbb{R} \to \mathbb{R}$ defined by $V_F(x) = V_F(-\infty, x]$.

Suppose that μ is a finite signed measure on $(\mathbb{R}, \mathscr{B}(\mathbb{R}))$. Define a function $F_\mu: \mathbb{R} \to \mathbb{R}$ by letting

$$F_\mu(x) = \mu((-\infty, x]) \tag{1}$$

hold at each x in $\mathbb{R}$. If $\{t_i\}_{i=0}^n$ is an increasing sequence of real numbers, then

$$\sum_{i=1}^n |F_\mu(t_i) - F_\mu(t_{i-1})| = \sum_{i=1}^n |\mu((t_{i-1}, t_i])| \leq |\mu|(\mathbb{R});$$

it follows that $V_{F_\mu}(-\infty, +\infty) \leq |\mu|(\mathbb{R})$ and hence that F_μ is of finite variation. It is easy to check that F_μ vanishes at $-\infty$ and is right-continuous (use Proposition 1.3.9 and the Jordan decomposition of μ). We will soon see that every right-continuous function of finite variation that vanishes at $-\infty$ arises from a finite signed measure in this way.

---

[4]Recall that a function $F: \mathbb{R} \to \mathbb{R}$ is said to *vanish at* $-\infty$ if $\lim_{x \to -\infty} F(x) = 0$.

It is easy to check that the function F_μ defined by (1) is continuous if and only if $\mu(\{x\}) = 0$ holds for each x in $\mathbb{R}$. In this case

$$\mu((a,b)) = \mu([a,b]) = \mu([a,b)) = \mu((a,b]) = F_\mu(b) - F_\mu(a)$$

holds whenever $a < b$.

Let us turn to some general properties of functions of finite variation.

Suppose that $F \colon \mathbb{R} \to \mathbb{R}$ is of finite variation. It is easy to check that F is bounded and that if $-\infty < a < b < +\infty$, then

$$V_F(-\infty, b] = V_F(-\infty, a] + V_F[a, b]. \tag{2}$$

Furthermore, if $b \in \mathbb{R}$, then

$$V_F(-\infty, b] = \lim_{a \to -\infty} V_F[a, b]; \tag{3}$$

to prove this, let ε be a positive number, choose an increasing sequence $\{t_i\}_{i=0}^n$ of numbers that belong to $(-\infty, b]$ and satisfy

$$\sum_{i=1}^n |F(t_i) - F(t_{i-1})| > V_F(-\infty, b] - \varepsilon,$$

and note that for each a that satisfies $a \le t_0$ we have

$$V_F(-\infty, b] - \varepsilon < V_F[a, b] \le V_F(-\infty, b].$$

A similar argument shows that if $a < c$ and if F is right-continuous at a, then

$$V_F[a, c] = \lim_{b \to a^+} V_F[b, c]. \tag{4}$$

Lemma 4.4.1. *Let F be a function of finite variation on $\mathbb{R}$. Then*

(a) *V_F is bounded and nondecreasing,*
(b) *V_F vanishes at $-\infty$, and*
(c) *if F is right-continuous, then V_F is right-continuous.*

Proof. Part (a) is clear. Equations (2) and (3) justify the calculation

$$\lim_{x \to -\infty} V_F(x) = \lim_{x \to -\infty} V_F(-\infty, x]$$

$$= \lim_{x \to -\infty} (V_F(-\infty, b] - V_F[x, b])$$

$$= V_F(-\infty, b] - V_F(-\infty, b] = 0,$$

and so part (b) is proved. A similar argument, using Eqs. (2) and (4), yields part (c). $\square$

Proposition 4.4.2. *Let F be a function of finite variation on $\mathbb{R}$. Then there are bounded nondecreasing functions F_1 and F_2 such that $F = F_1 - F_2$.*

Proof. It is easy to check that the functions defined by $F_1 = (V_F + F)/2$ and $F_2 = (V_F - F)/2$ have the required properties. □

Let $F: \mathbb{R} \to \mathbb{R}$ be of finite variation, and let F_1 and F_2 be the functions constructed in the proof of Proposition 4.4.2. Lemma 4.4.1 implies that if F is right-continuous, then F_1 and F_2 are right-continuous, and that if F vanishes at $-\infty$, then F_1 and F_2 vanish at $-\infty$.

Proposition 4.4.3. *Equation* (1) *defines a bijection* $\mu \mapsto F_\mu$ *between the set of all finite signed measures on* $(\mathbb{R}, \mathscr{B}(\mathbb{R}))$ *and the set of all right-continuous functions of finite variation that vanish at* $-\infty$.

Proof. We have already checked that F_μ is a right-continuous function of finite variation that vanishes at $-\infty$. If μ and ν are finite signed measures such that $F_\mu = F_\nu$ and if $\mu = \mu^+ - \mu^-$ and $\nu = \nu^+ - \nu^-$ are their Jordan decompositions, then $F_{\mu^+} - F_{\mu^-} = F_{\nu^+} - F_{\nu^-}$; since this implies that $F_{\mu^+} + F_{\nu^-} = F_{\nu^+} + F_{\mu^-}$, it follows from Proposition 1.3.10 that $\mu^+ + \nu^- = \nu^+ + \mu^-$ and hence that $\mu = \nu$. The injectivity of the map $\mu \mapsto F_\mu$ follows. The surjectivity follows from Proposition 1.3.10, Proposition 4.4.2, and the remarks following the proof of Proposition 4.4.2. □

A function $F: \mathbb{R} \to \mathbb{R}$ is *absolutely continuous* if for each positive number ε there is a positive number δ such that $\sum_i |F(t_i) - F(s_i)| < \varepsilon$ holds whenever $\{(s_i, t_i)\}$ is a finite sequence of disjoint open intervals for which $\sum_i (t_i - s_i) < \delta$.

It is clear that every absolutely continuous function is continuous and, in fact, uniformly continuous. There are, however, functions that are uniformly continuous and of finite variation, but are not absolutely continuous (see Exercise 3). It is easy to check that an absolutely continuous function is of finite variation on each closed bounded interval (see Exercise 5), but is not necessarily of finite variation on $\mathbb{R}$ (consider the function F defined by $F(x) = x$).

We turn to the relationship between absolute continuity for signed measures and absolute continuity for functions of a real variable.

Lemma 4.4.4. *If* $F: \mathbb{R} \to \mathbb{R}$ *is absolutely continuous and of finite variation, then* V_F *is absolutely continuous.*

Proof. Let ε be a positive number, and use the absolute continuity of F to choose a corresponding δ. If $\{(s_i, t_i)\}$ is a finite sequence of disjoint open intervals such that $\sum_i (t_i - s_i) < \delta$, then each finite sequence $\{(u_j, v_j)\}$ of disjoint open subintervals of $\cup_i (s_i, t_i)$ satisfies $\sum (v_j - u_j) < \delta$ and so satisfies $\sum_j |F(v_j) - F(u_j)| < \varepsilon$. Since the sequence $\{(u_j, v_j)\}$ can be chosen so as to make $\sum_j |F(v_j) - F(u_j)|$ arbitrarily close to $\sum_i V_F[s_i, t_i]$, we have

$$\sum_i |V_F(t_i) - V_F(s_i)| = \sum_i V_F[s_i, t_i] \le \varepsilon.$$

The absolute continuity of V_F follows. □

Proposition 4.4.5. *Let μ be a finite signed measure on $(\mathbb{R}, \mathscr{B}(\mathbb{R}))$, and let $F_\mu : \mathbb{R} \to \mathbb{R}$ be defined by* (1). *Then F_μ is absolutely continuous if and only if μ is absolutely continuous with respect to Lebesgue measure.*

Proof. First suppose that μ is absolutely continuous with respect to Lebesgue measure. Let ε be a positive number, and use Lemma 4.2.1 to choose a positive number δ such that $|\mu|(A) < \varepsilon$ holds whenever A is a Borel set that satisfies $\lambda(A) < \delta$. If $\{(s_i, t_i)\}$ is a finite sequence of disjoint open intervals such that $\sum_i (t_i - s_i) < \delta$, then $\lambda(\cup_i (s_i, t_i]) < \delta$, and so

$$\sum_i |F_\mu(t_i) - F_\mu(s_i)| = \sum_i |\mu((s_i, t_i])| \leq |\mu| \left(\bigcup_i (s_i, t_i] \right) < \varepsilon.$$

Hence F_μ is absolutely continuous.

Now suppose that F_μ is absolutely continuous. Then V_{F_μ} is absolutely continuous (Lemma 4.4.4), and so the functions F_1 and F_2 defined by $F_1 = (V_{F_\mu} + F_\mu)/2$ and $F_2 = (V_{F_\mu} - F_\mu)/2$ are absolutely continuous. Let μ_1 and μ_2 be the finite positive measures on $(\mathbb{R}, \mathscr{B}(\mathbb{R}))$ that correspond to F_1 and F_2. Since $F_\mu = F_1 - F_2$, it follows (Proposition 4.4.3) that $\mu = \mu_1 - \mu_2$; thus we need only show that $\mu_1 \ll \lambda$ and $\mu_2 \ll \lambda$. Let ε be a positive number, and let δ be a positive number such that

$$\sum_i |F_1(t_i) - F_1(s_i)| < \varepsilon \text{ holds whenever } \{(s_i, t_i)\} \text{ is a finite}$$
$$\text{sequence of disjoint open intervals such that } \sum_i (t_i - s_i) < \delta. \qquad (5)$$

Suppose that A is a Borel subset of $\mathbb{R}$ such that $\lambda(A) < \delta$, and use the regularity of Lebesgue measure to choose an open set U that includes A and satisfies $\lambda(U) < \delta$. Then U is the union of a sequence $\{(s_i, t_i)\}$ of disjoint open intervals (see Proposition C.4), and it follows from (5) that

$$\mu_1 \left(\bigcup_{i=1}^{n} (s_i, t_i) \right) = \sum_{i=1}^{n} (F_1(t_i) - F_1(s_i)) < \varepsilon$$

holds for each n. Hence $\mu_1(U) = \mu_1(\cup_{i=1}^{\infty}(s_i, t_i)) \leq \varepsilon$ (see Proposition 1.2.5), and so $\mu_1(A) \leq \varepsilon$. The absolute continuity of μ_1 now follows from Lemma 4.2.1. The case of μ_2 is similar, and so the proof is complete. $\qquad \square$

Proposition 4.4.6. *The functions $F : \mathbb{R} \to \mathbb{R}$ that can be written in the form*

$$F(x) = \int_{-\infty}^{x} f(t) \, dt \qquad (6)$$

for some f in $\mathscr{L}^1(\mathbb{R}, \mathscr{B}(\mathbb{R}), \lambda, \mathbb{R})$ are exactly the absolutely continuous functions of finite variation that vanish at $-\infty$.

Proof. First suppose that f belongs to $\mathscr{L}^1(\mathbb{R}, \mathscr{B}(\mathbb{R}), \lambda, \mathbb{R})$ and that F arises from f through (6). The signed measure μ defined by $\mu(A) = \int_A f \, d\lambda$ is absolutely continuous with respect to λ, and $F = F_\mu$; hence it follows from Propositions 4.4.3 and 4.4.5 that F is of finite variation, is absolutely continuous, and vanishes at $-\infty$.

Now suppose that $F : \mathbb{R} \to \mathbb{R}$ is of finite variation, is absolutely continuous, and vanishes at $-\infty$. Proposition 4.4.3 implies that there is a finite signed measure μ such that $F = F_\mu$, and Proposition 4.4.5 implies that $\mu \ll \lambda$. If $f = \frac{d\mu}{d\lambda}$, then (6) holds at each x in $\mathbb{R}$. $\square$

The study of absolute continuity for functions of a real variable, and in particular of Eq. (6), will be continued in Sect. 6.3.

Exercises

1. Suppose that $F : \mathbb{R} \to \mathbb{R}$ is defined by

$$F(x) = \begin{cases} 0 & \text{if } x \leq 0, \\ x \sin \frac{1}{x} & \text{if } x > 0. \end{cases}$$

   Find the closed bounded intervals $[a,b]$ for which $V_F[a,b]$ is finite.
2. Show that if $F : \mathbb{R} \to \mathbb{R}$ is of finite variation, then the limits $\lim_{x \to -\infty} F(x)$ and $\lim_{x \to +\infty} F(x)$ exist.
3. Let F be the Cantor function, extended so as to vanish on the interval $(-\infty, 0)$ and to have value 1 on the interval $(1, +\infty)$. Show directly (i.e., without using Proposition 4.4.5) that F is uniformly continuous but not absolutely continuous.
4. Show that if $F : \mathbb{R} \to \mathbb{R}$ is continuous and of finite variation, then $V_F : \mathbb{R} \to \mathbb{R}$ is continuous.
5. Show that if $F : \mathbb{R} \to \mathbb{R}$ is absolutely continuous, then F is of finite variation on each closed bounded interval. (Hint: Let δ be a positive number such that $\sum_i |F(t_i) - F(s_i)| < 1$ holds whenever $\{(s_i, t_i)\}$ is a finite sequence of disjoint open intervals such that $\sum_i (t_i - s_i) < \delta$, and let $[a,b]$ be a closed bounded interval. Show that if $\{u_i\}_{i=0}^n$ is a finite sequence such that

$$a \leq u_0 < u_1 < \ldots < u_n \leq b,$$

   then $\sum_{i=1}^n |F(u_i) - F(u_{i-1})| \leq (b-a)/\delta + 1$.)
6. Let μ be a finite signed measure on $(\mathbb{R}, \mathscr{B}(\mathbb{R}))$. Show that $V_{F_\mu}(-\infty, x] = |\mu|((-\infty, x])$ holds at each x in $\mathbb{R}$.

4.5 The Duals of the L^p Spaces

We return to the study, which we began in Sect. 3.5, of the duals of the L^p spaces. Let $(X, \mathscr{A}, \mu)$ be a measure space, let p satisfy $1 \leq p < +\infty$, and let q be defined by $1/p + 1/q = 1$. Recall that if f belongs to $\mathscr{L}^p(X, \mathscr{A}, \mu)$ (or to $\mathscr{L}^q(X, \mathscr{A}, \mu)$), then $\langle f \rangle$ is the coset in $L^p(X,\mathscr{A},\mu)$ (or in $L^q(X,\mathscr{A},\mu)$) to which f belongs. We have seen that each $\langle g \rangle$ in $L^q(X,\mathscr{A},\mu)$ induces a bounded linear functional $T_{\langle g \rangle}$ on $L^p(X,\mathscr{A},\mu)$ by means of the formula $T_{\langle g \rangle}(\langle f \rangle) = \int f g \, d\mu$ and that the operator T that takes $\langle g \rangle$ to $T_{\langle g \rangle}$ is an isometry of $L^q(X,\mathscr{A},\mu)$ into $(L^p(X,\mathscr{A},\mu))^*$ (Proposition 3.5.5). We now use the Radon–Nikodym theorem to show that in many situations the operator T is surjective and hence is an isometric isomorphism.

Theorem 4.5.1. *Let $(X,\mathscr{A},\mu)$ be a measure space, let p satisfy $1 \le p < +\infty$, and let q be defined by $1/p + 1/q = 1$. If $p = 1$ and μ is σ-finite, or if $1 < p < +\infty$ and μ is arbitrary, then the operator T defined above is an isometric isomorphism of $L^q(X,\mathscr{A},\mu)$ onto $(L^p(X,\mathscr{A},\mu))^*$.*

Proof. Since we know that T is an isometry (Proposition 3.5.5), we need only show that it is surjective.

Let F be an arbitrary element of $(L^p(X,\mathscr{A},\mu))^*$. First suppose that $\mu(X) < +\infty$ and that p satisfies $1 \le p < +\infty$. We define a function ν on the σ-algebra $\mathscr{A}$ by means of the formula $\nu(A) = F(\langle \chi_A \rangle)$. If $\{A_k\}$ is a sequence of disjoint sets in $\mathscr{A}$ and if $A = \cup_k A_k$, then the dominated convergence theorem implies that $\lim_n \|\chi_A - \sum_{k=1}^n \chi_{A_k}\|_p = 0$; since F is continuous and linear, this implies that $F(\langle \chi_A \rangle) = \sum_k F(\langle \chi_{A_k} \rangle)$ and hence that $\nu(A) = \sum_k \nu(A_k)$. Thus ν is countably additive and so is a finite signed or complex measure. It is clear that ν is absolutely continuous with respect to μ. Hence the Radon–Nikodym theorem (Theorem 4.2.4) provides a function g in $\mathscr{L}^1(X,\mathscr{A},\mu)$ that satisfies $\nu(A) = \int_A g \, d\mu$ for each A in $\mathscr{A}$. We will show that g belongs to $\mathscr{L}^q(X,\mathscr{A},\mu)$ and that $F(\langle f \rangle) = \int f g \, d\mu$ holds for each f in $\mathscr{L}^p(X,\mathscr{A},\mu)$.

For each positive integer n let $E_n = \{x \in X : |g(x)| \le n\}$. Then $g\chi_{E_n}$ is bounded and so belongs to $\mathscr{L}^q(X,\mathscr{A},\mu)$ (recall that μ is finite). Define a functional F_{E_n} on $L^p(X,\mathscr{A},\mu)$ by $F_{E_n}(\langle f \rangle) = F(\langle f\chi_{E_n} \rangle)$. Consider the relation

$$F_{E_n}(\langle f \rangle) = \int f g \chi_{E_n} \, d\mu. \tag{1}$$

If f is the characteristic function of an $\mathscr{A}$-measurable set A, then both sides of (1) are equal to $\nu(A \cap E_n)$); thus (1) holds if f is the characteristic function of an $\mathscr{A}$-measurable set and hence if f is an $\mathscr{A}$-measurable simple function. Since the $\mathscr{A}$-measurable simple functions determine a dense subspace of $L^p(X,\mathscr{A},\mu)$ (Proposition 3.4.2), Eq. (1) holds for all $\langle f \rangle$ in $L^p(X,\mathscr{A},\mu)$. It follows from Proposition 3.5.5 that

$$\|g\chi_{E_n}\|_q = \|F_{E_n}\| \le \|F\|.$$

If $q < +\infty$, then the monotone convergence theorem implies that $g \in \mathscr{L}^q(X,\mathscr{A},\mu)$ and $\|g\|_q \le \|F\|$. If $q = +\infty$, then (since $E = \cup_n E_n$) we have

$$\mu(\{x \in X : |g(x)| > \|F\|\}) = \lim_n \mu(\{x \in E_n : |g(x)| > \|F\|\}) = 0,$$

and we can redefine g so that it will be bounded, in fact satisfying $|g(x)| \le \|F\|$ at every x in X. Thus $\|g\|_q \le \|F\|$, whether q is finite or infinite. Furthermore, in both cases we can take limits in (1) as n approaches infinity and conclude that $F(\langle f \rangle) = \int fg\,d\mu$. With this the theorem is proved in the case of finite measures.

We need some notation in order to deal with the case where μ is not finite. Suppose that B belongs to $\mathscr{A}$. Let $\mathscr{A}_B$ be the σ-algebra on B consisting of those subsets of B that belong to $\mathscr{A}$, and let μ_B be the restriction of μ to $\mathscr{A}_B$. If f is a real- or complex-valued function on B, then we will denote by f' the function on X that agrees with f on B and vanishes outside B. The formula $F_B(\langle f \rangle) = F(\langle f' \rangle)$ defines a linear functional F_B on $L^p(B, \mathscr{A}_B, \mu_B)$; this functional satisfies $\|F_B\| \leq \|F\|$.

Now suppose that μ is σ-finite and that p satisfies $1 \leq p < +\infty$. Let $\{B_k\}$ be a sequence of disjoint sets that belong to $\mathscr{A}$, have finite measure under μ, and satisfy $X = \cup_k B_k$. According to the first part of this proof there is for each k a function g_k in $\mathscr{L}^q(B_k, \mathscr{A}_{B_k}, \mu_{B_k})$ that represents F_{B_k} on $L^p(B_k, \mathscr{A}_{B_k}, \mu_{B_k})$ and satisfies $\|g_k\|_q \leq \|F_{B_k}\|$. Define g on X so that it agrees on each B_k with g_k. It is not difficult to check (do so) that $g \in \mathscr{L}^q(X, \mathscr{A}, \mu)$ and that

$$F(\langle f \rangle) = \int fg\,d\mu$$

holds for each $\langle f \rangle$ in $L^p(X, \mathscr{A}, \mu)$.

Finally we turn to the case where μ is arbitrary. Now we assume that $1 < p < +\infty$ and hence that $1 < q < +\infty$. Let $\mathscr{S}$ be the collection of sets in $\mathscr{A}$ that are σ-finite under μ. Note that if $B \in \mathscr{S}$, then $(B, \mathscr{A}_B, \mu_B)$ is σ-finite, and so by what we have just proved, there is a function g in $\mathscr{L}^q(B, \mathscr{A}_B, \mu_B)$ such that

$$F_B(\langle f \rangle) = \int fg\,d\mu_B$$

holds for each $\langle f \rangle$ in $L^p(B, \mathscr{A}_B, \mu_B)$. Furthermore if B_1 and B_2 are disjoint sets in $\mathscr{S}$, then

$$\|F_{B_1 \cup B_2}\|^q = \|F_{B_1}\|^q + \|F_{B_2}\|^q; \tag{2}$$

to prove this, choose a function g in $\mathscr{L}^q(B_1 \cup B_2, \mathscr{A}_{B_1 \cup B_2}, \mu_{B_1 \cup B_2})$ that represents $F_{B_1 \cup B_2}$, and note that

$$\|F_{B_1 \cup B_2}\|^q = \int_{B_1 \cup B_2} |g|^q\,d\mu_{B_1 \cup B_2}$$

$$= \int_{B_1} |g|^q\,d\mu_{B_1} + \int_{B_2} |g|^q\,d\mu_{B_2} = \|F_{B_1}\|^q + \|F_{B_2}\|^q.$$

Now choose a sequence $\{C_n\}$ of sets in $\mathscr{S}$ such that

$$\lim_n \|F_{C_n}\| = \sup\{\|F_B\| : B \in \mathscr{S}\}.$$

Let $C = \cup_n C_n$. Then $C \in \mathscr{S}$,

$$\|F_C\| = \sup\{\|F_B\| : B \in \mathscr{S}\}, \tag{3}$$

and we can choose a function g_C in $\mathcal{L}^q(C, \mathcal{A}_C, \mu_C)$ such that

$$F_C(\langle f \rangle) = \int f g_C \, d\mu_C \tag{4}$$

holds for each $\langle f \rangle$ in $L^p(C, \mathcal{A}_C, \mu_C)$. Note that if f belongs to $\mathcal{L}^p(X, \mathcal{A}, \mu)$ and vanishes on C, then $F(\langle f \rangle) = 0$ (otherwise, if $D = \{x \in X : f(x) \neq 0\}$, then D would belong to $\mathcal{S}$ (Corollary 2.3.11) and would satisfy $F_D \neq 0$, and so in view of (2), $F_{C \cup D}$ would satisfy

$$\|F_{C \cup D}\|^q = \|F_C\|^q + \|F_D\|^q > \|F_C\|^q,$$

contradicting (3)). It follows from this and (4) that if g is the function on X that agrees with g_C on C and vanishes off C, then $g \in \mathcal{L}^q(X, \mathcal{A}, \mu)$ and

$$F(\langle f \rangle) = \int f g \, d\mu$$

holds for each $\langle f \rangle$ in $L^p(X, \mathcal{A}, \mu)$ (decompose f into the sum of a function that vanishes on C and a function that vanishes on C^c). Hence $F = T_{\langle g \rangle}$ and the proof of the surjectivity of T is complete.   □

Example 4.5.2. Let us consider an example that shows that the hypothesis of σ-finiteness cannot simply be omitted in Theorem 4.5.1 (see, however, Theorems 7.5.4 and 9.4.8). Let $X = \mathbb{R}$, let $\mathcal{A}$ be the σ-algebra consisting of those subsets A of $\mathbb{R}$ such that A or A^c is countable, and let μ be counting measure on $(X, \mathcal{A})$. Then $\mathcal{L}^1(X, \mathcal{A}, \mu)$ consists of those functions f on $\mathbb{R}$ that vanish outside a countable set and satisfy $\sum_x |f(x)| < +\infty$, and for such functions we have $\|f\|_1 = \sum_x |f(x)|$. Define a functional F on $L^1(X, \mathcal{A}, \mu)$ by $F(\langle f \rangle) = \sum_{x>0} f(x)$. Then F is continuous, and if g is a function that satisfies $F(\langle f \rangle) = \int f g \, d\mu$ for each f in $\mathcal{L}^1(X, \mathcal{A}, \mu)$, then g must be the characteristic function of the interval $(0, +\infty)$. However this function is not $\mathcal{A}$-measurable, and so the functional F is induced by no function in $\mathcal{L}^\infty(X, \mathcal{A}, \mu)$.   □

Exercises

1. Let V be a normed linear space, and let v and $v_1, v_2, \ldots$ belong to V. The sequence $\{v_n\}$ is said to *converge weakly* to v if $F(v) = \lim_n F(v_n)$ holds for each F in V^*.

   (a) Show that if $\{v_n\}$ converges to v in norm (that is, if $\lim_n \|v_n - v\| = 0$), then $\{v_n\}$ converges weakly to v.
   (b) Does the converse of part (a) hold if $V = L^2(\mathbb{R}, \mathcal{B}(\mathbb{R}), \lambda)$?

2. Let $(X, \mathscr{A}, \mu)$ be a measure space. Show that the formula $T_{\langle g \rangle}(\langle f \rangle) = \int fg \, d\mu$ defines an isometry T of $L^1(X, \mathscr{A}, \mu)$ into $(L^\infty(X, \mathscr{A}, \mu))^*$. (Thus we could have allowed p to be $+\infty$ in Proposition 3.5.5. See, however, the following exercise.)

3. (This exercise depends on Exercise 3.5.8, and hence on the Hahn–Banach theorem.) Let $(X, \mathscr{A}, \mu)$ be a finite measure space. Show that the conditions

   (i) the map T in Exercise 2 is surjective,
   (ii) $L^1(X, \mathscr{A}, \mu)$ is finite dimensional,
   (iii) $L^\infty(X, \mathscr{A}, \mu)$ is finite dimensional, and
   (iv) there is a finite σ-algebra $\mathscr{A}_0$ on X such that $\mathscr{A}_0 \subseteq \mathscr{A}$ and such that each set in $\mathscr{A}$ differs from a set in $\mathscr{A}_0$ by a μ-null set

   are equivalent. (Hint: To show that (i) implies (iv), assume that (iv) fails and use ideas from Exercise 3.5.8 to show that (i) fails.)

Notes

The basic facts about absolute continuity and singularity of measures are contained in essentially all books on measure and integration, while the results given in the last part of Sect. 4.1 and in Sect. 4.4 are sometimes omitted. See Chap. 10, on probability, for applications of most of these results.

The proof of the Radon–Nikodym theorem outlined in Exercise 4.2.11 is due to von Neumann (see [120, pp. 124–131]).

Chapter 5
Product Measures

In calculus courses one defines integrals over two- (or higher-) dimensional regions and then evaluates these integrals by applying the usual techniques of integration, one variable at a time. In this chapter we show that similar techniques work for the Lebesgue integral. More generally, given σ-finite measures μ and ν on spaces X and Y, we first define a natural product measure on the product space $X \times Y$ (Sect. 5.1). Then we look at how integrals with respect to this product measure can be evaluated in terms of integrals with respect to μ and ν over X and Y (Sect. 5.2). The chapter ends with a few applications (Sect. 5.3).

5.1 Constructions

Let $(X, \mathscr{A})$ and $(Y, \mathscr{B})$ be measurable spaces, and, as usual, let $X \times Y$ be the Cartesian product of the sets X and Y. A subset of $X \times Y$ is a *rectangle with measurable sides* if it has the form $A \times B$ for some A in $\mathscr{A}$ and some B in $\mathscr{B}$; the σ-algebra on $X \times Y$ generated by the collection of all rectangles with measurable sides is called the *product* of the σ-algebras $\mathscr{A}$ and $\mathscr{B}$ and is denoted by $\mathscr{A} \times \mathscr{B}$.

Example 5.1.1. Consider the space $\mathbb{R}^2$. This is, of course, a Cartesian product, the product of $\mathbb{R}$ with itself. Let us show that the product σ-algebra $\mathscr{B}(\mathbb{R}) \times \mathscr{B}(\mathbb{R})$ is equal to the σ-algebra $\mathscr{B}(\mathbb{R}^2)$ of Borel subsets of $\mathbb{R}^2$. Recall that $\mathscr{B}(\mathbb{R}^2)$ is generated by the collection of all sets of the form $(a, b] \times (c, d]$ (Proposition 1.1.5). Thus $\mathscr{B}(\mathbb{R}^2)$ is generated by a subfamily of the σ-algebra $\mathscr{B}(\mathbb{R}) \times \mathscr{B}(\mathbb{R})$ and so is included in $\mathscr{B}(\mathbb{R}) \times \mathscr{B}(\mathbb{R})$. We turn to the reverse inclusion. The projections π_1 and π_2 of $\mathbb{R}^2$ onto $\mathbb{R}$ defined by $\pi_1(x, y) = x$ and $\pi_2(x, y) = y$ are continuous and hence Borel measurable (Example 2.1.2(a)). It follows from this and the identity

$$A \times B = (A \times \mathbb{R}) \cap (\mathbb{R} \times B) = \pi_1^{-1}(A) \cap \pi_2^{-1}(B)$$

D.L. Cohn, *Measure Theory: Second Edition*, Birkhäuser Advanced
Texts Basler Lehrbücher, DOI 10.1007/978-1-4614-6956-8_5,
© Springer Science+Business Media, LLC 2013

that if A and B belong to $\mathscr{B}(\mathbb{R})$, then $A \times B$ belongs to $\mathscr{B}(\mathbb{R}^2)$. Since $\mathscr{B}(\mathbb{R}) \times \mathscr{B}(\mathbb{R})$ is the σ-algebra generated by the collection of all such rectangles $A \times B$, it must be included in $\mathscr{B}(\mathbb{R}^2)$. Thus $\mathscr{B}(\mathbb{R}) \times \mathscr{B}(\mathbb{R}) = \mathscr{B}(\mathbb{R}^2)$.   □

Let us introduce some terminology and notation. Suppose that X and Y are sets and that E is a subset of $X \times Y$. Then for each x in X and each y in Y the *sections* E_x and E^y are the subsets of Y and X given by

$$E_x = \{y \in Y : (x,y) \in E\}$$

and

$$E^y = \{x \in X : (x,y) \in E\}.$$

If f is a function on $X \times Y$, then the *sections* f_x and f^y are the functions on Y and X given by

$$f_x(y) = f(x,y)$$

and

$$f^y(x) = f(x,y).$$

Lemma 5.1.2. *Let* $(X, \mathscr{A})$ *and* $(Y, \mathscr{B})$ *be measurable spaces.*

(a) *If E is a subset of $X \times Y$ that belongs to $\mathscr{A} \times \mathscr{B}$, then each section E_x belongs to $\mathscr{B}$ and each section E^y belongs to $\mathscr{A}$.*

(b) *If f is an extended real-valued (or a complex-valued) $\mathscr{A} \times \mathscr{B}$-measurable function on $X \times Y$, then each section f_x is $\mathscr{B}$-measurable and each section f^y is $\mathscr{A}$-measurable.*

Proof. Suppose that x belongs to X, and let $\mathscr{F}$ be the collection of all subsets E of $X \times Y$ such that E_x belongs to $\mathscr{B}$. Then $\mathscr{F}$ contains all rectangles $A \times B$ for which $A \in \mathscr{A}$ and $B \in \mathscr{B}$ (note that $(A \times B)_x$ is either B or $\varnothing$). In particular, $X \times Y \in \mathscr{F}$. Furthermore, the identities $(E^c)_x = (E_x)^c$ and $(\cup_n E_n)_x = \cup_n((E_n)_x)$ imply that $\mathscr{F}$ is closed under complementation and under the formation of countable unions; thus $\mathscr{F}$ is a σ-algebra. It follows that $\mathscr{F}$ includes the σ-algebra $\mathscr{A} \times \mathscr{B}$ and hence that E_x belongs to $\mathscr{B}$ whenever E belongs to $\mathscr{A} \times \mathscr{B}$. A similar argument shows that E^y belongs to $\mathscr{A}$ whenever E belongs to $\mathscr{A} \times \mathscr{B}$. With this part (a) is proved.

Part (b) follows from part (a) and the identities $(f_x)^{-1}(D) = (f^{-1}(D))_x$ and $(f^y)^{-1}(D) = (f^{-1}(D))^y$.   □

Proposition 5.1.3. *Let* $(X, \mathscr{A}, \mu)$ *and* $(Y, \mathscr{B}, \nu)$ *be σ-finite measure spaces. If E belongs to the σ-algebra $\mathscr{A} \times \mathscr{B}$, then the function $x \mapsto \nu(E_x)$ is $\mathscr{A}$-measurable and the function $y \mapsto \mu(E^y)$ is $\mathscr{B}$-measurable.*

Proof. First suppose that the measure ν is finite. Let $\mathscr{F}$ be the class of those sets E in $\mathscr{A} \times \mathscr{B}$ for which the function $x \mapsto \nu(E_x)$ is $\mathscr{A}$-measurable (Lemma 5.1.2 implies that E_x belongs to $\mathscr{B}$, and hence that $\nu(E_x)$ is defined). If $A \in \mathscr{A}$ and $B \in \mathscr{B}$, then $\nu((A \times B)_x) = \nu(B)\chi_A(x)$, and so the rectangle $A \times B$ belongs to $\mathscr{F}$. In particular, the space $X \times Y$ belongs to $\mathscr{F}$. Note that if E and F are sets in $\mathscr{A} \times \mathscr{B}$ such that $E \subseteq F$, then $v((F-E)_x) = v(F_x) - v(E_x)$, and that if $\{E_n\}$ is an increasing sequence of sets in $\mathscr{A} \times \mathscr{B}$, then $v((\cup_n E_n)_x) = \lim_n v((E_n)_x)$; it follows that $\mathscr{F}$ is closed under the formation of proper differences and under the formation of unions of increasing sequences of sets. Thus $\mathscr{F}$ is a d-system (see Sect. 1.6). Since the family of rectangles with measurable sides is closed under the formation of finite intersections (note that

$$(A_1 \times B_1) \cap (A_2 \times B_2) = (A_1 \cap A_2) \times (B_1 \cap B_2)),$$

Theorem 1.6.2 implies that $\mathscr{F} = \mathscr{A} \times \mathscr{B}$. Thus $x \mapsto v(E_x)$ is measurable for each E in $\mathscr{A} \times \mathscr{B}$.

Now suppose that v is σ-infinite, and let $\{D_n\}$ be a sequence of disjoint subsets of Y that belong to $\mathscr{B}$, have finite measure under v, and satisfy $\cup_n D_n = Y$. Define finite measures $v_1, v_2, \ldots$ on $\mathscr{B}$ by letting $v_n(B) = v(B \cap D_n)$. According to what we have just proved, for each n the function $x \mapsto v_n(E_x)$ is $\mathscr{A}$-measurable; since $v(E_x) = \sum_n v_n(E_x)$ holds for each x, the measurability of $x \mapsto v(E_x)$ follows. The function $y \mapsto \mu(E^y)$ can be treated similarly, and so the proof is complete. $\square$

Theorem 5.1.4. *Let $(X, \mathscr{A}, \mu)$ and $(Y, \mathscr{B}, v)$ be σ-finite measure spaces. Then there is a unique measure $\mu \times v$ on the σ-algebra $\mathscr{A} \times \mathscr{B}$ such that*

$$(\mu \times v)(A \times B) = \mu(A)v(B)$$

holds for each A in $\mathscr{A}$ and B in $\mathscr{B}$. Furthermore, the measure under $\mu \times v$ of an arbitrary set E in $\mathscr{A} \times \mathscr{B}$ is given by

$$(\mu \times v)(E) = \int_X v(E_x)\,\mu(dx) = \int_Y \mu(E^y)\,v(dy). \tag{1}$$

The measure $\mu \times v$ is called the *product* of μ and v.

Proof. The measurability of $x \mapsto v(E_x)$ and $y \mapsto \mu(E^y)$ for each E in $\mathscr{A} \times \mathscr{B}$ follows from Proposition 5.1.3. Thus we can define functions $(\mu \times v)_1$ and $(\mu \times v)_2$ on $\mathscr{A} \times \mathscr{B}$ by $(\mu \times v)_1(E) = \int_X v(E_x)\,\mu(dx)$ and $(\mu \times v)_2(E) = \int_Y \mu(E^y)\,v(dy)$. It is clear that $(\mu \times v)_1(\varnothing) = (\mu \times v)_2(\varnothing) = 0$. If $\{E_n\}$ is a sequence of disjoint sets in $\mathscr{A} \times \mathscr{B}$, if $E = \cup_n E_n$, and if $x \in X$, then $\{(E_n)_x\}$ is a sequence of disjoint sets in $\mathscr{B}$ such that $E_x = \cup_n((E_n)_x)$ and hence such that $v(E_x) = \sum_n v((E_n)_x)$; thus Corollary 2.4.2 implies that

$$(\mu \times v)_1(E) = \int_X v(E_x)\,\mu(dx) = \sum_n \int_X v((E_n)_x)\,\mu(dx) = \sum_n (\mu \times v)_1(E_n),$$

and so $(\mu \times v)_1$ is countably additive. A similar argument shows that $(\mu \times v)_2$ is countably additive. It is easy to check that if $A \in \mathscr{A}$ and $B \in \mathscr{B}$, then

$$(\mu \times v)_1(A \times B) = \mu(A)v(B) = (\mu \times v)_2(A \times B).$$

Hence $(\mu \times v)_1$ and $(\mu \times v)_2$ are measures on $\mathscr{A} \times \mathscr{B}$ that have the required values on the rectangles with measurable sides.

The uniqueness of $\mu \times v$ follows from Corollary 1.6.4. Thus $(\mu \times v)_1 = (\mu \times v)_2$, and Eq. (1) holds for each E in $\mathscr{A} \times \mathscr{B}$. $\square$

Example 5.1.5. Let us look again at the space $\mathbb{R}^2$. We have already shown that $\mathscr{B}(\mathbb{R}^2) = \mathscr{B}(\mathbb{R}) \times \mathscr{B}(\mathbb{R})$. Let λ_1 be Lebesgue measure on the Borel subsets of $\mathbb{R}$, and let λ_2 be Lebesgue measure on the Borel subsets of $\mathbb{R}^2$. Each rectangle in $\mathbb{R}^2$ of the form $(a,b] \times (c,d]$ is assigned the same value, namely $(b-a)(d-c)$, by λ_2 and by $\lambda_1 \times \lambda_1$; thus Proposition 1.4.3 or Corollary 1.6.4 implies that $\lambda_2 = \lambda_1 \times \lambda_1$. With this we have a second construction of Lebesgue measure on $\mathbb{R}^2$. $\square$

Exercises

1. Use the results of Sect. 5.1 to give another solution to Exercise 1.4.1.
2. Let $(X, \mathscr{A}, \mu)$ and $(Y, \mathscr{B}, v)$ be σ-finite measure spaces, and let E belong to $\mathscr{A} \times \mathscr{B}$. Show that if μ-almost every section E_x has measure zero under v, then v-almost every section E^y has measure zero under μ.
3. Show that every $(d-1)$-dimensional hyperplane in $\mathbb{R}^d$ has zero d-dimensional Lebesgue measure (a $(d-1)$-dimensional hyperplane is a set that has the form $\{x \in \mathbb{R}^d : \sum_i a_i x_i = b\}$ for some b in $\mathbb{R}$ and some nonzero element $(a_1, \ldots, a_d)$ of $\mathbb{R}^d$).
4. Let $(X, \mathscr{A})$ and $(Y, \mathscr{B})$ be measurable spaces.

    (a) Use Proposition 2.6.2 to show that for each y in Y the function $x \mapsto (x,y)$ is measurable with respect to $\mathscr{A}$ and $\mathscr{A} \times \mathscr{B}$ and that for each x in X the function $y \mapsto (x,y)$ is measurable with respect to $\mathscr{B}$ and $\mathscr{A} \times \mathscr{B}$.
    (b) Use part (a) to give another proof of Lemma 5.1.2. (See Proposition 2.6.1.)

5. Let $\mathscr{M}_1$ be the σ-algebra of Lebesgue measurable subsets of $\mathbb{R}$, and let $\mathscr{M}_2$ be the σ-algebra of Lebesgue measurable subsets of $\mathbb{R}^2$. Show that $\mathscr{M}_2 \neq \mathscr{M}_1 \times \mathscr{M}_1$. (Hint: Which subsets of $\mathbb{R}$ can arise as sections of sets in $\mathscr{M}_2$?)
6. Let $(X, \mathscr{A})$ and $(Y, \mathscr{B})$ be measurable spaces, and let K be a kernel from $(X, \mathscr{A})$ to $(Y, \mathscr{B})$ such that $K(x,Y)$ is finite for each x in X (see Exercise 2.4.7).

    (a) Show that the formula $(x,E) \mapsto K(x,E_x)$ defines a kernel from $(X, \mathscr{A})$ to $(X \times Y, \mathscr{A} \times \mathscr{B})$.
    (b) Show that if μ is a measure on $(X, \mathscr{A})$, then

    $$E \mapsto \int K(x,E_x)\,\mu(dx)$$

    defines a measure on $\mathscr{A} \times \mathscr{B}$.
    (c) How can the existence of the product of a pair of finite measures be deduced from parts (a) and (b) of this exercise?

7. Let $(X, \mathscr{A})$ and $(Y, \mathscr{B})$ be measurable spaces. Show that if $C \in \mathscr{A} \times \mathscr{B}$, then the collection of sections $\{C_x : x \in X\}$ has at most the cardinality of the continuum. (Hint: See Exercise 1.1.7. Show that if

$$C \in \sigma(\{A_n \times B_n : n = 1, 2, \dots\})$$

and if x_1 and x_2 belong to exactly the same A_n's, then $C_{x_1} = C_{x_2}$. Next use the function $x \mapsto \{\chi_{A_n}(x)\}$ to map X into $\{0, 1\}^{\mathbb{N}}$, and note that $\{0, 1\}^{\mathbb{N}}$ has the cardinality of the continuum (see A.8).)

8. Show that if the cardinality of X is larger than that of the continuum and if $\mathscr{A}$ is a σ-algebra on X, then the diagonal in $X \times X$ (that is, the set $\{(x_1, x_2) \in X \times X : x_1 = x_2\}$) does not belong to $\mathscr{A} \times \mathscr{A}$. (Hint: Use Exercise 7.)

5.2 Fubini's Theorem

The following two theorems enable one to evaluate integrals with respect to product measures by evaluating iterated integrals.

Proposition 5.2.1 (Tonelli's Theorem). *Let $(X, \mathscr{A}, \mu)$ and $(Y, \mathscr{B}, \nu)$ be σ-finite measure spaces, and let $f : X \times Y \to [0, +\infty]$ be $\mathscr{A} \times \mathscr{B}$-measurable. Then*

(a) *the function $x \mapsto \int_Y f_x \, d\nu$ is $\mathscr{A}$-measurable and the function $y \mapsto \int_X f^y \, d\mu$ is $\mathscr{B}$-measurable, and*

(b) *f satisfies*

$$\int_{X \times Y} f \, d(\mu \times \nu) = \int_X \left(\int_Y f_x \, d\nu \right) \mu(dx) = \int_Y \left(\int_X f^y \, d\mu \right) \nu(dy). \quad (1)$$

Note that the functions f_x and f^y are nonnegative and measurable (Lemma 5.1.2); thus the expression $\int_Y f_x \, d\nu$ is defined for each x in X and the expression $\int_X f^y \, d\mu$ is defined for each y in Y. Note also that (1) can be reformulated as

$$\int_{X \times Y} f(x, y) \, d(\mu \times \nu)(x, y) = \int_X \left(\int_Y f(x, y) \, \nu(dy) \right) \mu(dx)$$

$$= \int_Y \left(\int_X f(x, y) \, \mu(dx) \right) \nu(dy).$$

Proof. First suppose that E belongs to $\mathscr{A} \times \mathscr{B}$ and that f is the characteristic function of E. Then the sections f_x and f^y are the characteristic functions of the sections E_x and E^y, and so the relations $\int f_x \, d\nu = \nu(E_x)$ and $\int f^y \, d\mu = \mu(E^y)$ hold for each x and y. Thus Proposition 5.1.3 and Theorem 5.1.4 imply that conclusions (a) and (b) hold if f is a characteristic function. The additivity and homogeneity of the integral now imply that they hold for nonnegative simple $\mathscr{A} \times \mathscr{B}$-measurable functions, and Proposition 2.1.5, Proposition 2.1.8, and Theorem 2.4.1 imply that they hold for arbitrary nonnegative $\mathscr{A} \times \mathscr{B}$-measurable functions. $\qquad \square$

Note that (1) is applicable to each nonnegative $\mathscr{A} \times \mathscr{B}$-measurable function, integrable or not; thus one can often determine whether an $\mathscr{A} \times \mathscr{B}$-measurable function f is integrable by using Proposition 5.2.1 to calculate $\int |f| \, d(\mu \times \nu)$.

Theorem 5.2.2 (Fubini's Theorem). *Let $(X, \mathscr{A}, \mu)$ and $(Y, \mathscr{B}, \nu)$ be σ-finite measure spaces, and let $f : X \times Y \to [-\infty, +\infty]$ be $\mathscr{A} \times \mathscr{B}$-measurable and $\mu \times \nu$-integrable. Then*

(a) *for μ-almost every x in X the section f_x is ν-integrable and for ν-almost every y in Y the section f^y is μ-integrable,*

(b) *the functions I_f and J_f defined by*

$$I_f(x) = \begin{cases} \int_Y f_x \, d\nu & \text{if } f_x \text{ is } \nu\text{-integrable,} \\ 0 & \text{otherwise} \end{cases}$$

and

$$J_f(y) = \begin{cases} \int_X f^y \, d\mu & \text{if } f^y \text{ is } \mu\text{-integrable,} \\ 0 & \text{otherwise} \end{cases}$$

belong to $\mathscr{L}^1(X, \mathscr{A}, \mu, \mathbb{R})$ and $\mathscr{L}^1(Y, \mathscr{B}, \text{and} \nu, \mathbb{R})$, respectively, and

(c) *the relation*

$$\int_{X \times Y} f \, d(\mu \times \nu) = \int_X I_f \, d\mu = \int_Y J_f \, d\nu$$

holds.

Note that part (c) of this theorem is just a precise way of rephrasing equation (1) in the case where f is integrable but not necessarily nonnegative.

Proof. Let f^+ and f^- be the positive and negative parts of f. Lemma 5.1.2 implies that the sections f_x, $(f^+)_x$, and $(f^-)_x$ are $\mathscr{B}$-measurable, and Proposition 5.2.1 implies that the functions $x \mapsto \int (f^+)_x \, d\nu$ and $x \mapsto \int (f^-)_x \, d\nu$ are $\mathscr{A}$-measurable and μ-integrable and hence that they are finite μ-almost everywhere (Corollary 2.3.14). Thus f_x is ν-integrable for almost every x. Let N be the set of those x for which $\int (f^+)_x \, d\nu = +\infty$ or $\int (f^-)_x \, d\nu = +\infty$. Then N belongs to $\mathscr{A}$, and $I_f(x)$ is equal to 0 if $x \in N$ and is equal to $\int (f^+)_x \, d\nu - \int (f^-)_x \, d\nu$ otherwise; consequently I_f belongs to $\mathscr{L}^1(X, \mathscr{A}, \mu, \mathbb{R})$. Propositions 5.2.1 and 2.3.9 now imply that

$$\int f \, d(\mu \times \nu) = \int f^+ \, d(\mu \times \nu) - \int f^- \, d(\mu \times \nu)$$

$$= \int \left(\int (f^+)_x \, d\nu \right) \mu(dx) - \int \left(\int (f^-)_x \, d\nu \right) \mu(dx)$$

$$= \int I_f \, d\mu.$$

Similar arguments apply to the functions f^y and J_f, and so the proof is complete.

$\square$

Of course we can deal with a complex-valued function on $X \times Y$ by separating it into its real and imaginary parts.

We briefly sketch the theory of products of a finite number of measure spaces. Let $(X_1, \mathscr{A}_1, \mu_1), \ldots, (X_n, \mathscr{A}_n, \mu_n)$ be σ-finite measure spaces. Then

$$\mathscr{A}_1 \times \cdots \times \mathscr{A}_n$$

is defined to be the σ-algebra on $X_1 \times \cdots \times X_n$ generated by the sets of the form $A_1 \times \cdots \times A_n$, where $A_i \in \mathscr{A}_i$ for $i = 1, \ldots, n$. It is not hard to check that if $1 \leq k < n$ and if we make the usual identification of $(X_1 \times \cdots \times X_k) \times (X_{k+1} \times \cdots \times X_n)$ with $X_1 \times \cdots \times X_n$, then

$$(\mathscr{A}_1 \times \cdots \times \mathscr{A}_k) \times (\mathscr{A}_{k+1} \times \cdots \times \mathscr{A}_n) = \mathscr{A}_1 \times \cdots \times \mathscr{A}_n.$$

Thus we can use Theorem 5.1.4 (applied $n - 1$ times) to construct a measure $\mu_1 \times \cdots \times \mu_n$ on $\mathscr{A}_1 \times \cdots \times \mathscr{A}_n$ that satisfies

$$(\mu_1 \times \cdots \times \mu_n)(A_1 \times \cdots \times A_n) = \mu_i(A_1) \cdots \mu_n(A_n)$$

whenever $A_i \in \mathscr{A}_i$ for $i = 1, \ldots, n$. Corollary 1.6.4 implies that the measure $\mu_1 \times \cdots \times \mu_n$ is unique. Integrals with respect to $\mu_1 \times \cdots \times \mu_n$ can be evaluated by repeated applications of Proposition 5.2.1 or Theorem 5.2.2.

Exercises

1. Let λ be Lebesgue measure on $(\mathbb{R}, \mathscr{B}(\mathbb{R}))$, let μ be counting measure on $(\mathbb{R}, \mathscr{B}(\mathbb{R}))$, and let $f \colon \mathbb{R}^2 \to \mathbb{R}$ be the characteristic function of the line $\{(x,y) \in \mathbb{R}^2 : y = x\}$. Show that

$$\int\int f(x,y)\,\mu(dy)\,\lambda(dx) \neq \int\int f(x,y)\,\lambda(dx)\,\mu(dy).$$

2. Suppose that $f \colon \mathbb{R}^2 \to \mathbb{R}$ is defined by

$$f(x,y) = \begin{cases} 1 & \text{if } x \geq 0 \text{ and } x \leq y < x+1, \\ -1 & \text{if } x \geq 0 \text{ and } x+1 \leq y < x+2, \\ 0 & \text{otherwise.} \end{cases}$$

Show that $\int\int f(x,y)\,\lambda(dy)\,\lambda(dx) \neq \int\int f(x,y)\,\lambda(dx)\,\lambda(dy)$. Why does this not contradict Theorem 5.2.2?

3. (a) Let μ be a measure on $(X, \mathscr{A})$. Show that if μ is σ-finite, then there are finite measures $\mu_1, \mu_2, \ldots$ on $(X, \mathscr{A})$ such that $\mu = \sum_n \mu_n$.

   (b) Show by example that the converse of part (a) does not hold.

(c) Let $(X, \mathscr{A}, \mu)$ and $(Y, \mathscr{B}, \nu)$ be measure spaces, and let $f: X \times Y \to [0, +\infty]$ be $\mathscr{A} \times \mathscr{B}$-measurable. Show that if μ and ν are sums of series of finite measures, then the functions $x \mapsto \int f(x, y) \nu(dy)$ and $y \mapsto \int f(x, y) \mu(dx)$ are measurable, and

$$\int \int f(x, y) \nu(dy) \mu(dx) = \int \int f(x, y) \mu(dx) \nu(dy).$$

4. Let $(X, \mathscr{A})$ and $(Y, \mathscr{B})$ be measurable spaces, let μ_1 and μ_2 be finite measures on $(X, \mathscr{A})$, and let ν_1 and ν_2 be finite measures on $(Y, \mathscr{B})$. Show that if $\mu_2 \ll \mu_1$ and $\nu_2 \ll \nu_1$, then $\mu_2 \times \nu_2 \ll \mu_1 \times \nu_1$. How are the various Radon–Nikodym derivatives related? (Hint: Do both parts at once, showing that $\mu_2 \times \nu_2$ can be computed by integrating an appropriate function with respect to $\mu_1 \times \nu_1$.)

5. Let $(X, \mathscr{A})$, $(Y, \mathscr{B})$, K, and μ be as in Exercise 5.1.6, and let ν be the measure on $(X \times Y, \mathscr{A} \times \mathscr{B})$ defined in part (b) of that exercise. Show that if $f: X \times Y \to [0, +\infty]$ is $\mathscr{A} \times \mathscr{B}$-measurable, then
   (a) $x \mapsto \int f(x, y) K(x, dy)$ is $\mathscr{A}$-measurable, and
   (b) $\int f d\nu = \int \int f(x, y) K(x, dy) \mu(dx)$.

6. Let $(X, \mathscr{A}, \mu)$ and $(Y, \mathscr{B}, \nu)$ be σ-finite measure spaces, let $(\mathscr{A} \times \mathscr{B})_{\mu \times \nu}$ be the completion of $\mathscr{A} \times \mathscr{B}$ under $\mu \times \nu$, and let $\overline{\mu \times \nu}$ be the completion of $\mu \times \nu$.

   (a) Suppose that $f: X \times Y \to [0, +\infty]$ is $(\mathscr{A} \times \mathscr{B})_{\mu \times \nu}$-measurable. Show that f_x is $\mathscr{B}_\nu$-measurable for μ-almost every x in X and that f^y is $\mathscr{A}_\mu$-measurable for ν-almost every y in Y. Show also that if $\int f(x, y) \overline{\nu}(dy)$ is defined to be 0 whenever f_x is not $\mathscr{B}_\nu$-measurable and $\int f(x, y) \overline{\mu}(dx)$ is defined to be 0 whenever f^y is not $\mathscr{A}_\mu$-measurable, then

   $$\int f d(\overline{\mu \times \nu}) = \int \int f(x, y) \overline{\nu}(dy) \overline{\mu}(dx) = \int \int f(x, y) \overline{\mu}(dx) \overline{\nu}(dy).$$

   (Hint: See Proposition 2.2.5.)
   (b) State and prove an analogous modification of Theorem 5.2.2.

5.3 Applications

We begin by noting a couple of easy-to-derive consequences of the theory of product measures.

Example 5.3.1. Let $(X, \mathscr{A}, \mu)$ be a σ-finite measure space, let λ be Lebesgue measure on $(\mathbb{R}, \mathscr{B}(\mathbb{R}))$, and let $f: X \to [0, +\infty]$ be $\mathscr{A}$-measurable. Let E be defined by

$$E = \{(x, y) \in X \times \mathbb{R} : 0 \le y < f(x)\};$$

in other words, E is the "region under the graph of f." Then E belongs to $\mathscr{A} \times \mathscr{B}(\mathbb{R})$ (check this), and so its measure under $\mu \times \lambda$ can be computed using Theorem 5.1.4. On the one hand,

$$(\mu \times \lambda)(E) = \int_X \lambda(E_x)\mu(dx) = \int_X f(x)\mu(dx),$$

while on the other,

$$(\mu \times \lambda)(E) = \int_{\mathbb{R}} \mu(E^y)\lambda(dy) = \int_0^\infty \mu(\{x \in X : f(x) > y\})\lambda(dy).$$

Thus we have the often useful relation

$$\int_X f(x)\mu(dx) = \int_0^\infty \mu(\{x \in X : f(x) > y\})dy. \qquad \square$$

Example 5.3.2. Next we use Fubini's theorem to derive a familiar result about double series. Let $\sum_{m,n} a_{m,n}$ be a double series, and let μ be counting measure on $\mathbb{N}$ (more precisely, on the σ-algebra of all subsets of $\mathbb{N}$). The series $\sum_{m,n} a_{m,n}$ is absolutely convergent if and only if the function $(m,n) \mapsto a_{m,n}$ is $\mu \times \mu$-integrable. Thus Fubini's theorem implies that if $\sum_{m,n} a_{m,n}$ is absolutely convergent, then $\sum_{m=1}^\infty \sum_{n=1}^\infty a_{m,n} = \sum_{n=1}^\infty \sum_{m=1}^\infty a_{m,n}$; in other words, the order of summation can be reversed for absolutely convergent series. See also Exercise 3. $\qquad \square$

Let us consider one version of integration by parts. Another version will be discussed in Sect. 6.3.

Proposition 5.3.3. *Let $F, G \colon \mathbb{R} \to \mathbb{R}$ be bounded nondecreasing right-continuous functions that vanish[1] at $-\infty$, let μ_F and μ_G be the measures they induce on $\mathscr{B}(\mathbb{R})$, and let a and b be real numbers such that $a < b$. Then*

$$\int_{[a,b]} \frac{F(x)+F(x-)}{2}\mu_G(dx) + \int_{[a,b]} \frac{G(x)+G(x-)}{2}\mu_F(dx) \qquad (1)$$

$$= F(b)G(b) - F(a-)G(a-).$$

Proof. Let S be the square $[a,b] \times [a,b]$, and let T_1 and T_2 be the triangular regions consisting of those points (x,y) in S for which $x \geq y$ and for which $x < y$, respectively. We compute the measure of S under $\mu_F \times \mu_G$ in two ways. On the one hand,

$$(\mu_F \times \mu_G)(S) = \mu_F([a,b])\mu_G([a,b]) = (F(b)-F(a-))(G(b)-G(a-)). \qquad (2)$$

---

[1] In other words, $\lim_{x\to-\infty} F(x) = \lim_{x\to-\infty} G(x) = 0$.

On the other hand, the relation $S = T_1 \cup T_2$ and Theorem 5.1.4 imply that

$$(\mu_F \times \mu_G)(S) = \int_{[a,b]} \mu_G([a,x]) \mu_F(dx) + \int_{[a,b]} \mu_F([a,y)) \mu_G(dy). \qquad (3)$$

If we replace the "dummy variable" y in this equation with x and express $\mu_G([a,x])$ and $\mu_F([a,x))$ in terms of G and F, then the right-hand side of (3) becomes

$$\int_{[a,b]} (G(x) - G(a-)) \mu_F(dx) + \int_{[a,b]} (F(x-) - F(a-)) \mu_G(dx).$$

Equating this to the expression on the right side of (2) and using a little algebra gives

$$\int_{[a,b]} G(x) \mu_F(dx) + \int_{[a,b]} F(x-) \mu_G(dx) = F(b)G(b) - F(a-)G(a-).$$

The functions F and G can be interchanged in this identity, yielding

$$\int_{[a,b]} G(x-) \mu_F(dx) + \int_{[a,b]} F(x) \mu_G(dx) = F(b)G(b) - F(a-)G(a-).$$

These two equations together imply Eq. (1).   □

See Exercises 4 and 5 for more information about Eq. (1).

Our last application of Fubini's theorem is to the convolution of functions in $\mathscr{L}^1(\mathbb{R}, \mathscr{B}(\mathbb{R}), \lambda)$.

Proposition 5.3.4. *Let f and g belong to $\mathscr{L}^1(\mathbb{R}, \mathscr{B}(\mathbb{R}), \lambda)$. Then*

(a) *for almost every x the function $t \mapsto f(x-t)g(t)$ belongs to $\mathscr{L}^1(\mathbb{R}, \mathscr{B}(\mathbb{R}), \lambda)$, and*

(b) *the function $f * g$ defined by*

$$f * g(x) = \begin{cases} \int f(x-t)g(t)\,dt & \text{if } t \mapsto f(x-t)g(t) \text{ is Lebesgue integrable,} \\ 0 & \text{otherwise} \end{cases}$$

belongs to $\mathscr{L}^1(\mathbb{R}, \mathscr{B}(\mathbb{R}), \lambda)$ and satisfies $\|f * g\|_1 \le \|f\|_1 \|g\|_1$.

Proof. We begin by checking that the function $(x,t) \mapsto f(x-t)g(t)$ is measurable with respect to $\mathscr{B}(\mathbb{R}^2)$, and hence (see Sect. 5.1) with respect to $\mathscr{B}(\mathbb{R}) \times \mathscr{B}(\mathbb{R})$. The function $(x,t) \mapsto f(x-t)$ is the composition of the continuous, and hence Borel measurable, function $(x,t) \mapsto x-t$ with the Borel measurable function f; thus it is Borel measurable (Proposition 2.6.1). A similar argument shows that $(x,t) \mapsto g(t)$ is Borel measurable. Consequently the function $(x,t) \mapsto f(x-t)g(t)$ is Borel measurable.

Thus we can use Proposition 5.2.1 and the translation invariance of Lebesgue measure (see the remarks at the end of Sect. 2.6) to justify the calculation

$$\int |f(x-t)g(t)| d(\lambda \times \lambda)(x,t) = \int \int |f(x-t)g(t)| \lambda(dx) \lambda(dt)$$

$$= \int \|f\|_1 |g(t)| \lambda(dt) = \|f\|_1 \|g\|_1. \qquad (4)$$

It follows that $(x,t) \mapsto f(x-t)g(t)$ belongs to

$$\mathcal{L}^1(\mathbb{R}^2, \mathcal{B}(\mathbb{R}) \times \mathcal{B}(\mathbb{R}), \lambda \times \lambda),$$

and so Fubini's theorem implies that $t \mapsto f(x-t)g(t)$ is integrable for almost every x. Since $|f * g(x)| \leq \int |f(x-t)g(t)| \lambda(dt)$ holds for each x in $\mathbb{R}$, Fubini's theorem and calculation (4) also imply that $f * g$ belongs to $\mathcal{L}^1(\mathbb{R}, \mathcal{B}(\mathbb{R}), \lambda)$ and that $\|f * g\|_1 \leq \|f\|_1 \|g\|_1$. □

The *convolution* of the functions f and g in $\mathcal{L}^1(\mathbb{R}, \mathcal{B}(\mathbb{R}), \lambda)$ is the function $f * g$ defined in part (b) of Proposition 5.3.4. Note that if $f_1, f_2, g_1,$ and g_2 belong to $\mathcal{L}^1(\mathbb{R}, \mathcal{B}(\mathbb{R}), \lambda)$ and if $f_1 = f_2$ and $g_1 = g_2$ hold λ-almost everywhere, then $(f_1 * g_1)(x) = (f_2 * g_2)(x)$ holds at each x in $\mathbb{R}$. Thus convolution, which we have defined as an operator that assigns a function in $\mathcal{L}^1(\mathbb{R}, \mathcal{B}(\mathbb{R}), \lambda)$ to each pair of functions in $\mathcal{L}^1(\mathbb{R}, \mathcal{B}(\mathbb{R}), \lambda)$, can be (and usually is) considered as an operator that assigns an element of $L^1(\mathbb{R}, \mathcal{B}(\mathbb{R}), \lambda)$ to each pair of elements of $L^1(\mathbb{R}, \mathcal{B}(\mathbb{R}), \lambda)$.

Convolution turns out to be a fundamental operation in harmonic analysis. Although we do not have space to develop its properties in detail, a few are presented in the exercises below. See Chap. 10 for convolutions in probability theory, and see Sect. 9.4 for convolutions in a much more general setting.

Exercises

1. Let μ be a σ-finite measure on $(X, \mathscr{A})$, and let $f, g \colon X \to [0, +\infty]$ be $\mathscr{A}$-measurable functions such that

$$\mu(\{x : f(x) > t\}) \leq \mu(\{x : g(x) > t\})$$

   holds for each positive t. Show that $\int f \, d\mu \leq \int g \, d\mu$.

2. Let μ be a σ-finite measure on $(X, \mathscr{A})$, let $f \colon X \to [0, +\infty]$ be $\mathscr{A}$-measurable, and let p satisfy $1 \leq p < +\infty$. Show that

$$\int f^p \, d\mu = \int_0^\infty pt^{p-1} \mu(\{x : f(x) > t\}) \, dt.$$

3. Let $\sum_{m,n} a_{m,n}$ be a double series whose terms are nonnegative. Show that $\sum_m \sum_n a_{m,n} = \sum_n \sum_m a_{m,n}$
    (a) by applying Proposition 5.2.1, and
    (b) by checking directly that $\sum_m \sum_n a_{m,n}$ and $\sum_n \sum_m a_{m,n}$ are both equal to

$$\sup \left\{ \sum_{(m,n) \in F} a_{m,n} : F \text{ is a finite subset of } \mathbb{N} \times \mathbb{N} \right\}.$$

(Note that we did not assume that the series involved are convergent.)

4. Show that if the functions F and G in Proposition 5.3.3 have no points of discontinuity in common, then Eq. (1) can be replaced with the equation

$$\int_{[a,b]} F(x)\, \mu_G(dx) + \int_{[a,b]} G(x)\, \mu_F(dx) = F(b)G(b) - F(a-)G(a-).$$

5. Show by example that formula (1) of Proposition 5.3.3 cannot in general be replaced with the formula in Exercise 4.

6. Show that if f and g belong to $\mathscr{L}^1(\mathbb{R}, \mathscr{B}(\mathbb{R}), \lambda)$, then $f * g = g * f$. (Hint: See the remarks at the end of Sect. 2.6.)

7. Show that if f and g belong to $\mathscr{L}^1(\mathbb{R}, \mathscr{B}(\mathbb{R}), \lambda)$ and if g is bounded, then $f * g$ is continuous. (Hint: See Exercise 3.4.5.)

8. Suppose that A is a Borel subset of $\mathbb{R}$ that satisfies $0 < \lambda(A) < +\infty$.

    (a) Show that the function $x \mapsto \lambda(A \cap (x + A))$ is continuous and is nonzero throughout some open interval about 0 (of course $x + A$ is the set $\{x + a : a \in A\}$). (Hint: Consider $\chi_{-A} * \chi_A$, where $-A = \{-a : a \in A\}$, and use Exercise 7.)

    (b) Use part (a) to give another proof of Proposition 1.4.10.

Notes

The theory of products of a finite number of σ-finite measure spaces, as given in this chapter, can be found in almost every book on measure and integration. The theory of products of an infinite number of measure spaces of total mass 1, as presented in Sect. 10.6, is needed for the study of probability and can be found in some books on measure theory and in most books on measure-theoretic probability.

The proof of Proposition 1.4.10 indicated in Exercise 5.3.8 was shown to me by Charles Rockland and (independently) by Lee Rubel.

Chapter 6
Differentiation

In this chapter we look at two aspects of the relationship between differentiation and integration. First, in Sect. 6.1, we look at changes of variables in d-dimensional integrals. Such changes of variables occur, for example, when one evaluates an integral over a region in $\mathbb{R}^2$ by converting to polar coordinates. Then, in Sects. 6.2 and 6.3, we look at some deeper aspects of differentiation theory, including the almost everywhere differentiability of monotone functions and of indefinite integrals and the relationship between Radon–Nikodym derivatives and differentiation theory. The Vitali covering theorem is an important tool for this. The discussion of differentiation theory will be resumed when we discuss the Henstock–Kurzweil integral in Appendix H.

6.1 Change of Variable in $\mathbb{R}^d$

In this section we deal with changes of variable in $\mathbb{R}^d$ and with their relation to Lebesgue measure. The main result is Theorem 6.1.7. Let us begin by recalling some definitions.

Let M_d be the set of all d by d matrices with real entries, and let D be a real-valued function on M_d. We will sometimes find it convenient to denote the columns of a d by d matrix A by $A_1, A_2, \ldots, A_d$ and to write $D(A_1, A_2, \ldots, A_d)$ in place of $D(A)$. The function D is *multilinear* if for each i and each choice of A_j (for $j \neq i$) the map $A_i \mapsto D(A_1, \ldots, A_d)$ is linear, is *alternating* if $D(A) = 0$ holds whenever two of the columns of A are equal, and is a *determinant* if it is multilinear, is alternating, and satisfies $D(I) = 1$ (here I is, of course, the d by d identity matrix).

We need to recall a few basic facts about determinants.

Lemma 6.1.1. *For each positive integer d there is a unique determinant on M_d.*

We follow the standard usage and use $\det(A)$ to denote the determinant of a matrix A.

D.L. Cohn, *Measure Theory: Second Edition*, Birkhäuser Advanced
Texts Basler Lehrbücher, DOI 10.1007/978-1-4614-6956-8_6,
© Springer Science+Business Media, LLC 2013

Lemma 6.1.2. *Let d be a positive integer, and let M_d be the set of all d by d matrices with real entries. Then*

(a) $\det(AB) = \det(A)\det(B)$ *holds for all A, B in M_d,*
(b) $\det(A)$ *is nonzero if and only if A is invertible,*
(c) $\det(A)$ *is a polynomial in the components of A, and*
(d) $\det(A^t) = \det(A)$*, where A^t is the transpose of A.*

Proofs of Lemmas 6.1.1 and 6.1.2 can be found in Halmos [53] and Hoffman and Kunze [61].

Recall that if $T: \mathbb{R}^d \to \mathbb{R}^d$ is linear, if A is the matrix of T with respect to some ordered basis of $\mathbb{R}^d$, and if B is the matrix of T with respect to some possibly different ordered basis of $\mathbb{R}^d$, then there is an invertible matrix U such that $A = UBU^{-1}$. It follows that $\det(A) = \det(U)\det(B)\det(U^{-1}) = \det(B)$. Thus $\det(T)$, the *determinant* of the linear operator T, can be defined to be the determinant of a matrix representing T; it does not matter which ordered basis is used to compute the matrix.

Let us prove the following special case of Theorem 6.1.7.

Proposition 6.1.3. *Let $T: \mathbb{R}^d \to \mathbb{R}^d$ be an invertible linear map. Then*

$$\lambda(T(B)) = |\det(T)|\lambda(B)$$

holds for each Borel subset B of $\mathbb{R}^d$.

Proof. The maps T and T^{-1} are continuous (check this) and hence measurable[1] with respect to $\mathscr{B}(\mathbb{R}^d)$ and $\mathscr{B}(\mathbb{R}^d)$; thus $T(B)$ is a Borel set if and only if B is a Borel set.

Since T is invertible, there exist linear maps $T_1, T_2, \ldots, T_n$ such that $T = T_1 \circ T_2 \circ \cdots \circ T_n$ and such that each T_k operates on a vector x in one of the following ways:

(a) one component of x is multiplied by a nonzero number, and the other components are left unchanged;
(b) two components of x are interchanged, and the other components are left unchanged;
(c) for some i and j the component x_i is replaced with $x_i + x_j$, while the other components of x are left unchanged

(see Exercise 1). In view of the relation $\det(T) = \det(T_1)\det(T_2)\ldots\det(T_n)$, it suffices to show that

$$\lambda(T_k(B)) = |\det(T_k)|\lambda(B) \tag{1}$$

holds for each k and each Borel set B.

---

[1]Since $T^{-1}(U)$ and $T(U)$ are open and hence Borel whenever U is a open subset of $\mathbb{R}^d$, the measurability of T and T^{-1} follows from Proposition 2.6.2.

First suppose that T_k arises through case (a) or case (b) above. Then it is easy to check that (1) holds if B is a cube with edges parallel to the coordinate axes and hence if B is an open set (use Lemma 1.4.2) or an arbitrary Borel set (use the regularity of λ).

Next suppose that T_k arises through case (c). Then there exist indices i and j such that if $x = (x_1, \ldots, x_d)$, then the ith component of $T_k(x)$ is $x_i + x_j$, while the other components of $T_k(x)$ agree with the corresponding components of x. Let us view $\mathbb{R}^d$ as the product of $\mathbb{R}$ (corresponding to the ith coordinate in $\mathbb{R}^d$) with $\mathbb{R}^{d-1}$ (corresponding to the remaining coordinates). Let B be a Borel subset of $\mathbb{R}^d$. It is easy to check that for each u in $\mathbb{R}^{d-1}$ the sections at u of B and of $T_k(B)$ are translates of one another and hence have the same Lebesgue measure. Thus it follows from the theory of product measures (in particular, from Theorem 5.1.4, an extension of Example 5.1.5 to $\mathbb{R}^d$, and the remarks at the end of Sect. 5.2) that $\lambda(B) = \lambda(T_k(B))$. Since $\det(T_k) = 1$ holds whenever T_k arises through case (c), the proof is complete.

$\square$

We will need the following standard facts about derivatives of vector-valued functions; proofs can be found in a number of advanced calculus or basic analysis texts,[2] and are sketched in Exercises 2, 3, and 4.

Let X and Y be Banach spaces, let U be an open subset of X, and let x_0 belong to U. A function $F: U \to Y$ is *differentiable* at x_0 if there is a continuous linear map $T: X \to Y$ such that

$$\lim_{x \to x_0} \frac{\|F(x) - F(x_0) - T(x - x_0)\|}{\|x - x_0\|} = 0. \tag{2}$$

It is easy to check that given x_0 and F, there is at most one such map T; it is called the *derivative* of F at x_0 and is denoted by $F'(x_0)$. It is also easy to check that if F is differentiable at x_0, then F is continuous at x_0. Furthermore, if $T: X \to Y$ is continuous and linear, then it is differentiable, with derivative T, at each point in X.

The *chain rule* now takes the following form.

Proposition 6.1.4. *Let X, Y, and Z be Banach spaces, and let U and V be open subsets of X and Y. If $x_0 \in U$, if $G: U \to Y$ is differentiable at x_0 and satisfies $G(U) \subseteq V$, and if $F: V \to Z$ is differentiable at $G(x_0)$, then $F \circ G$ is differentiable at x_0, and*

$$(F \circ G)'(x_0) = F'(G(x_0)) \circ G'(x_0).$$

A method for proving Proposition 6.1.4 is suggested in Exercise 5.

Let us now restrict our attention to the special case of the space $\mathbb{R}^d$. It will prove convenient to endow $\mathbb{R}^d$ with the norm $\|\cdot\|_\infty$ defined by

$$\|x\|_\infty = \max(|x_1|, |x_2|, \ldots, |x_d|)$$

---

[2]See Bartle [4], Hoffman [60], Loomis and Sternberg [85], Rudin [104], or Thomson, Bruckner, and Bruckner [117].

(here $x_1, x_2, \ldots, x_d$ are the components of the vector x). It is easy to check that the open sets and the continuous functions determined by $\|\cdot\|_\infty$ are the same as those determined by the usual norm $\|\cdot\|_2$ (see Exercise 6). If $\mathbb{R}^d$ is given the norm $\|\cdot\|_\infty$, if $T\colon \mathbb{R}^d \to \mathbb{R}^d$ is linear, and if (a_{ij}) is the matrix of T with respect to the usual ordered basis of $\mathbb{R}^d$, then T is continuous and its norm (see Sect. 3.5) is given by

$$\|T\| = \max_i \sum_{j=1}^d |a_{ij}| \tag{3}$$

(see Exercise 7).

Now let U be an open subset of $\mathbb{R}^d$, let F be a function from U to $\mathbb{R}^d$, and let $f_1, \ldots, f_d$ be the components of F; thus $F(x) = (f_1(x), \ldots, f_d(x))$ holds at each x in U. Then F is said to be a C^1 *function* (or to be *of class C^1*) if the partial derivatives $\partial f_i / \partial x_j, i, j = 1, \ldots, d$ exist and are continuous at each point in U.

We will need the following facts.

Lemma 6.1.5. *Let U be an open subset of $\mathbb{R}^d$, and let $F\colon U \to \mathbb{R}^d$ be a C^1 function. Then F is differentiable at each point in U, and the matrix of $F'(x)$ (with respect to the usual ordered basis of $\mathbb{R}^d$) is $(\partial f_i(x)/\partial x_j)$.*

Lemma 6.1.6. *Let U be an open subset of $\mathbb{R}^d$, and let $F\colon U \to \mathbb{R}^d$ be differentiable at each point in U. If x_0 and x_1, together with all the points on the line segment connecting them, belong to U and if $\|F'(x)\| \leq C$ holds at each point x on this line segment, then*

$$\|F(x_1) - F(x_0)\|_\infty \leq C\|x_1 - x_0\|_\infty.$$

See Exercises 8 and 9 for sketches of proofs of these lemmas.

The *Jacobian* J_F of the C^1 function F is defined by $J_F(x) = \det(F'(x))$. In view of Lemmas 6.1.2 and 6.1.5, the Jacobian of such a function is continuous and hence Borel measurable.

We turn to the main result of this section.

Theorem 6.1.7. *Let U and V be open subsets of $\mathbb{R}^d$, and let T be a bijection of U onto V such that T and T^{-1} are both of class C^1. Then each Borel subset B of U satisfies*

$$\lambda(T(B)) = \int_B |J_T(x)|\,\lambda(dx), \tag{4}$$

and each Borel measurable function $f\colon V \to \mathbb{R}$ satisfies

$$\int_V f\,d\lambda = \int_U f(T(x))|J_T(x)|\,\lambda(dx), \tag{5}$$

in the sense that if either of the integrals in (5) exists, then both exist and (5) holds.

Note that, in view of the identity $T^{-1}(T(x)) = x$, the chain rule implies that $(T^{-1})'(T(x)) \circ T'(x) = I$ holds at each x in U. Thus $T'(x)$ is invertible, and so $J_T(x)$ is nonzero, for each such x.

Note also that T and T^{-1} are Borel measurable (since they are continuous); thus a subset B of U is a Borel set if and only if $T(B)$ is Borel.

We need the following two lemmas for the proof of Theorem 6.1.7.

Lemma 6.1.8. *Let U be an open subset of $\mathbb{R}^d$, let $G: U \to \mathbb{R}^d$ be a differentiable function, let ε be a positive number, and let C be a cube that is a Borel set, is included in U, has edges parallel to the coordinate axes, and is such that*

$$\|G'(x) - I\| \le \varepsilon$$

holds at each x in C. Then the image $G(C)$ of C under G satisfies

$$\lambda^*(G(C)) \le (1+\varepsilon)^d \lambda(C).$$

Proof. Let x_0 be the center of C and let b be the length of the edges of C. Then each x in C satisfies $\|x - x_0\|_\infty \le b/2$, and so Lemma 6.1.6, applied to the function $x \mapsto G(x) - x$, implies that each x in C satisfies

$$\|(G(x) - x) - (G(x_0) - x_0)\|_\infty \le \varepsilon \|x - x_0\|_\infty$$

and hence satisfies

$$\|G(x) - G(x_0)\|_\infty \le (1+\varepsilon)\|x - x_0\|_\infty \le \frac{1}{2}(1+\varepsilon)b.$$

Thus $G(C)$ is a subset of the closed cube (with edges parallel to the coordinate axes) whose center is at $G(x_0)$ and whose edges are of length $(1+\varepsilon)b$. Since this cube has measure $(1+\varepsilon)^d b^d$, while C has measure b^d, the lemma follows. $\square$

Lemma 6.1.9. *Let U, V, and T be as in the statement of Theorem 6.1.7. Suppose that a is a positive number and that B is a Borel subset of U.*

(a) *If $|J_T(x)| \le a$ holds at each x in B, then $\lambda(T(B)) \le a\lambda(B)$.*
(b) *If $|J_T(x)| \ge a$ holds at each x in B, then $\lambda(T(B)) \ge a\lambda(B)$.*

Proof. First suppose that b is a positive number and that W is an open subset of U such that

(a) $\overline{W}$ is compact and included in U, and
(b) $|J_T(x)| < b$ holds at each x in W.

Let ε be a positive number. Since $\overline{W}$ is compact and T is of class C^1, the functions that take x to the components[3] of $T'(x)$—that is, to the partial derivatives of the

---

[3] Here we are dealing with the components of the matrices of these operators with respect to the usual ordered basis of $\mathbb{R}^d$.

components of T—are uniformly continuous on W (part (a) of Theorem C.12). A similar argument shows that the components of $(T'(x))^{-1}$ are bounded on W. Thus (see Eq. (3)) we can choose first a positive number M such that

$$\|(T'(x))^{-1}\| \leq M \tag{6}$$

holds at each x in W and then a positive number δ such that

$$\|T'(x) - T'(x_0)\| \leq \frac{\varepsilon}{M} \tag{7}$$

holds whenever x and x_0 belong to W and satisfy $\|x - x_0\| \leq \delta$.

According to Lemma 1.4.2 the set W is the union of a countable family $\{C_i\}$ of disjoint half-open cubes with edges parallel to the coordinate axes. By subdividing these cubes, if necessary, we can assume that each has edges of length at most 2δ. Let C be one of these cubes, let x_0 be its center, and define $G: U \to \mathbb{R}^d$ by

$$G = (T'(x_0))^{-1} \circ T.$$

The chain rule implies that for each x in U we have

$$G'(x) - I = (T'(x_0))^{-1} \circ T'(x) - I$$
$$= (T'(x_0))^{-1} \circ (T'(x) - T'(x_0)),$$

and so (6), (7), and Exercise 3.5.1 imply that

$$\|G'(x) - I\| \leq \|(T'(x_0))^{-1}\| \cdot \|T'(x) - T'(x_0)\|$$
$$\leq M \cdot \frac{\varepsilon}{M} = \varepsilon$$

holds at each x in C. It now follows from Lemma 6.1.8 that $\lambda(G(C)) \leq (1+\varepsilon)^d \lambda(C)$. If we use Proposition 6.1.3 and the fact that $T = T'(x_0) \circ G$, we find

$$\lambda(T(C)) = |\det(T'(x_0))|\lambda(G(C))$$
$$\leq b(1+\varepsilon)^d \lambda(C).$$

Since C was an arbitrary one of the cubes C_i, it follows that

$$\lambda(T(W)) = \sum_i \lambda(T(C_i))$$
$$\leq \sum_i b(1+\varepsilon)^d \lambda(C_i) = b(1+\varepsilon)^d \lambda(W)$$

holds for each ε, and hence that $\lambda(T(W)) \leq b\lambda(W)$.

Now suppose that W is an arbitrary open subset of U such that $|J_T(x)| < b$ holds at each x in W. We can choose an increasing sequence $\{W_n\}$ of open sets such that $W = \cup_n W_n$ and such that the closure of each W_n is compact and included in U (the details are left to the reader). Then each W_n satisfies $\lambda(T(W_n)) \leq b\lambda(W_n)$, and so we have

$$\lambda(T(W)) = \lim_n \lambda(T(W_n)) \leq \lim_n b\lambda(W_n) = b\lambda(W). \tag{8}$$

More generally, let B be a Borel subset of U such that $|J_T(x)| \leq a$ holds at each x in B. Let b be a number such that $a < b$. If W is an open subset of U that includes B and if W_b is defined by $W_b = \{x \in W : |J_T(x)| < b\}$, then $B \subseteq W_b$ and inequality (8) implies that

$$\lambda(T(B)) \leq \lambda(T(W_b)) \leq b\lambda(W_b) \leq b\lambda(W).$$

Since b can be made arbitrarily close to a and since λ is regular (Proposition 1.4.1), part (a) of the lemma follows.

We will prove part (b) by applying part (a) to the function $T^{-1}: V \to U$. If $|J_T(x)| \geq a$ holds at each x in B, then $|J_{T^{-1}}(y)| \leq 1/a$ holds at each y in $T(B)$, and so part (a) of the lemma implies that $\lambda(T^{-1}(T(B))) \leq (1/a)\lambda(T(B))$ or, equivalently, that $a\lambda(B) \leq \lambda(T(B))$. □

Proof of Theorem 6.1.7. First suppose that B is a Borel subset of U for which $\lambda(B)$ is finite. For each positive integer n define sets $B_{n,k}$, $k = 1, 2, \ldots$, by

$$B_{n,k} = \left\{ x \in B : \frac{k-1}{n} \leq |J_T(x)| < \frac{k}{n} \right\}.$$

It follows from Lemma 6.1.9 that

$$\frac{k-1}{n}\lambda(B_{n,k}) \leq \lambda(T(B_{n,k})) \leq \frac{k}{n}\lambda(B_{n,k}) \tag{9}$$

and from the definition of $B_{n,k}$ that

$$\frac{k-1}{n}\lambda(B_{n,k}) \leq \int_{B_{n,k}} |J_T(x)|\lambda(dx) \leq \frac{k}{n}\lambda(B_{n,k}). \tag{10}$$

We conclude from (9) and (10) that

$$\left| \lambda(T(B_{n,k})) - \int_{B_{n,k}} |J_T(x)|\lambda(dx) \right| \leq \frac{k}{n}\lambda(B_{n,k}) - \frac{k-1}{n}\lambda(B_{n,k}) = \frac{1}{n}\lambda(B_{n,k})$$

and, from this, since $B = \cup_k B_{n,k}$, that

$$\left| \lambda(T(B)) - \int_B |J_T(x)|\lambda(dx) \right| \leq \frac{1}{n}\lambda(B).$$

However, n is arbitrary and $\lambda(B)$ is finite, and so

$$\lambda(T(B)) = \int_B |J_T(x)|\lambda(dx).$$

Thus (4) is proved in the case where $\lambda(B)$ is finite.

If B is an arbitrary Borel subset of U, then it is the union of an increasing sequence $\{B_k\}$ of Borel sets of finite measure, and taking limits over k in the relation

$$\lambda(T(B_k)) = \int_{B_k} |J_T(x)| \lambda(dx)$$

yields

$$\lambda(T(B)) = \int_B |J_T(x)| \lambda(dx).$$

This completes the proof of (4).

We turn to the proof of (5). If f is the characteristic function of a Borel subset C of V and if $B = T^{-1}(C)$, then (5) reduces to (4). The linearity of the integral and the monotone convergence theorem now imply that (5) holds for all nonnegative Borel functions. The case of an arbitrary Borel function f reduces to this through the decomposition $f = f^+ - f^-$.   □

With more work it is possible to prove somewhat strengthened versions of Theorem 6.1.7 (see, for example, Theorem 8.26 in Rudin [105]). The version given here, however, seems adequate for most purposes.

Example 6.1.10. Let us apply Theorem 6.1.7 to polar coordinates in $\mathbb{R}^2$. Let R be a positive number, let

$$U = \{(r, \theta) : 0 < r < R \text{ and } 0 < \theta < 2\pi\},$$

let

$$V = \{(x, y) : x^2 + y^2 < R^2\},$$

and let V_0 be the set consisting of those points in V that do not lie on the nonnegative x-axis. Define $T : U \to \mathbb{R}^2$ by $T(r, \theta) = (r\cos\theta, r\sin\theta)$. Then T, U, and V_0 satisfy the hypotheses of Theorem 6.1.7. Furthermore, $J_T(r, \theta) = r$. Since V and V_0 differ only by a Lebesgue null set, each integrable function $f : V \to \mathbb{R}$ satisfies

$$\int_V f\, d\lambda = \int_{V_0} f\, d\lambda = \int_0^{2\pi} \int_0^R f(r\cos\theta, r\sin\theta) r\, dr\, d\theta.$$

This is, of course, the standard formula for the evaluation of integrals by means of polar coordinates.   □

Exercises

1. Show that if $T : \mathbb{R}^d \to \mathbb{R}^d$ is an invertible linear map, then T can be decomposed as specified in the second paragraph of the proof of Proposition 6.1.3. (Hint: Let

A be the matrix of T with respect to the usual basis of $\mathbb{R}^d$, and recall hows one can use Gaussian elimination to find the inverse of A by performing row operations on the d by $2d$ matrix $(A|I)$, consisting of A followed by the d by d identity matrix. How are linear maps satisfying conditions (a), (b), and (c) of the proof of Proposition 6.1.3 related to row operations?)

2. Show that if F is differentiable at x_0, then $F'(x_0)$ is unique. (Hint: Check that if S and T are both derivatives of F at x_0, then

$$\lim_{x \to x_0} \frac{\|(S-T)(x-x_0)\|}{\|x-x_0\|} = 0;$$

conclude that $S = T$.)

3. Show that if F is differentiable at x_0, then F is continuous at x_0. (Hint: Use Eq. (2), together with the continuity of $x \mapsto F'(x_0)(x)$, to verify that $\lim_{x \to x_0} \|F(x) - F(x_0)\| = 0$.)

4. Show that if $T: X \to Y$ is a continuous linear map from one Banach space to another, then T is differentiable, with derivative T, at each point in X. (Hint: Simplify the expression $T(x) - T(x_0) - T(x - x_0)$.)

5. Prove the chain rule, Proposition 6.1.4. (Hint: Let $y_0 = G(x_0)$ and define remainders R_{F,y_0} and R_{G,x_0} by

$$F(y) = F(y_0) + F'(y_0)(y - y_0) + R_{F,y_0}(y - y_0)$$

and

$$G(x) = G(x_0) + G'(x_0)(x - x_0) + R_{G,x_0}(x - x_0);$$

then compute $F(G(x))$ in terms of $F(G(x_0))$, $G'(x_0)$, $F'(G(x_0))$, R_{G,x_0}, R_{F,y_0}, and $x - x_0$. Consider the behavior of the remainders as x approaches x_0.)

6. Let $\|\cdot\|_2$ and $\|\cdot\|_\infty$ be the norms on $\mathbb{R}^d$ defined by $\|x\|_2 = (\sum_i x_i^2)^{1/2}$ and $\|x\|_\infty = \max_i |x_i|$.
   (a) Show that each x in $\mathbb{R}^d$ satisfies $\|x\|_\infty \le \|x\|_2 \le \sqrt{d}\,\|x\|_\infty$.
   (b) Use part (a) to show that the open sets determined by $\|\cdot\|_2$ are the same as those determined by $\|\cdot\|_\infty$.

7. Verify Eq. (3). (Hint: Suppose that $x \in \mathbb{R}^d$ and $y = T(x)$, and calculate an upper bound for $|y_i|$ in terms of $\|x\|_\infty$ and the elements of the matrix (a_{ij}). Also note how to construct a vector x that satisfies $\|x\|_\infty = 1$ and $\|T(x)\|_\infty = \max_i \sum_j |a_{ij}|$ by letting x be an appropriate sequence of 1's and -1's.)

8. Prove Lemma 6.1.5. (Hint: First consider the derivatives (as linear operators from $\mathbb{R}^d$ to $\mathbb{R}$) of the components f_i of F. Let x and x_0 belong to U, and define points u_j, for $j = 0, \ldots, d$, by letting the first j components of u_j agree with the corresponding components of x and letting the remaining components of u_j agree with the corresponding components of x_0. If x_0 is fixed and x is sufficiently close to x_0, then each u_i belongs to U. Use the formula $f_i(x) - f_i(x_0) =$

$\sum_{j=1}^{d}(f_i(u_j) - f_i(u_{j-1}))$, together with the mean value theorem (Theorem C.14), to show that there are points $v_1, \ldots, v_d$ such that[4]

$$f_i(x) - f_i(x_0) = \sum_{j=1}^{d}(\partial f_i(v_j)/\partial x_j)(x_j - x_{0,j}) \tag{11}$$

and such that for each j the point v_j lies on the line segment connecting u_{j-1} and u_j. Deduce the differentiability of f_i at x_0 and compute the matrix of $f_i'(x_0)$. Finally, turn to F.)

9. Prove Lemma 6.1.6. (Hint: Let $f_1, \ldots, f_d$ be the components of F. It is enough to show that $|f_i(x_1) - f_i(x_0)| \leq C\|x_1 - x_0\|_\infty$ holds for each i. Use the chain rule to compute the derivative of the function $t \mapsto f_i(x_0 + t(x_1 - x_0))$, and then use the mean value theorem (Theorem C.14) and Exercise 3.5.1 to obtain the required bound for $|f_i(x_1) - f_i(x_0)|$.)

6.2   Differentiation of Measures

Let $\mathscr{C}$ be the family consisting of those nondegenerate closed cubes in $\mathbb{R}^d$ whose edges are parallel to the coordinate axes. In other words, let $\mathscr{C}$ be the collection of all sets of the form

$$[a_1, b_1] \times [a_2, b_2] \times \cdots \times [a_d, b_d],$$

where $[a_1, b_1], \ldots, [a_d, b_d]$ are closed subintervals of $\mathbb{R}$ that have a common nonzero length. For each cube C in $\mathscr{C}$ let $e(C)$ be the length of the edges of C.

Suppose that A is a subset of $\mathbb{R}^d$. A *Vitali covering* of A is a subfamily $\mathscr{V}$ of $\mathscr{C}$ such that for each x in A and each positive number δ there is a cube C that belongs to $\mathscr{V}$, contains x, and satisfies $e(C) < \delta$.

The following fact about Vitali coverings forms the basis for our treatment of differentiation theory. The reader should note, however, that differentiation theory can also be based on the "rising sun lemma" of F. Riesz; see, for example, Chapter I of Riesz and Nagy [99].

Theorem 6.2.1 (Vitali Covering Theorem). *Let A be an arbitrary nonempty subset of $\mathbb{R}^d$, and let $\mathscr{V}$ be a Vitali covering of A. Then there is a finite or infinite sequence $\{C_n\}$ of disjoint sets that belong to $\mathscr{V}$ and are such that $\cup_n C_n$ contains λ-almost every point in A.*

Proof. First consider the case where the set A is bounded. Choose a bounded open subset U of $\mathbb{R}^d$ that includes A, and let $\mathscr{V}_0$ consist of those cubes in $\mathscr{V}$ that are included in U. It is clear that $\mathscr{V}_0$ is a Vitali covering of A. Let

$$\delta_1 = \sup\{e(C) : C \in \mathscr{V}_0\}.$$

---

[4]The symbols x_j and $x_{0,j}$ in (11) refer to the jth components of the vectors x and x_0.

Then δ_1 satisfies $0 < \delta_1 < +\infty$ (recall that A is nonempty and U is bounded), and we can choose a cube C_1 that belongs to $\mathcal{V}_0$ and satisfies $e(C_1) > \delta_1/2$. We continue this construction inductively, producing sequences $\{\delta_n\}$ and $\{C_n\}$ as follows. If $A \subseteq \bigcup_{k=1}^n C_k$, then the construction is complete. Otherwise there are points in A that lie outside $\bigcup_{k=1}^n C_k$, and so, since $\bigcup_{k=1}^n C_k$ is closed and $\mathcal{V}_0$ is a Vitali covering of A, there are cubes in $\mathcal{V}_0$ that are disjoint from $\bigcup_{k=1}^n C_k$. Thus the quantity δ_{n+1} defined by

$$\delta_{n+1} = \sup\{e(C) : C \in \mathcal{V}_0 \text{ and } C \cap (\bigcup_{k=1}^n C_k) = \varnothing\}$$

satisfies $0 < \delta_{n+1} < +\infty$, and we can choose a cube C_{n+1} in $\mathcal{V}_0$ that satisfies $e(C_{n+1}) > \delta_{n+1}/2$ and is disjoint from $\bigcup_{k=1}^n C_k$. This completes the induction step in the construction of the sequences $\{\delta_n\}$ and $\{C_n\}$.

If this construction terminates in a finite number, say N, of steps, then $A \subseteq \bigcup_{n=1}^N C_n$ and $\{C_n\}_{n=1}^N$ is the required sequence. We turn to the case in which the construction does not terminate.

Since the sets C_n are disjoint and included in the bounded set U, the series $\sum_n \lambda(C_n)$ must be convergent; thus $\lim_n \lambda(C_n) = 0$ and hence $\lim_n \delta_n = 0$. For each n let D_n be the cube in $\mathcal{C}$ with the same center as C_n but with edges 5 times as long as those of C_n. Then, since $\lambda(D_n) = 5^d \lambda(C_n)$, the series $\sum_n \lambda(D_n)$ is also convergent. We will show that

$$A - \bigcup_{n=1}^N C_n \subseteq \bigcup_{n=N+1}^\infty D_n \tag{1}$$

holds for each positive integer N. This inclusion implies that

$$\lambda^*(A - \bigcup_{n=1}^\infty C_n) \leq \lambda^*(A - \bigcup_{n=1}^N C_n) \leq \sum_{n=N+1}^\infty \lambda(D_n);$$

since the convergence of $\sum_{n=1}^\infty \lambda(D_n)$ implies that $\lim_N \sum_{n=N+1}^\infty \lambda(D_n) = 0$, it follows that $\lambda^*(A - \bigcup_{n=1}^\infty C_n) = 0$ and hence that $\{C_n\}$ is the required sequence.

We turn to the proof of (1). Suppose that x belongs to $A - \bigcup_{n=1}^N C_n$. Since $\bigcup_{n=1}^N C_n$ is closed and $\mathcal{V}_0$ is a Vitali covering of A, there are cubes in $\mathcal{V}_0$ that contain x and are disjoint from $\bigcup_{n=1}^N C_n$. Let C be such a cube. Then C meets $\bigcup_{n=1}^k C_n$ for some k, since otherwise we would have $e(C) \leq \delta_k$ for all k, contradicting $\lim_n \delta_n = 0$. Let k_0 be the smallest of those positive integers k for which C meets $\bigcup_{n=1}^k C_n$. Then $e(C) \leq \delta_{k_0}$ and $\delta_{k_0}/2 \leq e(C_{k_0})$, and it follows that $e(C) \leq 2e(C_{k_0})$. The definition of the sets D_n, the inequality $e(C) \leq 2e(C_{k_0})$, and the fact that $C \cap C_{k_0} \neq \varnothing$ together imply that $C \subseteq D_{k_0}$. Since C was chosen to be disjoint from $\bigcup_{n=1}^N C_n$, it follows that $k_0 \geq N+1$, and so

$$x \in C \subseteq D_{k_0} \subseteq \bigcup_{n=N+1}^\infty D_n.$$

Relation (1) follows, since x was an arbitrary element of $A - \bigcup_{n=1}^N C_n$. This completes the proof of the theorem in the case where A is bounded.

Now suppose that the set A is unbounded. Let $U_1, U_2, \ldots$ be disjoint bounded open subsets of $\mathbb{R}^d$ such that $\lambda(\mathbb{R}^d - (\bigcup_{k=1}^\infty U_k)) = 0$; for example, the open cubes whose edges have length 1 and whose corners have integer coordinates will do. For each k such that $A \cap U_k \neq \varnothing$ we can use the preceding argument to choose a sequence $\{C_{k,j}\}_j$ of disjoint cubes that belong to $\mathcal{V}$ and are such that $\cup_j C_{k,j}$ is included in U_k and contains almost every point in $A \cap U_k$. Merging the resulting sequences into one sequence completes the proof.   □

Let μ be a finite Borel measure on $\mathbb{R}^d$. Then $(\overline{D}\mu)(x)$, the *upper derivate of* μ at x, is defined by

$$(\overline{D}\mu)(x) = \limsup_{\varepsilon \downarrow 0} \left\{ \frac{\mu(C)}{\lambda(C)} : C \in \mathcal{C}, x \in C, \text{ and } e(C) < \varepsilon \right\}, \tag{2}$$

and $(\underline{D}\mu)(x)$, the *lower derivate of* μ at x, is defined by

$$(\underline{D}\mu)(x) = \liminf_{\varepsilon \downarrow 0} \left\{ \frac{\mu(C)}{\lambda(C)} : C \in \mathcal{C}, x \in C, \text{ and } e(C) < \varepsilon \right\}. \tag{3}$$

The *upper derivate* and the *lower derivate* of μ are the $[0, +\infty]$-valued functions $\overline{D}\mu$ and $\underline{D}\mu$ whose values at x are given by (2) and (3). The measure μ is *differentiable* at x if $(\overline{D}\mu)(x)$ and $(\underline{D}\mu)(x)$ are finite and equal, and at each such x the *derivative* $(D\mu)(x)$ *of* μ at x is defined by

$$(D\mu)(x) = (\overline{D}\mu)(x) = (\underline{D}\mu)(x). \tag{4}$$

The *derivative* of μ is the function $D\mu$ that is defined by (4) at each x at which μ is differentiable and is undefined elsewhere.

Lemma 6.2.2. *Let μ be a finite Borel measure on $\mathbb{R}^d$. Then $\overline{D}\mu$, $\underline{D}\mu$, and $D\mu$ are Borel measurable.*

Proof. Let $\mathcal{U}$ be the collection of all open cubes in $\mathbb{R}^d$ whose edges are parallel to the coordinate axes, and for each U in $\mathcal{U}$ let $e(U)$ be the length of the edges of U. Note that for each cube C in $\mathcal{C}$ there is a decreasing sequence $\{U_n\}$ of cubes in $\mathcal{U}$ for which $C = \cap_n U_n$ and hence (Proposition 1.2.5) for which $\mu(C)/\lambda(C) = \lim_n \mu(U_n)/\lambda(U_n)$. Likewise for each cube U in $\mathcal{U}$ there is an increasing sequence $\{C_n\}$ of cubes in $\mathcal{C}$ for which $U = \cup_n C_n$ and hence for which $\mu(U)/\lambda(U) = \lim_n \mu(C_n)/\lambda(C_n)$. It follows that $(\overline{D}\mu)(x)$ is given by

$$(\overline{D}\mu)(x) = \limsup_{\varepsilon \downarrow 0} \left\{ \frac{\mu(U)}{\lambda(U)} : U \in \mathcal{U}, x \in U, \text{ and } e(U) < \varepsilon \right\}.$$

For each positive ε let us define a function $s_\varepsilon : \mathbb{R}^d \to [0, \infty]$ whose value at x is the supremum considered above:

$$s_\varepsilon(x) = \sup \left\{ \frac{\mu(U)}{\lambda(U)} : U \in \mathcal{U}, x \in U, \text{ and } e(U) < \varepsilon \right\}.$$

Then for each a in $\mathbb{R}$ we have

$$\{x \in \mathbb{R}^d : s_\varepsilon(x) > a\} = \bigcup \left\{ U \in \mathscr{U} : e(U) < \varepsilon \text{ and } \frac{\mu(U)}{\lambda(U)} > a \right\},$$

and so s_ε is Borel measurable. If $\{\varepsilon_n\}$ is a sequence of numbers that decreases to 0, then $\overline{D}\mu$ is the pointwise limit of the sequence of functions $\{s_{\varepsilon_n}\}$ and so is Borel measurable. The measurability of $\underline{D}\mu$ can be proved in a similar way.

The measurability of $D\mu$ is a consequence of Proposition 2.1.3 and the measurability of $\overline{D}\mu$ and $\underline{D}\mu$.   □

The following theorem is the main result of this section.

Theorem 6.2.3. *Let μ be a finite Borel measure on $\mathbb{R}^d$. Then μ is differentiable at λ-almost every point in $\mathbb{R}^d$, and the function defined by*

$$x \mapsto \begin{cases} (D\mu)(x) & \text{if } \mu \text{ is differentiable at } x, \\ 0 & \text{otherwise} \end{cases}$$

is a Radon–Nikodym derivative of the absolutely continuous part of μ.

We will need the following two lemmas for the proof of Theorem 6.2.3.

Lemma 6.2.4. *Let μ be a finite Borel measure on $\mathbb{R}^d$, let a be a positive real number, and let A be a Borel subset of $\mathbb{R}^d$ such that $(\overline{D}\mu)(x) \geq a$ holds at each x in A. Then $\mu(A) \geq a\lambda(A)$.*

Proof. We can certainly assume that A is nonempty. Let U be an open set that includes A, let ε satisfy $0 < \varepsilon < a$, and let $\mathscr{V}$ be the family consisting of those cubes C in $\mathscr{C}$ that are included in U and satisfy $\mu(C) \geq (a-\varepsilon)\lambda(C)$. Since $(\overline{D}\mu)(x) \geq a$ holds at each x in A, the family $\mathscr{V}$ is a Vitali covering of A. Thus the Vitali covering theorem (Theorem 6.2.1) provides a sequence $\{C_n\}$ of disjoint cubes that belong to $\mathscr{V}$ and satisfy $\lambda(A - \cup_n C_n) = 0$. If we use the fact that the sets C_n are disjoint and included in U, the fact that each C_n satisfies $\mu(C_n) \geq (a-\varepsilon)\lambda(C_n)$, and finally the fact that $\lambda(A - \cup_n C_n) = 0$, we find

$$\mu(U) \geq \sum_n \mu(C_n) \geq \sum_n (a-\varepsilon)\lambda(C_n)$$

$$= (a-\varepsilon)\lambda\left(\bigcup_n C_n\right) \geq (a-\varepsilon)\lambda(A).$$

Since μ is regular (Proposition 1.5.6) and ε can be made arbitrarily close to 0, the inequality $\mu(A) \geq a\lambda(A)$ follows.   □

Lemma 6.2.5. *Let μ be a finite Borel measure on $\mathbb{R}^d$ that is absolutely continuous with respect to Lebesgue measure, let a be a positive real number, and let A be a Borel subset of $\mathbb{R}^d$ such that $(\underline{D}\mu)(x) \leq a$ holds at each x in A. Then $\mu(A) \leq a\lambda(A)$.*

Proof. We can again assume that A is not empty. Let U be an open set that includes A, and let ε be a positive number. Arguments similar to those used in the proof of the preceding lemma show that there is a sequence $\{C_n\}$ of disjoint closed cubes that are included in U, satisfy $\mu(C_n) \leq (a+\varepsilon)\lambda(C_n)$, and are such that $\cup_n C_n$ contains λ-almost every point in A. Since μ is absolutely continuous with respect to λ, the union of the sets C_n also contains μ-almost every point in A. It follows that

$$(a+\varepsilon)\lambda(U) \geq (a+\varepsilon)\sum_n \lambda(C_n) \geq \sum_n \mu(C_n) = \mu(\cup_n C_n) \geq \mu(A).$$

Since λ is regular (Proposition 1.4.1) and ε is arbitrary, it follows that $\mu(A) \leq a\lambda(A)$. $\qquad\qquad\square$

Proof of Theorem 6.2.3. We begin with the case where μ is singular with respect to Lebesgue measure. Let N be a Borel set such that $\lambda(N) = 0$ and $\mu(N^c) = 0$. For each n define a subset B_n of N^c by

$$B_n = \{x \in N^c : (\overline{D}\mu)(x) \geq 1/n\}.$$

Then Lemma 6.2.4 (with a equal to $1/n$) implies that

$$\lambda(B_n) \leq n\mu(B_n) \leq n\mu(N^c) = 0$$

holds for each n. Thus $\{x \in \mathbb{R}^d : (\overline{D}\mu)(x) > 0\}$, since it is a subset of $N \cup (\cup_n B_n)$, has Lebesgue measure 0; since also $0 \leq \underline{D}\mu \leq \overline{D}\mu$, it follows that μ is differentiable, with derivative 0, λ-almost everywhere.

Next let us consider the case where μ is absolutely continuous with respect to Lebesgue measure. We start by proving that in this case $\underline{D}\mu$ and $\overline{D}\mu$ are equal almost everywhere. For positive rational numbers p and q such that $p < q$, define $A(p,q)$ by

$$A(p,q) = \{x \in \mathbb{R}^d : (\underline{D}\mu)(x) \leq p < q \leq (\overline{D}\mu)(x)\}.$$

Lemmas 6.2.4 and 6.2.5 imply that

$$q\lambda(A(p,q)) \leq \mu(A(p,q)) \leq p\lambda(A(p,q));$$

it follows from this first that $\lambda(A(p,q))$ is finite and then, since $p < q$, that $\lambda(A(p,q)) = 0$. Since $(\underline{D}\mu)(x) \leq (\overline{D}\mu)(x)$ holds for every x, while

$$\{x \in \mathbb{R}^d : (\underline{D}\mu)(x) < (\overline{D}\mu)(x)\} = \bigcup_{p,q} A(p,q),$$

it follows that $\underline{D}\mu$ and $\overline{D}\mu$ are equal λ-almost everywhere. (Note that we have not yet shown that they are finite almost everywhere.)

We continue to assume that $\mu \ll \lambda$. Let f be a Radon–Nikodym derivative of μ with respect to λ. An easy modification of the argument in the preceding paragraph shows that $f \leq \underline{D}\mu$ holds λ-almost everywhere (use the fact that whenever a is a positive number and A is a Borel set such that $f(x) \geq a$ holds at each x in A, then $\mu(A) = \int_A f\,d\lambda \geq a\lambda(A)$). A similar argument shows that $f \geq \overline{D}\mu$ holds λ-a .e.

Since in addition f is finite almost everywhere, it follows that f, $\underline{D}\mu$, and $\overline{D}\mu$ are finite and equal almost everywhere and hence that μ is differentiable, with derivative f, almost everywhere.

Finally, let μ be an arbitrary finite Borel measure on $\mathbb{R}^d$, let $\mu = \mu_a + \mu_s$ be its Lebesgue decomposition, and let f be a Radon–Nikodym derivative of μ_a with respect to λ. Then

$$(D\mu)(x) = (D\mu_a)(x) + (D\mu_s)(x) = f(x) + 0 = f(x)$$

holds at almost every x, and the proof is complete. □

Let E be a Lebesgue measurable subset of $\mathbb{R}^d$. A point x in $\mathbb{R}^d$ is a *point of density* of E if for each positive ε there is a positive δ such that

$$\left| \frac{\lambda(E \cap C)}{\lambda(C)} - 1 \right| < \varepsilon$$

holds whenever C is a cube that belongs to $\mathscr{C}$, contains x, and satisfies $e(C) < \delta$. Less formally, x is a point of density of E if $\lim \lambda(E \cap C)/\lambda(C) = 1$, where the limit is taken as C approaches x (through the collection of cubes in $\mathscr{C}$ that contain x). A point x is a *point of dispersion* of E if it is a point of density of E^c. Equivalently, x is a point of dispersion of E if $\lim \lambda(E \cap C)/\lambda(C) = 0$ holds as the cube C approaches x.

Corollary 6.2.6 (Lebesgue Density Theorem). *Let E be a Lebesgue measurable subset of $\mathbb{R}^d$. Then λ-almost every point in E is a point of density of E, and λ-almost every point in E^c is a point of dispersion of E.*

Proof. First suppose that $\lambda(E) < +\infty$, and define a finite Borel measure μ on $\mathbb{R}^d$ by $\mu(A) = \lambda(A \cap E)$. Choose a Borel subset E_0 of E such that $\lambda(E - E_0) = 0$ (see Lemma 1.5.3). Since $\mu \ll \lambda$ and since χ_{E_0} is a Radon–Nikodym derivative of μ with respect to λ, Theorem 6.2.3 implies that almost every x in E satisfies $(D\mu)(x) = 1$ and so is a point of density of E.

If $\lambda(E)$ is infinite and if $\{E_n\}$ is a sequence of Lebesgue measurable sets of finite measure such that $E = \cup_n E_n$, then almost every point of E is a point of density of some E_n and so is a point of density of E. Finally, almost every point of E^c is a point of density of E^c and so is a point of dispersion of E. □

Exercises

1. Show that the union of an arbitrary family of closed cubes with edges parallel to the coordinate axes is Lebesgue measurable. (Hint: Use the Vitali covering theorem.)

2. Let I be the line segment in $\mathbb{R}^2$ that connects the points $(0,0)$ and $(1,1)$. Define a finite Borel measure μ on $\mathbb{R}^2$ by letting $\mu(A)$ be the one-dimensional Lebesgue

measure of $A \cap I$. (More precisely, let T be the map of the interval $[0, \sqrt{2}]$ onto I given by $T(t) = (t/\sqrt{2})(1,1)$, and define μ by $\mu(A) = \lambda(T^{-1}(A))$.) Find $\overline{D}\mu$ and $\underline{D}\mu$.

3. Let f be a nonnegative function in $\mathscr{L}^1(\mathbb{R}^d, \mathscr{B}(\mathbb{R}^d), \lambda, \mathbb{R})$, and let μ be the finite Borel measure on $\mathbb{R}^d$ given by $\mu(A) = \int_A f \, d\lambda$.
   (a) Show that $(D\mu)(x) = f(x)$ holds at each x at which f is continuous.
   (b) Show by example that the equation $(D\mu)(x) = f(x)$ need not hold at every x in $\mathbb{R}^d$.

4. Show by example that the assumption that $\mu \ll \lambda$ cannot be omitted in Lemma 6.2.5.

6.3   Differentiation of Functions

Let us apply the results of Sect. 6.2 to the differentiation of functions of a real variable. We begin with two lemmas.

Lemma 6.3.1. *Let μ be a finite Borel measure on $\mathbb{R}$, and let $F \colon \mathbb{R} \to \mathbb{R}$ be defined by $F(x) = \mu((-\infty, x])$. If μ is differentiable at x_0, then F is differentiable at x_0, and $F'(x_0) = (D\mu)(x_0)$.*

Proof. The differentiability of μ at x_0 implies that $\mu(\{x_0\}) = 0$ and hence that F is continuous at x_0. Thus $\frac{F(x) - F(x_0)}{x - x_0}$ is equal to $\frac{\mu((x_0, x])}{\lambda((x_0, x])}$ if $x_0 < x$ and to $\frac{\mu((x, x_0])}{\lambda((x, x_0])}$ if $x < x_0$. Now apply the definitions of $(D\mu)(x_0)$ and $F'(x_0)$ (note that the half-open interval $(x, x_0]$ causes no difficulty, since its measure is the limit of the measure of $[x + \frac{1}{n}, x_0]$ as n approaches infinity). $\qquad\square$

Lemma 6.3.2. *Let $F \colon \mathbb{R} \to \mathbb{R}$ be nondecreasing. Then*

(a) *the one-sided limits $F(x-)$ and $F(x+)$ exist at each x in $\mathbb{R}$,*
(b) *the set of points at which F fails to be continuous is at most countably infinite, and*
(c) *the function $G \colon \mathbb{R} \to \mathbb{R}$ defined by $G(x) = F(x+)$ is nondecreasing and right-continuous, and agrees with F at each point at which F is continuous.*

Proof. Since F is nondecreasing, the limits $F(x-)$ and $F(x+)$ exist and are given by $F(x-) = \sup\{F(t) : t < x\}$ and $F(x+) = \inf\{F(t) : t > x\}$. For each x we have $F(x-) \le F(x) \le F(x+)$, and so F is continuous at x if and only if $F(x-) = F(x+)$. Let D be the set of points at which F is not continuous, and for each x in D choose a rational number r_x that satisfies $F(x-) < r_x < F(x+)$. Then r_x and r_y are distinct whenever x and y are distinct elements of D, and the countability of D follows from the countability of $\mathbb{Q}$.

Now suppose that G is defined by $G(x) = F(x+)$. Then G satisfies the relation $G(x) = \inf\{F(t) : t > x\}$, which implies that G is nondecreasing and right-continuous. Since $F(x) = F(x+)$ holds if F is continuous at x, the proof is complete. $\qquad\square$

The following is one of the basic theorems of differentiation theory.

Theorem 6.3.3 (Lebesgue). *Let* $F: \mathbb{R} \to \mathbb{R}$ *be nondecreasing. Then* F *is differentiable* λ*-almost everywhere.*

Proof. First suppose that F is bounded, nondecreasing, and right-continuous, and that it vanishes at $-\infty$. Then there is a finite Borel measure μ such that $F(x) = \mu((-\infty,x])$ holds at each x in $\mathbb{R}$ (Proposition 1.3.10), and so Theorem 6.2.3 and Lemma 6.3.1 imply that F is differentiable almost everywhere.

Now remove the requirement that F be right-continuous, and define $G: \mathbb{R} \to \mathbb{R}$ by $G(x) = F(x+)$. Then G is right-continuous (Lemma 6.3.2) and so, by what we have just proved, differentiable almost everywhere. Note that F and G are continuous at the same points and they agree at each point at which they are continuous; furthermore, if $F(x_0) = G(x_0)$, then $\frac{F(x)-F(x_0)}{x-x_0}$ lies between $\frac{G(x)-G(x_0)}{x-x_0}$ and $\frac{G(x-)-G(x_0)}{x-x_0}$. Hence if G is differentiable at x_0, then F is differentiable at x_0, and $F'(x_0) = G'(x_0)$. The almost everywhere differentiability of F follows.

Finally, suppose that F is an arbitrary nondecreasing function. It is enough to prove that F is differentiable almost everywhere on an arbitrary bounded open interval (a,b). Since we can reduce this to the preceding case by considering the function

$$x \mapsto \begin{cases} 0 & \text{if } x \leq a, \\ F(x) - F(a) & \text{if } a < x < b, \\ F(b) - F(a) & \text{if } b \leq x, \end{cases}$$

the proof is complete. $\square$

Corollary 6.3.4. *Let* $F: \mathbb{R} \to \mathbb{R}$ *be of finite variation. Then* F *is differentiable* λ*-almost everywhere.*

Proof. Since each function of finite variation is the difference of two nondecreasing functions (Proposition 4.4.2), this is an immediate consequence of Theorem 6.3.3. $\square$

Proposition 6.3.5 (Fubini). *Let* $F_n: \mathbb{R} \to \mathbb{R}$, $n = 1, 2, \ldots$, *be nondecreasing functions such that the series* $\sum_n F_n(x)$ *converges at each x in* $\mathbb{R}$. *Define* $F: \mathbb{R} \to \mathbb{R}$ *by* $F(x) = \sum_n F_n(x)$. *Then* $F' = \sum_n F_n'$ *holds* λ*-almost everywhere.*

Proof. First suppose that the functions F_n, for $n = 1, 2, \ldots$, are bounded, nondecreasing, and right-continuous, that they vanish at $-\infty$, and that the function F (defined by $F(x) = \sum_n F_n(x)$) is bounded. Let μ_n, $n = 1, 2, \ldots$, be the finite Borel measures corresponding to the functions F_n, and define a Borel measure μ on $\mathbb{R}$ by $\mu(A) = \sum_n \mu_n(A)$ (check that μ is a measure). Since we are temporarily assuming that F is bounded and since $\mu((-\infty,x]) = F(x)$ holds at each x, the measure μ is finite.

For each n form the Lebesgue decomposition $\mu_n = \mu_{n,a} + \mu_{n,s}$ of μ_n with respect to Lebesgue measure,[5] let f_n be a Radon–Nikodym derivative of $\mu_{n,a}$ with respect to λ, and let N_n be a Borel set of Lebesgue measure zero on which $\mu_{n,s}$ is concentrated. It is easy to check that $\sum_n \mu_{n,s}$ is concentrated on $\cup_n N_n$ and that $\sum_n \mu_{n,a}(A) = \int_A \sum_n f_n \, d\lambda$ holds for each A in $\mathscr{B}(\mathbb{R})$. Thus the Lebesgue decomposition of μ is given by $\mu = (\sum_n \mu_{n,a}) + (\sum_n \mu_{n,s})$, and $\sum_n f_n$ is a Radon–Nikodym derivative of $\sum_n \mu_{n,a}$ with respect to λ. It now follows from Theorem 6.2.3 and Lemma 6.3.1 that

$$\sum_n F_n'(x) = \sum_n (D\mu_n)(x) = \sum_n f_n(x) = (D\mu)(x) = F'(x)$$

holds at almost every x in $\mathbb{R}$.

Arguments similar to those given in the second and third paragraphs of the proof of Theorem 6.3.3 allow one to reduce the proposition to the case just considered; the details are left to the reader.   □

Theorem 6.3.6 (Lebesgue). *Suppose that f belongs to $\mathscr{L}^1(\mathbb{R}, \mathscr{M}_{\lambda^*}, \lambda, \mathbb{R})$ and that $F: \mathbb{R} \to \mathbb{R}$ is defined by $F(x) = \int_{-\infty}^x f(t) \, dt$. Then F is differentiable, and its derivative is given by $F'(x) = f(x)$, at λ-almost every x in $\mathbb{R}$.*

Proof. First suppose that f is nonnegative, and define a finite Borel measure μ on $\mathbb{R}$ by $\mu(A) = \int_A f \, d\lambda$. Let f_0 be a Borel measurable function that agrees with f almost everywhere (see Proposition 2.2.5). Then Theorem 6.2.3 and Lemma 6.3.1 imply that

$$F'(x) = (D\mu)(x) = f_0(x) = f(x)$$

holds at almost every x, and so the proof is complete in the case where f is nonnegative.

An arbitrary f in $\mathscr{L}^1(\mathbb{R}, \mathscr{M}_{\lambda^*}, \lambda, \mathbb{R})$ can be dealt with through the decomposition $f = f^+ - f^-$.   □

We will often need to know that almost everywhere derivatives of reasonable functions are measurable. This is given by the following lemma.

Lemma 6.3.7. *Let $F: [a,b] \to \mathbb{R}$ be a Lebesgue measurable function that is differentiable almost everywhere. Suppose that $g: [a,b] \to \mathbb{R}$ satisfies $g(x) = F'(x)$ almost everywhere. Then g is Lebesgue measurable, as is F' (whose domain is the set where F is differentiable).*

Proof. Extend F to the interval $[a, +\infty)$ by letting $F(x)$ be equal to $F(b)$ if $x > b$. Since

$$g(x) = \overline{\lim_n} \, n\left(F\left(x + \frac{1}{n}\right) - F(x)\right)$$

---

[5] Thus $\mu_{n,a} \ll \lambda$ and $\mu_{n,s} \perp \lambda$.

holds at almost every x in $[a,b]$, it follows from Propositions 2.1.5 and 2.2.2 that g is Lebesgue measurable. Since the set of points where F is differentiable is the complement in $[a,b]$ of a Lebesgue null set, it follows that F' is also Lebesgue measurable. $\qquad\square$

We can now derive the following characterization of the absolutely continuous[6] functions on a closed bounded interval.

Corollary 6.3.8. *A function $F \colon [a,b] \to \mathbb{R}$ is absolutely continuous if and only if it is differentiable λ-almost everywhere, F' is integrable, and F can be reconstructed from its derivative through the formula*

$$F(x) = F(a) + \int_a^x F'(t)\,dt. \tag{1}$$

Proof. First suppose that F is absolutely continuous. Then F is also of finite variation (Exercise 4.4.5), and so Proposition 4.4.6, applied to the function

$$x \mapsto \begin{cases} 0 & \text{if } x \leq a, \\ F(x) - F(a) & \text{if } a < x < b, \\ F(b) - F(a) & \text{if } b \leq x, \end{cases}$$

provides a function f in $\mathscr{L}^1(\mathbb{R}, \mathscr{B}(\mathbb{R}), \lambda, \mathbb{R})$ such that

$$F(x) = F(a) + \int_a^x f(t)\,dt$$

holds at each x in $[a,b]$. Theorem 6.3.6 then implies that F is differentiable, with derivative given by $F'(x) = f(x)$, at almost every such x; hence (1) follows.

The other half of the proof is easy; Proposition 4.4.6 (see also Proposition 2.2.5 or Exercise 2) implies that each F that has an integrable derivative and satisfies (1) is absolutely continuous. $\qquad\square$

We are now in a position to prove the following version of integration by parts.

Corollary 6.3.9. *Let F and G be absolutely continuous functions on the interval $[a,b]$. Then*

$$F(b)G(b) - F(a)G(a) = \int_a^b F(t)G'(t)\,dt + \int_a^b F'(t)G(t)\,dt.$$

Proof. We begin by showing that the function FG is absolutely continuous. Since the functions F and G are continuous and the interval $[a,b]$ is compact, there are positive numbers M and N such that $|F(x)| \leq M$ and $|G(x)| \leq N$ hold at each x

---

[6]It is easy to modify the definition of absolute continuity for functions on $\mathbb{R}$ to make it apply to functions on $[a,b]$.

in $[a,b]$. Suppose that $\{(s_i,t_i)\}$ is a finite sequence of disjoint open subintervals of $[a,b]$. Then for each i we have

$$|F(t_i)G(t_i) - F(s_i)G(s_i)| \le |F(t_i) - F(s_i)||G(t_i)| + |F(s_i)||G(t_i) - G(s_i)|$$
$$\le N|F(t_i) - F(s_i)| + M|G(t_i) - G(s_i)|,$$

and so

$$\sum_i |F(t_i)G(t_i) - F(s_i)G(s_i)| \le N\sum_i |F(t_i) - F(s_i)| + M\sum_i |G(t_i) - G(s_i)|.$$

Since F and G are absolutely continuous, we can make the sums on the right side of this inequality small by making $\sum_i(t_i - s_i)$ small. The absolute continuity of FG follows.

Thus Corollary 6.3.8 can be applied to the function FG. Since FG' and $F'G$ are integrable (check this) and since

$$(FG)'(x) = F(x)G'(x) + F'(x)G(x)$$

holds at almost every x in $[a,b]$, the proof is complete.   □

Theorem 6.2.3 also implies the following strengthened version of Theorem 6.3.6 (see also Exercises 3 and 4).

Proposition 6.3.10. *Suppose that f belongs to $\mathscr{L}^1(\mathbb{R}, \mathscr{M}_{\lambda^*}, \lambda, \mathbb{R})$. Then*

$$\lim_I \frac{1}{\lambda(I)} \int_I |f(t) - f(x)|\, dt = 0 \tag{2}$$

holds at λ-almost every x in $\mathbb{R}$; here I is a closed interval that contains x, and the limit is taken as the length of I approaches zero.

Points x at which (2) holds are called *Lebesgue points*[7] of f, and the set of all Lebesgue points of f is called the *Lebesgue set* of f.

Proof. It is enough to choose an arbitrary bounded open interval (a,b) and to show that (2) holds at almost every x in (a,b).

Let us first suppose that the integrable function f is in fact Borel measurable. For each rational number r let μ_r be the finite Borel measure on $\mathbb{R}$ defined by

$$\mu_r(A) = \int_{A \cap (a,b)} |f(t) - r|\, dt.$$

---

[7]Some authors use the condition $\lim_{h \to 0^+} \frac{1}{h} \int_0^h |f(x+t) + f(x-t) - 2f(x)|\, dt = 0$ as the defining condition for being a Lebesgue point; of course each point that satisfies (2) is also a Lebesgue point in this sense.

Theorem 6.2.3 implies that there is a Lebesgue null set N_r such that $(D\mu_r)(x) = |f(x) - r|$ holds at each x in $(a,b) - N_r$. Let N be the Lebesgue null set $\cup_{r \in \mathbb{Q}} N_r$. Suppose that x belongs to $(a,b) - N$, that I is a closed subinterval of (a,b) that contains x, and that r is a rational number. Then

$$\int_I |f(t) - f(x)| \, dt \leq \int_I |f(t) - r| \, dt + \int_I |r - f(x)| \, dt,$$

and so if we divide the terms of this inequality by $\lambda(I)$ and let the length of I approach 0, we find

$$\overline{\lim_I} \frac{1}{\lambda(I)} \int_I |f(t) - f(x)| \, dt \leq (D\mu_r)(x) + |r - f(x)| = 2|f(x) - r|.$$

Since $|f(x) - r|$ can be made arbitrarily small by an appropriate choice of the rational number r, Eq. (2) follows.

In case f is Lebesgue measurable, rather than Borel measurable, we can complete the proof by applying the preceding argument to a Borel measurable function that agrees with f almost everywhere (see Proposition 2.2.5). □

It is of course of interest to have easily verified conditions that imply the absolute continuity of a function. One might conjecture that a continuous function on a closed bounded interval is absolutely continuous if it is differentiable almost everywhere, if it is differentiable almost everywhere and its derivative is integrable, or if it is differentiable everywhere. These conjectures all fail (see Exercises 5 and 6), but the following related result holds.

Theorem 6.3.11. *Let $F: [a,b] \to \mathbb{R}$ be a continuous function such that*

(a) *F is differentiable at all except countably many of the points in $[a,b]$, and*
(b) *F' is integrable.*

Then F is absolutely continuous, and so $F(x) = F(a) + \int_a^x F'(t) \, dt$ holds at each x in $[a,b]$.

Theorem 6.3.11 would fail if condition (b) were removed (see Exercise 6), and so condition (a) does not imply condition (b). There is, however, an analogue to Theorem 6.3.11 for the Henstock–Kurzweil integral in which condition (b) is not needed; see Exercise 23 in Appendix H.

For the proof we need the following definitions and lemmas.

A function $f: \mathbb{R} \to (-\infty, +\infty]$ is *lower semicontinuous* if for each x in $\mathbb{R}$ and each real number A such that $A < f(x)$ there is a positive number δ such that $A < f(t)$ holds whenever t satisfies $|t - x| < \delta$. A function $f: \mathbb{R} \to [-\infty, +\infty)$ is *upper semicontinuous* if $-f$ is lower semicontinuous. In other words, f is upper semicontinuous if for each x in $\mathbb{R}$ and each real number A such that $f(x) < A$ there is a positive number δ such that $f(t) < A$ holds whenever t satisfies $|t - x| < \delta$.

Of course, a function $f: \mathbb{R} \to \mathbb{R}$ is continuous if and only if it is both lower semicontinuous and upper semicontinuous. Furthermore, it is easy to check that

(a) a function $f: \mathbb{R} \to (-\infty, +\infty]$ is lower semicontinuous if and only if for each real number A, the set $\{x \in \mathbb{R} : A < f(x)\}$ is open,
(b) a function $f: \mathbb{R} \to [-\infty, +\infty)$ is upper semicontinuous if and only if for each real number A, the set $\{x \in \mathbb{R} : f(x) < A\}$ is open,
(c) if U is an open subset of $\mathbb{R}$, then the characteristic function χ_U is lower semicontinuous,
(d) if C is a closed subset of $\mathbb{R}$, then the characteristic function χ_C is upper semicontinuous,
(e) if f and g are lower semicontinuous, then $f + g$ is lower semicontinuous, and
(f) if $\{f_n\}$ is an increasing sequence of lower semicontinuous functions, then $\lim_n f_n$ is lower semicontinuous.

It follows (from (a) and (b)) that the upper semicontinuous functions and the lower semicontinuous functions are Borel measurable.

Lemma 6.3.12. *Let $f: [a,b] \to [-\infty, +\infty]$ be Lebesgue integrable. Then for each positive ε there is a lower semicontinuous function $g: \mathbb{R} \to (-\infty, +\infty]$ that is integrable on $[a,b]$ and satisfies*

(a) *$f(t) \le g(t)$ holds at each t in $[a,b]$, and*
(b) *$\int_a^b g(t)\, dt < \int_a^b f(t)\, dt + \varepsilon$.*

Proof. Let ε be a positive number. First suppose that f is nonnegative. There is a nondecreasing sequence $\{f_n\}$ of nonnegative simple measurable functions such that $f = \lim_n f_n$ (Proposition 2.1.8), and so we can find Lebesgue measurable sets A_k, $k = 1, 2, \ldots$, and positive real numbers a_k such that $f = \sum_k a_k \chi_{A_k}$ (write each $f_n - f_{n-1}$ as a sum of positive multiples of characteristic functions). Use the regularity of Lebesgue measure (Proposition 1.4.1) to choose open sets U_k, $k = 1, 2, \ldots$, that include the corresponding A_k's and satisfy $\sum_k a_k \lambda(U_k) < \sum_k a_k \lambda(A_k) + \varepsilon/2$. Then the formula $f^\infty = \sum_k a_k \chi_{U_k}$ defines a lower semicontinuous function f^∞ that satisfies

$$\int_a^b f^\infty(t)\, dt < \int \sum_k a_k \chi_{A_k}\, dt + \varepsilon/2 = \int_a^b f(t)\, dt + \varepsilon/2$$

and is such that $f(t) \le f^\infty(t)$ holds for each t in $[a,b]$.

Now suppose that f is an arbitrary integrable function on $[a,b]$. For each n define a function h_n by $h_n(x) = \max(f(x), -n)$. The dominated convergence theorem implies that $\int_a^b f(t)\, dt = \lim_n \int_a^b h_n(t)\, dt$ and hence that we can choose a positive integer N such that $\int_a^b h_N(t)\, dt < \int_a^b f(t)\, dt + \varepsilon/2$. If we apply the argument in the preceding paragraph to the nonnegative function $h_N + N$, producing the lower semicontinuous function f^∞, then the required function g is given by $g = f^\infty - N\chi_{[a,b]}$. $\square$

Lemma 6.3.13. *Let $H: [a,b] \to \mathbb{R}$ be continuous, and let C be a countable subset of $[a,b]$. Suppose that for each x in $[a,b) - C$ there is a positive number δ_x such that $H(t) > H(x)$ holds at each t in the interval $(x, x+\delta_x)$. Then H is strictly increasing.*

Proof. It suffices to prove that H is nondecreasing (why?), and for this it is enough to show that numbers x_1 and x_2 in $[a,b]$ that satisfy $x_1 < x_2$ and $H(x_1) > H(x_2)$ do not exist. Assume that such numbers do exist, and for each y between $H(x_1)$ and $H(x_2)$ define a number x_y by

$$x_y = \sup\{x \in [x_1, x_2] : H(x) \geq y\}.$$

It is easy to check that each x_y satisfies $H(x_y) = y$ and belongs to the countable exceptional set C. Since there are uncountably many such points x_y, we have reached a contradiction, and the proof is complete. $\qquad\qquad\square$

Proof of Theorem 6.3.11. Suppose that the function F satisfies the hypotheses of Theorem 6.3.11 and that C is a countable subset of $[a,b)$ such that F is differentiable at each point of $[a,b) - C$. Let ε be a positive number. Lemma 6.3.12 (applied to the function that agrees with F' where F is differentiable and that vanishes elsewhere) provides a lower semicontinuous function g such that $F'(t) \leq g(t)$ holds at each t in $[a,b) - C$ and such that $\int_a^b g(t)\,dt < \int_a^b F'(t)\,dt + \varepsilon$. By adding a small positive continuous function to g, if necessary, we can assume that $F'(t) < g(t)$ holds at each t in $[a,b) - C$. Define $G: [a,b] \to \mathbb{R}$ by $G(x) = F(a) + \int_a^x g(t)\,dt$. The lower semicontinuity of g implies that

$$\lim_{y \downarrow x} \frac{G(y) - G(x)}{y - x} \geq g(x)$$

holds at each x in $[a,b)$. Thus

$$\lim_{y \downarrow x} \frac{(G(y) - F(y)) - (G(x) - F(x))}{y - x} \geq g(x) - F'(x) > 0$$

holds at each x in $[a,b) - C$, and so Lemma 6.3.13 implies that $G - F$ is nondecreasing. Since furthermore $G(a) = F(a)$, it follows that $F \leq G$. This and the inequality $\int_a^b g(t)\,dt < \int_a^b F'(t)\,dt + \varepsilon$ imply that

$$F(x) \leq G(x) = F(a) + \int_a^x g(t)\,dt$$

$$= F(a) + \int_a^x F'(t)\,dt + \int_a^x (g(t) - F'(t))\,dt$$

$$\leq F(a) + \int_a^x F'(t)\,dt + \varepsilon$$

holds at each x in $[a,b]$. Since ε was arbitrary,

$$F(x) \leq F(a) + \int_a^x F'(t)\,dt$$

must hold at each such x. The reverse inequality can be proved by applying the same argument to $-F$, and so the proof is complete.   □

Exercises

1. Let $F\colon \mathbb{R} \to \mathbb{R}$ be nondecreasing. Show that if ε is a positive number, then each bounded interval contains only a finite number of values x such that $F(x+) - F(x-) \geq \varepsilon$. Use this observation to give a second proof of part (b) of Lemma 6.3.2.
2. Prove the following modified version of Lemma 6.3.7: If $F\colon [a,b] \to \mathbb{R}$ is continuous and if D is the set consisting of those points in $[a,b]$ at which F is differentiable, then $D \in \mathscr{B}(\mathbb{R})$ and F' (as a function from D to $\mathbb{R}$) is Borel measurable.
3. Derive Theorem 6.3.6 from Proposition 6.3.10.
4. Let f and F be as in Theorem 6.3.6. Show by example that there can be points x that are not Lebesgue points of f, but are such that $F'(x)$ exists and is equal to $f(x)$.
5. Show that the Cantor function provides a counterexample to two of the three conjectures suggested just before the statement of Theorem 6.3.11.
6. Define $F\colon [0,1] \to \mathbb{R}$ by

$$F(x) = \begin{cases} 0 & \text{if } x = 0, \\ x^2 \sin\dfrac{1}{x^2} & \text{if } 0 < x \leq 1. \end{cases}$$

Show that F is differentiable everywhere on $[0,1]$ but is not absolutely continuous.
7. Show that there is a strictly increasing continuous function $F\colon [0,1] \to \mathbb{R}$ such that $F'(x) = 0$ holds at λ-almost every x in $[0,1]$. (Hint: Let F be the sum of a suitable series of functions, and use Proposition 6.3.5.)

Notes

The proof of Theorem 6.1.7 presented here was inspired by one given by A.M. Gleason in some unpublished notes on advanced calculus.

Munroe [92], Rudin [105], and Wheeden and Zygmund [127] carry the study of the differentiation of measures and functions a bit farther than it is taken here. See Bruckner [21], Bruckner [22], de Guzmán [33], Hayes and Pauc [56], Kölzow [74], and Saks [106] for more advanced treatments of differentiation theory.

The proof of Theorem 6.3.11 given here is taken from Walker [123].

Chapter 7
Measures on Locally Compact Spaces

Let $\mathscr{K}(\mathbb{R})$ be the vector space consisting of those continuous functions $f\colon \mathbb{R} \to \mathbb{R}$ that have compact support—that is, for which the set $\{x \in \mathbb{R} : f(x) \neq 0\}$ has a compact closure. Then $f \mapsto \int_{\mathbb{R}} f\, d\lambda$ defines a positive[1] linear functional on $\mathscr{K}(\mathbb{R})$. More generally, if μ is a measure on $(\mathbb{R}, \mathscr{B}(\mathbb{R}))$ that has finite values on the compact subsets of $\mathbb{R}$, then $f \mapsto \int_{\mathbb{R}} f\, d\mu$ defines a positive linear functional on $\mathscr{K}(\mathbb{R})$. According to a special case of the Riesz representation theorem (see Theorem 7.2.8), the converse also holds: for every positive linear functional $I\colon \mathscr{K}(\mathbb{R}) \to \mathbb{R}$ there is a Borel measure μ on $\mathbb{R}$ that is finite on compact sets and represents I, in the sense that $I(f) = \int_{\mathbb{R}} f\, d\mu$ holds for each f in $\mathscr{K}(\mathbb{R})$.

This chapter is devoted to the Riesz representation theorem and related results. The first section (Sect. 7.1) contains some basic facts about locally compact Hausdorff spaces, the spaces that provide the natural setting for the Riesz representation theorem, while the second section (Sect. 7.2) gives a proof of the Riesz representation theorem. The next two sections (Sects. 7.3 and 7.4) contain some useful and relatively basic related material. The results of Sects. 7.5 and 7.6 are needed for dealing with large locally compact Hausdorff spaces; for relatively small locally compact Hausdorff spaces (those that have a countable base), Proposition 7.6.2 is the only result from these sections that one really needs (see also Proposition 7.2.5 and Theorems 4.5.1 and 5.2.2).

The Daniell–Stone integral gives another way to deal with integration on locally compact Hausdorff spaces. A measure-theoretic version of the basic result of the Daniell–Stone theory is given by Theorem 7.7.4; the general Daniell–Stone setup is outlined in the exercises at the end of Sect. 7.7.

---

[1] Recall that a linear functional I on a vector space of functions is *positive* if $I(f) \geq 0$ holds for each nonnegative function f in the domain of I.

D.L. Cohn, *Measure Theory: Second Edition*, Birkhäuser Advanced Texts Basler Lehrbücher, DOI 10.1007/978-1-4614-6956-8_7, © Springer Science+Business Media, LLC 2013

7.1 Locally Compact Spaces

In this chapter we will be dealing with measures and integrals on locally compact Hausdorff spaces. This first section contains a summary of some of the necessary topological facts and constructions; the main development begins in Sect. 7.2.

Recall that a topological space is *locally compact* if each of its points has an open neighborhood whose closure is compact.

Examples 7.1.1. Examples of locally compact Hausdorff spaces include the Euclidean spaces $\mathbb{R}^d$, spaces with the discrete[2] topology (for example, the set $\mathbb{Z}$ of all integers), and compact Hausdorff spaces. The space ℓ^2 of sequences $\{x_n\}$ such that $\sum_n x_n^2 < +\infty$, with the topology given by the norm $\{x_n\} \mapsto (\sum_n x_n^2)^{1/2}$, is not locally compact (inside each open ball there is an infinite sequence that has no convergent subsequence; see item D.39 in Appendix D). $\quad\square$

The following elementary proposition will be a basic tool for what follows.

Proposition 7.1.2. *Let X be a Hausdorff space, and let K and L be disjoint compact subsets of X. Then there are disjoint open subsets U and V of X such that $K \subseteq U$ and $L \subseteq V$.*

Proof. We can assume that K and L are both nonempty (otherwise we could use $\varnothing$ as one of our open sets and X as the other). Let us begin with the case where K contains exactly one point, say x. For each y in L there is a pair U_y, V_y of disjoint open sets such that $x \in U_y$ and $y \in V_y$ (recall that X is Hausdorff). Since L is compact, there is a finite family $y_1, \ldots, y_n$ such that the sets $V_{y_1}, \ldots, V_{y_n}$ cover L. The sets U and V defined by $U = \cap_{i=1}^n U_{y_i}$ and $V = \cup_{i=1}^n V_{y_i}$ are then the required open sets.

Next consider the case where K has more than one element. We have just shown that for each x in K there are disjoint open sets, say U_x and V_x, such that $x \in U_x$ and $L \subseteq V_x$. Since K is compact, there is a finite family $x_1, \ldots, x_k$ such that $U_{x_1}, \ldots, U_{x_k}$ cover K. The proof is complete if we define U and V by $U = \cup_{i=1}^k U_{x_i}$ and $V = \cap_{i=1}^k V_{x_i}$. $\quad\square$

The sets U and V just constructed are said to *separate* the sets K and L.

Let us note several useful results (Propositions 7.1.3–7.1.6), each of which makes at least indirect use of Proposition 7.1.2.

Proposition 7.1.3. *Let X be a locally compact Hausdorff space, let x be a point in X, and let U be an open neighborhood of x. Then x has an open neighborhood whose closure is compact and included in U.*

Proof. Since X is locally compact, there is an open neighborhood of x, say W, whose closure is compact. By replacing W with $W \cap U$, we can assume that W is included in U. The difficulty is that $\overline{W}$ may extend outside U. Use Proposition 7.1.2

---

[2]A topological space is *discrete* (or has the *discrete topology*) if each of its subsets is open.

to choose disjoint open sets V_1 and V_2 that separate the compact sets $\{x\}$ and $\overline{W} - W$. The closure of $V_1 \cap W$ is then compact and included in W and hence in U; thus $V_1 \cap W$ is the required open neighborhood of x.   □

Proposition 7.1.4. *Let X be a locally compact Hausdorff space, let K be a compact subset of X, and let U be an open subset of X that includes K. Then there is an open subset V of X that has a compact closure and satisfies $K \subseteq V \subseteq \overline{V} \subseteq U$.*

Proof. Proposition 7.1.3 implies that each point in K has an open neighborhood whose closure is compact and included in U. Since K is compact, some finite collection of these neighborhoods covers K. Let V be the union of the sets in such a finite collection; then V is the required set.   □

A subset of a topological space X is a G_δ if it is the intersection of a sequence of open subsets of X and is an F_σ if it is the union of a sequence of closed subsets of X.

Proposition 7.1.5. *Let X be a locally compact Hausdorff space that has a countable base for its topology. Then each open subset of X is an F_σ and is in fact the union of a sequence of compact sets. Likewise, each closed subset of X is a G_δ.*

Proof. Suppose that $\mathscr{U}$ is a countable base for the topology of X. Let U be an open subset of X, and let $\mathscr{U}_U$ be the collection of those sets V in $\mathscr{U}$ for which $\overline{V}$ is compact and included in U. Proposition 7.1.3 implies that U is the union of the closures of the sets in $\mathscr{U}_U$ (the open neighborhoods provided by Proposition 7.1.3 can be replaced with smaller sets that belong to $\mathscr{U}$). Thus U is the union of a countable collection of sets that are closed and, in fact, compact.

Now suppose that C is a closed subset of X. Then C^c is open and so is the union of a sequence $\{F_n\}$ of closed sets. Consequently C is the intersection of the open sets $F_n^c, n = 1, 2, \ldots$.   □

Recall that a topological space (or a subset thereof) is *σ-compact* if it is the union of a countable collection of compact sets.

Proposition 7.1.6. *Every locally compact Hausdorff space that has a countable base for its topology is σ-compact.*

Proof. Since a topological space is an open subset of itself, this proposition is an immediate corollary of Proposition 7.1.5.   □

We turn to the continuous functions on a locally compact Hausdorff space.

Recall that a topological space is *normal* if it is Hausdorff and each pair of disjoint closed subsets of it can be separated by a pair of disjoint open sets.

Proposition 7.1.7. *Every compact Hausdorff space is normal.*

Proof. Note that every closed subset of a compact space is compact, and use Proposition 7.1.2   □

We will need the following standard result. A proof of it is sketched in Exercise 5.

Theorem 7.1.8 (Urysohn's Lemma). *Let X be a normal topological space, and let E and F be disjoint closed subsets of X. Then there is a continuous function* $f: X \to [0,1]$ *such that* $f(x) = 0$ *holds at each x in E and* $f(x) = 1$ *holds at each x in F.*

Let f be a continuous real- or complex-valued function on a topological space X. The *support* of f, written $\text{supp}(f)$, is the closure of $\{x \in X : f(x) \neq 0\}$. In case X is a locally compact Hausdorff space, we will denote by $\mathcal{K}(X)$ the set of those continuous functions $f: X \to \mathbb{R}$ for which $\text{supp}(f)$ is compact. Likewise, we will denote by $\mathcal{K}^{\mathbb{C}}(X)$ the set of those continuous functions $f: X \to \mathbb{C}$ for which $\text{supp}(f)$ is compact.

It is clear that $\mathcal{K}(X)$ and $\mathcal{K}^{\mathbb{C}}(X)$ are vector spaces over $\mathbb{R}$ and $\mathbb{C}$, respectively, and that each function in $\mathcal{K}(X)$ or in $\mathcal{K}^{\mathbb{C}}(X)$ is bounded (recall that continuous functions are bounded on compact sets).

The following fact about $\mathcal{K}(X)$ is essential for the development of measure theory on locally compact Hausdorff spaces.

Proposition 7.1.9. *Let X be a locally compact Hausdorff space, let K be a compact subset of X, and let U be an open subset of X that includes K. Then there is a function f that belongs to* $\mathcal{K}(X)$, *satisfies* $\chi_K \leq f \leq \chi_U$, *and is such that* $\text{supp}(f) \subseteq U$.

Proof. Use Proposition 7.1.4 to choose an open set V that has a compact closure and satisfies $K \subseteq V \subseteq \overline{V} \subseteq U$. According to Urysohn's lemma (applied to the compact Hausdorff space $\overline{V}$), there is a continuous function $g: \overline{V} \to [0,1]$ such that $g(x) = 1$ holds at each x in K and $g(x) = 0$ holds at each x in $\overline{V} - V$. Now define the function $f: X \to [0,1]$ by requiring that f agree with g on $\overline{V}$ and vanish outside $\overline{V}$. The continuity of f follows from D.6 (note that f is continuous on $\overline{V}$ and is constant, and hence continuous, on $X - V$). The support of f is included in $\overline{V}$ and so is compact and included in U.   □

Next we derive two consequences of Proposition 7.1.9 (Propositions 7.1.11 and 7.1.12); they will be needed in Sects. 7.2 and 7.3, respectively. Let us begin with the following lemma.

Lemma 7.1.10. *Let X be a Hausdorff space, let K be a compact subset of X, and let* U_1 *and* U_2 *be open subsets of X such that* $K \subseteq U_1 \cup U_2$. *Then there are compact sets* K_1 *and* K_2 *such that* $K = K_1 \cup K_2$, $K_1 \subseteq U_1$, *and* $K_2 \subseteq U_2$.

Proof. Let $L_1 = K - U_1$ and $L_2 = K - U_2$. Then L_1 and L_2 are disjoint and compact, and so according to Proposition 7.1.2 they can be separated by disjoint open sets, say by V_1 and V_2. If we define K_1 and K_2 by $K_1 = K - V_1$ and $K_2 = K - V_2$, then K_1 and K_2 are compact, are included in U_1 and U_2, respectively, and have K as their union.   □

Proposition 7.1.11. *Let X be a locally compact Hausdorff space, let f belong to* $\mathcal{K}(X)$, *and let* $U_1, \ldots, U_n$ *be open subsets of X such that* $\text{supp}(f) \subseteq \bigcup_{i=1}^{n} U_i$. *Then there are functions* $f_1, \ldots, f_n$ *in* $\mathcal{K}(X)$ *such that* $f = f_1 + f_2 + \cdots + f_n$ *and such* *that for each i the support of f_i is included in U_i. Furthermore, if the function f is nonnegative, then the functions $f_i, \ldots, f_n$ can be chosen so that all are nonnegative.*

Proof. In case $n = 1$ we need only let f_1 be f. So we can begin by supposing that $n = 2$. Use Lemma 7.1.10 to construct compact sets K_1 and K_2 such that $K_1 \subseteq U_1$, $K_2 \subseteq U_2$, and $\text{supp}(f) = K_1 \cup K_2$, and then use Proposition 7.1.9 to construct functions h_1 and h_2 that belong to $\mathscr{K}(X)$ and satisfy $\chi_{K_i} \leq h_i \leq \chi_{U_i}$ and $\text{supp}(h_i) \subseteq U_i$ for $i = 1, 2$. Define functions g_1 and g_2 by $g_1 = h_1$ and $g_2 = h_2 - (h_1 \wedge h_2)$. Then g_1 and g_2 are non-negative, their supports are included in U_1 and U_2, respectively, and they satisfy

$$g_1(x) + g_2(x) = (h_1 \vee h_2)(x) = 1$$

at each x in $\text{supp}(f)$. We can complete the proof in the case where $n = 2$ by defining f_1 and f_2 to be fg_1 and fg_2.

The general case can be dealt with by induction: use what we have just proved to write f as the sum of two functions, having supports included in $\cup_{i=1}^{n-1} U_i$ and U_n, respectively, and then use the induction hypothesis to decompose the first of these functions into the sum of $n - 1$ suitable functions. □

Proposition 7.1.12. *Let X be a locally compact Hausdorff space, let $K_1, \ldots, K_n$ be disjoint compact subsets of X, and let $\alpha_1, \ldots, \alpha_n$ be real (or complex) numbers. Then there is a function f that belongs to $\mathscr{K}(X)$ (or to $\mathscr{K}^{\mathbb{C}}(X)$) and satisfies*

(a) $f(x) = \alpha_i$ *if* $x \in K_i$, $i = 1, \ldots, n$, *and*
(b) $\|f\|_\infty = \max\{|a_1|, \ldots, |a_n|\}$.

Proof. We begin by constructing disjoint open sets $U_1, \ldots, U_n$ such that $K_i \subseteq U_i$ holds for each i. If $n = 2$, such sets are provided by Proposition 7.1.2. The general case follows by induction: use Proposition 7.1.2 to choose disjoint open sets V_1 and V_2 that separate $\cup_{i=1}^{n-1} K_i$ from K_n, use the induction hypothesis to choose disjoint open sets $W_1, \ldots, W_{n-1}$ that separate $K_1, \ldots, K_{n-1}$ from one another, and then define $U_1, \ldots, U_n$ to be $V_1 \cap W_1, \ldots, V_1 \cap W_{n-1}, V_2$.

Next we use Proposition 7.1.9 to choose functions $f_1, \ldots, f_n$ that belong to $\mathscr{K}(X)$ and satisfy $\chi_{K_i} \leq f_i \leq \chi_{U_i}$ for $i = 1, \ldots, n$. The required function f is now given by $f = \sum_{i=1}^{n} \alpha_i f_i$. □

We will have use for the *one-point compactification* of a locally compact Hausdorff space X; it is constructed as follows. The underlying set X^* consists of the points in X, together with one additional point, called the *point at infinity*. The open subsets of X^* are, by definition, the open subsets of X, together with the complements (with respect to X^*) of the compact subsets of X. It is not hard to check that this does define a topology on X^* and that the topology induced by it on the subspace X is the original topology on X; the details are left to the reader. We need to verify that X^* is a compact Hausdorff space. Let us begin by checking that X^* is Hausdorff. Suppose that x and y are distinct points in X^*. If both points belong to X, then they can be separated with sets that are open in X and hence in X^*. If one of these points, say y, is the point at infinity and if we choose an open neighborhood U of x whose closure (in X) is compact, then U and $X^* - \overline{U}$ are disjoint open neighborhoods in X^* of x and y. Hence X^* is Hausdorff. We turn to the compactness of X^*. Let $\mathscr{U}$ be an open cover of X^*. The point at infinity must belong to some set in $\mathscr{U}$, say U_0. Then $X^* - U_0$ is a compact subset of X and so is covered by some finite subfamily of $\mathscr{U}$. The sets in this subfamily, together with U_0, form a finite cover of X^*. Thus X^* is compact.

The remaining results in this section will be used in a few exercises, but otherwise they will not be used until Chap. 8. They do, however, provide some perspective on the spaces considered here.

Proposition 7.1.13. *A compact Hausdorff space is metrizable if and only if there is a countable base for its topology.*

Proof. First suppose that X is compact and metrizable. Then X is separable (Corollary D.40) and so has a countable base (see D.32).

Now suppose that X is a compact Hausdorff space that has a countable base, and let $\mathscr{U}$ be such a base. The space X is normal (Proposition 7.1.7), and so for each pair of sets U, V that belong to $\mathscr{U}$ and satisfy $\overline{U} \cap \overline{V} = \varnothing$ there is, by Urysohn's lemma, a continuous function $f \colon X \to [0,1]$ that vanishes on $\overline{U}$ and has value 1 everywhere on $\overline{V}$. Form a sequence $\{f_n\}$ by choosing one such function for each such pair of sets. Our next step is to check that this sequence of functions separates the points of X, and for this it is enough to show that for each pair x, y of distinct points in X there are sets U and V that belong to $\mathscr{U}$, have disjoint closures, and contain x and y, respectively. To construct such sets, choose disjoint open neighborhoods W_1 and W_2 of x and y, and use Proposition 7.1.3 to choose open sets U and V such that $x \in U \subseteq \overline{U} \subseteq W_1$ and $y \in V \subseteq \overline{V} \subseteq W_2$. By making U and V a bit smaller, if necessary, we can assume that they belong to $\mathscr{U}$. Thus the required sets U and V exist, and the sequence $\{f_n\}$ separates the points of X.

Define a function $d \colon X \times X \to \mathbb{R}$ by setting

$$d(x,y) = \sum_n \frac{1}{2^n} |f_n(x) - f_n(y)|.$$

It is easy to use the fact that the functions $f_1, f_2, \ldots$ separate the points of X to check that d is a metric on the set X and to use the fact that the functions $f_1, f_2, \ldots$ are continuous (with respect to the original topology on X) to check that the topology induced by d is weaker than the original topology. Since the original topology makes X a compact space, while the topology induced by d is weaker and Hausdorff, the two topologies must be the same (see D.17). Thus the original topology on X is metrizable and in fact is metrized by d. $\square$

Our next task is to prove that each locally compact Hausdorff space that has a countable base is metrizable. For this we need the following lemma.

Lemma 7.1.14. *Let X be a locally compact Hausdorff space. If there is a countable base for the topology of X, then there is a countable base for the topology of the one-point compactification of X.*

Proof. Let $\mathscr{U}$ be a countable base for the topology of X, and let $\mathscr{U}_0$ be the collection of those sets V in $\mathscr{U}$ for which $\overline{V}$ is compact. Arrange the sets in $\mathscr{U}_0$ in a sequence, say $\{V_k\}$. Then $X = \cup_k V_k$, and so for each compact subset K of X there is a positive integer n such that $K \subseteq \cup_{i=1}^{n} V_k$. Thus if U is an open neighborhood in X^* of the point at infinity and if $K = X^* - U$, then there is a positive integer n such that $K \subseteq \cup_{k=1}^{n} V_k$ and hence such that $X^* - \overline{(\cup_{k=1}^{n} V_k)} \subseteq U$. It follows that the sets in $\mathscr{U}$, together with the sets $X^* - \overline{(\cup_{k=1}^{n} V_k)}$, $n = 1, 2, \ldots$, form a countable base for the topology of X^*. $\qquad\square$

Proposition 7.1.15. *Each locally compact Hausdorff space that has a countable base for its topology is metrizable.*

Proof. Let X be a locally compact Hausdorff space whose topology has a countable base. Proposition 7.1.13 and Lemma 7.1.14 imply that the one-point compactification X^* of X is metrizable. Then X, as a subspace of the metrizable space X^*, is metrizable. $\qquad\square$

A locally compact Hausdorff space can be metrizable without having a countable base; see Exercise 1.

Exercises

1. (a) Show that each discrete topological space is metrizable and locally compact.
   (b) Conclude that there are metrizable locally compact Hausdorff spaces that are not second countable.
2. Let X be a locally compact Hausdorff space, and let Y be a subspace of X. Show that if Y is open or closed (as a subset of X), then Y is locally compact.
3. Let X be a locally compact Hausdorff space, and let Y be a subspace of X. Show that Y is locally compact if and only if $Y = U \cap F$ for some open subset U and some closed subset F of X. (Hint: See Exercise 2. Also show that if Y is locally compact, then Y is an open subset of $\overline{Y}$ (of course $\overline{Y}$ is the closure of Y in X and is to be given the topology it inherits from X).)
4. Find all continuous functions $f : \mathbb{Q} \to \mathbb{R}$ such that $\mathrm{supp}(f)$ is compact.
5. Prove Urysohn's lemma, Theorem 7.1.8. (Hint: Let D be the set of all dyadic rationals in the interval $(0, 1)$ (that is, let D be the set of all numbers of the form $m/2^n$, where m and n are positive integers and $m < 2^n$). Use the normality of X first to choose an open set $U_{1/2}$ such that $E \subseteq U_{1/2} \subseteq \overline{U_{1/2}} \subseteq F^c$ and then to choose open sets $U_{1/4}$ and $U_{3/4}$ such that $E \subseteq U_{1/4} \subseteq \overline{U_{1/4}} \subseteq U_{1/2}$ and $\overline{U_{1/2}} \subseteq U_{3/4} \subseteq \overline{U_{3/4}} \subseteq F^c$. Continue inductively, producing an indexed family $\{U_r\}_{r \in D}$ of open subsets of X such that

$$E \subseteq U_r \subseteq \overline{U_r} \subseteq U_s \subseteq \overline{U_s} \subseteq F^c$$

holds whenever r and s belong to D and satisfy $r < s$. Define a function $f : X \to \mathbb{R}$ by

$$f(x) = \begin{cases} 1 & \text{if } x \notin \bigcup_r U_r, \\ \inf\{r : x \in U_r\} & \text{otherwise,} \end{cases}$$

and check that it has the required properties.)

6. Prove the Tietze extension theorem: if X is a normal topological space, if E is a closed subspace of X, and if $f : E \to \mathbb{R}$ is bounded and continuous, then there is a bounded continuous function $g : X \to \mathbb{R}$ whose restriction to E is f. (Hint: Check that we can assume that $f(E) \subseteq [-1, 1]$. Use Urysohn's lemma to choose a continuous function $g_1 : X \to [-1/3, 1/3]$ such that $g_1(x) = -1/3$ if $x \in \{x \in E : f(x) \leq -1/3\}$ and $g_1(x) = 1/3$ if $x \in \{x \in E : f(x) \geq 1/3\}$. Show that $|f(x) - g_1(x)| \leq 2/3$ holds at each x in E. Continue inductively, choosing continuous functions $g_2, g_3, \ldots$ such that $|g_n(x)| \leq 2^{n-1}/3^n$ holds at each x in X and $|f(x) - (g_1 + \cdots + g_n)(x)| \leq (2/3)^n$ holds at each x in E. Then define g by $g = \sum_{n=1}^{\infty} g_n$.)

7. Let X be a compact Hausdorff space that contains at least two points, and let I be an uncountable set. Show that the product space X^I (which is of course compact[3] and Hausdorff) does not have a countable base. (Hint: Use D.11 to show that if X^I has a countable base and if $\mathscr{U}$ is the base for X^I constructed in D.19, then some countable subset of $\mathscr{U}$ is a base for X^I. Then show that no countable subfamily of $\mathscr{U}$ can be a base for X^I.)

8. Let X be the set consisting of those step functions $f : [0, 1] \to [0, 1]$ such that

   (i) each value of f is rational, and
   (ii) each jump in the graph of $y = f(x)$ occurs at a rational value of x.

   Show that X is a countable dense subset of the product space $[0, 1]^{[0,1]}$. Conclude that a compact Hausdorff space can be separable without being second countable. (See Exercise 7.)

9. Let X be a second countable compact Hausdorff space (in other words, a compact metrizable space), and let $C(X)$ be the vector space of all real-valued continuous functions on X. Give $C(X)$ the norm $\|\cdot\|_\infty$ defined by $\|f\|_\infty = \sup\{|f(x)| : x \in X\}$. Show that $C(X)$ is separable. (Hint: We saw in the proof of Proposition 7.1.13 that one can choose a countable collection S of continuous functions on X such that S separates the points of X. The Stone–Weierstrass theorem (Theorem D.22) implies that the polynomials in the functions belonging to S form a dense subset of $C(X)$. Those polynomials that have rational coefficients form a countable dense subset of $C(X)$.)

10. Let Ω be the smallest uncountable ordinal, let X be the set of all ordinal numbers α such that $\alpha \leq \Omega$, and let Y be the set of all ordinal numbers α such that $\alpha < \Omega$

---

[3]See Theorem D.20.

(thus Y consists of the countable ordinals). Give X and Y the order topology (see D.24). Show that

(a) X is a compact Hausdorff space, and

(b) Y is a locally compact Hausdorff space.

(Hint: Use transfinite induction to show that for each α in X the set $\{\beta \in X : \beta \leq \alpha\}$ is compact.)

7.2   The Riesz Representation Theorem

Let X be a Hausdorff topological space. Then $\mathscr{B}(X)$, the *Borel σ-algebra* on X, is the σ-algebra generated by the open subsets of X; the *Borel subsets* of X are those that belong to $\mathscr{B}(X)$. Note that $\mathscr{B}(X)$ is also the σ-algebra generated by the closed subsets of X.

We will need the following two elementary facts about the Borel subsets of Hausdorff spaces.

Lemma 7.2.1. *Let X and Y be Hausdorff topological spaces, and let $f \colon X \to Y$ be continuous. Then f is Borel measurable (that is, measurable with respect to $\mathscr{B}(X)$ and $\mathscr{B}(Y)$).*

Proof. The continuity of f implies that if U is an open subset of Y, then $f^{-1}(U)$ is an open and hence a Borel subset of X. Since the collection of open subsets of Y generates $\mathscr{B}(Y)$, the measurability of f follows from Proposition 2.6.2.   □

Lemma 7.2.2. *Let X be a Hausdorff topological space, and let Y be a subspace of X. Then*

$$\mathscr{B}(Y) = \{A : \text{there is a set } B \text{ in } \mathscr{B}(X) \text{ such that } A = B \cap Y\}.$$

Proof. Let $\mathscr{B}(X)_Y$ denote the collection of subsets of Y that have the form $B \cap Y$ for some B in $\mathscr{B}(X)$. We need to show that $\mathscr{B}(Y) = \mathscr{B}(X)_Y$. Let f be the standard injection of Y into X (in other words, let $f(y) = y$ hold at each y in Y). Then f is continuous and hence measurable with respect to $\mathscr{B}(Y)$ and $\mathscr{B}(X)$. Since $f^{-1}(B) = B \cap Y$ holds for each subset B of X, the measurability of f implies that $\mathscr{B}(X)_Y \subseteq \mathscr{B}(Y)$. On the other hand, it is easy to check that $\mathscr{B}(X)_Y$ is a σ-algebra on Y that contains all the open subsets of Y and hence includes $\mathscr{B}(Y)$. With this we have shown that $B(Y) = \mathscr{B}(X)_Y$.   □

We turn to terminology for measures. Let X be a Hausdorff topological space. A *Borel measure* on X is a measure whose domain is $\mathscr{B}(X)$. Suppose that $\mathscr{A}$ is a σ-algebra on X such that $\mathscr{B}(X) \subseteq \mathscr{A}$. A positive measure μ on $\mathscr{A}$ is *regular* if

(a) each compact subset K of X satisfies $\mu(K) < +\infty$,

(b) each set A in $\mathscr{A}$ satisfies

$$\mu(A) = \inf\{\mu(U) : A \subseteq U \text{ and } U \text{ is open}\}, \text{ and}$$

(c) each open subset U of X satisfies

$$\mu(U) = \sup\{\mu(K) : K \subseteq U \text{ and } K \text{ is compact}\}.$$

A *regular Borel measure* on X is a regular measure whose domain is $\mathscr{B}(X)$. A measure that satisfies condition (b) is often called *outer regular*, and a measure that satisfies condition (c), *inner regular*.

We have already seen that Lebesgue measure on $\mathbb{R}^d$ is regular (Proposition 1.4.1) and that every finite Borel measure on $\mathbb{R}^d$ is regular (Proposition 1.5.6).

The regularity of a measure allows many approximations and calculations that would be impossible without it. In particular, various linear functionals can be represented in a useful way with regular measures; see Theorems 7.2.8 and 7.3.6.

On certain rather complicated locally compact Hausdorff spaces there exist finite Borel measures that are not regular; see Exercise 7. However, for a locally compact Hausdorff space that has a countable base, we have the following result.

Proposition 7.2.3. *Let X be a locally compact Hausdorff space that has a countable base, and let μ be a Borel measure on X that is finite on compact sets. Then μ is regular.*

Proof. First consider the inner regularity of μ. Let U be an open subset of X. Proposition 7.1.5 implies that U is the union of a sequence $\{K_j\}$ of compact sets, and Proposition 1.2.5 then implies that $\mu(U) = \lim_n \mu(\cup_{j=1}^n K_j)$. The inner regularity of μ follows.

We will use the following reformulation of Lemma 1.5.7 in our proof of the outer regularity of μ.

Lemma 7.2.4. *Let X be a Hausdorff space in which each open set is an F_σ, and let μ be a finite Borel measure on X. Then each Borel subset A of X satisfies*

$$\mu(A) = \inf\{\mu(U) : A \subseteq U \text{ and } U \text{ is open}\} \tag{1}$$

and

$$\mu(A) = \sup\{\mu(F) : F \subseteq A \text{ and } F \text{ is closed}\}. \tag{2}$$

Proof. The arguments used to prove Lemma 1.5.7 also prove this lemma; the details will not be repeated. $\square$

Let us continue with the proof of Proposition 7.2.3. We still need to check the outer regularity of μ. In order to apply Lemma 7.2.4, we will consider certain finite measures that are closely related to μ. Let $\{U_n\}$ be a sequence of open sets such that $X = \cup_n U_n$ and such that $\mu(U_n) < +\infty$ holds for each n (for instance, take a countable base $\mathscr{U}$ for X and arrange in a sequence those sets U in $\mathscr{U}$ for which $\overline{U}$ is compact). For each n define a Borel measure μ_n on X by $\mu_n(A) = \mu(A \cap U_n)$. The measures μ_n are finite, and so Proposition 7.1.5 and Lemma 7.2.4 imply that they are outer regular. Hence if A belongs to $\mathscr{B}(X)$ and if ε is a positive number, then for each n there is an open set V_n that includes A and satisfies $\mu_n(V_n) < \mu_n(A) + \varepsilon/2^n$.

Consequently $\mu((U_n \cap V_n) - A) < \varepsilon/2^n$. The set V defined by $V = \cup_n(U_n \cap V_n)$ is open, includes A, and satisfies

$$\mu(V - A) \leq \sum_n \mu((U_n \cap V_n) - A) < \varepsilon.$$

Hence $\mu(V) \leq \mu(A) + \varepsilon$, and the outer regularity of μ follows. $\square$

Proposition 7.2.5. *Let X be a locally compact Hausdorff space that has a countable base. Then every regular measure on X is σ-finite.*

Proof. The space X is, according to Proposition 7.1.6, the union of a sequence of compact sets. Since the measure of a compact set is finite under a regular measure, the proposition follows. $\square$

The following proposition enables one to approximate many sets from within by compact sets.

Proposition 7.2.6. *Let X be a Hausdorff space, let $\mathscr{A}$ be a σ-algebra on X that includes $\mathscr{B}(X)$, and let μ be a regular measure on $\mathscr{A}$. If A belongs to $\mathscr{A}$ and is σ-finite under μ, then*

$$\mu(A) = \sup\{\mu(K) : K \subseteq A \text{ and } K \text{ is compact}\}. \tag{3}$$

Proof. First suppose that $\mu(A) < +\infty$. Let ε be a positive number, and use the regularity of μ first to choose an open set V such that $A \subseteq V$ and $\mu(V) < \mu(A) + \varepsilon$ and then to choose a compact subset L of V such that $\mu(L) > \mu(V) - \varepsilon$. Since $\mu(V - A) < \varepsilon$, we can choose an open set W that includes $V - A$ and satisfies $\mu(W) < \varepsilon$. The set $L - W$ is then a compact subset of A, and it satisfies

$$\mu(L - W) = \mu(L) - \mu(L \cap W) > \mu(V) - 2\varepsilon \geq \mu(A) - 2\varepsilon.$$

Since ε is arbitrary, relation (3) follows in the case where $\mu(A)$ is finite.

In the case where $\mu(A) = +\infty$, we can suppose that $A = \cup_n A_n$, where for each n we have $A_n \in \mathscr{A}$ and $\mu(A_n) < +\infty$. For each positive number α, we need to construct a compact subset K of A such that $\mu(K) > \alpha$. We can construct such a set by first choosing N large enough that $\mu(\cup_{n=1}^N A_n) > \alpha$ and then using the construction in the first part of the proof to produce an appropriate compact subset of $\cup_{n=1}^N A_n$. $\square$

Let X be a locally compact Hausdorff space. Recall that $\mathscr{K}(X)$ is the vector space consisting of all real-valued functions on X that are continuous and have compact support. We will study the relationship between regular measures on X and linear functionals on $\mathscr{K}(X)$. The first thing to note is that each function in $\mathscr{K}(X)$ is integrable with respect to each regular measure on X (each such function is measurable (Lemma 7.2.1) and so, since it is bounded and vanishes outside a set that is compact and hence of finite measure, is integrable). It follows that if μ is a regular Borel measure on X, then $f \mapsto \int f \, d\mu$ defines a linear functional on $\mathscr{K}(X)$. Two questions arise immediately. Can several regular Borel measures induce the same functional? Which functionals arise in this way? Both of these questions will be answered in Theorem 7.2.8.

For dealing with such questions the concept of positivity for linear functionals is essential. A linear functional I on $\mathcal{K}(X)$ is *positive* if for each nonnegative f in $\mathcal{K}(X)$ we have $I(f) \geq 0$. Note that if μ is a regular Borel measure on X, then the functional $f \mapsto \int f \, d\mu$ is positive. Note also that a positive linear functional I on $\mathcal{K}(X)$ is order preserving, in the sense that if f and g belong to $\mathcal{K}(X)$ and satisfy $f \leq g$, then $I(f) \leq I(g)$ (if $f \leq g$, then $g - f$ is nonnegative, and we have $I(g) - I(f) = I(g - f) \geq 0$).

Let U be an open subset of the locally compact Hausdorff space X. We will often deal with functions f that belong to $\mathcal{K}(X)$ and satisfy

$$0 \leq f \leq \chi_U. \tag{4}$$

Among the functions f in $\mathcal{K}(X)$ that satisfy (4), those that also satisfy $\mathrm{supp}(f) \subseteq U$ are especially nice to deal with; accordingly we will write $f \prec U$ to indicate that f satisfies both (4) and the relation $\mathrm{supp}(f) \subseteq U$.

Lemma 7.2.7. *Let X be a locally compact Hausdorff space, and let μ be a regular Borel measure on X. If U is an open subset of X, then*

$$\mu(U) = \sup\left\{ \int f \, d\mu : f \in \mathcal{K}(X) \text{ and } 0 \leq f \leq \chi_U \right\}$$

$$= \sup\left\{ \int f \, d\mu : f \in \mathcal{K}(X) \text{ and } f \prec U \right\}.$$

Proof. It is clear that $\mu(U)$ is at least as large as the first supremum and that the first supremum is at least as large as the second. So it is enough to prove that

$$\mu(U) \leq \sup\left\{ \int f \, d\mu : f \in \mathcal{K}(X) \text{ and } f \prec U \right\}.$$

Let α be a number that satisfies $\alpha < \mu(U)$, and use the regularity of μ to choose a compact subset K of U such that $\alpha < \mu(K)$. Proposition 7.1.9 provides a function f in $\mathcal{K}(X)$ that satisfies $\chi_K \leq f$ and $f \prec U$. Then $\alpha < \int f \, d\mu$, and so

$$\alpha < \sup\left\{ \int f \, d\mu : f \in \mathcal{K}(X) \text{ and } f \prec U \right\}.$$

Since α was an arbitrary number less than $\mu(U)$, the proof is complete. □

We are now in a position to prove the main result of this section.

Theorem 7.2.8 (Riesz Representation Theorem). *Let X be a locally compact Hausdorff space, and let I be a positive linear functional on $\mathcal{K}(X)$. Then there is a unique regular Borel measure μ on X such that*

$$I(f) = \int f \, d\mu$$

holds for each f in $\mathcal{K}(X)$.

Proof. We first prove the uniqueness of μ. Suppose that μ and ν are regular Borel measures on X such that

$$\int f\,d\mu = \int f\,d\nu = I(f)$$

holds for each f in $\mathscr{K}(X)$. It follows from Lemma 7.2.7 that $\mu(U) = \nu(U)$ holds for each open subset U of X and then from the outer regularity of μ and ν that $\mu(A) = \nu(A)$ holds for each Borel subset A of X. Thus μ and ν are equal, and the uniqueness is proved.

We turn to the construction of a measure representing the functional I. Lemma 7.2.7 and condition (b) in the definition of regularity suggest how to proceed. Define a function μ^* on the open subsets of X by

$$\mu^*(U) = \sup\{I(f) : f \in \mathscr{K}(X) \text{ and } f \prec U\}, \tag{5}$$

and then extend it to all subsets of X by

$$\mu^*(A) = \inf\{\mu^*(U) : U \text{ is open and } A \subseteq U\} \tag{6}$$

(it is easy to check that Eq. (6) is consistent with Eq. (5), in the sense that an open set is assigned the same value by both). We will presently see that the required measure μ can be obtained by restricting μ^* to $\mathscr{B}(X)$.

The rest of the proof of Theorem 7.2.8 will be given by Proposition 7.2.9, Lemma 7.2.10, and Proposition 7.2.11.

Proposition 7.2.9. *Let X and I be as in the statement of Theorem 7.2.8, and let μ^* be defined by (5) and (6). Then μ^* is an outer measure on X, and every Borel subset of X is μ^*-measurable.*

Proof. The relation $\mu^*(\varnothing) = 0$ and the monotonicity of μ^* are clear. We need to check the countable subadditivity of μ^*. First suppose that $\{U_n\}$ is a sequence of open subsets of X; we will verify that

$$\mu^*\left(\bigcup_n U_n\right) \leq \sum_n \mu^*(U_n). \tag{7}$$

Let f be a function that belongs to $\mathscr{K}(X)$ and satisfies $f \prec \cup_n U_n$. Then $\mathrm{supp}(f)$ is a compact subset of $\cup_n U_n$, and so there is a positive integer N such that $\mathrm{supp}(f) \subseteq \cup_{n=1}^N U_n$. Proposition 7.1.11 implies that f is the sum of functions $f_1, \ldots, f_N$ that belong to $\mathscr{K}(X)$ and satisfy $f_n \prec U_n$ for $n = 1, \ldots, N$. It follows that

$$I(f) = \sum_{n=1}^N I(f_n) \leq \sum_{n=1}^N \mu^*(U_n) \leq \sum_{n=1}^\infty \mu^*(U_n).$$

This and Eq. (5) yield inequality (7).

Now suppose that $\{A_n\}$ is an arbitrary sequence of subsets of X. The inequality $\mu^*(\cup_n A_n) \leq \sum_n \mu^*(A_n)$ is clear if $\sum_n \mu^*(A_n) = +\infty$. So suppose that

$\sum_n \mu^*(A_n) < +\infty$, let ε be a positive number, and for each n use (6) to choose an open set U_n that includes A_n and satisfies $\mu^*(U_n) \leq \mu^*(A_n) + \varepsilon/2^n$. Then (see inequality (7))

$$\mu^*(\cup_n A_n) \leq \mu^*(\cup_n U_n) \leq \sum_{n=1}^{\infty} \mu^*(U_n) \leq \sum_{n=1}^{\infty} \mu^*(A_n) + \varepsilon.$$

Since ε is arbitrary, the relation $\mu^*(\cup_n A_n) \leq \sum_n \mu^*(A_n)$ follows. Thus μ^* is countably subadditive and so is an outer measure.

Since the family of μ^*-measurable sets is a σ-algebra, we can show that every Borel subset of X is μ^*-measurable by checking that each open subset of X is μ^*-measurable. So let U be an open subset of X. According to the discussion preceding Proposition 1.3.5, we can prove that U is μ^*-measurable by showing that

$$\mu^*(A) \geq \mu^*(A \cap U) + \mu^*(A \cap U^c) \tag{8}$$

holds for each subset A of X that satisfies $\mu^*(A) < +\infty$. Let A be such a set, let ε be a positive number, and use (6) to choose an open set V that includes A and satisfies $\mu^*(V) < \mu^*(A) + \varepsilon$. If we show that

$$\mu^*(V) > \mu^*(V \cap U) + \mu^*(V \cap U^c) - 2\varepsilon, \tag{9}$$

it will follow that

$$\mu^*(A) + \varepsilon > \mu^*(A \cap U) + \mu^*(A \cap U^c) - 2\varepsilon,$$

and, since ε is arbitrary, that (8) holds. So let us verify (9). Choose a function f_1 in $\mathcal{K}(X)$ that satisfies $f_1 \prec V \cap U$ and $I(f_1) > \mu^*(V \cap U) - \varepsilon$, let $K = \text{supp}(f_1)$, and then choose a function f_2 in $\mathcal{K}(X)$ that satisfies $f_2 \prec V \cap K^c$ and $I(f_2) > \mu^*(V \cap K^c) - \varepsilon$. (This would be a good time to draw a sketch of the sets involved here.) Since $f_1 + f_2 \prec V$ and $V \cap U^c \subseteq V \cap K^c$, we have

$$\mu^*(V) \geq I(f_1 + f_2) > \mu^*(V \cap U) + \mu^*(V \cap U^c) - 2\varepsilon.$$

Thus (9) holds and proof of Proposition 7.2.9 is complete. $\qquad\qquad\square$

Lemma 7.2.10. *Let X and I be as in the statement of Theorem 7.2.8, and let μ^* be defined by (5) and (6). Suppose that A is a subset of X and that f belongs to $\mathcal{K}(X)$. If $\chi_A \leq f$, then $\mu^*(A) \leq I(f)$, while if $0 \leq f \leq \chi_A$ and if A is compact,[4] then $I(f) \leq \mu^*(A)$.*

Proof. First assume that $\chi_A \leq f$. Let ε satisfy $0 < \varepsilon < 1$, and define U_ε by $U_\varepsilon = \{x \in X : f(x) > 1 - \varepsilon\}$. Then U_ε is open, and each g in $\mathcal{K}(X)$ that satisfies $g \leq \chi_{U_\varepsilon}$ also satisfies $g \leq \frac{1}{1-\varepsilon} f$; hence (5) implies that $\mu^*(U_\varepsilon) \leq \frac{1}{1-\varepsilon} I(f)$. Since $A \subseteq U_\varepsilon$ and since ε can be made arbitrarily close to 0, it follows that $\mu^*(A) \leq I(f)$.

---

[4]The assumption that A is compact simplifies the proof, but is not actually necessary.

Now suppose that $0 \leq f \leq \chi_A$ and that A is compact. Let U be an open set that includes A. Then $f \prec U$ and so (5) implies that $I(f) \leq \mu^*(U)$. Since U was an arbitrary open set that includes A, (6) implies that $I(f) \leq \mu^*(A)$. □

Proposition 7.2.11. *Let X and I be as in the statement of Theorem 7.2.8, let μ^* be defined by (5) and (6), let μ be the restriction of μ^* to $\mathscr{B}(X)$, and let μ_1 be the restriction of μ^* to the σ-algebra $\mathscr{M}_{\mu^*}$ of μ^*-measurable sets. Then μ and μ_1 are regular measures, and*

$$\int f \, d\mu = \int f \, d\mu_1 = I(f)$$

holds for each f in $\mathscr{K}(X)$.

Proof. Theorem 1.3.6 implies that μ_1 is a measure on $\mathscr{M}_{\mu^*}$ and, since $\mathscr{B}(X) \subseteq \mathscr{M}_{\mu^*}$ (Proposition 7.2.9), that μ is a measure on $\mathscr{B}(X)$. Since for each compact subset K of X there is a function f that belongs to $\mathscr{K}(X)$ and satisfies $\chi_K \leq f$ (Proposition 7.1.9), the first part of Lemma 7.2.10 implies that μ and μ_1 are finite on compact sets. The outer regularity of μ and μ_1 follows from (6), and the inner regularity of these measures follows from (5) and the second part of Lemma 7.2.10 (where we let A be the support of f).

We turn to the identity $I(f) = \int f \, d\mu = \int f \, d\mu_1$. Since each function in $\mathscr{K}(X)$ is the difference of two nonnegative functions in $\mathscr{K}(X)$, we can restrict our attention to the nonnegative functions in $\mathscr{K}(X)$. Let f be such a function. Let ε be a positive number, and for each positive integer n define a function $f_n : X \to \mathbb{R}$ by

$$f_n(x) = \begin{cases} 0 & \text{if } f(x) \leq (n-1)\varepsilon, \\ f(x) - (n-1)\varepsilon & \text{if } (n-1)\varepsilon < f(x) \leq n\varepsilon, \\ \varepsilon & \text{if } n\varepsilon < f(x). \end{cases} \quad (10)$$

(See Fig. 7.1 below.) Then each f_n belongs to $\mathscr{K}(X)$, $f = \sum_n f_n$, and there is a positive integer N such that $f_n = 0$ if $n > N$. Let $K_0 = \text{supp}(f)$ and for each positive integer n let $K_n = \{x \in X : f(x) \geq n\varepsilon\}$. Then $\varepsilon \chi_{K_n} \leq f_n \leq \varepsilon \chi_{K_{n-1}}$ holds for each n, and so Lemma 7.2.10 and the basic properties of the integral imply that $\varepsilon \mu(K_n) \leq I(f_n) \leq \varepsilon \mu(K_{n-1})$ and $\varepsilon \mu(K_n) \leq \int f_n \, d\mu \leq \varepsilon \mu(K_{n-1})$ hold for each n. Since $f = \sum_{n=1}^{N} f_n$, the relations

$$\sum_{n=1}^{N} \varepsilon \mu(K_n) \leq I(f) \leq \sum_{n=0}^{N-1} \varepsilon \mu(K_n)$$

and

$$\sum_{n=1}^{N} \varepsilon \mu(K_n) \leq \int f \, d\mu \leq \sum_{n=0}^{N-1} \varepsilon \mu(K_n)$$

follow. Thus $I(f)$ and $\int f \, d\mu$ both lie in the interval $[\sum_{n=1}^{N} \varepsilon \mu(K_n), \sum_{n=0}^{N-1} \varepsilon \mu(K_n)]$, which has length $\varepsilon \mu(K_0) - \varepsilon \mu(K_N)$. Since ε is arbitrary and this length is at most

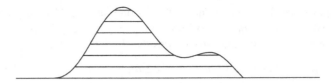

Fig. 7.1 Decomposing f as $\sum f_n$ (see Eq. (10))

$\varepsilon\mu(K_0)$, $I(f)$ and $\int f\,d\mu$ must be equal. It is clear that $\int f\,d\mu_1 = \int f\,d\mu$, and so the proof of Proposition 7.2.11, and hence that of Theorem 7.2.8, is complete.   □

Exercises

1. Let X be a locally compact Hausdorff space, let $\mathscr{A}$ be a σ-algebra on X that includes $\mathscr{B}(X)$, and let μ be a regular measure on $(X, \mathscr{A})$. Show that if A belongs to $\mathscr{A}$ and is σ-finite under μ, then for each positive ε there is an open set U that includes A and satisfies $\mu(U - A) < \varepsilon$. (Be sure to consider the case where $\mu(A)$ is infinite.)
2. Let X be a locally compact Hausdorff space, let $\mathscr{A}$ be a σ-algebra on X that includes $\mathscr{B}(X)$, and let μ be a regular measure on $(X, \mathscr{A})$. Show that the completion of μ is regular.
3. Let X be a locally compact Hausdorff space, let $\mathscr{A}$ be a σ-algebra on X that includes $\mathscr{B}(X)$, and let μ be a regular measure on $(X, \mathscr{A})$. Show that if A belongs to $\mathscr{A}$ and is σ-finite under μ, then there are sets E and F in $\mathscr{B}(X)$ such that $E \subseteq A \subseteq F$ and $\mu(F - E) = 0$. (In particular, if μ is σ-finite, then $\mathscr{A}$ is included in the completion of $\mathscr{B}(X)$ under the restriction of μ to $\mathscr{B}(X)$.)
4. Let us construct a topological space X by letting the underlying set be $\mathbb{R}^2$ and declaring that the open subsets U of X are those for which each section of the form U_x is an open subset of $\mathbb{R}$.

   (a) Show that X is locally compact and Hausdorff.
   (b) Characterize the functions $f : X \to \mathbb{R}$ that belong to $\mathscr{K}(X)$ in terms of their sections f_x.
   (c) Show that the formula

   $$I(f) = \sum_x \int f_x\,d\lambda$$

   (where λ is Lebesgue measure on $\mathbb{R}$) defines a positive linear functional on $\mathscr{K}(X)$ and that the regular Borel measure associated to I by the Riesz representation theorem is the restriction to $\mathscr{B}(X)$ of the measure defined in Exercise 3.3.6. (We will see in Exercise 9.4.12 that the σ-algebra $\mathscr{A}$ in Exercise 3.3.6 is strictly larger than $\mathscr{B}(X)$.)
   (d) Show that if μ is the regular Borel measure on X that corresponds to I, then

   $$\mu(A) = \sup\{\mu(K) : K \subseteq A \text{ and } K \text{ is compact}\}$$

   fails for some Borel subset A of X.

5. Let X be a locally compact Hausdorff space, let $\mathscr{A}$ be a σ-algebra on X that includes $\mathscr{B}(X)$, and let μ be a regular measure on $(X, \mathscr{A})$. Define $\mu^{\bullet}$ as in Exercise 1.2.8.
   (a) Show that
$$\mu^{\bullet}(A) = \sup\{\mu^{\bullet}(K) : K \subseteq A \text{ and } K \text{ is compact}\}$$
   holds for each A in $\mathscr{A}$. (In particular, $\mu^{\bullet}$ is inner regular.)
   (b) Show that the conditions
   (i) $\mu^{\bullet}$ is regular,
   (ii) $\mu^{\bullet} = \mu$, and
   (iii) every locally μ-null set in $\mathscr{A}$ is μ-null
   are equivalent.
6. Let X be a locally compact Hausdorff space, and let μ be a regular Borel measure on X. Suppose that $\mu(\{x\}) = 0$ holds for each x in X. Show that if B is a Borel subset of X such that $\mu(B) < +\infty$ and if a is a real number such that $0 < a < \mu(B)$, then there is a Borel subset A of B that satisfies $\mu(A) = a$. Can the Borel set A be replaced with a compact set?
7. Let Y be the collection of all countable ordinals, with the order topology (see Exercise 7.1.10).

   (a) Show that a subset A of Y is uncountable if and only if for each countable ordinal α there is an ordinal β that belongs to A and satisfies $\beta > \alpha$.
   (b) Show that if $\{C_n\}$ is a sequence of uncountable closed subsets of Y, then $\cap_n C_n$ is an uncountable closed set. (Hint: Use part (a); show that if $\{\alpha_k\}$ is an increasing sequence of countable ordinals such that each C_n contains infinitely many terms of $\{\alpha_k\}$, then $\lim_k \alpha_k$ exists and belongs to $\cap_n C_n$.)
   (c) Show that if $A \in \mathscr{B}(Y)$, then exactly one of A and A^c includes an uncountable closed subset of Y.
   (d) Suppose that we define a function μ on $\mathscr{B}(Y)$ by letting $\mu(A)$ be 1 if A includes an uncountable closed set and letting $\mu(A)$ be 0 otherwise. Show that μ is a Borel measure that is not regular. Find the regular Borel measure μ' on Y that satisfies $\int f \, d\mu' = \int f \, d\mu$ for each f in $\mathscr{K}(Y)$.
   (e) Let X be the collection of all ordinal numbers that are less than or equal to the first uncountable ordinal, and give X the order topology (again see Exercise 7.1.10). Show that the formula $\nu(A) = \mu(A \cap Y)$ defines a non-regular Borel measure ν on X. Find the regular Borel measure ν' on X that satisfies $\int f \, d\nu' = \int f \, d\nu$ for each f in $\mathscr{K}(X)$.

8. Let X be a compact Hausdorff space, and let $C(X)$ be the set of all real-valued continuous functions on X. Then $\mathscr{B}_0(X)$, the *Baire σ-algebra* on X, is the smallest σ-algebra on X that makes each function in $C(X)$ measurable; the sets that belong to $\mathscr{B}_0(X)$ are called the *Baire subsets* of X. A *Baire measure* on X is a *finite* measure on $(X, \mathscr{B}_0(X))$.

   (a) Show that $\mathscr{B}_0(X)$ is the σ-algebra generated by the closed G_δ's in X. (Hint: Check that if $f \in C(X)$ and if $a \in \mathbb{R}$, then $\{x : f(x) \leq a\}$ is a closed G_δ, and use Proposition 7.1.9 to check that every closed G_δ in X arises in this way.)

(b) Show that if the compact Hausdorff space X is second countable, then $\mathscr{B}_0(X) = \mathscr{B}(X)$.

9. Show that if μ is a Baire measure on a compact Hausdorff space X, then μ is regular, in the sense that

$$\mu(A) = \inf\{\mu(U) : A \subseteq U, U \text{ is open, and } U \in \mathscr{B}_0(X)\}$$

holds for each set A in $\mathscr{B}_0(X)$ and

$$\mu(U) = \sup\{\mu(K) : K \subseteq U, K \text{ is compact, and } K \in \mathscr{B}_0(X)\}$$

holds for each open set U in $\mathscr{B}_0(X)$. (Hint: Modify the proof of Lemma 1.5.7; show that the collection of Baire sets that can be approximated from above by open Baire sets and from below by compact Baire sets is a σ-algebra and that this σ-algebra contains all the closed G_δ's in X. See Exercise 8.)

10. Let X be a compact Hausdorff space, and let I be a positive linear functional on $C(X)$ (note that since X is compact, $C(X) = \mathscr{K}(X)$). Show that there is a unique Baire measure μ on X such that $I(f) = \int f \, d\mu$ holds for each f in $C(X)$. (Hint: First check that the restriction to $\mathscr{B}_0(X)$ of the measure given by Theorem 7.2.8 works. Then modify the part of the proof of Theorem 7.2.8 that deals with uniqueness; see Exercise 9.)

11. Let X be a compact Hausdorff space. Show that if K is a closed Baire subset of X, then K is a G_δ. (Hint: Use Exercise 1.1.7 to choose a sequence $\{f_n\}$ of functions in $C(X)$ such that K belongs to the smallest σ-algebra making $f_1, f_2,$ ... measurable; then define $F: X \to \mathbb{R}^{\mathbb{N}}$ by letting F take x to the sequence $\{f_n(x)\}$. Show that $F(K)$ is a compact subset of the second countable space $\mathbb{R}^{\mathbb{N}}$ and so is a G_δ; then check that $K = F^{-1}(F(K))$ (see Exercise 2.6.5), and conclude that K is a G_δ in X.)

12. Let I be the interval $[0, 1]$, and let X be the product space I^I, with the product topology (here the interval $[0, 1]$, when considered as a factor in the product space, is to have its usual topology). Thus each element x of X is an indexed family $\{x_i\}_{i \in I}$ of elements of I.

   (a) Show that if f belongs to $C(X)$, then $f(x)$ depends on only countably many of the components x_i of x (in other words, for each f in $C(X)$ there is a countable subset C of I such that if $x_i = y_i$ holds for each i in C, then $f(x) = f(y)$). (Hint: First consider the case where f is a polynomial in the components of x, and then use the Stone–Weierstrass theorem.)

   (b) Show that if $A \in \mathscr{B}_0(X)$, then $\chi_A(x)$ depends on only countably many of the components of x. (Hint: Check that the collection of sets A such that $\chi_A(x)$ depends on only countably many of the components of x is a σ-algebra.)

   (c) Show that if $f: X \to \mathbb{R}$ is $\mathscr{B}_0(X)$-measurable, then $f(x)$ depends on only countably many of the components of x.

   (d) Show that the one-element subsets of X belong to $\mathscr{B}(X)$ but not to $\mathscr{B}_0(X)$. Conclude that $\mathscr{B}_0(X) \neq \mathscr{B}(X)$.

13. Let X be the space of all ordinals less than or equal to the first uncountable ordinal, and let X have the order topology (see Exercise 7.1.10). Find $\mathscr{B}_0(X)$, the Baire σ-algebra on X. Is $\mathscr{B}_0(X)$ equal to $\mathscr{B}(X)$?

7.3 Signed and Complex Measures; Duality

This section is devoted to regularity for finite signed and complex Borel measures. The main result is a measure-theoretic representation for the duals of certain Banach spaces of continuous functions.

Let X be a locally compact Hausdorff space, and let f be a real- or complex-valued continuous function on X. Then f is said to *vanish at infinity* if for every positive number ε there is a compact subset K of X such that $|f(x)| < \varepsilon$ holds at each x outside K. We will denote by $C_0(X)$ the set of all real-valued continuous functions on X that vanish at infinity and by $C_0^{\mathbb{C}}(X)$ the set of all complex-valued continuous functions on X that vanish at infinity.

Examples 7.3.1. Note that a continuous function f on $\mathbb{R}$ vanishes at infinity if and only if $\lim_{x \to -\infty} f(x) = 0$ and $\lim_{x \to +\infty} f(x) = 0$. Note also that every continuous function on a compact Hausdorff space vanishes at infinity. See Exercises 1, 2, and 9 for some more examples, and see Exercise 3 for another characterization of the continuous functions that vanish at infinity.   □

Of course, $C_0(X)$ and $C_0^{\mathbb{C}}(X)$ are vector spaces over $\mathbb{R}$ and $\mathbb{C}$, respectively. These spaces are normed spaces: each continuous function that vanishes at infinity is bounded (since a continuous function is bounded on a compact set), and so the formula

$$\|f\|_\infty = \sup\{|f(x)| : x \in X\}$$

defines norms on $C_0(X)$ and $C_0^{\mathbb{C}}(X)$.

Proposition 7.3.2. *Let X be a locally compact Hausdorff space. Then $\mathscr{K}(X)$ and $\mathscr{K}^{\mathbb{C}}(X)$ are dense subspaces of $C_0(X)$ and $C_0^{\mathbb{C}}(X)$, respectively.*

Proof. It is clear that $\mathscr{K}(X)$ and $\mathscr{K}^{\mathbb{C}}(X)$ are linear subspaces of $C_0(X)$ and $C_0^{\mathbb{C}}(X)$. We need only show that they are dense. Suppose that f belongs to $C_0(X)$ or to $C_0^{\mathbb{C}}(X)$ and that ε is a positive number. Choose a compact set K such that $|f(x)| \le \varepsilon$ holds at each x outside K, and use Proposition 7.1.9 to choose a function $g : X \to [0, 1]$ that belongs to $\mathscr{K}(X)$ and satisfies $g(x) = 1$ at each x in K. Let $h = fg$. Then h belongs to $\mathscr{K}(X)$ or to $\mathscr{K}^{\mathbb{C}}(X)$ and satisfies $\|f - h\|_\infty \le \varepsilon$. Since ε is arbitrary, the proof is complete.   □

Proposition 7.3.3. *Let X be a locally compact Hausdorff space. Then $C_0(X)$ and $C_0^{\mathbb{C}}(X)$ are Banach spaces.*

Proof. The only issue is the completeness of these spaces. So let $\{f_n\}$ be a Cauchy sequence in one of them. A standard argument (see the proof given in Sect. 3.2 of the completeness of $C[a,b]$) shows that there is a continuous function f such that $\{f_n\}$ converges uniformly to f. We need only check that f vanishes at infinity. Let ε be a positive number, choose a positive integer n such that $|f(x) - f_n(x)| < \varepsilon$ holds at each x in X, and use the fact that f_n vanishes at infinity to choose a compact set K such that $|f_n(x)| < \varepsilon$ holds at each x outside K. Then

$$|f(x)| \leq |f(x) - f_n(x)| + |f_n(x)| < 2\varepsilon$$

holds at each x outside K, and since ε is arbitrary, the proof is complete.   □

Let X be a locally compact Hausdorff space. A finite signed or complex measure μ on $(X, \mathscr{B}(X))$ is *regular* if its variation $|\mu|$ is regular (in the sense of Sect. 7.2). It is convenient to note the following equivalent formulations of regularity.

Proposition 7.3.4. *Let X be a locally compact Hausdorff space, and let μ be a finite signed or complex measure on $(X, \mathscr{B}(X))$. Then the conditions*

(a) *μ is regular,*
(b) *each of the positive measures appearing in the Jordan decomposition of μ is regular, and*
(c) *μ is a linear combination of finite positive regular Borel measures*

are equivalent.

Proof. Suppose that condition (a) holds, and let μ' be one of the measures appearing in the Jordan decomposition of μ. Then μ' satisfies $\mu' \leq |\mu|$ (consider how μ' arises from a Hahn decomposition). Thus if $A \in \mathscr{B}(X)$, if ε is a positive number, and if U is an open set that includes A and satisfies $|\mu|(U) < |\mu|(A) + \varepsilon$, then $\mu'(U - A) \leq |\mu|(U - A) < \varepsilon$, and so

$$\mu'(U) = \mu'(A) + \mu'(U - A) < \mu'(A) + \varepsilon.$$

The outer regularity of μ' follows. The inner regularity of μ' can be proved in a similar manner. Hence condition (a) implies condition (b).

Condition (b) certainly implies condition (c).

The proof that (c) implies (a) is similar to the proof that (a) implies (b) and makes use of the fact that if $\mu = \alpha_1 \mu_1 + \cdots + \alpha_n \mu_n$, where each α_i is a real or complex number and each μ_i is positive, then $|\mu| \leq |\alpha_1| \mu_1 + \cdots + |\alpha_n| \mu_n$.   □

Regularity makes possible the following approximation (see also Exercise 5).

Lemma 7.3.5. *Let X be a locally compact Hausdorff space, and let μ be a finite signed or complex regular Borel measure on X. Then for each A in $\mathscr{B}(X)$ and each positive number ε there is a compact subset K of A such that $|\mu(A) - \mu(B)| < \varepsilon$ holds whenever B is a Borel set that satisfies $K \subseteq B \subseteq A$.*

Proof. Let A and ε be as in the statement of the lemma. Use the regularity of $|\mu|$ and Proposition 7.2.6 to choose a compact subset K of A such that $|\mu|(A - K) < \varepsilon$. Then each Borel set B that satisfies $K \subseteq B \subseteq A$ also satisfies

$$|\mu(A) - \mu(B)| = |\mu(A - B)| \leq |\mu|(A - B) \leq |\mu|(A - K) < \varepsilon,$$

which completes the proof of the lemma.   □

Let X be a locally compact Hausdorff space. We will denote by $M_r(X, \mathbb{R})$ the set of all finite signed regular Borel measures on X and by $M_r(X, \mathbb{C})$ the set of all complex regular Borel measures on X. It is easy to check that $M_r(X, \mathbb{R})$ and $M_r(X, \mathbb{C})$ are linear subspaces of the vector spaces $M(X, \mathscr{B}(X), \mathbb{R})$ and $M(X, \mathscr{B}(X), \mathbb{C})$ of all finite signed or complex measures on $(X, \mathscr{B}(X))$. These larger spaces are Banach spaces under the total variation norm (Proposition 4.1.8). Moreover $M_r(X, \mathbb{R})$ and $M_r(X, \mathbb{C})$ are closed subspaces of $M(X, \mathscr{B}(X), \mathbb{R})$ and $M(X, \mathscr{B}(X), \mathbb{C})$ (to check this, note that if μ is regular, if $\|\mu - v\| < \varepsilon$, and if A is a Borel set and U is an open set chosen so that $A \subseteq U$ and $|\mu|(U - A) < \varepsilon$, then

$$|v|(U - A) \leq \|v - \mu\| + |\mu|(U - A) < 2\varepsilon).$$

It follows that $M_r(X, \mathbb{R})$ and $M_r(X, \mathbb{C})$ are themselves Banach spaces under the total variation norm.

Recall (see Sect. 4.1) that if $(X, \mathscr{A})$ is a measurable space, if f is a bounded $\mathscr{A}$-measurable function on X, and if μ is a finite signed measure on $(X, \mathscr{A})$ with Jordan decomposition $\mu = \mu^+ - \mu^-$, then the integral of f with respect to μ is defined by

$$\int f \, d\mu = \int f \, d\mu^+ - \int f \, d\mu^-.$$

Likewise, if μ is a complex measure with Jordan decomposition $\mu = \mu_1 - \mu_2 + i\mu_3 - i\mu_4$, then

$$\int f \, d\mu = \int f \, d\mu_1 - \int f \, d\mu_2 + i \int f \, d\mu_3 - i \int f \, d\mu_4.$$

Theorem 7.3.6. *Let X be a locally compact Hausdorff space. Then the map that takes the finite signed (or complex) regular Borel measure μ to the functional $f \mapsto \int f \, d\mu$ is an isometric isomorphism of the Banach space $M_r(X, \mathbb{R})$ (or $M_r(X, \mathbb{C})$) onto the dual of the Banach space $C_0(X)$ (or $C_0^{\mathbb{C}}(X)$).*

Proof. For each finite signed regular Borel measure μ on X define a functional Φ_μ on $C_0(X)$ by $\Phi_\mu(f) = \int f \, d\mu$. It is easy to see that Φ_μ is a linear functional on $C_0(X)$ and that

$$|\Phi_\mu(f)| \leq \|f\|_\infty \|\mu\|$$

holds for each f and μ (see the discussion at the end of Sect. 4.1). Thus Φ_μ is continuous and its norm satisfies

$$\|\Phi_\mu\| \leq \|\mu\|. \tag{1}$$

Moreover, $\mu \mapsto \Phi_\mu$ defines a linear map Φ from $M_r(X, \mathbb{R})$ to the dual of $C_0(X)$. Analogous results hold for complex measures and complex-valued functions.

We need to show that Φ is norm preserving and surjective. Let us begin with the first of these tasks. In view of (1), it is enough to show that

$$\|\Phi_\mu\| \geq \|\mu\| \tag{2}$$

holds for each μ. So let μ belong to $M_r(X, \mathbb{R})$ or to $M_r(X, \mathbb{C})$, and let ε be a positive number. We can assume that $\|\mu\| \neq 0$. According to the definition of $\|\mu\|$, we can choose a finite partition of X into Borel sets A_j, $j = 1, \ldots, n$, such that $\sum_{j=1}^{n} |\mu(A_j)| > \|\mu\| - \varepsilon$. Now choose compact subsets $K_1, \ldots, K_n$ of $A_1, \ldots, A_n$ such that

$$\|\mu\| - \varepsilon < \sum_j |\mu(K_j)| \leq \sum_j |\mu|(K_j)$$

(see Lemma 7.3.5). We can assume that $\mu(K_j) \neq 0$ holds for each j. Choose a continuous function f that has compact support (and hence vanishes at infinity), satisfies $\|f\|_\infty \leq 1$, and is such that $f(x) = \overline{\mu(K_j)}/|\mu(K_j)|$ holds for each j and each x in K_j (see Proposition 7.1.12). Let $K = \cup_j K_j$. Then $\int_K f \, d\mu = \sum_j |\mu(K_j| > \|\mu\| - \varepsilon$, while $|\int_{K^c} f \, d\mu| \leq |\mu|(K^c) < \varepsilon$. It follows that $|\int f \, d\mu| > \|\mu\| - 2\varepsilon$. Since f satisfies $\|f\|_\infty \leq 1$ and ε is arbitrary, relation (2) follows. Thus Φ is norm preserving.

We turn to the surjectivity of Φ. First consider the case of real-valued functions and measures. Suppose that L is a continuous linear functional on $C_0(X)$ that is positive, in the sense that $L(f) \geq 0$ holds for each nonnegative f in $C_0(X)$. The restriction of L to $\mathscr{K}(X)$ is also positive, and so the Riesz representation theorem (Theorem 7.2.8) provides a regular Borel measure μ on X such that $L(f) = \int f \, d\mu$ holds for each f in $\mathscr{K}(X)$. Lemma 7.2.7 implies that

$$\mu(X) = \sup\{L(f) : f \in \mathscr{K}(X) \text{ and } 0 \leq f \leq 1\},$$

and hence that $\mu(X) \leq \|L\|$; in particular, μ is finite. Note that so far we have only proved that $L(f) = \Phi_\mu(f)$ holds when f belongs to the subspace $\mathscr{K}(X)$ of $C_0(X)$. However, since $\mathscr{K}(X)$ is dense in $C_0(X)$ (Proposition 7.3.2), while L and Φ_μ are continuous, the equality of L and Φ_μ on $C_0(X)$ follows. With this we have proved that each positive continuous linear functional on $C_0(X)$ is of the form Φ_μ.

We need the following lemma to complete the proof of Theorem 7.3.6.

Lemma 7.3.7. *Let X be a locally compact Hausdorff space. Then for each continuous linear functional L on $C_0(X)$ there are positive continuous linear functionals L_+ and L_- on $C_0(X)$ such that $L = L_+ - L_-$.*

Proof. For each nonnegative f in $C_0(X)$ define $L_+(f)$ by

$$L_+(f) = \sup\{L(g) : g \in C_0(X) \text{ and } 0 \leq g \leq f\}. \tag{3}$$

The relation

$$|L(g)| \leq \|L\|\|g\|_\infty \leq \|L\|\|f\|_\infty,$$

which is valid if $0 \leq g \leq f$, implies that the supremum involved in the definition of $L_+(f)$ is finite and in fact that

$$L_+(f) \leq \|L\|\|f\|_\infty. \tag{4}$$

We need to check that if $t \geq 0$ and if f, f_1, and f_2 are nonnegative functions in $C_0(X)$, then

$$0 \leq L_+(f),$$

$$L_+(tf) = tL_+(f), \text{ and}$$

$$L_+(f_1 + f_2) = L_+(f_1) + L_+(f_2).$$

The first two of these properties are easy to check, and so we turn to the third. If g_1 and g_2 belong to $C_0(X)$ and satisfy $0 \leq g_1 \leq f_1$ and $0 \leq g_2 \leq f_2$, then $0 \leq g_1 + g_2 \leq f_1 + f_2$, and so

$$L(g_1) + L(g_2) = L(g_1 + g_2) \leq L_+(f_1 + f_2).$$

Since g_1 and g_2 can be chosen so as to make $L(g_1) + L(g_2)$ arbitrarily close to $L_+(f_1) + L_+(f_2)$, the inequality

$$L_+(f_1) + L_+(f_2) \leq L_+(f_1 + f_2)$$

follows. Now consider the reverse inequality. Suppose that g belongs to $C_0(X)$ and satisfies $0 \leq g \leq f_1 + f_2$, and define functions g_1 and g_2 by $g_1 = g \wedge f_1$ and $g_2 = g - g_1$. Then g_1 and g_2 belong to $C_0(X)$ and satisfy $0 \leq g_1 \leq f_1$ and $0 \leq g_2 \leq f_2$, and so

$$L(g) = L(g_1) + L(g_2) \leq L_+(f_1) + L_+(f_2).$$

Since g can be chosen so as to make $L(g)$ arbitrarily close to $L_+(f_1 + f_2)$, the inequality

$$L_+(f_1 + f_2) \leq L_+(f_1) + L_+(f_2)$$

follows. With this the third of our properties is proved.

Now use the formula

$$L_+(f) = L_+(f^+) - L_+(f^-), \tag{5}$$

where f^+ and f^- are the positive and negative parts of f, to extend the definition of L_+ to all of $C_0(X)$. By imitating some arguments used in the construction of the integral (see Lemma 2.3.5 and Proposition 2.3.6), the reader can show that L_+ is a linear functional on $C_0(X)$. The positivity of L_+ is clear. Relations (4) and (5), together with the positivity of L_+, imply that $|L_+(f)| \leq \|L\| \|f\|_\infty$ and hence that L_+ is continuous.

Define a functional L_- on $C_0(X)$ by $L_- = L_+ - L$. The linearity and continuity of L_- are immediate. Its positivity follows from its definition and the fact that $L_+(f) \geq L(f)$ holds for each nonnegative f (let $g = f$ in relation (3)). Since $L = L_+ - L_-$, the proof of the lemma is complete. $\qquad\square$

Let us return to the proof of Theorem 7.3.6. Since we have already checked that each positive continuous linear functional on $C_0(X)$ is of the form Φ_μ, the surjectivity of $\Phi: M_r(X, \mathbb{R}) \to C_0(X)^*$ follows from Lemma 7.3.7. The extension to the case of complex-valued functions and measures is easy: if $L \in C_0^{\mathbb{C}}(X)^*$, then there are functionals L_1 and L_2 in $C_0(X)^*$ such that $L(f) = L_1(f) + iL_2(f)$ for each f in $C_0(X)$ (that is, for each *real-valued* f in $C_0^{\mathbb{C}}(X)$), and so if μ_1 and μ_2 are the finite signed regular Borel measures that represent L_1 and L_2, then $\mu_1 + i\mu_2$ is a complex regular Borel measure that represents L. $\qquad\square$

We close this section by turning to those finite signed or complex measures ν on $(X, \mathcal{B}(X))$ that have the form $\nu(A) = \int_A f \, d\mu$ for some μ-integrable f (here μ is a positive regular Borel measure on X). Two questions arise: Does the regularity of ν follow from the regularity of μ? Can such measures ν be characterized by a version of the Radon–Nikodym theorem, even if μ is not σ-finite? The next two propositions answer these questions.

These results will be used only in Sect. 9.4.

Proposition 7.3.8. *Let X be a locally compact Hausdorff space, let μ be a regular Borel measure on X, let f belong to $\mathscr{L}^1(X, \mathcal{B}(X), \mu)$, and let ν be the finite signed or complex measure on $(X, \mathcal{B}(X))$ defined by $\nu(A) = \int_A f \, d\mu$. Then ν is regular.*

Proof. For each f in $\mathscr{L}^1(X, \mathcal{B}(X), \mu)$ define a finite signed or complex measure ν_f on $(X, \mathcal{B}(X))$ by $\nu_f(A) = \int_A f \, d\mu$. Let us deal first with the case where f is the characteristic function of a Borel set B for which $\mu(B) < +\infty$. In this case ν_f is the positive measure given by $\nu_f(A) = \mu(A \cap B)$, and for each A in $\mathcal{B}(X)$ Proposition 7.2.6, applied to the measure μ and the set $A \cap B$, implies that

$$\nu_f(A) = \sup\{\nu_f(K) : K \subseteq A \text{ and } K \text{ is compact }\}. \qquad (6)$$

Thus ν_f is inner regular. The outer regularity of ν_f follows if for each A in $\mathcal{B}(X)$ we use (6) (with A replaced by A^c) to approximate A^c from below by compact sets and hence to approximate A from above by open sets.

We can use the regularity of ν_f for such characteristic functions to conclude first that ν_f is regular if f is simple and integrable and then that ν_f is regular if f is an arbitrary integrable function (see Propositions 3.4.2 and 4.2.5). $\qquad\square$

Proposition 7.3.9. *Let X be a locally compact Hausdorff space, let μ be a regular Borel measure on X, and let ν be a finite signed or complex regular Borel measure on X. Then the conditions*

(a) there is a function f in $\mathcal{L}^1(X,\mathcal{B}(X),\mu)$ such that $v(A) = \int_A f\,d\mu$ holds for
   each A in $\mathcal{B}(X)$,
(b) v is absolutely continuous with respect to μ (each Borel subset A of X that
   satisfies $\mu(A) = 0$ also satisfies $v(A) = 0$), and
(c) each compact subset K of X that satisfies $\mu(K) = 0$ also satisfies $v(K) = 0$

are equivalent.

Proof. It is clear that condition (a) implies condition (b) and that condition (b)
implies condition (c).

   If $A \in \mathcal{B}(X)$, then Lemma 7.3.5 implies that for every positive ε there is a
compact subset K of A such that $|v(A) - v(K)| < \varepsilon$. Consequently if condition (c)
holds and if A satisfies $\mu(A) = 0$, then A must also satisfy $v(A) = 0$. Thus condition
(c) implies condition (b).

   Next suppose that condition (b) holds. The difficulty in using the Radon–
Nikodym theorem (Theorem 4.2.4) to derive condition (a) is that we are not
assuming that μ is σ-finite. We take care of this as follows. Use the regularity of
v to choose an increasing sequence $\{K_n\}$ of compact sets such that $\lim_n |v|(K_n) =$
$|v|(X)$. Then, because of the regularity of μ, $\mu(K_n)$ is finite for each n, and so the
measure μ_0 defined by $\mu_0(A) = \mu(A \cap (\cup_n K_n))$ is σ-finite. Since v is absolutely
continuous with respect to μ and since $|v|(X - (\cup_n K_n)) = 0$, v is also absolutely
continuous with respect to μ_0. Thus the Radon–Nikodym theorem provides a
function f in $\mathcal{L}^1(X,\mathcal{B}(X),\mu_0)$ such that $v(A) = \int_A f\,d\mu_0$ holds for each A in $\mathcal{B}(X)$.
If we modify f so that it vanishes outside $\cup_n K_n$, then $v(A) = \int_A f\,d\mu$ holds for each
A in $\mathcal{B}(X)$. With this we have shown that condition (b) implies condition (a). □

Proposition 7.3.10. *Let X be a locally compact Hausdorff space, and let μ be a
regular Borel measure on X. For each f in $\mathcal{L}^1(X,\mathcal{B}(X),\mu)$ define a finite signed or
complex measure v_f on $(X,\mathcal{B}(X))$ by means of the formula $v_f(A) = \int_A f\,d\mu$. Then
the map $f \mapsto v_f$ induces a linear isometry of $L^1(X,\mathcal{B}(X),\mu)$ onto the subspace of
$M_r(X,\mathbb{R})$ (or of $M_r(X,\mathbb{C})$) that consists of those v that are absolutely continuous
with respect to μ.*

Proof. The proposition is an immediate consequence of Propositions 7.3.8 and 7.3.9
and the fact that $\|v_f\| = \int |f|\,d\mu$ (see Proposition 4.2.5). □

Exercises

1. Describe $\mathcal{K}(X)$ and $C_0(X)$ rather explicitly in the case where X is the space
   $\{(x_1,x_2) \in \mathbb{R}^2 : (x_1,x_2) \neq (0,0)\}$.
2. Give an example of a continuous function $f \colon \mathbb{R}^2 \to \mathbb{R}$ that does not belong to
   $C_0(\mathbb{R}^2)$ but satisfies $\lim_{t\to\infty} f(tx_1,tx_2) = 0$ for each nonzero (x_1,x_2) in $\mathbb{R}^2$.

3. Let X be a locally compact Hausdorff space, let X^* be its one-point compact-ification, and let x_∞ be the point at infinity. Show that a function $f: X \to \mathbb{R}$ belongs to $C_0(X)$ if and only if the function $f^*: X^* \to \mathbb{R}$ defined by

$$f^*(x) = \begin{cases} f(x) & \text{if } x \in X, \\ 0 & \text{if } x = x_\infty, \end{cases}$$

   is continuous.

4. Show that the decomposition $L = L_+ - L_-$ given in the proof of Lemma 7.3.7 is minimal, in the sense that if $L = L_1 - L_2$ is another decomposition of L into a difference of positive linear functionals, then $L_1(f) \geq L_+(f)$ and $L_2(f) \geq L_-(f)$ hold for each nonnegative f in $C_0(X)$.

5. Prove the converse of Lemma 7.3.5: if X is a locally compact Hausdorff space, if μ is a finite signed or complex measure on $(X, \mathcal{B}(X))$, and if μ satisfies the conclusion of Lemma 7.3.5, then μ is regular.

6. Let X be a locally compact Hausdorff space, and let μ be a regular Borel measure on X such that $\mu(X) = +\infty$. Show that there is a nonnegative function f in $C_0(X)$ such that $\int f \, d\mu = +\infty$.

7. Let X be a locally compact Hausdorff space. Show that each positive linear functional on $C_0(X)$ is continuous.

8. Show that if X is a second countable locally compact Hausdorff space, then $C_0(X)$ is separable. (Hint: Use Exercises 7.1.9 and 7.3.3.)

9. Let Y be the space of all countable ordinals, with the order topology (see Exercise 7.1.10). Show that Y is not compact, but $C_0(Y) = \mathcal{K}(Y)$.

10. Let X be a compact Hausdorff space, let $\mathcal{B}_0(X)$ be the Baire σ-algebra on X (see Exercise 7.2.8), and let $C(X)$ be the space of all continuous real-(or complex-) valued functions on X. Give $C(X)$ the norm $\| \cdot \|_\infty$ defined by $\|f\|_\infty = \sup\{|f(x)| : x \in X\}$. Show that the map that assigns to a finite signed (or complex) measure μ on $(X, \mathcal{B}_0(X))$ the functional $f \mapsto \int f \, d\mu$ is an isometric isomorphism of $M(X, \mathcal{B}_0(X), \mathbb{R})$ (or of $M(X, \mathcal{B}_0(X), \mathbb{C})$) onto $C(X)^*$. (Hint: Modify the proof of Theorem 7.3.6; see Exercises 7.2.9 and 7.2.10.)

7.4 Additional Properties of Regular Measures

This section is devoted to several useful facts about regular measures.

Proposition 7.4.1. *Let X be a locally compact Hausdorff space, let $\mathcal{A}$ be a σ-algebra on X that includes $\mathcal{B}(X)$, and let μ be a regular measure[5] on $(X, \mathcal{A})$. Then the union of all the open subsets of X that have measure zero under μ is itself an open set that has measure zero under μ.*

---

[5]Note that μ is a positive measure, since its specification has no modifier such as "signed" or "complex."

Proof. Let $\mathscr{U}$ be the collection of all open subsets of X that have measure zero under μ, and let U be the union of the sets in $\mathscr{U}$. Then U is open and so belongs to $\mathscr{A}$. If K is a compact subset of U, then K can be covered by a finite collection U_1, $U_2, \dots, U_n$ of sets that belong to $\mathscr{U}$, and so we have

$$\mu(K) \leq \sum_{i=1}^{n} \mu(U_i) = 0.$$

This and the inner regularity of μ imply that $\mu(U) = 0$. $\qquad\square$

Let us continue for a moment with X and μ having the same meaning as in the statement of Proposition 7.4.1. Then X has a largest open subset of μ-measure zero, namely the union of all its open subsets of μ-measure zero. The complement of this open set is called the *support* of μ and is denoted by $\mathrm{supp}(\mu)$. Of course $\mathrm{supp}(\mu)$ is the smallest closed set whose complement has measure zero under μ. Furthermore, a point x belongs to $\mathrm{supp}(\mu)$ if and only if every open neighborhood of x has positive measure under μ.

If μ is a finite signed or complex regular Borel measure on a locally compact Hausdorff space, then its *support* is defined to be the support of its variation $|\mu|$.

Examples 7.4.2. It is easy to check that if, as usual, λ is Lebesgue measure on $\mathbb{R}$, then $\mathrm{supp}(\lambda) = \mathbb{R}$. At the other extreme, if δ_x is the point mass on $(\mathbb{R}, \mathscr{B}(\mathbb{R}))$ concentrated at x, then $\mathrm{supp}(\delta_x) = \{x\}$. See Exercises 1 through 5 for more information about supports. $\qquad\square$

We turn to two theorems that deal with the approximation of measurable functions by continuous functions. These results are often useful, since continuous functions are in many ways easier to handle than are measurable functions.

Proposition 7.4.3. *Let X be a locally compact Hausdorff space, let $\mathscr{A}$ be a σ-algebra on X that includes $\mathscr{B}(X)$, and let μ be a regular measure on $(X, \mathscr{A})$. Suppose that $1 \leq p < +\infty$. Then $\mathscr{K}(X)$ is a dense subspace of $\mathscr{L}^p(X, \mathscr{A}, \mu, \mathbb{R})$ and so determines a dense subspace of $L^p(X, \mathscr{A}, \mu, \mathbb{R})$.*

Note that Proposition 7.4.3 is a generalization of Proposition 3.4.4.

Proof. It is clear that $\mathscr{K}(X) \subseteq \mathscr{L}^p(X, \mathscr{A}, \mu, \mathbb{R})$. Since the simple functions in $\mathscr{L}^p(X, \mathscr{A}, \mu, \mathbb{R})$ are dense in $\mathscr{L}^p(X, \mathscr{A}, \mu, \mathbb{R})$ (Proposition 3.4.2), it suffices to show that if A belongs to $\mathscr{A}$ and has finite measure under μ, then there are functions f in $\mathscr{K}(X)$ that make $\|\chi_A - f\|_p$ arbitrarily small.

So let A be as specified above, and let ε be a positive number. Use the outer regularity of μ to choose an open set U that includes A and satisfies $\mu(U) < \mu(A) + \varepsilon$, and use Proposition 7.2.6 to choose a compact set K that is included in A and satisfies $\mu(K) > \mu(A) - \varepsilon$. Let f belong to $\mathscr{K}(X)$ and satisfy $\chi_K \leq f \leq \chi_U$ (see Proposition 7.1.9). Then $|\chi_A - f| \leq \chi_U - \chi_K$, and so

$$\|\chi_A - f\|_p \leq \|\chi_U - \chi_K\|_p = (\mu(U - K))^{1/p} < (2\varepsilon)^{1/p};$$

since $(2\varepsilon)^{1/p}$ can be made arbitrarily small by a suitable choice of ε, the proof is complete. $\qquad\square$

Theorem 7.4.4 (Lusin's Theorem).   *Let X be a locally compact Hausdorff space, let $\mathscr{A}$ be a σ-algebra on X that includes $\mathscr{B}(X)$, let μ be a regular measure on $(X, \mathscr{A})$, and let $f : X \to \mathbb{R}$ be $\mathscr{A}$-measurable. If A belongs to $\mathscr{A}$ and satisfies $\mu(A) < +\infty$ and if ε is a positive number, then there is a compact subset K of A such that $\mu(A - K) < \varepsilon$ and such that the restriction of f to K is continuous. Moreover, there is a function g in $\mathscr{K}(X)$ that agrees with f at each point in K; if $A \neq \varnothing$ and f is bounded on A, then the function g can be chosen so that*

$$\sup\{|g(x)| : x \in X\} \leq \sup\{|f(x)| : x \in A\}. \tag{1}$$

Proof. First suppose that f has only countably many values, say $a_1, a_2, \ldots,$ and that these values are attained on the sets $A_1, A_2, \ldots$. Use Proposition 1.2.5 to choose a positive integer n such that $\mu(A - (\cup_{i=1}^{n} A_i)) < \varepsilon/2$, and then use Proposition 7.2.6 to choose compact subsets $K_1, \ldots, K_n$ of $A \cap A_1, \ldots, A \cap A_n$ that satisfy $\sum_{i=1}^{n} \mu((A \cap A_i) - K_i) < \varepsilon/2$. Let $K = \cup_{i=1}^{n} K_i$. Then K is a compact subset of A, and

$$\mu(A - K) = \mu(A - (\cup_{i=1}^{n} A_i)) + \sum_{i=1}^{n} \mu((A \cap A_i) - K_i) < \varepsilon/2 + \varepsilon/2 = \varepsilon.$$

Furthermore, since f is constant on each K_i, its restriction to K is continuous (see D.6). Thus K is the required set.

Now let f be an arbitrary $\mathscr{A}$-measurable function. Then f is the uniform limit of a sequence $\{f_n\}$ of functions, each of which is $\mathscr{A}$-measurable and has only countably many values (for example, f_n might be defined by letting $f_n(x)$ be k/n, where k is the integer that satisfies $k/n \leq f(x) < (k+1)/n$). According to what we have just proved, for each n there is a compact subset K_n of A such that $\mu(A - K_n) < \varepsilon/2^n$ and such that the restriction of f_n to K_n is continuous. Let $K = \cap_n K_n$. Then K is a compact subset of A,

$$\mu(A - K) \leq \sum_n \mu(A - K_n) < \sum_n \varepsilon/2^n = \varepsilon,$$

and f, as the uniform limit of the functions f_n, each of which is continuous on K, is itself continuous on K. With this the first part of the theorem is proved.

We turn to the construction of a function g in $\mathscr{K}(X)$ that agrees with f on K. The one-point compactification X^* of X is normal (Proposition 7.1.7), and so the Tietze extension theorem (Exercise 7.1.6) provides a continuous function $h^* : X^* \to \mathbb{R}$ that agrees with f on K. Let $g : X \to \mathbb{R}$ be the product hp, where h is the restriction of h^* to X and p is a function that belongs to $\mathscr{K}(X)$ and satisfies $p(x) = 1$ at each x in K (Proposition 7.1.9). Then g belongs to $\mathscr{K}(X)$ and agrees with f on K. In order to make sure that g satisfies inequality (1), let $B = \sup\{|f(x)| : x \in A\}$, define $\varphi : \mathbb{R} \to \mathbb{R}$ by

$$\varphi(t) = \begin{cases} -B & \text{if } t < -B, \\ t & \text{if } -B \le t \le B, \\ B & \text{if } B < t, \end{cases}$$

and replace g with $\varphi \circ g$.   □

Note that Proposition 7.4.3 and Theorem 7.4.4 can be extended to apply to complex-valued functions. Everything except for inequality (1) can be proved by dealing with real and imaginary parts separately. For the proof of (1), let $B = \sup\{|f(x)| : x \in A\}$, and define $\varphi \colon \mathbb{C} \to \mathbb{C}$ by

$$\varphi(t) = \begin{cases} t & \text{if } |t| \le B, \\ \frac{t}{|t|}B & \text{if } |t| > B. \end{cases}$$

Then φ is continuous, and, as before, the function g can be replaced with $\varphi \circ g$.

The reader should note that in certain cases Lusin's theorem can be used to characterize measurable functions (see Exercise 7.5.2). In fact, Bourbaki defines a function to be measurable if it satisfies the conclusion of Lusin's theorem.

For our next result we need to recall two definitions. Let X be a topological space. A function $f \colon X \to (-\infty, +\infty]$ is *lower semicontinuous* if for each x in X and each real number A that satisfies $A < f(x)$ there is an open neighborhood V of x such that $A < f(t)$ holds at each t in V. It is easy to see that f is lower semicontinuous if and only if for each real number A the set $\{x \in X : A < f(x)\}$ is open. It follows that the supremum of a collection of continuous (or lower semicontinuous) functions is lower semicontinuous and that each lower semicontinuous function on a Hausdorff space is Borel measurable.

Now suppose that X is an arbitrary set and that $\mathscr{H}$ is a family of $[-\infty, +\infty]$-valued functions on X. Then $\mathscr{H}$ is *directed upward* if for each pair h_1, h_2 of functions in $\mathscr{H}$ there is a function h in $\mathscr{H}$ that satisfies $h_1 \le h$ and $h_2 \le h$. Note that if $\mathscr{H}$ is directed upward and if $h_1, \dots, h_n$ belong to $\mathscr{H}$, then there is a function h in $\mathscr{H}$ that satisfies $h_i \le h$ for $i = 1, \dots, n$.

Proposition 7.4.5. *Let X be a locally compact Hausdorff space, let $\mathscr{A}$ be a σ-algebra on X that includes $\mathscr{B}(X)$, and let μ be a regular measure on $(X, \mathscr{A})$. Suppose that $f \colon X \to [0, +\infty]$ is lower semicontinuous and that $\mathscr{H}$ is a family of nonnegative lower semicontinuous functions that is directed upward and satisfies*

$$f(x) = \sup\{h(x) : h \in \mathscr{H}\} \tag{2}$$

at each x in X. Then

$$\int f \, d\mu = \sup\{\int h \, d\mu : h \in \mathscr{H}\}.$$

Proof. Certainly $\int h\,d\mu \le \int f\,d\mu$ holds whenever h belongs to $\mathcal{H}$. Thus we need only show that for each real number A that satisfies $A < \int f\,d\mu$ there is a function h that belongs to $\mathcal{H}$ and satisfies $A < \int h\,d\mu$. So let A be a real number (which we will hold fixed) that satisfies $A < \int f\,d\mu$.

We begin by approximating f with simple functions in the following way. For each positive integer n define open sets $U_{n,i}$, $i = 1, \ldots, n2^n$, by

$$U_{n,i} = \{x \in X : f(x) > i/2^n\},$$

and then define a function $f_n \colon X \to \mathbb{R}$ by

$$f_n = \frac{1}{2^n} \sum_{i=1}^{n2^n} \chi_{U_{n,i}}.$$

Each f_n is Borel measurable and hence $\mathscr{A}$-measurable. It is easy to check that $f_n(x) = 0$ if $f(x) = 0$, that $f_n(x) = i/2^n$ if $0 < f(x) \le n$ and i is the integer that satisfies $i/2^n < f(x) \le (i+1)/2^n$, and that $f_n(x) = n$ if $n < f(x)$. Consequently $\{f_n\}$ is a nondecreasing sequence of nonnegative functions for which $f(x) = \lim_n f_n(x)$ holds at each x in X, and so the monotone convergence theorem (Theorem 2.4.1) implies that $\int f\,d\mu = \lim_n \int f_n\,d\mu$. Hence we can choose a positive integer N such that $A < \int f_N\,d\mu$. The plan now is to choose a function g that satisfies $A < \int g\,d\mu$ but is a bit more convenient than f_N and then to choose a function h in $\mathcal{H}$ that is at least as large as g.

Since $\int f_N\,d\mu = (1/2^N)\sum_i \mu(U_{N,i})$, we can use the regularity of μ to get compact subsets K_i of $U_{N,i}$, $i = 1, \ldots, N2^N$, such that $A < (1/2^N)\sum_i \mu(K_i)$. Let $g = (1/2^N)\sum_i \chi_{K_i}$.

Note that $g(x) \le f_N(x) < f(x)$ holds at each x for which $f(x) > 0$ and hence at each x in $\cup_1^{N2^N} K_i$. Thus (see also (2)) for each x in $\cup_1^{N2^N} K_i$ there is a function h_x in $\mathcal{H}$ such that $g(x) < h_x(x)$. Since h_x is lower semicontinuous and g is a positive multiple of a finite sum of characteristic functions of compact (and hence closed) sets, we can choose an open neighborhood U_x of x such that $g(t) < h_x(t)$ holds at each t in U_x. Carrying this out for each x in $\cup_1^{N2^N} K_i$ gives an open cover of $\cup_1^{N2^N} K_i$; since $\cup_1^{N2^N} K_i$ is compact, we can get first a finite subcover $U_{x_1}, \ldots, U_{x_m}$ of $\cup_1^{N2^N} K_i$ and then a function h in $\mathcal{H}$ such that $h_{x_j} \le h$ holds for $j = 1, \ldots, m$ (recall that $\mathcal{H}$ is directed upward). The function h satisfies $g \le h$ and so satisfies

$$A < \frac{1}{2^N} \sum_i \mu(K_i) = \int g\,d\mu \le \int h\,d\mu.$$

Thus we have produced the required function h, and the proof is complete. $\square$

Exercises

1. Let $\{a_n\}$ be a sequence of positive real numbers such that $\sum_n a_n < +\infty$, let $\{x_n\}$ be an arbitrary sequence of real numbers, and let μ be the measure on $(\mathbb{R}, \mathscr{B}(\mathbb{R}))$ defined by $\mu = \sum_n a_n \delta_{x_n}$. Find $\mathrm{supp}(\mu)$.

2. Construct a finite signed regular Borel measure μ on $\mathbb{R}$ such that $\mathrm{supp}(\mu^+)$ and $\mathrm{supp}(\mu^-)$ are both equal to $\mathbb{R}$.

3. Let X be a locally compact Hausdorff space, and let μ be a regular Borel measure on X. Show that a point x in X belongs to $\mathrm{supp}(\mu)$ if and only if every nonnegative function f in $\mathscr{K}(X)$ that satisfies $f(x) > 0$ also satisfies $\int f\, d\mu > 0$.

4. Let X be an uncountable space that has the discrete topology (and so is locally compact), and let X^* be the one-point compactification of X. Show that there is no regular Borel measure μ on X^* such that $\mathrm{supp}(\mu) = X^*$.

5. Let X and Y be as in Exercise 7.1.10.
   (a) Show that there is no regular Borel measure μ on X such that $\mathrm{supp}(\mu) = X$.
   (b) Is there a regular Borel measure μ on Y such that $\mathrm{supp}(\mu) = Y$?

6. Give a proof of Lusin's theorem that does not depend on the Tietze extension theorem. (Hint: Construct real-valued $\mathscr{A}$-measurable functions $f_1, f_2, \ldots$ such that each f_n has only countably many values and such that $|f_n(x) - f(x)| < 1/2^n$ holds for each n and x. Show that by applying part of the argument in the first paragraph of the proof of Theorem 7.4.4 to the functions f_1, $f_2 - f_1$, $f_3 - f_2$, $\ldots$ and then using Proposition 7.1.12, we can construct functions $g_1, g_2, \ldots$ in $\mathscr{K}(X)$ such that $\sum_n g_n$ belongs to $C_0(X)$ and agrees with f on a suitably large compact subset of A. Then modify $\sum_n g_n$ so that it belongs to $\mathscr{K}(X)$ and satisfies inequality (1).)

7. Let X be a topological space and let A be a subset of X. Show that χ_A is lower semicontinuous if and only if A is open.

8. Let X be a topological space and let $f: X \to (-\infty, +\infty]$ be lower semicontinuous. Show that if K is a nonempty compact subset of X, then
   (a) f is bounded below on K, and
   (b) there is a point x_0 in K such that $f(x_0) = \inf\{f(x) : x \in K\}$.

9. Let X be a locally compact Hausdorff space, and let f be a nonnegative lower semicontinuous function on X. Show that

$$f(x) = \sup\{g(x) : g \in \mathscr{K}(X) \text{ and } 0 \le g \le f\}$$

holds at each x in X.

10. Show by example that in Proposition 7.4.5 we can not replace the assumption that the functions in $\mathscr{H}$ are lower semicontinuous with the assumption that they are Borel measurable. (Hint: Let $X = \mathbb{R}$, let μ be Lebesgue measure, let f be the constant function 1, and choose $\mathscr{H}$ in such a way that $\int h\, d\mu = 0$ holds for each h in $\mathscr{H}$.)

7.5 The μ^*-Measurable Sets and the Dual of L^1

Let X be a locally compact Hausdorff space, and let I be a positive linear functional on $\mathscr{K}(X)$. In Sect. 7.2 we constructed an outer measure μ^* on X by using the equation

$$\mu^*(U) = \sup\{I(f) : f \in \mathscr{K}(X) \text{ and } f \prec U\}, \tag{1}$$

to define the outer measure of the open subsets of X, and then using the equation

$$\mu^*(A) = \inf\{\mu^*(U) : U \text{ is open and } A \subseteq U\} \tag{2}$$

to extend μ^* to all the subsets of X. Let $\mathscr{M}_{\mu^*}$ be the σ-algebra of μ^*-measurable sets. We showed that $\mathscr{B}(X) \subseteq \mathscr{M}_{\mu^*}$ and that the restrictions μ and μ_1 of μ^* to $\mathscr{B}(X)$ and to $\mathscr{M}_{\mu^*}$ are regular measures such that

$$\int f \, d\mu = \int f \, d\mu_1 = I(f)$$

holds for each f in $\mathscr{K}(X)$.

Although the Borel measure μ is appropriate for most purposes, its extension μ_1 is occasionally useful (see Theorem 7.5.4 and Exercise 2). In this section we will study a few of the properties of μ_1 and of $\mathscr{M}_{\mu^*}$.

Proposition 7.5.1. *Let X be a locally compact Hausdorff space, and let μ^* and $\mathscr{M}_{\mu^*}$ be as in the introduction to this section. If B is a subset of X, then the conditions*

(a) $B \in \mathscr{M}_{\mu^*}$,
(b) $B \cap U \in \mathscr{M}_{\mu^*}$ *whenever U is an open subset of X for which $\mu^*(U)$ is finite, and*
(c) $B \cap K \in \mathscr{M}_{\mu^*}$ *whenever K is a compact subset of X*

are equivalent.

Proof. Since the open subsets of X and the compact subsets of X belong to $\mathscr{M}_{\mu^*}$, condition (a) implies conditions (b) and (c).

Next assume that condition (b) holds. According to the discussion preceding Proposition 1.3.5, we can prove that B is μ^*-measurable by showing that

$$\mu^*(A) \geq \mu^*(A \cap B) + \mu^*(A \cap B^c) \tag{3}$$

holds for each subset A of X that satisfies $\mu^*(A) < +\infty$. So let A be such a set, and let U be an open set that includes A and satisfies $\mu^*(U) < +\infty$. Then condition (b) says that $U \cap B$ is μ^*-measurable, and so

$$\mu^*(U) = \mu^*(U \cap B) + \mu^*(U \cap B^c) \geq \mu^*(A \cap B) + \mu^*(A \cap B^c).$$

Since U can be chosen so as to make $\mu^*(U)$ arbitrarily close to $\mu^*(A)$, inequality (3) follows. With this the proof that (b) implies (a) is complete.

Finally, suppose that condition (c) holds. We will show that condition (b) follows. Let U be an open set such that $\mu^*(U) < +\infty$, and choose a sequence $\{K_n\}$ of compact subsets of U such that $\mu^*(U) = \sup_n \mu^*(K_n)$. Then on the one hand, condition (c) says that each $B \cap K_n$ belongs to $\mathscr{M}_{\mu^*}$, while on the other hand, $B \cap (U - \cup_n K_n)$, as a subset of $U - \cup_n K_n$, has μ^*-measure 0 and so belongs to $\mathscr{M}_{\mu^*}$. Since $B \cap U$ is the union of these sets, it also belongs to $\mathscr{M}_{\mu^*}$ and condition (b) follows. $\square$

The following lemma is needed for our proof of Proposition 7.5.3, which is an important technical fact about the σ-algebra $\mathscr{M}_{\mu^*}$ of μ^*-measurable sets.

Lemma 7.5.2. *Let X be a locally compact Hausdorff space, let $\mathscr{A}$ be a σ-algebra on X that includes $\mathscr{B}(X)$, and let μ be a regular measure on $(X, \mathscr{A})$. If K is a compact subset of X such that $\mu(K) > 0$, then there is a compact subset K_0 of K such that $\mu(K_0) = \mu(K)$ and such that each open subset U of X that meets K_0 satisfies $\mu(U \cap K_0) > 0$.*

Proof. The proof here is very similar to that of Proposition 7.4.1: here we let U be the union of the open sets V such that $\mu(V \cap K) = 0$, and we check that every compact subset of $U \cap K$ has measure zero. It then follows from Proposition 7.2.6 that $\mu(U \cap K) = 0$, and so we can let K_0 be $K \cap U^c$. $\square$

Proposition 7.5.3. *Let X be a locally compact Hausdorff space, and let μ^*, $\mathscr{M}_{\mu^*}$, and μ_1 be as in the introduction to this section. Then there is a disjoint family $\mathscr{C}_0$ of compact subsets of X such that*

(a) *if $K \in \mathscr{C}_0$, then $\mu_1(K) > 0$,*
(b) *if U is open, if $K \in \mathscr{C}_0$, and if $U \cap K \neq \varnothing$, then $\mu_1(U \cap K) > 0$,*
(c) *if $A \in \mathscr{M}_{\mu^*}$ and if $\mu_1(A) < +\infty$, then $A \cap K \neq \varnothing$ for only countably many sets K in $\mathscr{C}_0$, and*

$$\mu_1(A) = \sum_K \mu_1(A \cap K),$$

(d) *a subset A of X belongs to $\mathscr{M}_{\mu^*}$ if and only if for each K in $\mathscr{C}_0$ the set $A \cap K$ belongs to $\mathscr{M}_{\mu^*}$, and*
(e) *a function $f : X \to \mathbb{R}$ is $\mathscr{M}_{\mu^*}$-measurable if and only if for each K in $\mathscr{C}_0$ the function $f \chi_K$ is $\mathscr{M}_{\mu^*}$-measurable.*

Proof. Let Ξ be the collection of all families $\mathscr{C}$ of compact subsets of X such that

  (i) the sets in $\mathscr{C}$ are disjoint from one another,
 (ii) if $K \in \mathscr{C}$, then $\mu_1(K) > 0$, and
(iii) if U is open, if $K \in \mathscr{C}$, and if $U \cap K \neq \varnothing$, then $\mu_1(U \cap K) > 0$.

Note that Ξ contains $\varnothing$ and so is nonempty, and that Ξ is partially ordered by inclusion. Furthermore, if Ξ_0 is a linearly ordered subcollection of Ξ, then $\cup \Xi_0$ belongs to Ξ and so is an upper bound for Ξ_0. Hence Zorn's lemma (see Theorem A.13) implies that Ξ has a maximal element.

Let $\mathscr{C}_0$ be a maximal element of Ξ. We will check that $\mathscr{C}_0$ satisfies properties (a) through (e). Properties (a) and (b) are immediate.

We turn to property (c). Suppose that A belongs to $\mathscr{M}_{\mu^*}$ and satisfies $\mu_1(A) < +\infty$, and use (2) to choose an open set U such that $A \subseteq U$ and $\mu_1(U) < +\infty$. Then each set K in $\mathscr{C}_0$ that meets A also meets U and so, by property (b), satisfies $\mu_1(U \cap K) > 0$. Since $\mu_1(U) < +\infty$ and the sets in $\mathscr{C}_0$ are disjoint from one another, there can for each n be only finitely many sets K in $\mathscr{C}_0$ such that $\mu_1(U \cap K) > 1/n$ and hence only countably many sets K in $\mathscr{C}_0$ such that $\mu_1(U \cap K) > 0$. Since $A \subseteq U$, it follows that only countably many of the sets in $\mathscr{C}_0$ meet A.

Now consider the second half of (c). To prove that $\mu_1(A) = \sum_K \mu_1(A \cap K)$, where K ranges over those sets in $\mathscr{C}_0$ that meet A, we need only show that $A - (\cup_K(A \cap K))$ has μ_1-measure zero. But if that set had positive measure, then according to Proposition 7.2.6 and Lemma 7.5.2, it would include a compact subset K that would satisfy $\mu_1(K) > 0$ and $\mu_1(U \cap K) > 0$ for each open set U such that $U \cap K \neq \varnothing$. Such a set K would be disjoint from all the sets in $\mathscr{C}_0$. This cannot happen, however, since it would contradict the maximality of the family $\mathscr{C}_0$. With this the proof of property (c) is complete.

To begin the proof of property (d), suppose that A is a set such that $A \cap K \in \mathscr{M}_{\mu^*}$ holds for each K in $\mathscr{C}_0$. According to Proposition 7.5.1, it is enough to show that $A \cap L \in \mathscr{M}_{\mu^*}$ for an arbitrary compact subset L of X. So let L be such a set. Part (c) of the current proposition says that L meets only countably many of the sets in $\mathscr{C}_0$ and that $\mu_1(L - \cup_n K_n) = 0$, where $\{K_n\}$ is the collection of sets in $\mathscr{C}_0$ that meet L. Thus $A \cap L$ is the union of the countable collection of sets of the form $A \cap K_n \cap L$, together with a subset of the μ_1-null set $L - \cup_n K_n$. Since all these sets are μ^*-measurable, the measurability of A follows and half of property (d) is proved. The converse half is immediate.

Property (e) follows from property (d), since for each Borel subset B of $\mathbb{R}$ and each K in $\mathscr{C}_0$ we have $f^{-1}(B) \cap K = (f \chi_K)^{-1}(B) \cap K$. $\square$

Let us turn to an application of the preceding result. Suppose that $(X, \mathscr{A}, \mu)$ is an arbitrary measure space and that T is the map from $L^\infty(X, \mathscr{A}, \mu)$ to $(L^1(X, \mathscr{A}, \mu))^*$ that associates to each $\langle g \rangle$ in $L^\infty(X, \mathscr{A}, \mu)$ the functional $T_{\langle g \rangle}$ defined by

$$T_{\langle g \rangle}(\langle f \rangle) = \int fg \, d\mu \tag{4}$$

(see Sect. 3.5). Recall that T is an isometric isomorphism of $L^\infty(X, \mathscr{A}, \mu)$ onto a subspace of $(L^1(X, \mathscr{A}, \mu))^*$ (Proposition 3.5.5). Recall also that T is surjective if $(X, \mathscr{A}, \mu)$ is σ-finite but fails to be surjective in some other situations (see Theorem 4.5.1 and the example at the end of Sect. 4.5). We now use Proposition 7.5.3 to show that the map T is surjective for a large class of not necessarily σ-finite spaces.

Theorem 7.5.4. *Let X be a locally compact Hausdorff space, and let μ^*, $\mathscr{M}_{\mu^*}$, and μ_1 be as in the introduction to this section. Then the map T given by $T_{\langle g \rangle}(\langle f \rangle) = \int fg \, d\mu_1$ is an isometric isomorphism of $L^\infty(X, \mathscr{M}_{\mu^*}, \mu_1)$ onto $(L^1(X, \mathscr{M}_{\mu^*}, \mu_1))^*$.*

Proof. In view of the preceding discussion, only the surjectivity of T needs to be checked. Let F belong to $(L^1(X, \mathcal{M}_{\mu^*}, \mu_1))^*$, and let $\mathcal{C}_0$ be a disjoint family of compact subsets of X for which properties (a) through (e) of Proposition 7.5.3 hold. For each K in $\mathcal{C}_0$ consider the measure space $(K, \mathcal{M}_K, \mu_K)$, where $\mathcal{M}_K$ is the σ-algebra consisting of those subsets of K that belong to $\mathcal{M}_{\mu^*}$ and μ_K is the restriction of μ_1 to $\mathcal{M}_K$. Let F_K be the functional on $L^1(K, \mathcal{M}_K, \mu_K)$ defined by $F_K(\langle f \rangle) = F(\langle f' \rangle)$, where f' is the function on X that agrees on K with f and that vanishes outside K. Since μ_K is a finite measure, there is (Theorem 4.5.1) an $\mathcal{M}_K$-measurable function g_k on K such that

$$\sup\{|g_K(x)| : x \in K\} = \|F_K\| \leq \|F\| \tag{5}$$

and such that $F_K(\langle f \rangle) = \int_K f g_K \, d\mu_K$ holds for each $\langle f \rangle$ in $L^1(K, \mathcal{M}_K, \mu_K)$. For each K in $\mathcal{C}_0$ choose such a function g_K. Let g be the function on X that vanishes outside $\cup \mathcal{C}_0$ and that for each K in $\mathcal{C}_0$ agrees with g_K on K. It follows from part (e) of Proposition 7.5.3 and inequality (5) that $g \in \mathcal{L}^\infty(X, \mathcal{M}_{\mu^*}, \mu_1)$.

Let us check that $F = T_{\langle g \rangle}$. It is clear that if f is a member of $\mathcal{L}^1(X, \mathcal{M}_{\mu^*}, \mu_1)$ that vanishes outside some K in $\mathcal{C}_0$, then $F(\langle f \rangle) = T_{\langle g \rangle}(\langle f \rangle)$. If f is an arbitrary function in $\mathcal{L}^1(X, \mathcal{M}_{\mu^*}, \mu_1)$, then f vanishes outside the union of a sequence of sets of finite measure (Corollary 2.3.11); thus according to part (c) of Proposition 7.5.3, there is a sequence $\{K_n\}$ of sets in $\mathcal{C}_0$ such that f vanishes almost everywhere outside $\cup_n K_n$. Since the functionals F and $T_{\langle g \rangle}$ agree on each $\langle f \chi_{K_n} \rangle$ and since $\lim_N \|f - \sum_{n=1}^N f \chi_{K_n}\|_1 = 0$, it follows that $F(\langle f \rangle) = T_{\langle g \rangle}(\langle f \rangle)$. Thus $F = T_{\langle g \rangle}$, and the proof is complete. $\square$

It is natural to ask whether in Theorem 7.5.4 the measure space $(X, \mathcal{M}_{\mu^*}, \mu_1)$ can be replaced with $(X, \mathcal{B}(X), \mu)$. This change can of course be made if μ_1 and μ are σ-finite and can also be made in certain other situations (see Theorem 9.4.8); it cannot be made in general (see Fremlin [47]).

We are now in a position to sketch the relationship of the treatment of integration on locally compact Hausdorff spaces given here to that given by Bourbaki (see [18]).

Let X be a locally compact Hausdorff space and let $\mathcal{I}_+(X)$ be the set of all $[0, +\infty]$-valued lower semicontinuous functions on X. Suppose that I is a positive linear functional on $\mathcal{K}(X)$ (in Bourbaki's terminology, I is a *positive Radon measure* on X). Bourbaki defines a function $I^* : \mathcal{I}_+(X) \to [0, +\infty]$ by

$$I^*(f) = \sup\{I(g) : g \in \mathcal{K}(X) \text{ and } 0 \leq g \leq f\}$$

and then uses the formula

$$I^*(f) = \inf\{I^*(h) : h \in \mathcal{I}_+(X) \text{ and } f \leq h\}$$

to extend I^* to the set of all $[0, +\infty]$-valued functions on X. He checks that I^* satisfies

$$I^*(f + g) \leq I^*(f) + I^*(g) \tag{6}$$

and

$$I^*(af) = aI^*(f) \tag{7}$$

for all $f, g: X \to [0, +\infty]$ and all a in $[0, +\infty)$. Of course, $I^*(0) = 0$. It follows that the set $\mathscr{F}^1$ of functions $f: X \to \mathbb{R}$ for which $I^*(|f|) < +\infty$ is a vector space over $\mathbb{R}$ and that the function $N_1: \mathscr{F}^1 \to \mathbb{R}$ defined by $N_1(f) = I^*(|f|)$ is a seminorm on $\mathscr{F}^1$. Bourbaki then defines $\mathscr{L}^1(X, I)$ to be the closure of $\mathscr{K}(X)$ in[6] $\mathscr{F}^1$ (of course $\mathscr{F}^1$ is given the topology determined by N_1), and extends I from $\mathscr{K}(X)$ to $\mathscr{L}^1(X, I)$ by letting

$$I(f) = \lim_n I(f_n) \tag{8}$$

hold whenever $\{f_n\}$ is a sequence of functions in $\mathscr{K}(X)$ for which $\lim_n N_1(f_n - f) = 0$ (check that the limit in (8) exists and depends only on f). He calls the functions that belong to $\mathscr{L}^1(X, I)$ *I-integrable*, and he calls the extension of I to $\mathscr{L}^1(X, I)$ the *integral*; he often writes[7] $\int f \, dI$ in place of $I(f)$. He calls a function $f: X \to \mathbb{R}$ *I-measurable* if for each compact subset K of X and each positive number ε there is a compact subset L of K that satisfies $I^*(\chi_{K-L}) < \varepsilon$ and is such that the restriction of f to L is continuous. Furthermore, he calls a subset A of X *I-integrable* if χ_A is *I*-integrable and *I-measurable* if χ_A is *I*-measurable.

The following theorem shows how these concepts are related to those treated earlier in this chapter.

Theorem 7.5.5. *Let X be a locally compact Hausdorff space, let I be a positive linear functional on $\mathscr{K}(X)$, let $\mathscr{L}^1(X, I)$ be as defined in the preceding paragraphs, and let μ^*, $\mathscr{M}_{\mu^*}$, and μ_1 be as defined at the beginning of this section. Then*

(a) $\mathscr{L}^1(X, I) = \mathscr{L}^1(X, \mathscr{M}_{\mu^*}, \mu_1, \mathbb{R})$,
(b) $\int f \, dI = \int f \, d\mu_1$ *holds for each f in $\mathscr{L}^1(X, I)$*,
(c) *a subset A of X is I-measurable if and only if it belongs to $\mathscr{M}_{\mu^*}$, and*
(d) *a function $f: X \to \mathbb{R}$ is I-measurable if and only if it is $\mathscr{M}_{\mu^*}$-measurable.*

Proof (A Sketch). It follows from Proposition 7.4.5 and Exercise 7.4.9 that

$$I^*(f) = \int f \, d\mu_1 \tag{9}$$

holds for each f in $\mathscr{I}_+(X)$ and then from (9), together with the additivity and homogeneity of the integral, that (6) and (7) hold for all $f, g: X \to [0, +\infty]$ and

---

[6]Note that $I^*(|f|) = I(|f|) < +\infty$ holds for each f in $\mathscr{K}(X)$ and hence that $\mathscr{K}(X)$ is included in $\mathscr{F}^1$.

[7]Actually, he usually calls his positive linear functional μ, and he writes $\mu(f)$ and $\int f \, d\mu$, rather than $I(f)$ and $\int f \, dI$; such notation will not be used in this book, since we have been using μ to denote a measure.

all a in $[0, +\infty)$.[8] Consequently $\mathscr{F}^1$ is a vector space and N_1 is a seminorm on it. The reader should check that

$$I^*(|f|) = \int |f| \, d\mu_1 \tag{10}$$

holds for each f in $\mathscr{L}^1(X, \mathscr{M}_{\mu^*}, \mu_1, \mathbb{R})$ (use (9) and an appropriate extension of Lemma 6.3.12 to the case of functions on locally compact Hausdorff spaces).

In view of (10), Proposition 7.4.3 implies that $\mathscr{L}^1(X, \mathscr{M}_{\mu^*}, \mu_1, \mathbb{R})$ is included in $\mathscr{L}^1(X, I)$ and that $\int f \, d\mu_1 = \int f \, dI$ holds for each f in $\mathscr{L}^1(X, \mathscr{M}_{\mu^*}, \mu_1, \mathbb{R})$. The reverse inclusion is left to the reader (use (9) to show that if $f \in \mathscr{L}^1(X, I)$, then there is a sequence $\{f_n\}$ in $\mathscr{K}(X)$ that converges to f almost everywhere with respect to μ_1 and satisfies $\lim_n N_1(f - f_n) = 0$; then use the completeness of $L^1(X, \mathscr{M}_{\mu^*}, \mu_1, \mathbb{R})$). With this parts (a) and (b) of the theorem are proved.

Part (d) follows from Lusin's theorem (Theorem 7.4.4) and Exercise 2. Finally, part (c) is a special case of part (d).   □

Note that if I is a positive linear functional on $\mathscr{K}(X)$, then

> for each compact subset K of X there is a number c_K such that $|I(f)| \leq c_K \|f\|_\infty$ holds whenever f belongs to $\mathscr{K}(X)$ and satisfies $\mathrm{supp}(f) \subseteq K$ $\tag{11}$

(choose a function g that belongs to $\mathscr{K}(X)$ and satisfies $\chi_K \leq g$, and let c_K be $I(g)$). Bourbaki calls a (not necessarily positive) linear functional I on $\mathscr{K}(X)$ a *Radon measure* on X if it satisfies (11). Since each difference of positive linear functionals on $\mathscr{K}(X)$ satisfies (11), each such difference is a Radon measure. The proof of Lemma 7.3.7 can be modified so as to show that every Radon measure on X is the difference of positive Radon measures on X (that is, of positive linear functionals on $\mathscr{K}(X)$). Thus the set of Radon measures on X is the vector space generated by the set of positive linear functionals on $\mathscr{K}(X)$.

Note that the formula

$$I(f) = \int_0^{+\infty} f(x) \, \lambda(dx) - \int_{-\infty}^0 f(x) \, \lambda(dx)$$

defines a Radon measure on $\mathbb{R}$; this Radon measure cannot be represented in terms of integration with respect to a signed measure on $(\mathbb{R}, \mathscr{B}(\mathbb{R}))$ (recall that the positive and negative parts of a signed measure cannot both be infinite). See, however, Exercise 6.

---

[8]Bourbaki develops integration theory without first developing measure theory; his proofs, for example, of (6) and (7) are therefore quite different from those given here.

Exercises

1. Show that the assumption that $A \in \mathscr{M}_{\mu^*}$ can be omitted from part (c) of Proposition 7.5.3; that is, show that if $\mu^*(A) < +\infty$, then $A \cap K \neq \varnothing$ holds for only countably many of the sets K in $\mathscr{C}_0$, and

$$\mu^*(A) = \sum_K \mu^*(A \cap K).$$

2. Let X, μ^*, and μ_1 be as in the introduction to this section, and let f be a real-valued function on X. Suppose that for each compact subset K of X and each positive ε there is a compact subset L of K such that

   (i) $\mu_1(K - L) < \varepsilon$, and
   (ii) the restriction of f to L is continuous.

   Show that f is $\mathscr{M}_{\mu^*}$-measurable. (Note that this is a sort of converse to Lusin's theorem and that it explains one of the remarks following the proof of Theorem 7.4.4.)
3. Let X be a locally compact Hausdorff space, let $\mathscr{A}$ be a σ-algebra on X that includes $\mathscr{B}(X)$, and let ν be a regular measure on $(X, \mathscr{A})$. Define a positive linear functional I on $\mathscr{K}(X)$ by $I(f) = \int f \, d\nu$. Show that if μ^*, $\mathscr{M}_{\mu^*}$, and μ_1 are associated to I as in this section, then $\mathscr{A} \subseteq \mathscr{M}_{\mu^*}$ and ν is the restriction of μ_1 to $\mathscr{A}$.
4. Show by example that the assumption of σ-finiteness can not be omitted in Exercise 7.2.3. (Hint: See Exercise 7.2.4.)
5. Let X, I, μ^*, $\mathscr{M}_{\mu^*}$, μ, and μ_1 be as in the introduction to this section. Suppose that $1 \leq p < +\infty$.
   (a) Show that if $f \in \mathscr{L}^p(X, \mathscr{M}_{\mu^*}, \mu_1)$, then there is a function that belongs to $\mathscr{L}^p(X, \mathscr{B}(X), \mu)$ and agrees with f μ-almost everywhere.
   (b) Conclude that $L^p(X, \mathscr{M}_{\mu^*}, \mu_1)$ and $L^p(X, \mathscr{B}(X), \mu)$ are isometrically isomorphic to one another.
6. Show that if I is a Radon measure on the locally compact Hausdorff space X, then there are regular Borel measures μ_1 and μ_2 on X such that $I(f) = \int f \, d\mu_1 - \int f \, d\mu_2$ holds for each f in $\mathscr{K}(X)$.

7.6 Products of Locally Compact Spaces

This section is devoted to the study of products of regular Borel measures on locally compact Hausdorff spaces. In Chap. 5 we proved that if μ and ν are σ-finite measures on measurable spaces $(X, \mathscr{A})$ and $(Y, \mathscr{B})$, then there is a unique measure $\mu \times \nu$ on $(X \times Y, \mathscr{A} \times \mathscr{B})$ such that $(\mu \times \nu)(A \times B) = \mu(A)\nu(B)$ holds for each A in $\mathscr{A}$ and each B in $\mathscr{B}$. Now assume that X and Y are locally compact Hausdorff spaces. Then $X \times Y$ is a locally compact Hausdorff space, and it would be convenient if for each pair of regular Borel measures on X and Y, the constructions in Chap. 5 gave a regular Borel measure on $X \times Y$. However, two problems arise. First, regular Borel measures can fail to be σ-finite, and so the earlier theory can fail to apply. Second, the product σ-algebra $\mathscr{B}(X) \times \mathscr{B}(Y)$ can fail to contain all the Borel subsets of $X \times Y$ (see Exercise 5.1.8), in which case no measure on $\mathscr{B}(X) \times \mathscr{B}(Y)$ can be regular.

We will begin by proving that these difficulties cannot arise if the spaces X and Y have countable bases for their topologies; then we will turn to a theory of product measures that is suitable for Borel measures on arbitrary locally compact Hausdorff spaces. Lemma 7.6.1 and Proposition 7.6.2 suffice for most applications. The remaining parts of this section will be used only in Chap. 9 and should be skipped by most readers.

Let us recall some notation. Suppose that X and Y are sets and that E is a subset of $X \times Y$. For each x in X and each y in Y the sections E_x and E^y are the subsets of Y and X given by

$$E_x = \{y \in Y : (x,y) \in E\}$$

and

$$E^y = \{x \in X : (x,y) \in E\}.$$

Likewise, if f is a function whose domain is $X \times Y$, then for each x in X and each y in Y the sections f_x and f^y are the functions on Y and X defined by

$$f_x(y) = f(x,y)$$

and

$$f^y(x) = f(x,y).$$

The following lemma summarizes some useful elementary facts.

Lemma 7.6.1. *Let X and Y be Hausdorff topological spaces, and let $X \times Y$ be their product. Then*

(a) *the product σ-algebra $\mathscr{B}(X) \times \mathscr{B}(Y)$ is included in $\mathscr{B}(X \times Y)$,*
(b) *if $E \in \mathscr{B}(X \times Y)$, then for each x in X the section E_x belongs to $\mathscr{B}(Y)$, and for each y in Y the section E^y belongs to $\mathscr{B}(X)$, and*
(c) *if $f : X \times Y \to \mathbb{R}$ is $\mathscr{B}(X \times Y)$-measurable, then for each x in X the section f_x is $\mathscr{B}(Y)$-measurable, and for each y in Y the section f^y is $\mathscr{B}(X)$-measurable.*

Proof. The projection π_1 of $X \times Y$ onto X is continuous and so is measurable with respect to $\mathscr{B}(X \times Y)$ and $\mathscr{B}(X)$ (Lemma 7.2.1). Likewise, the projection π_2 of $X \times Y$ onto Y is measurable with respect to $\mathscr{B}(X \times Y)$ and $\mathscr{B}(Y)$. Note that if $A \subseteq X$ and $B \subseteq Y$, then

$$A \times B = (A \times Y) \cap (X \times B) = \pi_1^{-1}(A) \cap \pi_2^{-1}(B).$$

Hence if $A \in \mathcal{B}(X)$ and $B \in \mathcal{B}(Y)$, then $A \times B \in \mathcal{B}(X \times Y)$. Since $\mathcal{B}(X) \times \mathcal{B}(Y)$ is the σ-algebra generated by the collection of all such rectangles $A \times B$, it follows that $\mathcal{B}(X) \times \mathcal{B}(Y) \subseteq \mathcal{B}(X \times Y)$. Thus part (a) is proved.

To check the first assertion in part (b), suppose that x belongs to X, and define $g \colon Y \to X \times Y$ by $g(y) = (x, y)$. Then g is continuous and so is measurable with respect to $\mathcal{B}(Y)$ and $\mathcal{B}(X \times Y)$. Each subset E of $X \times Y$ satisfies $E_x = g^{-1}(E)$; hence if $E \in \mathcal{B}(X \times Y)$, then $E_x \in \mathcal{B}(Y)$. The second assertion in part (b) is proved in the same way.

Part (c) follows from part (b) and the fact that if $B \subseteq \mathbb{R}$, then $(f_x)^{-1}(B) = (f^{-1}(B))_x$ and $(f^y)^{-1}(B) = (f^{-1}(B))^y$. □

Now we prove that the difficulties mentioned in the introduction to this section do not occur if each of the spaces X and Y has a countable base.

Proposition 7.6.2. *Let X and Y be locally compact Hausdorff spaces that have countable bases for their topologies. Then $\mathcal{B}(X \times Y) = \mathcal{B}(X) \times \mathcal{B}(Y)$. Furthermore, if μ and ν are regular Borel measures on X and Y, respectively, then μ and ν are σ-finite, and $\mu \times \nu$ is a regular Borel measure on $X \times Y$.*

Proof. Lemma 7.6.1 implies that $\mathcal{B}(X) \times \mathcal{B}(Y) \subseteq \mathcal{B}(X \times Y)$. We turn to the reverse inclusion. Let $\mathcal{U}$ and $\mathcal{V}$ be countable bases for X and Y, and let $\mathcal{W}$ be the collection of rectangles of the form $U \times V$, where $U \in \mathcal{U}$ and $V \in \mathcal{V}$. Then $\mathcal{W}$ is a countable base for $X \times Y$ and is included in $\mathcal{B}(X) \times \mathcal{B}(Y)$. Each open subset of $X \times Y$ is the union of a (necessarily countable) subfamily of the base $\mathcal{W}$ and so belongs to $\mathcal{B}(X) \times \mathcal{B}(Y)$. Since $\mathcal{B}(X \times Y)$ is generated by the open subsets of $X \times Y$, it follows that $\mathcal{B}(X \times Y) \subseteq \mathcal{B}(X) \times \mathcal{B}(Y)$. Thus $\mathcal{B}(X \times Y) = \mathcal{B}(X) \times \mathcal{B}(Y)$.

Now suppose that μ and ν are regular Borel measures on X and Y, respectively. Then μ and ν are σ-finite (Proposition 7.2.5), and so the constructions of Chap. 5 provide a unique product measure $\mu \times \nu$ on $\mathcal{B}(X) \times \mathcal{B}(Y)$. Since $\mathcal{B}(X) \times \mathcal{B}(Y) = \mathcal{B}(X \times Y)$, the measure $\mu \times \nu$ is a Borel measure. If K is a compact subset of $X \times Y$ and if K_1 and K_2 are the projections of K on X and Y, respectively, then K_1 and K_2 are compact, and so

$$(\mu \times \nu)(K) \leq (\mu \times \nu)(K_1 \times K_2) = \mu(K_1)\nu(K_2) < +\infty.$$

Thus $\mu \times \nu$ is finite on the compact subsets of $X \times Y$. Since there is a countable base for $X \times Y$ (for example, the base $\mathcal{W}$ defined above), Proposition 7.2.3 implies that $\mu \times \nu$ is regular. □

Now let X and Y be arbitrary locally compact Hausdorff spaces, and let μ and ν be regular Borel measures on X and Y, respectively. As we noted in the introduction to this section, μ and ν can fail to be σ-finite, and the σ-algebra $\mathcal{B}(X) \times \mathcal{B}(Y)$ can fail to contain all the sets in $\mathcal{B}(X \times Y)$. Suppose, however, that we could prove that for each f in $\mathcal{K}(X \times Y)$ the iterated integrals $\int_X \int_Y f(x, y) \, \nu(dy) \, \mu(dx)$ and $\int_Y \int_X f(x, y) \, \mu(dx) \, \nu(dy)$ exist and are equal. We could then proceed in two steps, first defining a positive linear functional I on $\mathcal{K}(X \times Y)$ by letting $I(f)$ be the common value of these iterated integrals and then using the Riesz representation theorem to obtain the corresponding regular Borel measure on $X \times Y$. This is indeed the course that we will follow. The following propositions contain the necessary details.

Lemma 7.6.3. *Suppose that S and T are topological spaces, that T is compact, and that $f \colon S \times T \to \mathbb{R}$ is continuous. Then for each s_0 in S and each positive number ε there is an open neighborhood U of s_0 such that $|f(s,t) - f(s_0,t)| < \varepsilon$ holds for each s in U and each t in T.*

Proof. Suppose that s_0 belongs to S and that ε is a positive number. For each t in T choose open neighborhoods U_t of s_0 and V_t of t such that if $(s,t') \in U_t \times V_t$, then $|f(s,t') - f(s_0,t)| < \varepsilon/2$. It follows that if $s \in U_t$ and $t' \in V_t$, then

$$|f(s,t') - f(s_0,t')| \leq |f(s,t') - f(s_0,t)| + |f(s_0,t) - f(s_0,t')|$$
$$< \varepsilon/2 + \varepsilon/2 = \varepsilon.$$

Since T is compact, we can choose a finite collection $t_1, \ldots, t_n$ of points in T such that the neighborhoods $V_{t_1}, \ldots, V_{t_n}$ cover T. Then $\cap_{i=1}^n U_{t_i}$ is the required neighborhood of s_0. $\qquad\qquad\square$

Proposition 7.6.4. *Let X and Y be locally compact Hausdorff spaces, let μ and ν be regular Borel measures on X and Y, respectively, and let f belong to $\mathcal{K}(X \times Y)$. Then*

(a) *for each x in X and each y in Y the sections f_x and f^y belong to $\mathcal{K}(Y)$ and $\mathcal{K}(X)$, respectively,*
(b) *the functions*

$$x \mapsto \int_Y f(x,y)\, \nu(dy)$$

and

$$y \mapsto \int_X f(x,y)\, \mu(dx)$$

belong to $\mathcal{K}(X)$ and $\mathcal{K}(Y)$, respectively, and
(c) *$\int_X \int_Y f(x,y)\, \nu(dy)\, \mu(dx) = \int_Y \int_X f(x,y)\, \mu(dx)\, \nu(dy)$.*

Proof. Let f belong to $\mathcal{K}(X \times Y)$, let K be the support of f, and let K_1 and K_2 be the projections of K on X and Y, respectively. Then K_1 and K_2 are compact.

If $x \in X$, then the section f_x is continuous, since it results from composing the continuous function $y \mapsto (x,y)$ with the continuous function f. The support of f_x is included in K_2 and so is compact. Thus $f_x \in \mathcal{K}(Y)$. A similar argument shows that $f^y \in \mathcal{K}(X)$.

It follows that the integrals in part (b) exist. We now check that the function $x \mapsto \int_Y f(x,y)\, \nu(dy)$ is continuous. Let x_0 belong to X and let ε be a positive number. According to Lemma 7.6.3, applied to the space $X \times K_2$, there is an open neighborhood U of x_0 such that if $x \in U$ and $y \in K_2$, then $|f(x,y) - f(x_0,y)| < \varepsilon$. Hence if $x \in U$, then

$$\left| \int_Y f(x,y)\, \nu(dy) - \int_Y f(x_0,y)\, \nu(dy) \right|$$

$$\leq \int_{K_2} |f(x,y) - f(x_0,y)|\, \nu(dy) \leq \varepsilon \nu(K_2).$$

Since ε was arbitrary, the continuity of $x \mapsto \int_Y f(x,y)\, \nu(dy)$ follows. In addition this function vanishes outside K_1, and so it belongs to $\mathscr{K}(X)$. A similar argument shows that the function $y \mapsto \int_X f(x,y)\, \mu(dx)$ belongs to $\mathscr{K}(Y)$.

We turn to part (c). Parts (a) and (b) imply that the integrals involved here exist. We prove that they are equal by approximating f with simpler functions. Let ε be an arbitrary positive number. For each x in K_1 choose a neighborhood U_x of x such that if $x' \in U_x$ and $y \in K_2$, then $|f(x',y) - f(x,y)| < \varepsilon$ (see Lemma 7.6.3). The set K_1 is compact, and so there exist points $x_1, \ldots, x_n$ in K_1 such that the sets $U_{x_1}, \ldots, U_{x_n}$ cover K_1. Now use these sets to construct disjoint Borel sets $A_1, \ldots, A_n$ such that $K_1 = \cup_i A_i$ and such that $A_i \subseteq U_{x_i}$ holds for $i = 1, \ldots, n$. Define $g \colon X \times Y \to \mathbb{R}$ by $g(x,y) = \sum_{i=1}^n \chi_{A_i}(x) f(x_i,y)$. The functions f and g vanish outside $K_1 \times K_2$ and satisfy $|f(x,y) - g(x,y)| < \varepsilon$ at each (x,y) in $K_1 \times K_2$; hence they satisfy

$$\left| \int_Y \int_X f(x,y)\, \mu(dx)\, \nu(dy) - \int_Y \int_X g(x,y)\, \mu(dx)\, \nu(dy) \right| \leq \varepsilon \mu(K_1) \nu(K_2)$$

and

$$\left| \int_X \int_Y f(x,y)\, \nu(dy)\, \mu(dx) - \int_X \int_Y g(x,y)\, \nu(dy)\, \mu(dx) \right| \leq \varepsilon \mu(K_1) \nu(K_2).$$

The two iterated integrals of g are both equal to $\sum \mu(A_i) \int f(x_i,y)\, \nu(dy)$; thus they are equal to each other, and so

$$\left| \int_Y \int_X f(x,y)\, \mu(dx)\, \nu(dy) - \int_X \int_Y f(x,y)\, \nu(dy)\, \mu(dx) \right| \leq 2\varepsilon \mu(K_1) \nu(K_2).$$

Since ε is arbitrary, the proof is complete. $\square$

Let X and Y be locally compact Hausdorff spaces, and let μ and ν be regular Borel measures on X and Y, respectively. As promised earlier, we define $I \colon \mathscr{K}(X \times Y) \to \mathbb{R}$ by letting $I(f)$ be the common value of the iterated integrals $\int_X \int_Y f(x,y)\, \nu(dy)\, \mu(dx)$ and $\int_Y \int_X f(x,y)\, \mu(dx)\, \nu(dy)$. The *regular Borel product* of μ and ν is the regular Borel measure on $X \times Y$ induced by the functional I via the Riesz representation theorem. This measure will be denoted by $\mu \times \nu$.

Proposition 7.6.5. *Let X and Y be locally compact Hausdorff spaces, let μ and ν be regular Borel measures on X and Y, respectively, and let $\mu \times \nu$ be the regular Borel product of μ and ν. If U is an open subset of $X \times Y$, then*

(a) *the functions $x \mapsto \nu(U_x)$ and $y \mapsto \mu(U^y)$ are lower semicontinuous and hence Borel measurable, and*

(b) *$(\mu \times \nu)(U) = \int_X \nu(U_x)\,\mu(dx) = \int_Y \mu(U^y)\,\nu(dy)$.*

Proof. Of course U_x and U^y are open sets and therefore Borel sets. Let

$$\mathscr{F} = \{f \in \mathscr{K}(X \times Y) : 0 \le f \le \chi_U\},$$

and for each x in X and y in Y define sets $\mathscr{F}_x$ and $\mathscr{F}^y$ by

$$\mathscr{F}_x = \{f_x : f \in \mathscr{F}\} \text{ and}$$
$$\mathscr{F}^y = \{f^y : f \in \mathscr{F}\}.$$

Then $\mathscr{F}_x$ and $\mathscr{F}^y$ are included in $\mathscr{K}(Y)$ and $\mathscr{K}(X)$, respectively, are directed upward, and have χ_{U_x} and χ_{U^y} as their suprema. Since these characteristic functions are lower semicontinuous, Proposition 7.4.5 implies that

$$\nu(U_x) = \sup\left\{ \int f_x\, d\nu : f_x \in \mathscr{F}_x \right\} \tag{1}$$

holds for each x in X and that

$$\mu(U^y) = \sup\left\{ \int f^y\, d\mu : f^y \in \mathscr{F}^y \right\} \tag{2}$$

holds for each y in Y. Thus the functions $x \mapsto \nu(U_x)$ and $y \mapsto \mu(U^y)$ are suprema of collections of continuous functions (see part (b) of Proposition 7.6.4), and so are lower semicontinuous.

The first half of part (b) will follow, once we check the calculation

$$(\mu \times \nu)(U) = \sup_{f \in \mathscr{F}} \int_X \int_Y f(x,y)\, \nu(dy)\,\mu(dx)$$

$$= \int_X \left(\sup_{f \in \mathscr{F}} \int f_x\, d\nu \right) \mu(dx)$$

$$= \int_X \nu(U_x)\,\mu(dx);$$

here the first equality is a consequence of Lemma 7.2.7 and the definition of the functional I, the second a consequence of Proposition 7.4.5, and the third a consequence of Eq. (1). The other half of part (b) is proved in a similar way. $\square$

Corollary 7.6.6. *Let X, Y, μ, ν, and $\mu \times \nu$ be as in Proposition 7.6.5. If E is a Borel subset of $X \times Y$ that is included in a rectangle whose sides are Borel sets that are σ-finite under μ and ν, respectively, then*

(a) *the functions $x \mapsto \nu(E_x)$ and $y \mapsto \mu(E^y)$ are Borel measurable, and*
(b) $(\mu \times \nu)(E) = \int_X \nu(E_x)\mu(dx) = \int_Y \mu(E^y)\nu(dy)$.

Proof. We begin with Borel sets that are included in rectangles whose sides are Borel sets of finite measure. So let A and B be Borel subsets of X and Y that satisfy $\mu(A) < +\infty$ and $\nu(B) < +\infty$. Use the regularity of μ and ν to choose open sets U and V that include A and B and satisfy $\mu(U) < +\infty$ and $\nu(V) < +\infty$. Let $W = U \times V$ and let $\mathscr{S}$ consist of those Borel subsets D of $X \times Y$ for which the functions $x \mapsto \nu((D \cap W)_x)$ and $y \mapsto \mu((D \cap W)^y)$ are Borel measurable and for which the identity

$$(\mu \times \nu)(D \cap W) = \int_X \nu((D \cap W)_x)\mu(dx)$$
$$= \int_Y \mu((D \cap W)^y)\nu(dy)$$

holds (according to Lemma 7.6.1, the sections $(D \cap W)_x$ and $(D \cap W)^y$ are Borel sets, and so these formulas make sense). According to Proposition 7.6.5, $\mathscr{S}$ contains all the open subsets of $X \times Y$. It is easy to check that

$$\text{if } D_1, D_2 \in \mathscr{S} \text{ and if } D_1 \subseteq D_2, \text{ then } D_2 - D_1 \in \mathscr{S}, \text{ and} \tag{3}$$

$$\text{if } D_1, D_2, \cdots \in \mathscr{S} \text{ and if } D_1 \subseteq D_2 \subseteq \ldots, \text{ then } \cup_n D_n \in \mathscr{S}. \tag{4}$$

Thus $\mathscr{S}$ is a d-system (see Sect. 1.6) that includes the π-system made up of the open subsets of $X \times Y$, and so Theorem 1.6.2 implies that $\mathscr{B}(X \times Y) \subseteq \mathscr{S}$. Thus if E is a Borel set that is included in $A \times B$, then E satisfies the conclusions of the corollary. Since a Borel set that is included in a rectangle with σ-finite sides is the union of an increasing sequence of Borel sets that are included in rectangles with sides of finite measure, the corollary follows. $\square$

Theorem 7.6.7. *Let X and Y be locally compact Hausdorff spaces, let μ and ν be regular Borel measures on X and Y, respectively, and let $\mu \times \nu$ be the regular Borel product of μ and ν. If f belongs to $\mathscr{L}^1(X \times Y, \mathscr{B}(X \times Y), \mu \times \nu)$ and vanishes outside a rectangle whose sides are Borel sets that are σ-finite under μ and ν, respectively, then*

(a) $f_x \in \mathscr{L}^1(Y, \mathscr{B}(Y), \nu)$ *for μ-almost every x, and $f^y \in \mathscr{L}^1(X, \mathscr{B}(X), \mu)$ for ν-almost every y,*
(b) *the functions*

$$x \mapsto \begin{cases} \int f_x \, d\nu & \text{if } f_x \in \mathscr{L}^1(Y, \mathscr{B}(Y), \nu), \\ 0 & \text{otherwise,} \end{cases}$$

and

$$y \mapsto \begin{cases} \int f^y d\mu & \text{if } f^y \in \mathscr{L}^1(X,\mathscr{B}(X),\mu), \\ 0 & \text{otherwise,} \end{cases}$$

belong to $\mathscr{L}^1(X,\mathscr{B}(X),\mu)$ *and* $\mathscr{L}^1(Y,\mathscr{B}(Y),v)$, *respectively, and*
(c) $\int f d(\mu \times v) = \int_X \int_Y f(x,y) v(dy)\mu(dx) = \int_Y \int_X f(x,y)\mu(dx)v(dy)$.

Proof. Let $\mathscr{F}$ be the collection of all functions in $\mathscr{L}^1(X \times Y, \mathscr{B}(X \times Y), \mu \times v)$ that vanish outside a rectangle with σ-finite sides. Corollary 7.6.6 implies that if f is a characteristic function that belongs to $\mathscr{F}$, then f satisfies the conclusions of the theorem (the finiteness of $\int f_x dv$ and $\int f^y d\mu$ for almost all x and y follows from Corollary 2.3.14). The linearity of the integral and the monotone convergence theorem imply that the same is true first for nonnegative simple functions in $\mathscr{F}$, then for nonnegative functions in $\mathscr{F}$, and finally for arbitrary functions in $\mathscr{F}$.   □

See Exercises 3 and 4 for some techniques for computing $\int |f| d(\mu \times v)$ and hence for determining whether f is $(\mu \times v)$-integrable.

The reader should note several things about the hypotheses of Corollary 7.6.6 and Theorem 7.6.7:

(a) Corollary 7.6.6 would fail if E were only assumed to be a Borel subset of $X \times Y$; see Exercise 1.
(b) Corollary 7.6.6 would also fail if the Borel set E were only assumed to be σ-finite (or even of finite measure) under $\mu \times v$; see Exercise 2.
(c) We will see that if μ and v are Haar measures on locally compact groups X and Y, then each Borel subset E of $X \times Y$ that satisfies $(\mu \times v)(E) < +\infty$ is included in a rectangle with σ-finite sides, and each integrable function on $X \times Y$ vanishes outside a rectangle with σ-finite sides (this follows from Lemma 9.4.2, applied to the group $X \times Y$).
(d) See Exercise 6 for an alternate version of Corollary 7.6.6 and Theorem 7.6.7.

Exercises

1. Show that the conclusions of Corollary 7.6.6 can fail if E is an arbitrary Borel (or even closed) subset of $X \times Y$. More precisely, show that part (a) can fail and that part (b) can fail even in cases where part (a) holds. (Hint: Let X be $\mathbb{R}$ with its usual topology, let Y be $\mathbb{R}$ with the discrete topology, let μ be Lebesgue measure on $(X,\mathscr{B}(X))$, let v be counting measure on $(Y,\mathscr{B}(Y))$, and let E be a suitable subset of $\{(x,y) : x = y\}$.)
2. Let X be $\mathbb{R}$ with its usual topology, let Y be $\mathbb{R}$ with the discrete topology, let μ be a point mass on $(X,\mathscr{B}(X))$, and let v be counting measure on $(Y,\mathscr{B}(Y))$. Suppose that A is a non-Borel subset of $\mathbb{R}$, and define E to be the set of all pairs (x,x) for which $x \in A$. Show that E is a Borel subset of $X \times Y$ that has finite measure under $\mu \times v$, but for which the conclusion of Corollary 7.6.6 fails.

3. Let X, Y, μ, ν, and $\mu \times \nu$ be as in Proposition 7.6.5. Show that if $f\colon X \times Y \to$
   $[0, +\infty]$ is lower semicontinuous, then
   (a) $x \mapsto \int f(x,y)\,\nu(dy)$ and $y \mapsto \int f(x,y)\,\mu(dx)$ are Borel measurable, and
   (b) $\int f\, d(\mu \times \nu) = \int \int f(x,y)\,\nu(dy)\mu(dx) = \int \int f(x,y)\,\mu(dx)\nu(dy)$.
4. Show that the conclusions of Exercise 3 also hold if f is a nonnegative Borel
   measurable function that vanishes outside a Borel rectangle with σ-finite sides.
5. Show that the Baire σ-algebras on compact Hausdorff spaces (see Exercise
   7.2.8) behave "properly" under the formation of products, in the sense that
   $\mathscr{B}_0(X \times Y) = \mathscr{B}_0(X) \times \mathscr{B}_0(Y)$. (Hint: Use the Stone–Weierstrass theorem (Theorem D.22) to show that each function in $C(X \times Y)$ can be uniformly approximated
   by functions of the form $(x,y) \mapsto \sum_i f_i(x)g_i(y)$, where the sum is finite, each f_i
   belongs to $C(X)$, and each g_i belongs to $C(Y)$.)
6. Let X, Y, μ, ν, and $\mu \times \nu$ be as in Proposition 7.6.5, and consider the outer
   measures μ^*, ν^*, and $(\mu \times \nu)^*$ and measures μ_1, ν_1, and $(\mu \times \nu)_1$ that are
   associated to μ, ν, and $\mu \times \nu$ as in Sect. 7.5.
   (a) Show that if E belongs to $\mathscr{M}_{(\mu \times \nu)^*}$ and satisfies $(\mu \times \nu)_1(E) = 0$, then
       $\nu^*(E_x) = 0$ holds for μ_1-almost every x in X and $\mu^*(E^y) = 0$ holds for ν_1-
       almost every y in Y. (Note that E is not assumed to be included in a rectangle
       with σ-finite sides.)
   (b) Prove modifications of Corollary 7.6.6, Theorem 7.6.7, and Exercise 4 that
       apply to $\mathscr{M}_{(\mu \times \nu)^*}$-measurable, rather than Borel measurable, functions. Your
       modification of Theorem 7.6.7 should not contain the assumption that f
       vanishes outside a rectangle with σ-finite sides. (Hint: Replace $(X, \mathscr{B}(X), \mu)$
       and $(Y, \mathscr{B}(Y), \nu)$ with $(X, \mathscr{M}_{\mu^*}, \mu_1)$ and $(Y, \mathscr{M}_{\nu^*}, \nu_1)$; see Exercises 7.2.3
       and 5.2.6.)

7.7   The Daniell–Stone Integral

There is an alternate approach to integration theory, due to Daniell [32] and Stone
[114], in which one does not begin with a measure but rather with a positive
linear functional on a vector space of functions. One extends this functional to a
larger collection of functions, proves analogues of the monotone and dominated
convergence theorems for the extended functional, and finally shows that the
extended functional can be viewed as integration with respect to a measure.

Exercises 3 through 36 at the end of this section contain an outline of these
classical results. I hope that I have arranged these exercises in such a way that
the student can supply the missing details without too much trouble. In the body
of this section we simply give an argument due to Kindler [70] (see also Zaanen
[130]) that shows that the functionals considered by Daniell and Stone in fact
correspond to integration with respect to measures. This theorem does not, of course,
give the entire Daniell–Stone theory, but it does provide what is needed for many
applications.

We turn to some basic definitions. Let X be a nonempty set. Recall that for real-valued (or $[-\infty, +\infty]$-valued) functions f and g on X, the functions $f \vee g$ and $f \wedge g$ are defined by

$$(f \vee g)(x) = \max(f(x), g(x))$$

and

$$(f \wedge g)(x) = \min(f(x), g(x)).$$

A *vector lattice* on X is a vector space V of real-valued functions on X that is closed under the operations $\wedge$ and $\vee$. A vector lattice V satisfies *Stone's condition* if

$$f \wedge 1 \in V \text{ whenever } f \in V. \tag{1}$$

(Here 1 is the constant function whose value is 1 at every point in X. Note that the constant function 1 may or may not belong to V.)

A linear functional L on a vector lattice V is an *elementary integral* if it is positive (that is, $L(f) \geq 0$ holds for every nonnegative function f in V) and satisfies

$$\lim_n L(f_n) = 0 \text{ for every sequence } \{f_n\} \text{ in } V \text{ that decreases pointwise to } 0. \tag{2}$$

We have the following basic facts about elementary integrals.

Lemma 7.7.1. *Suppose that L is an elementary integral on the vector lattice V and that f and f_1, f_2, $\ldots$, are nonnegative functions in V.*

(a) *If the sequence $\{f_n\}$ increases to f, then $L(f) = \lim_n L(f_n)$.*
(b) *If $f = \sum_n f_n$, then $L(f) = \sum_n L(f_n)$.*
(c) *If $f \leq \sum_n f_n$, then $L(f) \leq \sum_n L(f_n)$.*

Proof. Part (a) follows from condition (2), applied to the sequence $\{f - f_n\}_{n=1}^{\infty}$. Then parts (b) and (c) are consequences of part (a), applied to the sequences $\{\sum_{i=1}^{n} f_i\}_{n=1}^{\infty}$ and $\{(\sum_{i=1}^{n} f_i) \wedge f\}_{n=1}^{\infty}$. $\square$

The following lemma is often useful for verifying condition (2).

Lemma 7.7.2 (Dini's Theorem). *Suppose that X is a closed bounded subinterval of $\mathbb{R}$ (or, more generally, a compact Hausdorff space). Let $\{f_n\}$ be a sequence of nonnegative continuous functions on X that decreases to 0 (in the sense that $\{f_n(x)\}$ decreases to 0 for each x in X). Then the sequence $\{f_n\}$ converges uniformly to 0.*

Proof. We need to show that for each positive ε there is a positive integer N such that $\|f_n\|_{\infty} \leq \varepsilon$ holds whenever $n \geq N$.

So suppose that ε is a positive number. For each x in X choose a positive integer n_x such that $f_{n_x}(x) < \varepsilon$, and then use the continuity of f_{n_x} to choose an open neighborhood U_x of x such that $f_{n_x}(t) < \varepsilon$ holds for all t in U_x. The family $\{U_x\}_{x \in X}$ is an open cover of X, and so the compactness of X gives a finite subcover U_{x_i}, where $i = 1, \ldots, k$, of X. Let N be the maximum of n_{x_i}, for $i = 1, \ldots, k$. If $x \in X$, then $x \in U_{x_i}$ for some i, and so

$$0 \leq f_n(x) \leq f_{n_i}(x) < \varepsilon$$

holds for all n that satisfy $n \geq N$. Since this estimate is valid for every x in X, we have $\|f_n\|_\infty \leq \varepsilon$ and the proof is complete. $\square$

Examples 7.7.3.

(a) Let $[a,b]$ be a closed bounded subinterval of $\mathbb{R}$ and let $C([a,b])$ be the set of all continuous real-valued functions on $[a,b]$. Then $C([a,b])$ is a vector lattice that satisfies Stone's condition. Suppose we define a functional $L\colon C([a,b]) \to \mathbb{R}$ by letting L be the Riemann integral: $L(f) = \int_a^b f$. Dini's theorem implies that L satisfies condition (2) and so is an elementary integral.

(b) Let X be a locally compact Hausdorff space, and (as in Sect. 7.1) let $\mathscr{K}(X)$ be the set of all continuous functions $f\colon X \to \mathbb{R}$ for which the support of f is compact. Then $\mathscr{K}(X)$ is a vector lattice that satisfies Stone's condition. (Note that the constant function 1 does not belong to $\mathscr{K}(X)$ if X is not compact.) If L is a positive linear functional on $\mathscr{K}(X)$, then L is an elementary integral (again use Dini's theorem to check that L satisfies condition (2)).

(c) The set of all differentiable real-valued functions on $\mathbb{R}$ is a vector space, but not a vector lattice.

(d) Let V be the set of all constant multiples of the function $f\colon [0,1] \to \mathbb{R}$ defined by $f(x) = x$. Then V is a vector lattice, but it does not satisfy Stone's condition.

(e) Let V be the set of all continuous functions $f\colon [0,+\infty) \to \mathbb{R}$ such that $\lim_{x \to +\infty} f(x)$ exists, and define $L\colon V \to \mathbb{R}$ by $L(f) = \lim_{x \to +\infty} f(x)$. Then V is a vector lattice that satisfies Stone's condition, and L is a positive linear functional that does *not* satisfy condition (2)—consider, for example, the sequence $\{f_n\}$ defined by

$$f_n(x) = \begin{cases} 0 & \text{if } x < n, \\ x-n & \text{if } n \leq x < n+1, \text{ and} \\ 1 & \text{otherwise.} \end{cases}$$
$\square$

Before we look at the main theorem of this section, it will be convenient to look at a slight generalization of the concept of a σ-algebra. So let X be a set. A collection $\mathscr{R}$ of subsets of X is a σ-*ring* on X if

(a) $\varnothing$ belongs to $\mathscr{R}$,

(b) for all sets A, B that belong to $\mathscr{R}$, the set $A - B$ belongs to $\mathscr{R}$,

(c) for each infinite sequence $\{A_i\}$ of sets that belong to $\mathscr{R}$, the set $\cup_{i=1}^\infty A_i$ belongs to $\mathscr{R}$, and

(d) for each infinite sequence $\{A_i\}$ of sets that belong to $\mathscr{R}$, the set $\cap_{i=1}^\infty A_i$ belongs to $\mathscr{R}$.

Of course, every σ-algebra is a σ-ring. If X is an uncountable set, then the set of all countable subsets of X is a σ-ring but not a σ-algebra. It is sometimes useful to deal with σ-rings when one wants to deal only with sets that are in some sense not too large.

Here are a few properties of σ-rings; their proofs are left for the reader:

(a) If $\mathscr{F}$ is a collection of subsets of a set X, then there is a smallest σ-ring on X that includes $\mathscr{F}$.
(b) If $\mathscr{R}$ is a σ-ring on a set X, then the collection of subsets A of X such that either A or A^c belongs to $\mathscr{R}$ is a σ-algebra on X; it is in fact the σ-algebra $\sigma(\mathscr{R})$ generated by $\mathscr{R}$.
(c) Suppose that μ_0 is a measure on a σ-ring $\mathscr{R}$ (i.e., a countably additive $[0,+\infty]$-valued function on $\mathscr{R}$ such that $\mu_0(\varnothing) = 0$). Let $\mathscr{A}$ be the σ-algebra generated by $\mathscr{R}$. Then the function $\mu : \mathscr{A} \to [0,+\infty]$ defined by

$$\mu(A) = \begin{cases} \mu_0(A) & \text{if } A \in \mathscr{R}, \text{ and} \\ +\infty & \text{if } A \in \mathscr{A} - \mathscr{R} \end{cases}$$

is a measure on $\mathscr{A}$.

Now suppose that X is a set and that V is a vector lattice on X. Let $\mathscr{F}$ be the collection of sets of the form $\{x \in X : f(x) > B\}$, where f ranges over V and B ranges over the positive reals. Let $\mathscr{R}$ be the smallest σ-ring on X that includes $\mathscr{F}$, and let $\mathscr{A}$ be the smallest σ-algebra on X that includes $\mathscr{F}$. It is easy to check that $\mathscr{A}$ is the smallest σ-algebra on X that makes each function in V measurable.

The following theorem is the main result of this section.

Theorem 7.7.4. *Let X be a set, let V be a vector lattice on X that satisfies Stone's condition, let L be an elementary integral on V, and let $\mathscr{R}$ and $\mathscr{A}$ be as defined above. Then there is a measure μ on $(X, \mathscr{A})$ such that $L(f) = \int f \, d\mu$ holds for each f in V. The restriction of this measure to $\mathscr{R}$ is unique, in the sense that if μ_1 and μ_2 are measures on $(X, \mathscr{A})$ such that $\int f \, d\mu_1 = L(f) = \int f \, d\mu_2$ holds for all f in V, then $\mu_1(A) = \mu_2(A)$ holds for all A in $\mathscr{R}$.*

The uniqueness assertion in this theorem may seem rather weak, since it involves only sets in $\mathscr{R}$. Note, however, that if μ is a measure on $\mathscr{A}$ that represents L, if f is a nonnegative function in V, and if we let $A_{n,i} = \{x : i/2^n < f(x) \le (i+1)/2^n\}$ for each n and i, then the sequence $\{\frac{i}{2^n} \sum_{i=0}^{n2^n-1} \frac{i}{2^n} \chi_{A_{n,i}}\}_{n=1}^{\infty}$ increases pointwise to f, and so

$$L(f) = \int f \, d\mu = \lim_n \int \sum_{i=0}^{n2^n-1} \frac{i}{2^n} \chi_{A_{n,i}} \, d\mu = \lim_n \sum_{i=0}^{n2^n-1} \frac{i}{2^n} \mu(A_{n,i}).$$

Thus the sets in $\mathscr{R}$ are the only ones needed for computing $\int f \, d\mu$. Also see Exercise 2.

For functions f and g in V let $[f,g)$ be the subset of $X \times \mathbb{R}$ given by

$$[f,g) = \{(x,t) \in X \times \mathbb{R} : f(x) \le t < g(x)\}$$

(be careful: we are not assuming that $f \le g$). Note that if f is a nonnegative function in V, then $[0,f)$ can be interpreted as the region under the graph of f. Let $\mathscr{I}$ be the collection of all such sets $[f,g)$, and let $\mathscr{B}$ to be the smallest σ-algebra on $X \times \mathbb{R}$ that includes $\mathscr{I}$. We will begin the proof of Theorem 7.7.4 by constructing a measure

v on $(X \times \mathbb{R}, \mathscr{B})$ such that $v([f,g)) = L(g - f)$ holds whenever f and g belong to V and satisfy $f \leq g$. Next we will define a measure μ on $(X, \mathscr{A})$ that satisfies $\mu(A) = v(A \times [0,1))$ for each A in $\mathscr{R}$, and finally we will show that μ satisfies $L(f) = \int f \, d\mu$ for each f in V.

Here are a few basic facts about $\mathscr{I}$.

Lemma 7.7.5. *Suppose that V is a vector lattice of functions, that L is a positive linear functional on V, and that $\mathscr{I}$ is as defined above.*

(a) *If $I \in \mathscr{I}$, then there exist functions f and g in V such that $f \leq g$ and $I = [f,g)$.*

(b) *If the member I of $\mathscr{I}$ can be written in the form $[f_1, g_1)$ and in the form $[f_2, g_2)$, where $f_1 \leq g_1$ and $f_2 \leq g_2$, then $g_1 - f_1 = g_2 - f_2$, and so $L(g_1 - f_1) = L(g_2 - f_2)$.*

(c) *If I_1 and I_2 belong to $\mathscr{I}$, then $I_1 \cap I_2$ also belongs to $\mathscr{I}$.*

(d) *If I_1 and I_2 belong to $\mathscr{I}$, then there are disjoint sets I' and I'' in $\mathscr{I}$ such that $I_1 \cap I_2^c = I' \cup I''$ and hence such that $I_1 = (I_1 \cap I_2) \cup I' \cup I''$.*

Proof. For part (a), note that if $f_0, g_0 \in V$ and if we let $f = f_0 \wedge g_0$ and $g = g_0$, then $f \leq g$ and $[f,g) = [f_0, g_0)$. For part (b), note that if the section I_x is nonempty, then $f_1(x) = f_2(x)$ and $g_1(x) = g_2(x)$, while if the section I_x is empty, then $f_1(x) = g_1(x)$ and $f_2(x) = g_2(x)$. In either case we have $g_1(x) - f_1(x) = g_2(x) - f_2(x)$, and so $L(g_1 - f_1) = L(g_2 - f_2)$. Part (c) follows from the calculation $[f_1, g_1) \cap [f_2, g_2) = [f_1 \vee f_2, g_1 \wedge g_2)$. Finally, if $I_1 = [f_1, g_1)$ and $I_2 = [f_2, g_2)$, where $f_2 \leq g_2$, then $I_1 \cap I_2^c = [f_1, g_1 \wedge f_2) \cup [f_1 \vee g_2, g_1)$, from which part (d) follows. $\qquad\square$

In view of part (b) of Lemma 7.7.5, we can define a function $L_{\mathscr{I}} : \mathscr{I} \to \mathbb{R}$ by

$$L_{\mathscr{I}}(I) = L(g - f),$$

where f and g are elements of V such that $f \leq g$ and $I = [f,g)$.

Lemma 7.7.6. *Suppose that I and I_1, I_2, ... are members of $\mathscr{I}$.*

(a) *If the sets I_n, $n = 1, 2, \ldots$, are disjoint and if $I = \cup_n I_n$, then $L_{\mathscr{I}}(I) = \sum_n L_{\mathscr{I}}(I_n)$.*

(b) *If $I \subseteq \cup_n I_n$, then $L_{\mathscr{I}}(I) \leq \sum_n L_{\mathscr{I}}(I_n)$.*

Proof. Suppose that $I = \cup_n I_n$, and let $I = [f,g)$ and $I_n = [f_n, g_n)$, $n = 1, 2, \ldots$, where $f \leq g$ and $f_n \leq g_n$, $n = 1, 2, \ldots$. For each x in X the sections of these sets at x satisfy $I_x = \cup_n (I_n)_x$, and so the countable additivity of Lebesgue measure implies that

$$g(x) - f(x) = \lambda(I_x) = \sum_n \lambda((I_n)_x) = \sum_n (g_n(x) - f_n(x)).$$

It follows from Lemma 7.7.1 that $L_{\mathscr{I}}(I) = \sum_n L_{\mathscr{I}}(I_n)$, and so the proof of part (a) is complete. Part (b) can be proved with a similar argument. $\qquad\square$

Proof of Theorem 7.7.4. We define a function v^* on the subsets of $X \times \mathbb{R}$ by letting $v^*(A)$ be the infimum of the set of sums of the form $\sum_i L_{\mathscr{I}}(I_i)$, where $\{I_i\}$ is a sequence in $\mathscr{I}$ such that $A \subseteq \cup_i I_i$. (Of course, $v^*(A) = +\infty$ if there is no sequence $\{I_i\}$ such that $A \subseteq \cup_i I_i$.)

Lemma 7.7.7. *Let v^* be as defined above. Then*

(a) v^* *is an outer measure on* X,
(b) *every set in* $\mathscr{I}$ *is* v^*-*measurable, and*
(c) *if* $I \in \mathscr{I}$, *then* $v^*(I) = L_{\mathscr{I}}(I)$.

Proof. It is immediate that v^* is an outer measure. Now suppose that $I \in \mathscr{I}$. We can show that I is v^*-measurable by checking that

$$v^*(A) \geq v^*(A \cap I) + v^*(A \cap I^c)$$

holds for each subset A of $X \times \mathbb{R}$ such that $v^*(A) < +\infty$ (see Sect. 1.3 and, in particular, the discussion of inequality (1) in that section). So suppose that A is such a set, that ε is a positive number, and that $\{I_n\}$ is a sequence of sets in $\mathscr{I}$ such that $A \subseteq \cup_n I_n$ and

$$v^*(A) + \varepsilon > \sum_n L_{\mathscr{I}}(I_n).$$

According to part (d) of Lemma 7.7.5, for each n there are sets I'_n and I''_n such that $I_n \cap I$, I'_n, and I''_n are disjoint and have I_n as their union. Thus

$$L_{\mathscr{I}}(I_n) = L_{\mathscr{I}}(I_n \cap I) + L_{\mathscr{I}}(I'_n) + L_{\mathscr{I}}(I''_n);$$

since $A \cap I \subseteq \cup_n(I_n \cap I)$ and $A \cap I^c \subseteq \cup_n(I'_n \cup I''_n)$, we have

$$
\begin{aligned}
v^*(A) + \varepsilon &> \sum_n L_{\mathscr{I}}(I_n) \\
&= \sum_n L_{\mathscr{I}}(I_n \cap I) + \sum_n (L_{\mathscr{I}}(I'_n) + L_{\mathscr{I}}(I''_n)) \\
&\geq v^*(A \cap I) + v^*(A \cap I^c).
\end{aligned}
$$

Thus I is measurable. Each I in $\mathscr{I}$ certainly satisfies $v^*(I) \leq L_{\mathscr{I}}(I)$. The reverse inequality follows from part (b) of Lemma 7.7.6, and with that the proof of Lemma 7.7.7 is complete. $\square$

We return to the proof of Theorem 7.7.4. Let f be a nonnegative function in V, let B be a positive real number, and for each n define f_n by

$$f_n = 1 \wedge n(f - (f \wedge B)).$$

Since V is a vector lattice that satisfies Stone's condition, each f_n belongs to V. For each positive number C, the sequence $\{Cf_n\}$ is increasing and converges pointwise to $C\chi_{\{f>B\}}$, and the sets $[0, Cf_n)$ increase to the set $\{f > B\} \times [0, C)$. Let us consider three consequences of this.

First, for each f in V and each positive number B we have $\{f > B\} \times [0, 1) \in \mathscr{B}$ (recall that $\mathscr{B}$ is the σ-algebra generated by the family $\mathscr{I}$). It follows that each A in $\mathscr{R}$ satisfies $A \times [0, 1) \in \mathscr{B}$ and hence that $\mu(A) = v(A \times [0, 1))$ defines a measure on $\mathscr{R}$. As we noted earlier in this section, we can extend μ to a measure on the σ-algebra $\mathscr{A}$ by letting $\mu(A) = +\infty$ if A belongs to $\mathscr{A}$ but not to $\mathscr{R}$.

Second, the fact that the sets $[0,Cf_n)$ increase to $\{f > B\} \times [0,C)$ implies that

$$v(\{f > B\} \times [0,C)) = \lim_n v([0,Cf_n)) = \lim_n L(Cf_n)$$
$$= C \lim_n L(f_n) = C \lim_n v([0,f_n))$$
$$= Cv(\{f > B\} \times [0,1));$$

that is,

$$v(\{f > B\} \times [0,C)) = C\mu(\{f > B\}). \tag{3}$$

Finally, for each n we have $f_n \le \chi_{\{f>B\}} \le f/B$ and so $L(f_n) \le L(f)/B$. Since $v(\{f > B\} \times [0,C)) = \lim_n CL(f_n) \le CL(f)/B$, it follows that the values in (3) are finite.

Now let n and i range over the positive integers. If we apply (3) twice, once with $B = i/2^n$ and $C = i/2^n$ and once with $B = (i+1)/2^n$ and $C = i/2^n$, we find that

$$v\left(\left\{\frac{i}{2^n} < f \le \frac{i+1}{2^n}\right\} \times [0,i/2^n)\right) = \frac{i}{2^n}\mu\left(\left\{\frac{i}{2^n} < f \le \frac{i+1}{2^n}\right\}\right)$$

and hence that

$$v\left(\bigcup_{i=1}^{n2^n}\left\{\frac{i}{2^n} < f \le \frac{i+1}{2^n}\right\} \times [0,i/2^n)\right) = \sum_{i=1}^{n2^n}\frac{i}{2^n}\mu\left(\left\{\frac{i}{2^n} < f \le \frac{i+1}{2^n}\right\}\right). \tag{4}$$

The countable additivity of v implies that the left side of (4) approaches $v([0,f))$ as n approaches infinity, while the monotone convergence theorem implies that the right side approaches $\int f\,d\mu$. With this we have

$$L(f) = v([0,f)) = \int f\,d\mu.$$

Since each f in V can be separated into its positive and negative parts, we have $L(f) = \int f\,d\mu$ for each f in V, and the construction of μ is complete.

We turn to the uniqueness. Let μ_1 and μ_2 be measures on $\mathscr{A}$ such that $\int f\,d\mu_1 = L(f) = \int f\,d\mu_2$ holds for all f in V. Suppose that $f_1, \ldots, f_k$ belong to V, that $B_1, \ldots, B_k$ are positive numbers, and that $A = \cap_i\{f_i > B_i\}$. For each n let

$$g_n = \wedge_{i=1}^k (1 \wedge n(f_i - (f_i \wedge B_i))).$$

Then each g_n belongs to V and is nonnegative, and the sequence $\{g_n\}$ increases to χ_A. Hence

$$\mu_1(A) = \lim_n \int g_n\,d\mu_1 = \lim_n L(g_n) = \lim_n \int g_n\,d\mu_2 = \mu_2(A).$$

Now fix f_1 and B_1, and let $\mathscr{F}_{f_1,B_1}$ be the collection of all subsets of $\{f_1 > B_1\}$ that have the form $\cap_i\{f_i > B_i\}$. Then μ_1 and μ_2 agree on the π-system $\mathscr{F}_{f_1,B_1}$, and so it follows from Corollary 1.6.3 that μ_1 and μ_2 agree on the σ-algebra on $\{f_1 > B_1\}$ that $\mathscr{F}_{f_1,B_1}$ generates. However, it is easy to check that this σ-algebra is just the collection of all subsets of $\{f_1 > B_1\}$ that belong to $\mathscr{R}$. Finally, every set in $\mathscr{R}$ is included in a countable union of sets of the form $\{f > B\}$, and so μ_1 and μ_2 agree on $\mathscr{R}$. $\qquad\square$

Exercises

1. (a) Let X be the interval $[-1, 1]$. Find (i.e., describe precisely) the smallest vector lattice V on X that contains the function $x \mapsto x$.
   (b) Does V satisfy Stone's condition?
2. Let $X = \mathbb{R}$ and let V be the set of those continuous functions $f: X \to \mathbb{R}$ whose support is compact and included in $(0, +\infty)$. Define L on V by letting $L(f)$ be the Riemann integral of f, and let $\mathscr{A}$ and $\mathscr{R}$ be as in Theorem 7.7.4.
   (a) What sets do $\mathscr{A}$ and $\mathscr{R}$ contain? (Your answer should relate these families of sets to the Borel or the Lebesgue measurable subsets of $\mathbb{R}$.)
   (b) Give two measures on $\mathscr{A}$ that represent the functional L.

The following exercises contain an outline of the usual development of the Daniell–Stone integral. As noted at the beginning of this section, this development is not based on measure theory. Thus solutions to Exercises 3 through 32 should not contain any references to the earlier chapters of this book. Sigma-algebras, measurable functions, measures, and the Lebesgue integral appear first in Exercises 33–36

Suppose that V is a vector lattice of functions on a set X, that V satisfies Stone's condition, and that L is an elementary integral on V (i.e., it is a positive linear functional on V that satisfies relation (2)). Let $V^\bullet$ be the set of all $(-\infty, +\infty]$-valued functions on X that are pointwise limits of increasing sequences of functions in V, and define $L^\bullet: V^\bullet \to (-\infty, +\infty]$ by $L^\bullet(f) = \lim_n L(f_n)$, where $\{f_n\}$ is an increasing sequence of functions in V that converges pointwise to f (see Exercise 3). Likewise, let $V_\bullet$ be the set of all $[-\infty, +\infty)$-valued functions on X that are pointwise limits of decreasing sequences of functions in V, and define $L_\bullet: V_\bullet \to [-\infty, +\infty)$ by $L_\bullet(f) = \lim_n L(f_n)$, where $\{f_n\}$ is an decreasing sequence of functions in V that converges pointwise to f.

3. Show that $L^\bullet$ and $L_\bullet$ are well defined on $V^\bullet$ and $V_\bullet$. For example, to show that $L^\bullet$ is well defined, one needs to show that if $f \in V^\bullet$ and if $\{f_n\}$ and $\{g_n\}$ are sequences of functions in V that increase to f, then $\lim_n L(f_n) = \lim_n L(g_n)$. (Hint: Start by showing that if $g \in V$ and if $\{f_n\}$ is an increasing sequence in V such that $g(x) \le \lim_n f_n(x)$ holds for each x, then $L(g) \le \lim L(f_n)$.)
4. Show that $f \in V_\bullet$ if and only if there is a function g in $V^\bullet$ such that $f = -g$ and that in this case $L_\bullet(f) = -L^\bullet(g)$.

5. Suppose that $f_1, f_2 \in V^\bullet$ and that α is a nonnegative real number. Show that
   (a) $f_1 \wedge f_2, f_1 \vee f_2 \in V^\bullet$,
   (b) $f_1 + f_2 \in V^\bullet$ and $L^\bullet(f_1 + f_2) = L^\bullet(f_1) + L^\bullet(f_2)$, and
   (c) $\alpha f_1 \in V^\bullet$ and $L^\bullet(\alpha f_1) = \alpha L^\bullet(f_1)$.
6. Show that if $f_1, f_2 \in V^\bullet$ and $f_1 \le f_2$, then $L^\bullet(f_1) \le L^\bullet(f_2)$.
7. (a) Show that if $g \in V_\bullet$ and $h \in V^\bullet$, then $h - g \in V^\bullet$ and $L^\bullet(h - g) = L^\bullet(h) - L_\bullet(g)$. In particular, all the differences involved here are defined (that is, neither $+\infty$ or $-\infty$ is ever subtracted from itself here).
   (b) Conclude that if $g \in V_\bullet$, $h \in V^\bullet$, and $g \le h$, then $L_\bullet(g) \le L^\bullet(h)$.
8. Show that if $\{f_n\}$ is a sequence of functions in $V^\bullet$ and if $\{f_n\}$ increases to f, then $f \in V^\bullet$ and $L^\bullet(f) = \lim_n L^\bullet(f_n)$. (Hint: Use some ideas from the proof of Theorem 2.4.1.)

Suppose that f is an arbitrary $[-\infty, +\infty]$-valued function on X. Define $\overline{L}(f)$ and $\underline{L}(f)$ by

$$\overline{L}(f) = \inf\{L^\bullet(h) : h \in V^\bullet \text{ and } f \le h\}$$

and

$$\underline{L}(f) = \sup\{L_\bullet(g) : g \in V_\bullet \text{ and } g \le f\}$$

(of course, $\overline{L}(f) = +\infty$ if there is no h in $V^\bullet$ such that $f \le h$, and $\underline{L}(f) = -\infty$ if there is no g in $V_\bullet$ such that $g \le f$). For each f we have $\underline{L}(f) \le \overline{L}(f)$ (see Exercise 9). A function $f\colon X \to [-\infty, +\infty]$ is *L-summable*, or simply *summable*, if $\underline{L}(f)$ and $\overline{L}(f)$ are finite and equal. We define L_1 on the collection of summable functions by letting $L_1(f)$ be the common value of $\underline{L}(f)$ and $\overline{L}(f)$.

9. Show that each $f\colon X \to [-\infty, +\infty]$ satisfies $\underline{L}(f) \le \overline{L}(f)$.
10. Show that $f\colon X \to [-\infty, +\infty]$ is L-summable if and only if for every positive ε there exist g in $V_\bullet$ and h in $V^\bullet$ such that $g \le f \le h$ and $L^\bullet(h - g) < \varepsilon$. (Hint: See Exercise 7.)
11. Show that if $f \in V$, then f is summable and $L_1(f) = L(f)$. Thus L_1 is an extension of L.
12. Show that if f_1 and f_2 are $\mathbb{R}$-valued summable functions and $\alpha \in \mathbb{R}$, then
    (a) $f_1 + f_2$ is summable and $L_1(f_1 + f_2) = L_1(f_1) + L_1(f_2)$, and
    (b) αf_1 is summable and $L_1(\alpha f_1) = \alpha L_1(f_1)$.
13. Show that if f_1 and f_2 are $[0, +\infty]$-valued summable functions, then $f_1 + f_2$ is summable and $L_1(f_1 + f_2) = L_1(f_1) + L_1(f_2)$.
14. Show that if f_1 and f_2 are $[-\infty, +\infty]$-valued summable functions, then $f_1 \wedge f_2$ and $f_1 \vee f_2$ are summable.
15. (a) Generalize part (a) of Exercise 12 to the case where f_1 and f_2 are summable $[-\infty, +\infty]$-valued functions such that $f_1(x) + f_2(x)$ is defined for each x (i.e., such that for no x is this sum of the form $+\infty + (-\infty)$ or $-\infty + (+\infty)$).
    (b) Show that a function $f\colon X \to [-\infty, +\infty]$ is summable if and only if f^+ and f^- are summable and that in such cases $L_1(f) = L_1(f^+) - L_1(f^-)$.

16. Show that if $\{f_n\}$ is a sequence of $[0, +\infty]$-valued summable functions, if $\{f_n\}$ increases pointwise to f, and if $\sup_n L_1(f_n) < +\infty$, then f is summable and $L_1(f) = \lim_n L_1(f_n)$. (Hint: It might be useful to verify and use the fact that if f_1 and f_2 are nonnegative summable functions such that $f_1 \leq f_2$, if ε_1 and ε_2 are positive numbers, and if g_1 and g_2 belong to $V^\bullet$ and satisfy $f_i \leq g_i$ and $L^\bullet(g_i) < L_1(f_i) + \varepsilon_i$ for $i = 1, 2$, then $L^\bullet(g_1 \vee g_2) < L_1(f_2) + \varepsilon_1 + \varepsilon_2$.)

17. Suppose that $\{f_n\}$ is a sequence of $\mathbb{R}$-valued summable functions, that $f(x) = \lim_n f_n(x)$ holds for all x, and that h is a $[0, +\infty)$-valued summable function such that $|f_n(x)| \leq h(x)$ holds for all n and all x. Show that f is summable and that $L_1(f) = \lim_n L_1(f_n)$.

18. Show that if f is a $[-\infty, +\infty]$-valued summable function, then $f \wedge 1$ is summable. (Hint: See Exercise 10.)

19. Show that if f is a $[0, +\infty]$-valued summable function, if α is a positive real number, and if $A = \{x \in X : f(x) > \alpha\}$, then χ_A is summable and $L_1(\alpha \chi_A) \leq L_1(f)$. (Hint: Use Exercise 18. Reduce the question to one involving only $[0, +\infty)$-valued functions. Check that if f is $[0, +\infty)$-valued, then the sequence $\{f_n\}$, where $f_n = \alpha \wedge n(f - (f \wedge \alpha))$, increases to $\alpha \chi_A$.)

A subset A of X is *L-negligible* or *L-null* if χ_A is summable and $L_1(\chi_A) = 0$. A property of points x in X is said to hold *L-almost everywhere* if the set of points at which it fails is an *L*-negligible set.

20. Show that a subset A of X is *L*-negligible if and only if for every ε there is a function f in $V^\bullet$ such that $\chi_A \leq f$ and $L^\bullet(f) < \varepsilon$.

21. Show that each subset of an *L*-negligible set is *L*-negligible.

22. Show that the union of a countable collection of *L*-negligible sets is *L*-negligible.

23. Suppose that f_1 and f_2 are $[-\infty, +\infty]$-valued functions that are equal *L*-almost everywhere. Show that if one of these functions is summable, then both are summable and $L_1(f_1) = L_1(f_2)$.

24. Show that if f is a $[-\infty, +\infty]$-valued summable function, then $\{x \in X : |f(x)| = +\infty\}$ is *L*-negligible.

25. Reformulate Exercises 16 and 17 by allowing the appropriate hypotheses to hold only *L*-almost everywhere and (in the case of Exercise 17) by allowing the functions involved to have infinite values on *L*-negligible sets. Prove your reformulated versions.

A function $f : X \to [-\infty, +\infty]$ is called *L-measurable*, or simply *measurable*, if $(g \vee f) \wedge h$ is summable for every choice of functions g and h, where g is a non-positive summable function and h is a nonnegative summable function. A subset A of X is *L-measurable* if χ_A is a measurable function. Let $\mathcal{M}$ be the collection of all *L*-measurable subsets of X.

26. Show that every summable function is measurable.

27. Show that
    (a) a $[0, +\infty]$-valued function f is measurable if and only if for each nonnegative summable h the function $f \wedge h$ is also summable, and
    (b) a $[-\infty, +\infty]$-valued function f is measurable if and only if f^+ and f^- are measurable.
28. Show that the constant function 1 is measurable.
29. Let f_1 and f_2 be $[-\infty, +\infty]$-valued functions that are equal L-almost everywhere. Show that f_1 is measurable if and only if f_2 is measurable.
30. Show that if f_1 and f_2 are $[-\infty, +\infty]$-valued measurable functions, then $f_1 \wedge f_2$ and $f_1 \vee f_2$ are measurable.
31. Show that if $\{f_n\}$ is a sequence of $[-\infty, +\infty]$-valued measurable functions and if $f(x) = \lim_n f_n(x)$ holds for almost every x in X, then f is measurable.
32. Show that if f_1 and f_2 are $\mathbb{R}$-valued measurable functions and if $\alpha \in \mathbb{R}$, then $f_1 + f_2$ and αf_1 are measurable.
33. Show that the collection $\mathcal{M}$ of L-measurable sets is a σ-algebra.
34. Show that a function $f: X \to [-\infty, +\infty]$ is L-measurable if and only if it is measurable (in the sense of Chap. 2) with respect to the σ-algebra $\mathcal{M}$.

    Define a function $\mu: \mathcal{M} \to [0, +\infty]$ by

    $$\mu(A) = \begin{cases} L_1(\chi_A) & \text{if } \chi_A \text{ is summable, and} \\ +\infty & \text{if } \chi_A \text{ is measurable but not summable.} \end{cases}$$

35. Show that μ is a measure on $\mathcal{M}$.
36. Show that a function $f: X \to [-\infty, +\infty]$ is L-summable if and only if it is $\mathcal{M}$-measurable and μ-integrable and that then $L_1(f) = \int f \, d\mu$.
37. Let $[a, b]$ be a closed bounded interval and let L be the Riemann integral on $C[a, b]$, as in Example 7.7.3(a). Show that the L-summable functions on $[a, b]$ are exactly the Lebesgue measurable functions on $[a, b]$ that are Lebesgue integrable, and that $L_1(f) = \int f \, d\lambda$ holds for each such function f.
38. Let V and L be as in Exercise 2. Characterize the set of L-summable functions in terms of concepts from the Lebesgue theory. Be very precise.

Notes

The historical notes in Chapter III of Hewitt and Ross [58] contain a nice summary of the history of integration theory on locally compact Hausdorff spaces.

The reader who wants to see another elementary treatment of integration on locally compact Hausdorff spaces might find Halmos [54], Hewitt and Stromberg [59], Rudin [105], or Hewitt and Ross [58] useful. He or she would also do well to look up the paper of Kakutani [67].

The definition given here for the collection of Borel subsets of X agrees with that given by Hewitt and Ross [58], Hewitt and Stromberg [59], and Rudin [105]; it agrees with that given by Halmos [54] only when X is σ-compact. The definition given in Exercise 7.2.8 for the σ-algebra $\mathscr{B}_0(X)$ of Baire subsets of a compact Hausdorff space X is a special case of that given by Halmos (Halmos considers σ-rings, in addition to σ-algebras, and so is able to give a definition of $\mathscr{B}_0(X)$ that can reasonably be applied to an arbitrary locally compact Hausdorff space X).

Bourbaki [18] and Hewitt and Ross [58] deal with the μ^*-, ν^*-, and $(\mu \times \nu)^*$-measurable sets, rather than with the Borel sets, when considering product measures. Proposition 7.6.5, Corollary 7.6.6, and Theorem 7.6.7 were suggested by de Leeuw [34]. (See also Godfrey and Sion [51] and Bledsoe and Wilks [13].) Point (b) in the discussion at the end of Sect. 7.6, and also Exercise 7.6.2, come from suggestions made by Roy Johnson.

See Loomis [84], Riesz and Nagy [99], Royden [102], or Taylor [116] for more details on the Daniell–Stone version of integration theory. Taylor's exposition is especially clear and detailed. The Daniell treatment of integration theory can also be used to prove a version of the Riesz representation theorem; this is done, for instance, by Loomis [84] and Royden [102].

Chapter 8
Polish Spaces and Analytic Sets

The Borel subsets of a complete separable metric space have a number of interesting and useful characteristics. For example, if A and B are uncountable Borel subsets of complete separable metric spaces, then A and B are Borel isomorphic—that is, there is a bijection $f: A \to B$ such that f and f^{-1} are both Borel measurable. A related result says that if A is a Borel subset of a complete separable metric space, if Y is a complete separable metric space, and if $f: A \to Y$ is injective and Borel measurable, then $f(A)$ is a Borel subset of Y. If the function f is not injective, then $f(A)$ may not be a Borel set, but it will be μ-measurable for every finite Borel measure μ on Y (that is, there will be Borel subsets B_1 and B_2 of Y such that $B_1 \subseteq f(A) \subseteq B_2$ and $\mu(B_2 - B_1) = 0$).

This chapter is devoted to proving such results and to showing the context in which they arise.

8.1 Polish Spaces

A *Polish* space is a separable topological space that can be metrized using a complete metric. This section contains a number of elementary properties of Polish spaces. In Sects. 8.3 through 8.6 we will use these properties, plus the concept of an analytic set (defined in Sect. 8.2), to derive some deep and useful results about measurable sets and functions.

There are many topological spaces that are Polish, but have no complete metric that is particularly natural or simple. Furthermore, many constructions and facts of interest in measure theory depend on the existence of a complete metric, but not on the choice of a particular metric. It has thus become rather common to deal with the class of Polish spaces, rather than with the class of complete separable metric spaces.

D.L. Cohn, *Measure Theory: Second Edition*, Birkhäuser Advanced
Texts Basler Lehrbücher, DOI 10.1007/978-1-4614-6956-8_8,
© Springer Science+Business Media, LLC 2013

Examples 8.1.1.

(a) For each d the space $\mathbb{R}^d$, with its usual topology, is Polish.
(b) More generally, each separable Banach space, with the topology induced by its norm, is Polish.
(c) Each compact metrizable space is Polish (see Theorem D.39 and Corollary D.40). It amounts to the same thing to say that each compact Hausdorff space that has a countable base is Polish (see Proposition 7.1.13).

$\square$

We need the following results before we look at some additional examples.

Proposition 8.1.2. *Each closed subspace, and each open subspace, of a Polish space is Polish.*

Proof. Let X be a Polish space. According to D.33, every subspace of X is separable. Hence we need only check that the closed subspaces and the open subspaces of X can be metrized by means of complete metrics.

Let d be a complete metric for X. If F is a closed subspace of X, then the restriction of d to F is a complete metric for F. Hence each closed subspace of X is Polish.

Now suppose that U is an open subspace of X. We can assume that $U \neq X$. Recall that $d(x, U^c)$, the distance between x and U^c, is defined by

$$d(x, U^c) = \inf\{d(x, z) : z \in U^c\}$$

(see D.27). It is easy to see that

$$d_0(x, y) = d(x, y) + \left| \frac{1}{d(x, U^c)} - \frac{1}{d(y, U^c)} \right|$$

defines a metric d_0 on the set U; we will check that d_0 metrizes the topology that U inherits as a subspace of X and then that U is complete under d_0.

The function $x \mapsto d(x, U^c)$ is continuous (again see D.27), from which it follows that if x and $x_1, x_2, \ldots$ belong to U, then the sequence $\{x_n\}$ converges to x with respect to d if and only if it converges to x with respect to d_0. Thus d_0 metrizes the topology of U.

We turn to the completeness of U under d_0. A sequence $\{x_n\}$ that is Cauchy under d_0 is also Cauchy under d, and so converges under d to a point x of X. The point x belongs to U, since otherwise we would have $\lim_n d(x_n, U^c) = 0$, which would imply that

$$\overline{\lim_{m,n}} d_0(x_m, x_n) = +\infty,$$

contradicting the assumption that $\{x_n\}$ is Cauchy under d_0. It now follows that $\{x_n\}$ also converges to x under d_0, and the completeness of U under d_0 follows. $\square$

For the next results we need to recall a technique for constructing bounded metrics. Suppose that d is a metric on a set X. It is easy to check that the formula

$$d_0(x,y) = \min(1, d(x,y)) \tag{1}$$

defines a metric on X and that $d(x,y) = d_0(x,y)$ holds whenever x and y are such that $d(x,y)$ (or $d_0(x,y)$) is less than 1. It follows that d and d_0 determine the same topology on X and that X is complete under d_0 if and only if it is complete under d.

Recall that the *disjoint union* $\sum_\alpha X_\alpha$ of an indexed collection $\{X_\alpha\}$ of topological spaces is defined by letting the underlying set $\sum_\alpha X_\alpha$ be the disjoint union[1] of the X_α's and then declaring that a subset of $\sum_\alpha X_\alpha$ is open if and only if for each α its intersection with X_α is an open subset of X_α.

Proposition 8.1.3. *The disjoint union of a finite or infinite sequence of Polish spaces is Polish.*

Proof. Let $X_1, X_2, \ldots$ be Polish spaces, and let $\sum_n X_n$ be their disjoint union. For each n let D_n be a countable dense subset of X_n and let d_n be a complete metric on X_n. We can assume that $d_n(x,y) \le 1$ holds for each n and for all x and y in X_n (see Eq. (1)). Then $\sum_n D_n$ is a countable dense subset of $\sum_n X_n$, and

$$d(x,y) = \begin{cases} d_n(x,y) & \text{if } x, y \in X_n \text{ for some } n, \\ 1 & \text{if } x \in X_m \text{ and } y \in X_n, \text{ where } m \ne n \end{cases}$$

defines a complete metric that metrizes $\sum_n X_n$. $\square$

Proposition 8.1.4. *The product of a finite or infinite sequence of Polish spaces is Polish.*

Proof. Let $X_1, X_2, \ldots$ be a finite or infinite sequence of Polish spaces. We can assume that no X_n is empty. For each n let d_n be a complete metric that metrizes X_n and satisfies $d_n(x,y) \le 1$ for all x and y in X_n (see Eq. (1)). For points x and y in $\prod_n X_n$, with coordinates $x_1, x_2, \ldots$ and $y_1, y_2, \ldots$, respectively, let

$$d(x,y) = \sum_n \frac{1}{2^n} d_n(x_n, y_n).$$

It is easy to check that this defines a metric d on $\prod_n X_n$, that d metrizes the product topology on $\prod_n X_n$, and that $\prod_n X_n$ is complete under d.

---

[1]Let $\{Y_\alpha\}$ be an indexed collection of sets such that

(a) for each α the set Y_α has the same cardinality as the set X_α, and
(b) Y_{α_1} and Y_{α_2} are disjoint if $\alpha_1 \ne \alpha_2$

(for instance, one might let Y_α be $X_\alpha \times \{\alpha\}$). The *disjoint union* of the X_α's is defined to be the union of the Y_α's. (One generally thinks of the Y_α's as being identified with the corresponding X_α's.)

To prove the separability of $\prod_n X_n$, it is enough to construct a countable base for $\prod_n X_n$ (see D.10). For each n choose a countable base $\mathcal{U}_n$ for X_n (see D.32). Then the collection of subsets of $\prod_n X_n$ that have the form

$$U_1 \times \cdots \times U_N \times X_{N+1} \times X_{N+2} \times \cdots$$

for some N and some choice of sets U_n in $\mathcal{U}_n$, $n = 1, \ldots, N$, is the required base for $\prod_n X_n$.   □

Proposition 8.1.5. *Let X be a Polish space. Then a subspace of X is Polish if and only if it is a G_δ in X.*

Proof. First let $\{U_n\}$ be a sequence of open subsets of X and let $Y = \cap_n U_n$. Each U_n is Polish (Proposition 8.1.2), as is the product $\prod_n U_n$ (Proposition 8.1.4). Let Δ be the subset of $\prod_n U_n$ defined by

$$\Delta = \left\{ \{u_n\} \in \prod_n U_n : u_j = u_k \text{ for all } j, k \right\}.$$

Then Δ is a closed subset of $\prod_n U_n$, and so is Polish. Furthermore Y is homeomorphic to Δ via the map that takes an element y of Y to the sequence each term of which is y. Hence Y is Polish.

We turn to the converse. So suppose that Y is a subspace of X that is Polish. Let d be a metric for the topology of X, and let d_0 be a complete metric for the topology of Y. For each n let V_n be the union of those open subsets W of X that have diameter at most $1/n$ under d and for which $W \cap Y$ is nonempty and has diameter at most $1/n$ under d_0. Since d and d_0 induce the same topology on Y, every point in Y belongs to each V_n. Let us show that

$$Y = \overline{Y} \cap (\cap_n V_n). \tag{2}$$

We just noted that $Y \subseteq V_n$ holds for each n, and so, we have $Y \subseteq \overline{Y} \cap (\cap_n V_n)$. We turn to the reverse inclusion. Suppose that $x \in \overline{Y} \cap (\cap_n V_n)$. Since $x \in \cap_n V_n$, we can choose a sequence $\{W_n\}$ of open neighborhoods of x such that for each n the sets W_n and $Y \cap W_n$ have diameters (under d and d_0, respectively) at most $1/n$. Since $x \in \overline{Y}$, our sets W_n satisfy $W_n \cap Y \neq \varnothing$ for each n. Thus we can form a sequence $\{x_n\}$ by choosing (for each n) a point x_n in $W_n \cap Y$. Our conditions on the diameters of the sets W_n under d and d_0 imply that $\{x_n\}$ converges to x with respect to d and that it is a Cauchy sequence (in Y) with respect to d_0. Thus there is a point y in Y to which $\{x_n\}$ converges under d_0. Since d and d_0 metrize the same topology on Y, it follows $\{x_n\}$ also converges to y under d and hence that $x = y \in Y$. Thus $\overline{Y} \cap (\cap_n V_n) \subseteq Y$ and the proof of (2) is complete. Since each closed subset of X (in particular, $\overline{Y}$) is a G_δ in X (see D.28), relation (2) implies that Y is a G_δ in X.   □

Examples 8.1.6.

(a) Let X be a locally compact Hausdorff space that has a countable base for its topology. Its one-point compactification X^* also has a countable base (Lemma 7.1.14) and so is Polish (Example 8.1.1(c)). Proposition 8.1.2 now implies that X, as an open subset of X^*, is Polish.

(b) The space $\mathbb{N}^{\mathbb{N}}$ is, according to Proposition 8.1.4, Polish. We will often denote this space by $\mathcal{N}$. Its elements are, of course, sequences of positive integers. A typical such sequence will generally be denoted by $\{n_i\}$ or by $\mathbf{n}$ (the boldface $\mathbf{n}$ is a useful substitute for $\{n_i\}$ in complicated expressions).

For positive integers k and $n_1, \ldots, n_k$ we will denote by $\mathcal{N}(n_1, \ldots, n_k)$ the set of those elements $\{m_i\}$ of $\mathcal{N}$ that satisfy $m_i = n_i$ for $i = 1, \ldots, k$. It is easy to check that the family of all such sets is a countable base for $\mathcal{N}$. It is also easy to check that the collection of those elements $\{m_i\}$ of $\mathcal{N}$ that are eventually constant (that is, for which there is a positive integer k such that $m_i = m_k$ holds whenever $i > k$) is a countable dense subset of $\mathcal{N}$.

(c) Next consider the space $\mathcal{I}$ of irrational numbers in the interval $(0, 1)$, together with the topology it inherits from $\mathbb{R}$. The complement of $\mathcal{I}$ in $\mathbb{R}$ is an F_σ, and so $\mathcal{I}$ is a G_δ; thus Proposition 8.1.5 implies that $\mathcal{I}$ is Polish. It can be shown that $\mathcal{I}$ is homeomorphic to $\mathcal{N}$ (see Exercise 3 in Sect. 8.2).

(d) The space $\mathbb{Q}$ of rational numbers is not Polish (see Exercise 2).

(e) The space $\{0, 1\}^{\mathbb{N}}$, which consists of all sequences of zeroes and ones, is Polish (Proposition 8.1.4 or Example 8.1.1(c)). It can be shown that this space is homeomorphic to the Cantor set (see Exercise 1). $\qquad\square$

The spaces $\mathcal{N}$ and $\{0, 1\}^{\mathbb{N}}$ turn out to be very important in the development of the theory of Polish spaces and analytic sets.

We turn to some basic facts about the Borel subsets of Polish spaces.

Let $(X_1, \mathscr{A}_1)$, $(X_2, \mathscr{A}_2)$, ... be measurable spaces. The *product* of these measurable spaces is the measurable space $(\prod_n X_n, \prod_n \mathscr{A}_n)$ where $\prod_n \mathscr{A}_n$ is the σ-algebra on $\prod_n X_n$ that is generated by the sets that have the form

$$A_1 \times A_2 \times \cdots \times A_N \times X_{N+1} \times X_{N+2} \times \ldots \tag{3}$$

for some positive integer N and some choice of A_n in $\mathscr{A}_n$, $n = 1, \ldots, N$. For each i let π_i be the projection of $\prod_n X_n$ onto X_i. Then

$$\pi_i^{-1}(A) = X_1 \times \cdots \times X_{i-1} \times A \times X_{i+1} \times \ldots$$

holds for each subset A of X_i, and so π_i is measurable with respect to $\prod_n \mathscr{A}_n$ and $\mathscr{A}_i$. The set in display (3) is equal to $\cap_{i=1}^N \pi_i^{-1}(A_i)$; hence $\prod_n \mathscr{A}_n$ is the smallest σ-algebra on $\prod_n X_n$ that makes all the projections π_i measurable.

Proposition 8.1.7. *Let X_1, X_2, ... be a finite or infinite sequence of separable metrizable spaces. Then $\mathscr{B}(\prod_n X_n) = \prod_n \mathscr{B}(X_n)$.*

Proof. For each i consider the projection π_i of $\prod_n X_n$ onto X_i. Each such projection is continuous and so is measurable (Lemma 7.2.1) with respect to $\mathscr{B}(\prod_n X_n)$ and $\mathscr{B}(X_i)$. Since $\prod_n \mathscr{B}(X_n)$ is the smallest σ-algebra on $\prod_n X_n$ that makes these projections measurable (see the remarks above), it follows that $\prod_n \mathscr{B}(X_n) \subseteq \mathscr{B}(\prod_n X_n)$.

We turn to the reverse inclusion. For each n choose a countable base $\mathcal{U}_n$ for X_n (see D.32), and then let $\mathcal{U}$ be the collection of sets that have the form

$$U_1 \times \cdots \times U_N \times X_{N+1} \times \cdots$$

for some positive integer N and some choice of sets U_n in $\mathcal{U}_n$, for $n = 1, \ldots, N$. Then $\mathcal{U}$ is a countable base for $\prod_n X_n$, and $\mathcal{U} \subseteq \prod_n \mathscr{B}(X_n)$. Since each open subset of $\prod_n X_n$ is the union of a (necessarily countable) subfamily of $\mathcal{U}$, it follows that $\mathscr{B}(\prod_n X_n) \subseteq \prod_n \mathscr{B}(X_n)$. Thus $\mathscr{B}(\prod_n X_n) = \prod_n \mathscr{B}(X_n)$, and the proof is complete. $\square$

Let X and Y be sets, and let f be a function from X to Y. The *graph* of f, denoted by $\mathrm{gr}(f)$, is defined by

$$\mathrm{gr}(f) = \{(x,y) \in X \times Y : y = f(x)\}.$$

Proposition 8.1.8. *Let X and Y be separable metrizable spaces, and let $f : X \to Y$ be Borel measurable. Then the graph of f is a Borel subset of $X \times Y$.*

Proof. Let $F : X \times Y \to Y \times Y$ be the map that takes (x,y) to $(f(x),y)$. The Borel measurability of f implies that if $A, B \in \mathscr{B}(Y)$, then $F^{-1}(A \times B) \in \mathscr{B}(X) \times \mathscr{B}(Y)$; hence F is measurable with respect to $\mathscr{B}(X) \times \mathscr{B}(Y)$ and $\mathscr{B}(Y) \times \mathscr{B}(Y)$ (Proposition 2.6.2) and so with respect to $\mathscr{B}(X \times Y)$ and $\mathscr{B}(Y \times Y)$ (Proposition 8.1.7). Let $\Delta = \{(y_1, y_2) \in Y \times Y : y_1 = y_2\}$. Then Δ is a closed subset of $Y \times Y$ and $\mathrm{gr}(f) = F^{-1}(\Delta)$. It follows that $\mathrm{gr}(f)$ is a Borel subset of $X \times Y$. $\square$

Lemma 8.1.9. *Let $(X, \mathscr{A})$ be a measurable space, and let Y be a metrizable topological space. Then a function $f : X \to Y$ is measurable with respect to $\mathscr{A}$ and $\mathscr{B}(Y)$ if and only if for each continuous function $g : Y \to \mathbb{R}$ the function $g \circ f$ is $\mathscr{A}$-measurable.*

Proof. If f is measurable with respect to $\mathscr{A}$ and $\mathscr{B}(Y)$, then the measurability of $g \circ f$ for each continuous g follows from the measurability of g (Lemma 7.2.1), together with Proposition 2.6.1.

Now assume that for each continuous $g : Y \to \mathbb{R}$ the function $g \circ f$ is $\mathscr{A}$-measurable, and let d be a metric that metrizes Y. Suppose that U is an open subset of Y. Then there is a continuous function $g_U : Y \to \mathbb{R}$ such that

$$U = \{y \in Y : g_U(y) > 0\}$$

(if $U \neq Y$, define g_U by $g_U(y) = d(y, U^c)$; otherwise, let g_U be the constant function 1). The set $f^{-1}(U)$ is equal to

$$\{x \in X : (g_U \circ f)(x) > 0\}$$

and so belongs to $\mathscr{A}$. Since U was an arbitrary open subset of X, the measurability of f follows (Proposition 2.6.2). $\square$

Proposition 8.1.10. *Let $(X, \mathscr{A})$ be a measurable space, let Y be a metrizable topological space, and for each positive integer n let $f_n \colon X \to Y$ be measurable with respect to $\mathscr{A}$ and $\mathscr{B}(Y)$. If $\lim_n f_n(x)$ exists for each x in X, then the function $f \colon X \to Y$ given by $f(x) = \lim_n f_n(x)$ is measurable with respect to $\mathscr{A}$ and $\mathscr{B}(Y)$.*

Proof. Note that if $g \colon Y \to \mathbb{R}$ is continuous, then $g(f(x)) = \lim_n g(f_n(x))$ holds for each x in X. The proposition is now an immediate consequence of Lemma 8.1.9 and Proposition 2.1.5. □

Proposition 8.1.11. *Let $(X, \mathscr{A})$ be a measurable space, let Y be a Polish space, and for each positive integer n let $f_n \colon X \to Y$ be measurable with respect to $\mathscr{A}$ and $\mathscr{B}(Y)$. Let $C = \{x \in X : \lim_n f_n(x) \text{ exists}\}$. Then $C \in \mathscr{A}$. Furthermore, the map $f \colon C \to Y$ defined by $f(x) = \lim_n f_n(x)$ is measurable with respect to $\mathscr{A}$ and $\mathscr{B}(Y)$.*

Proof. Let d be a complete metric for Y. Then C is the set consisting of those x in X for which $\{f_n(x)\}$ is a Cauchy sequence in Y. For each positive integer n the set $\{(y_1, y_2) \in Y \times Y : d(y_1, y_2) < 1/n\}$ is an open subset of $Y \times Y$ and so belongs to $\mathscr{B}(Y) \times \mathscr{B}(Y)$ (Proposition 8.1.7). Thus for each i, j, and n the set $C(i, j, n)$ defined by

$$C(i, j, n) = \left\{ x \in X : d(f_i(x), f_j(x)) < \frac{1}{n} \right\}$$

belongs to $\mathscr{A}$. Since

$$C = \bigcap_n \bigcup_k \bigcap_{i \geq k} \bigcap_{j \geq k} C(i, j, n),$$

it follows that $C \in \mathscr{A}$. The measurability of f is now a consequence of Proposition 8.1.10, applied to the spaces $(C, \mathscr{A}_C)$ and Y (here $\mathscr{A}_C$ is the trace of $\mathscr{A}$ on C; see Exercise 1.5.11). □

We conclude this section with the following useful fact about measures on Polish spaces.

Proposition 8.1.12. *Every finite Borel measure on a Polish space is regular.*

Proof. Let X be a Polish space, let d be a complete metric for X, and let μ be a finite Borel measure on X. We can assume that X is not empty. Since each open subset of X is an F_σ in X (see D.28), Lemma 7.2.4 implies that each Borel subset A of X satisfies

$$\mu(A) = \inf\{\mu(U) : A \subseteq U \text{ and } U \text{ is open}\} \tag{4}$$

and

$$\mu(A) = \sup\{\mu(F) : F \subseteq A \text{ and } F \text{ is closed}\}. \tag{5}$$

We will strengthen (5) by showing that each Borel subset A of X satisfies

$$\mu(A) = \sup\{\mu(K) : K \subseteq A \text{ and } K \text{ is compact}\}. \tag{6}$$

First consider the case where $A = X$. Let $\{x_k\}$ be a sequence whose terms form a dense subset of X, and let ε be a positive number. For each positive integer n use Proposition 1.2.5 and the fact that X is the union of the open balls $B(x_k, 1/n)$, $k = 1$, $2, \ldots$, to choose a positive integer k_n such that

$$\mu\left(\bigcup_{k=1}^{k_n} B(x_k, 1/n)\right) > \mu(X) - \varepsilon/2^n.$$

Let $K = \bigcap_n \bigcup_{k=1}^{k_n} \overline{B(x_k, 1/n)}$. Then K is complete and totally bounded under the restriction of d to K, and so is compact (Theorem D.39). Furthermore

$$\mu(K^c) \leq \sum_n \mu\left(\left(\bigcup_{k=1}^{k_n} B(x_k, 1/n)\right)^c\right) < \sum_n \varepsilon/2^n = \varepsilon,$$

and so $\mu(K) > \mu(X) - \varepsilon$. Since ε is arbitrary, (6) follows in the case where $A = X$.

Now let A be an arbitrary Borel subset of X, and let ε be a positive number. Choose a compact set K such that $\mu(K) > \mu(X) - \varepsilon$, and use (5) to choose a closed subset F of A such that $\mu(F) > \mu(A) - \varepsilon$. Then $K \cap F$ is a compact subset of A, and $\mu(K \cap F) > \mu(A) - 2\varepsilon$. Since ε is arbitrary, A must satisfy (6). Thus μ is regular. $\qquad\square$

Exercises

1. Show that the map that takes the sequence $\{n_k\}$ to the number $\sum_k 2n_k/3^k$ is a homeomorphism of $\{0,1\}^{\mathbb{N}}$ onto the Cantor set.
2. Show that the set $\mathbb{Q}$ of rational numbers, with the topology it inherits as a subspace of $\mathbb{R}$, is not Polish. (Hint: Use the Baire category theorem, Theorem D.37.)
3. Let $(X, \mathscr{A})$ be a measurable space, let Y be a separable metrizable space, and let $f, g : X \to Y$ be measurable with respect to $\mathscr{A}$ and $\mathscr{B}(Y)$. Show that $\{x \in X : f(x) = g(x)\}$ belongs to $\mathscr{A}$.
4. Suppose that $\{X_n\}$ is a sequence of nonempty separable metrizable spaces and that, for each n, D_n is a countable dense subset of X_n. Give (rather explicitly) a countable dense subset of $\prod_n X_n$.
5. Let X be a Polish space, let $\{U_n\}$ be a sequence of open subsets of X, and let d be a complete metric for X. Construct a complete metric for $\bigcap_n U_n$; show directly that it has the required properties. (Hint: Examine the proofs of Propositions 8.1.2, 8.1.4, and 8.1.5.)
6. Let $C[0, +\infty)$ be the set of all continuous real-valued functions on the interval $[0, +\infty)$.

(a) Show that the formula

$$d(f,g) = \sup\{1 \wedge |f(t) - g(t)| : t \in [0, +\infty)\}$$

defines a metric on $C[0, +\infty)$.
(b) Suppose that f and $f_1, f_2, \ldots$ belong to $C[0, +\infty)$. Show that $\{f_k\}$ converges to f with respect to the metric in part (a) if and only if it converges to f uniformly on $[0, +\infty)$.
(c) Show that $C[0, +\infty)$, when endowed with the topology determined by the metric in part (a), is not separable and hence not Polish.

7. (a) Show that the formula

$$d(f,g) = \sum_n \frac{1}{2^n} \sup\{1 \wedge |f(t) - g(t)| : t \in [0, n]\}$$

defines a metric on the set $C[0, +\infty)$ (see Exercise 6).
(b) Suppose that f and $f_1, f_2, \ldots$ belong to $C[0, +\infty)$. Show that $\{f_k\}$ converges to f with respect to the metric in part (a) if and only if it converges to f uniformly on each compact subset of $[0, +\infty)$.
(c) Show that $C[0, +\infty)$ is complete and separable under the metric defined in part (a). (Hint: See Exercise 7.1.9.)

8. Prove Proposition 8.1.10 directly, without using continuous functions.

9. Suppose that in Proposition 8.1.11 the space Y were only required to be separable and metrizable. Show by example that the set C would not need to belong to $\mathscr{A}$.

10. Show that every finite Borel measure on $\mathbb{Q}$ is regular. (Recall that $\mathbb{Q}$ is not Polish; see Exercise 2.)

11. Show by example that a finite Borel measure on a separable metrizable space can fail to be regular. (Hint: Suppose that X is a subset of $\mathbb{R}$ that satisfies $\lambda^*(X) < +\infty$ but is not Lebesgue measurable. Consider the measure on $(X, \mathscr{B}(X))$ that results when the construction of Exercise 1.5.11 is applied to Lebesgue measure.)

12. Show that every separable metrizable space is homeomorphic to a subspace of the product space $[0, 1]^{\mathbb{N}}$ and that every Polish space is homeomorphic to a G_δ in $[0, 1]^{\mathbb{N}}$. (Hint: Let d be a metric for the separable metrizable space X, and let $\{x_n\}$ be a sequence whose terms form a dense subset of X. Consider the map from X to $[0, 1]^{\mathbb{N}}$ that takes the point x to the sequence whose nth term is $\min(1, d(x, x_n))$.)

13. Let $(X, \mathscr{A})$ be a measurable space, let Y be a Polish space, let A be a subset of X that might not belong to $\mathscr{A}$, and let $\mathscr{A}_A$ be the trace of $\mathscr{A}$ on A (see Exercise 1.5.11). Show that if $f : A \to Y$ is measurable with respect to $\mathscr{A}_A$ and $\mathscr{B}(Y)$, then f has an extension $F : X \to Y$ that is measurable with respect to $\mathscr{A}$ and $\mathscr{B}(Y)$.

14. Give a counterexample that shows that the metrizability of Y cannot be omitted in Proposition 8.1.10. (Hint: Let $(X, \mathscr{A})$ be $([0,1], \mathscr{B}([0,1]))$ and let Y be $[0,1]^{[0,1]}$ with the product topology. For each n let $f_n \colon X \to Y$ be the function that takes x to the element of Y (i.e., to the function from $[0,1]$ to $[0,1]$) given by $t \mapsto \max(0, 1 - n|t - x|)$.)

8.2   Analytic Sets

Let X be a Polish space. A subset A of X is *analytic* if there is a Polish space Z and a continuous function $f \colon Z \to X$ such that $f(Z) = A$.

We will soon see that every Borel subset of a Polish space is analytic, but that there are analytic sets that are not Borel.

Analytic sets are useful tools for the study of Borel sets and Borel measurable functions (see Sect. 8.3); they also possess measurability properties that make them useful in their own right (see Sects. 8.4 and 8.5). This section contains a few elementary properties of analytic sets, some techniques for constructing those continuous maps that will be needed later in this chapter, and a construction that provides an analytic set that is not Borel. The reader might well skip from Proposition 8.2.9 to Sect. 8.3 at a first reading, returning for the remaining results as they are needed.

Proposition 8.2.1. *Let X be a Polish space. Then each open subset, and each closed subset, of X is analytic.*

Proof. This is an immediate consequence of Proposition 8.1.2, together with the continuity of the standard injection of a subspace of X into X.   □

Proposition 8.2.2. *Let X be a Polish space, and let $A_1, A_2, \ldots$ be analytic subsets of X. Then $\cup_k A_k$ and $\cap_k A_k$ are analytic.*

Proof. For each k choose a Polish space Z_k and a continuous function $f_k \colon Z_k \to X$ such that $f(Z_k) = A_k$. Let Z be the disjoint union of the spaces $Z_1, Z_2, \ldots$, and define $f \colon Z \to X$ so that for each k it agrees on Z_k with f_k. Then Z is a Polish space (Proposition 8.1.3), f is a continuous function, and $f(Z) = \cup_k A_k$; hence $\cup_k A_k$ is analytic.

Next form the product space $\prod_k Z_k$, and let Δ consist of those sequences $\{z_k\}$ in $\prod_k Z_k$ such that $f_i(z_i) = f_j(z_j)$ holds for all i and j. Then Δ is a closed subspace of $\prod_k Z_k$ and so is Polish (Propositions 8.1.2 and 8.1.4). The set $\cap_k A_k$ is the image of Δ under the continuous function that takes the sequence $\{z_k\}$ to the point $f_1(z_1)$; hence it is analytic.   □

It should be noted that the complement of an analytic set is not necessarily analytic. In fact, the complement of an analytic set A is analytic if and only if A is Borel (see Proposition 8.2.3 and Corollary 8.3.3).

Proposition 8.2.3. *Let X be a Polish space. Then each Borel subset of X is analytic.*

The proof will depend on the following lemma. Because of later applications, this lemma is given in a slightly stronger form than is needed here.

Lemma 8.2.4. *Let X be a Hausdorff topological space. Then $\mathscr{B}(X)$ is the smallest family of subsets of X that*

(a) *contains the open and the closed subsets of X,*
(b) *is closed under the formation of countable intersections, and*
(c) *is closed under the formation of countable disjoint unions.*

Note that closure under complementation is not one of the conditions listed in Lemma 8.2.4.

Proof. Let $\mathscr{S}$ be the smallest collection of subsets of X that satisfies conditions (a), (b), and (c) of the lemma (why does such a smallest collection exist?), and let $\mathscr{S}_0 = \{A : A \in \mathscr{S} \text{ and } A^c \in \mathscr{S}\}$. It is clear that $\mathscr{S}_0 \subseteq \mathscr{S} \subseteq \mathscr{B}(X)$. Thus if we show that $\mathscr{S}_0$ is a σ-algebra that contains each open subset of X, it will follow that $\mathscr{S}_0 = \mathscr{S} = \mathscr{B}(X)$, and the proof will be complete.

It is immediate that $\mathscr{S}_0$ contains the open subsets of X and is closed under complementation. Now suppose that $\{A_n\}$ is a sequence of sets in $\mathscr{S}_0$. Then $\cup_n A_n$ is the union of the sets

$$A_1, A_1^c \cap A_2, A_1^c \cap A_2^c \cap A_3, \ldots;$$

these sets are disjoint and belong to $\mathscr{S}$, and so $\cup_n A_n$ must also belong to $\mathscr{S}$. Furthermore $(\cup_n A_n)^c$ is the intersection of a sequence (namely $\{A_n^c\}$) of sets in $\mathscr{S}$, and so belongs to $\mathscr{S}$. Consequently $\cup_n A_n$ belongs to $\mathscr{S}_0$. It follows that $\mathscr{S}_0$ is closed under the formation of countable unions. With this we have shown that $\mathscr{S}_0$ is a σ-algebra that contains the open subsets of X, and the proof of Lemma 8.2.4 is complete. $\square$

Proof of Proposition 8.2.3. Since the collection of analytic subsets of X satisfies conditions (a), (b), and (c) of Lemma 8.2.4 (see Propositions 8.2.1 and 8.2.2), it must include $\mathscr{B}(X)$. $\square$

Proposition 8.2.5. *Let $X_1, X_2, \ldots$ be a finite or infinite sequence of Polish spaces, and for each k let A_k be an analytic subset of X_k. Then $\prod_k A_k$ is an analytic subset of $\prod_k X_k$.*

Proof. If some A_k is empty, then $\prod_k A_k$ is empty and so is an analytic set. Otherwise for each k choose a Polish space Z_k and a continuous function $f_k : Z_k \to X_k$ such that $f_k(Z_k) = A_k$. Define a function $f : \prod_k Z_k \to \prod_k X_k$ by $f(\{z_k\}) = \{f_k(z_k)\}$. Then $\prod_k Z_k$ is Polish, f is continuous, and $f(\prod_k Z_k) = \prod_k A_k$. Thus $\prod_k A_k$ is analytic. $\square$

Proposition 8.2.6. *Let X and Y be Polish spaces, let A be an analytic subset of X, and let $f : A \to Y$ be Borel measurable (that is, measurable with respect to $\mathscr{B}(A)$ and $\mathscr{B}(Y)$). If A_1 and A_2 are analytic subsets of X and Y, respectively, then $f(A \cap A_1)$ and $f^{-1}(A_2)$ are analytic subsets of Y and X, respectively.*

Proof. Let π_Y be the projection of $X \times Y$ onto Y. Proposition 8.1.8 implies that $\mathrm{gr}(f) \in \mathscr{B}(A \times Y)$, and Lemma 7.2.2 then implies that there is a Borel subset B of $X \times Y$ such that $\mathrm{gr}(f) = B \cap (A \times Y)$. Hence $\mathrm{gr}(f) \cap (A_1 \times Y)$ is an analytic subset of $X \times Y$ (Propositions 8.2.2, 8.2.3, and 8.2.5) and so is the image of a Polish space (say Z) under a continuous map (say h). It follows that $f(A \cap A_1)$, since it is the projection of $\mathrm{gr}(f) \cap (A_1 \times Y)$ on Y, is the image of Z under the continuous map $\pi_Y \circ h$ and so is analytic. A similar argument shows that $f^{-1}(A_2)$ is analytic (note that it is the projection of $\mathrm{gr}(f) \cap (X \times A_2)$ on X). $\qquad\square$

We turn to the construction of some continuous functions that are useful in the study of Borel and analytic sets.

Proposition 8.2.7. *Each nonempty Polish space is the image of $\mathscr{N}$ under a continuous function.*

Proof. Let X be a nonempty Polish space, and let d be a complete metric for X. We begin by constructing a family $\{C(n_1, \ldots, n_k)\}$ of subsets of X, indexed by the set of all finite sequences $(n_1, \ldots, n_k)$ of positive integers, in such a way that

(a) $C(n_1, \ldots, n_k)$ is closed and nonempty,
(b) the diameter of $C(n_1, \ldots, n_k)$ is at most $1/k$,
(c) $C(n_1, \ldots, n_{k-1}) = \cup_{n_k} C(n_1, \ldots, n_k)$, and
(d) $X = \cup_{n_1} C(n_1)$.

We do this by induction on k.

First, suppose that $k = 1$, and let $\{x_{n_1}\}_{n_1=1}^{\infty}$ be a sequence whose terms form a dense subset of X. For each n_1 in $\mathbb{N}$ define $C(n_1)$ to be the closed ball with center x_{n_1} and radius $1/2$. Certainly each $C(n_1)$ is closed and nonempty and has diameter at most 1. Furthermore, $X = \cup_{n_1} C(n_1)$.

Now suppose that $k > 1$ and that $C(n_1, \ldots, n_{k-1})$ has already been chosen. It is easy to use a modification of the construction of the $C(n_1)$'s, now applied to $C(n_1, \ldots, n_{k-1})$ rather than to X, to produce sets $C(n_1, \ldots, n_k)$, $n_k = 1, 2, \ldots$, that satisfy conditions (a) through (c). With this, the inductive step in our construction is complete.

We turn to the construction of a continuous function that maps $\mathscr{N}$ onto X. Let $\mathbf{n} = \{n_k\}$ be an element of $\mathscr{N}$. It follows from (a), (b), and (c) above that $C(n_1)$, $C(n_1, n_2)$, $\ldots$ is a decreasing sequence of nonempty closed subsets of X whose diameters approach 0. Thus there is a unique element in the intersection of these sets (see Theorem D.35), and we can define a function $f : \mathscr{N} \to X$ by letting $f(\mathbf{n})$ be the unique member of $\cap_k C(n_1, \ldots, n_k)$. Note that if $\mathbf{m}$ and $\mathbf{n}$ are elements of $\mathscr{N}$ such that $m_i = n_i$ holds for $i = 1, \ldots, k$, then $d(f(\mathbf{m}), f(\mathbf{n})) \leq 1/k$. It follows that f is continuous. Finally, (c) and (d) above imply that for each x in X there is an element $\mathbf{n} = \{n_k\}$ of $\mathscr{N}$ such that $x \in \cap_k C(n_1, \ldots, n_k)$ and hence such that $x = f(\mathbf{n})$; thus f is surjective. $\qquad\square$

Corollary 8.2.8. *Each nonempty analytic subset of a Polish space is the image of $\mathscr{N}$ under some continuous function.*

Proof. If A is the image of the Polish space Z under the continuous function f and if Z is the image of $\mathcal{N}$ under the continuous function g (Proposition 8.2.7), then A is the image of $\mathcal{N}$ under $f \circ g$. $\square$

Proposition 8.2.9. *Let X be a Polish space. A subset A of X is analytic if and only if there is a closed subset of $\mathcal{N} \times X$ whose projection on X is A.*

Proof. The projection on X of a closed subset of $\mathcal{N} \times X$ is the image of a Polish space (see Propositions 8.1.2 and 8.1.4) under a continuous function (the projection), and so is analytic.

Now suppose that A is an analytic subset of X. If A is empty, then it is the projection of the empty subset of $\mathcal{N} \times X$. Otherwise there is a continuous function $f \colon \mathcal{N} \to X$ such that $f(\mathcal{N}) = A$ (Corollary 8.2.8). Then $\mathrm{gr}(f)$ is a closed subset of $\mathcal{N} \times X$ whose projection on X is A. $\square$

While the preceding material is fundamental, the following results will be used only occasionally in this book. The reader who does Exercises 1 and 5 and replaces the proof for Theorem 8.3.6 given below with the one sketched in Exercise 8.3.5 can skip everything from here through Corollary 8.2.14.

We need to recall a definition and a few facts before proving Proposition 8.2.10. A topological space is *zero dimensional* if its topology has a base that consists of sets that are both open and closed. Among the zero-dimensional spaces are the space of all rational numbers, the space of all irrational numbers, and each space that has a discrete topology. Note that a subspace of a zero-dimensional space is zero dimensional, that a product of zero-dimensional spaces is zero dimensional, and that the disjoint union of a collection of zero-dimensional spaces is zero dimensional. In particular, the spaces $\mathcal{N}$ and $\{0,1\}^{\mathbb{N}}$ are products of zero-dimensional spaces, and so are zero dimensional.

Proposition 8.2.10. *Each Borel subset of a Polish space is the image under a continuous injective map of some zero-dimensional Polish space.*

Proof. We begin by showing that each Polish space is the image under a continuous injective map of some zero-dimensional Polish space. First consider the interval $[0,1]$. It is the image of the space $\{0,1\}^{\mathbb{N}}$ under the map $F \colon \{0,1\}^{\mathbb{N}} \to [0,1]$ that takes the sequence $\{x_k\}$ to the number $\sum_k (x_k/2^k)$. Each number in $[0,1)$ that has two binary expansions (that is, each number in $(0,1)$ that is of the form $m/2^n$ for some m and n) is the image under F of two elements of $\{0,1\}^{\mathbb{N}}$; the remaining members of $[0,1]$ are images of only one element of $\{0,1\}^{\mathbb{N}}$. Thus if we remove a suitable countably infinite subset from $\{0,1\}^{\mathbb{N}}$, the remaining points form a space Z such that the restriction of F to Z is a bijection of Z onto $[0,1]$. Note that F is continuous, that Z is zero dimensional (it is a subspace of the zero-dimensional space $\{0,1\}^{\mathbb{N}}$), and that Z is Polish (its complement in $\{0,1\}^{\mathbb{N}}$ is countable, and so it is a G_δ in $\{0,1\}^{\mathbb{N}}$). Hence $[0,1]$ is the image of a zero-dimensional Polish space under a continuous injective map.

It follows that $[0,1]^{\mathbb{N}}$ is the image of the zero-dimensional Polish space $Z^{\mathbb{N}}$ under a continuous injective map.

Now suppose that X is an arbitrary Polish space. Recall (see Exercise 8.1.12) that there is a homeomorphism G of X onto a G_δ in $[0,1]^{\mathbb{N}}$. Let H be a continuous injective map of $Z^{\mathbb{N}}$ onto $[0,1]^{\mathbb{N}}$. Since $G(X)$ is a G_δ in $[0,1]^{\mathbb{N}}$, it follows that $H^{-1}(G(X))$ is a G_δ in $Z^{\mathbb{N}}$, and so is Polish. Let H_0 be the restriction of H to $H^{-1}(G(X))$. Then X is the image of the zero-dimensional Polish space $H^{-1}(G(X))$ under the continuous injective map $G^{-1} \circ H_0$.

We turn to the Borel subsets of X. Let $\mathscr{F}$ consist of those Borel subsets B of X for which there is a zero-dimensional Polish space Y and a continuous injective map $f: Y \to X$ such that $f(Y) = B$. According to the first part of this proof, $\mathscr{F}$ contains the open and the closed subsets of X (see Proposition 8.1.2), and an easy modification of the proof of Proposition 8.2.2 shows that $\mathscr{F}$ is closed under the formation of countable intersections and under the formation of countable disjoint unions. Thus Lemma 8.2.4 implies that $\mathscr{F} = \mathscr{B}(X)$.   □

See Theorem 8.3.7 for a rather powerful result that implies the converse of Proposition 8.2.10.

Let us make some preparations for the proof of our next major result, Proposition 8.2.13.

Lemma 8.2.11. *Let X be a zero-dimensional separable metric space, let U be an open and non-compact subset of X, and let ε be a positive number. Then U is the union of a countably infinite family of disjoint sets, each of which is nonempty, open, closed, and of diameter at most ε.*

Proof. Since U is open and not compact, there is a family $\mathscr{U}$ of open sets whose union is U, but that has no finite subfamily whose union is U. Let $\mathscr{V}$ be the collection of all subsets of X that are open, closed, of diameter at most ε, and included in some member of $\mathscr{U}$. Since X is zero dimensional, the set U is the union of the family $\mathscr{V}$. According to D.11, there is a countable subfamily $\mathscr{V}_0$ of $\mathscr{V}$ whose union is U. List the sets in $\mathscr{V}_0$ in a sequence $V_1, V_2, \ldots$, and consider the nonempty sets that appear in the sequence

$$V_1,\ V_1^c \cap V_2,\ V_1^c \cap V_2^c \cap V_3, \ldots.$$

These sets are open, closed, disjoint, and of diameter at most ε, and their union is U. There are infinitely many of them, since otherwise there would be a finite subfamily of $\mathscr{U}$ that would cover U.   □

Let X be a topological space and let A be a subset of X, possibly the entire space X. A point x of X is a *condensation point* of A if every open neighborhood of x contains uncountably many points of A.

Lemma 8.2.12. *Let X be a separable metrizable space, and let C be the set of condensation points of X. Then C is closed, and C^c is countable.*

Proof. Let $\mathscr{U}$ be a countable base for X (see D.32). Then x fails to belong to C if and only if there is a countable open set that belongs to $\mathscr{U}$ and contains x. Hence C^c is the union of a countable collection of countable open sets, and so C^c itself is countable and open.   □

Proposition 8.2.13. *Let X be a Polish space, and let B be an uncountable Borel subset of X. Then there is a continuous injective map $f: \mathcal{N} \to X$ such that $f(\mathcal{N}) \subseteq B$ and such that $B - f(\mathcal{N})$ is countable.*

Proof. According to Proposition 8.2.10, there exist a zero-dimensional Polish space Z and a continuous injective map $g: Z \to X$ such that $g(Z) = B$. Thus it will suffice to construct a continuous injective map $h: \mathcal{N} \to Z$ such that $Z - h(\mathcal{N})$ is countable, and then to define f to be $g \circ h$.

Let Z_0 be the collection of all points in Z that are condensation points of Z. Then Z_0 is Polish (Lemma 8.2.12) and zero dimensional. Since Z_0^c is countable, every point in Z_0 is a condensation point of Z_0 (and not just a condensation point of Z).

Suppose that d is a complete metric that metrizes Z_0. For each k we construct a family of sets, indexed by $\mathbb{N}^k$, as follows. Let us begin with the case where $k = 1$. Apply Lemma 8.2.11 to the space Z_0, letting ε be 1 and letting U consist of the points that remain when one point is removed from Z_0 (this is to guarantee that U is not compact). The sets provided by Lemma 8.2.11, say $A(n_1), n_1 = 1, 2, \ldots$, are disjoint, nonempty, open, closed, and of diameter at most 1, each of them consists entirely of condensation points of itself, and the union of these sets is Z_0 less a single point. We can repeat this construction over and over, for each k and $n_1, \ldots, n_{k-1}$ producing sets $A(n_1, \ldots, n_k), n_k = 1, 2, \ldots$, that are disjoint, nonempty, open, closed, and of diameter at most $1/k$, and are such that $\cup_{n_k} A(n_1, \ldots, n_k)$ is $A(n_1, \ldots, n_{k-1})$ less a single point.

Define $h: \mathcal{N} \to Z$ by letting $h(\mathbf{n})$ be the unique point in $\cap_k A(n_1, \ldots, n_k)$ (Theorem D.35). It is easy to check that h is continuous and injective and that $Z_0 - h(\mathcal{N})$ is the countably infinite set consisting of the points removed from Z_0 during the construction of the sets $A(n_1, \ldots, n_k)$. It follows that $Z - h(\mathcal{N})$ is countable. Thus the construction of h, and so of f, is complete. $\qquad\square$

The following is an interesting and well-known consequence of Proposition 8.2.13 (see also Exercise 1).

Corollary 8.2.14. *Each uncountable Borel subset of a Polish space includes a subset that is homeomorphic to $\{0, 1\}^{\mathbb{N}}$.*

Proof. Let X be a Polish space, and let A be an uncountable Borel subset of X. Proposition 8.2.13 provides a continuous injective map $f: \mathcal{N} \to X$ such that $f(\mathcal{N}) \subseteq A$. If we regard $\{0, 1\}^{\mathbb{N}}$ as a subspace of $\mathcal{N}$ in the natural way, then the restriction of f to $\{0, 1\}^{\mathbb{N}}$ is a homeomorphism of $\{0, 1\}^{\mathbb{N}}$ onto the subset $f(\{0, 1\}^{\mathbb{N}})$ of A (see D.17). $\qquad\square$

Let X be a set, and let $\mathscr{F}$ be a family of subsets of X. A subset A of $\mathcal{N} \times X$ is *universal* for $\mathscr{F}$ if the collection of sections $\{A_{\mathbf{n}} : \mathbf{n} \in \mathcal{N}\}$ is equal to $\mathscr{F}$.

Our goal now is to show that if X is Polish, then there is an analytic subset of $\mathcal{N} \times X$ that is universal for the class of analytic subsets of X. We will use such a universal set to construct an analytic set that is not a Borel set.

Lemma 8.2.15. *Let X be a separable metrizable space. Then there is an open subset of $\mathcal{N} \times X$ that is universal for the collection of open subsets of X and a closed subset of $\mathcal{N} \times X$ that is universal for the collection of closed subsets of X.*

Proof. Let $\mathscr{U}$ be a countable base for X, and let $\{U_n\}$ be an infinite sequence whose terms are the sets in $\mathscr{U}$, together with the empty set (the sequence may have repeated terms). Define a subset W of $\mathscr{N} \times X$ by

$$W = \{(\mathbf{n}, x) : x \in U_{n_k} \text{ for some } k\}$$

(recall that $\mathbf{n}$ is an abbreviation for $\{n_k\}$). For each n and each k the set $W(k, n)$ defined by

$$W(k, n) = \{\mathbf{n} \in \mathscr{N} : n_k = n\} \times U_n$$

is open, and so W, since it is equal to $\cup_k \cup_n W(k, n)$, is also open. For each $\mathbf{n}$ in $\mathscr{N}$ the section $W_{\mathbf{n}}$ is given by

$$W_{\mathbf{n}} = \bigcup_k U_{n_k};$$

hence W is universal for the collection of open subsets of X (recall the definition of the sequence $\{U_n\}$).

The complement of W is a closed subset of $\mathscr{N} \times X$ and is universal for the class of closed subsets of X. $\square$

Proposition 8.2.16. *Let X be a Polish space. Then there is an analytic subset of $\mathscr{N} \times X$ that is universal for the collection of analytic subsets of X.*

Proof. Use Lemma 8.2.15, applied to the space $\mathscr{N} \times X$, to choose a closed subset F of $\mathscr{N} \times \mathscr{N} \times X$ that is universal for the collection of closed subsets of $\mathscr{N} \times X$. Let A be the image of F under the map $(\mathbf{m}, \mathbf{n}, x) \mapsto (\mathbf{m}, x)$. Then A is analytic, and it is easy to check that for each $\mathbf{m}$ in $\mathscr{N}$ the section $A_{\mathbf{m}}$ is the projection on X of the corresponding section $F_{\mathbf{m}}$ of F. Since F is universal for the collection of closed subsets of $\mathscr{N} \times X$, Proposition 8.2.9 implies that the analytic subsets of X are exactly the projections on X of the sections $F_{\mathbf{m}}$. Thus A is universal for the collection of analytic subsets of X. $\square$

Corollary 8.2.17. *There is an analytic subset of $\mathscr{N}$ that is not a Borel set.*

Proof. According to Proposition 8.2.16, there is an analytic subset A of $\mathscr{N} \times \mathscr{N}$ that is universal for the collection of analytic subsets of $\mathscr{N}$. Let $S = \{\mathbf{n} \in \mathscr{N} : (\mathbf{n}, \mathbf{n}) \in A\}$. Then S is analytic, since it is the projection on $\mathscr{N}$ of the intersection of A with the diagonal $\{(\mathbf{m}, \mathbf{n}) \in \mathscr{N} \times \mathscr{N} : \mathbf{m} = \mathbf{n}\}$. Now suppose that S is a Borel set. Then S^c is a Borel set, and so is analytic (Proposition 8.2.3). Thus, since A is universal, there is an element $\mathbf{n}_0$ of $\mathscr{N}$ such that $S^c = A_{\mathbf{n}_0}$. Let us consider whether $\mathbf{n}_0$ belongs to S or to S^c. If $\mathbf{n}_0 \in S$, then by the definition of S we have $(\mathbf{n}_0, \mathbf{n}_0) \in A$ and so $\mathbf{n}_0 \in A_{\mathbf{n}_0} = S^c$, which is impossible. A similar argument shows that if $\mathbf{n}_0 \in S^c$, then $\mathbf{n}_0 \in S$. In either case we have a contradiction, and so we must reject the assumption that S is a Borel set. $\square$

One can use Corollary 8.2.17 to show that each uncountable Polish space has an analytic subset that is not a Borel set; see Exercise 6.

Exercises

1. (a) Let A be an uncountable analytic subset of the Polish space X. Show that A has a subset that is homeomorphic to $\{0,1\}^{\mathbb{N}}$. (Hint: Let $f\colon \mathcal{N} \to X$ be a continuous function such that $f(\mathcal{N}) = A$. Choose a subset S of $\mathcal{N}$ such that the restriction of f to S is a bijection of S onto A (why does such a set exist?), and let S_0 consist of the points in S that are condensation points of S. Modify the proof of Proposition 8.2.7 so as to produce a continuous function $g\colon \{0,1\}^{\mathbb{N}} \to \mathcal{N}$ such that $f \circ g\colon \{0,1\}^{\mathbb{N}} \to X$ is injective.)

   (b) Conclude that each uncountable analytic subset of a Polish space has the cardinality of the continuum.

2. Let X be an uncountable Polish space. Show that the collection of analytic subsets of X and the collection of Borel subsets of X have the cardinality of the continuum. (Hint: Use Proposition 8.2.9 or 8.2.16.)

3. (a) Let X be a nonempty zero-dimensional Polish space such that each nonempty open subset of X is uncountable and not compact. Show that X is homeomorphic to $\mathcal{N}$. (Hint: Modify the proof of Proposition 8.2.7, and use Lemma 8.2.11.)

   (b) Conclude that the space $\mathcal{I}$ of irrational numbers in the interval $(0,1)$ is homeomorphic to $\mathcal{N}$.

4. Show that each nonempty Polish space is the image of $\mathcal{N}$ under a continuous open[2] map. (Hint: Modify the construction of the sets $C(n_1,\dots,n_k)$ in the proof of Proposition 8.2.7, replacing condition (a) with the requirement that each $C(n_1,\dots,n_k)$ be nonempty and open and adding the requirement that for each $n_1, \dots, n_k, n_{k+1}$ the closure of $C(n_1,\dots,n_{k+1})$ be included in $C(n_1,\dots,n_k)$.)

5. Show that if the phrase "zero-dimensional" is omitted from the statement of Proposition 8.2.10, then a much simpler proof can be given. (Hint: Use Lemma 8.2.4.)

6. Show that if X is an uncountable Polish space, then there is an analytic subset of X that is not a Borel set. (Hint: Use Proposition 8.2.13 and Corollary 8.2.17. One can avoid Proposition 8.2.13 by using Theorem 8.3.6.)

7. In this and the following two exercises, we study a generalization of the sequences $\mathscr{F}$, $\mathscr{F}_\sigma$, $\mathscr{F}_{\sigma\delta}$, ... and $\mathscr{G}$, $\mathscr{G}_\delta$, $\mathscr{G}_{\delta\sigma}$, ... introduced in Sect. 1.1. Suppose that X is a metrizable space. For each countable ordinal α we define collections $\mathscr{F}_\alpha(X)$ and $\mathscr{G}_\alpha(X)$ of subsets of X as follows. Let $\mathscr{F}_0(X)$ be the collection of all closed subsets of X, and let $\mathscr{G}_0(X)$ be the collection of all open subsets of X. Once $\mathscr{F}_\alpha(X)$ and $\mathscr{G}_\alpha(X)$ are defined, let $\mathscr{F}_{\alpha+1}(X)$ and $\mathscr{G}_{\alpha+1}(X)$ be given by[3]

---

[2] Suppose that X and Y are topological spaces. A function $f\colon X \to Y$ is *open* if for each open subset U of X the set $f(U)$ is an open subset of Y.

[3] Recall that each ordinal α can be written in a unique way in the form $\alpha = \beta + n$, where β is either zero or a limit ordinal and where n is finite. The ordinal α is called *even* if n is even and *odd* if n is odd.

$$\mathscr{F}_{\alpha+1}(X) = \begin{cases} (\mathscr{F}_\alpha(X))_\sigma & \text{if } \alpha \text{ is even,} \\ (\mathscr{F}_\alpha(X))_\delta & \text{if } \alpha \text{ is odd,} \end{cases}$$

and

$$\mathscr{G}_{\alpha+1}(X) = \begin{cases} (\mathscr{G}_\alpha(X))_\delta & \text{if } \alpha \text{ is even,} \\ (\mathscr{G}_\alpha(X))_\sigma & \text{if } \alpha \text{ is odd.} \end{cases}$$

Finally, if α is a limit ordinal, let $\mathscr{F}_\alpha(X)$ be $(\cup_{\beta<\alpha}\mathscr{F}_\beta(X))_\delta$ and let $\mathscr{G}_\alpha(X)$ be $(\cup_{\beta<\alpha}\mathscr{G}_\beta(X))_\sigma$.

(a) Show that for each α the sets that belong to $\mathscr{G}_\alpha(X)$ are exactly those whose complements belong to $\mathscr{F}_\alpha(X)$.

(b) Show that for each α and each A in $\mathscr{G}_\alpha(X)$ (or in $\mathscr{F}_\alpha(X)$) the set A^c belongs to $\mathscr{G}_{\alpha+1}(X)$ (or to $\mathscr{F}_{\alpha+1}(X)$).

(c) Show that $\mathscr{B}(X) = \cup_\alpha \mathscr{G}_\alpha(X) = \cup_\alpha \mathscr{F}_\alpha(X)$.

(d) Suppose that Y is also a metrizable space and that $f: X \to Y$ is continuous. Show that for each α and each A in $\mathscr{G}_\alpha(Y)$ (or in $\mathscr{F}_\alpha(Y)$) the set $f^{-1}(A)$ belongs to $\mathscr{G}_\alpha(X)$ (or to $\mathscr{F}_\alpha(X)$).

8. Suppose that X is an uncountable Polish space. We already know that the collection of Borel subsets of X has the cardinality of the continuum (see Exercise 2). Here you are not to use that result, but rather to use transfinite induction to show that each $\mathscr{G}_\alpha(X)$ has the cardinality of the continuum, that each $\mathscr{F}_\alpha(X)$ has the cardinality of the continuum, and that $\mathscr{B}(X)$ has the cardinality of the continuum.

9. (a) Show that if X is a Polish space, then for each countable ordinal α there is a set in $\mathscr{G}_\alpha(\mathcal{N} \times X)$ (or in $\mathscr{F}_\alpha(\mathcal{N} \times X)$) that is universal for $\mathscr{G}_\alpha(X)$ (or for $\mathscr{F}_\alpha(X)$). (Hint: Use transfinite induction. Lemma 8.2.15 provides a beginning. Next suppose that $\alpha > 0$. Let $\varphi: \mathcal{N} \to \mathcal{N}^\mathbb{N}$ be a continuous surjection, and let $\varphi_k(\mathbf{n})$, $k = 1, 2, \ldots$, be the components (in $\mathcal{N}$) of the element $\varphi(\mathbf{n})$ of $\mathcal{N}^\mathbb{N}$. If α is a limit ordinal, let $\{\alpha_k\}$ be an enumeration of the ordinals less than α; otherwise, let $\{\alpha_k\}$ be the sequence each of whose terms is the immediate predecessor of α. For each k choose a set A_k in $\mathscr{G}_{\alpha_k}(\mathcal{N} \times X)$ that is universal for $\mathscr{G}_{\alpha_k}(X)$; then define sets $B_1, B_2, \ldots$ by

$$B_k = \{(\mathbf{n}, x) \in \mathcal{N} \times X : (\varphi_k(\mathbf{n}), x) \in A_k\}.$$

Show that the set B defined by

$$B = \begin{cases} \cup_k B_k & \text{if } \alpha \text{ is even,} \\ \cap_k B_k & \text{if } \alpha \text{ is odd,} \end{cases}$$

belongs to $\mathscr{G}_\alpha(\mathcal{N} \times X)$ and is universal for $\mathscr{G}_\alpha(X)$. Finally, use part (a) of Exercise 7.)

(b) Show that there is no set in $\mathscr{B}(\mathcal{N} \times \mathcal{N})$ that is universal for $\mathscr{B}(\mathcal{N})$. (Hint: Modify the proof of Corollary 8.2.17.)

(c) Suppose that X is an uncountable Polish space and that there is a bijection $F: \mathcal{N} \to X$ such that both F and F^{-1} are Borel measurable (such a bijection always exists; see Theorem 8.3.6). Show that there is no set in $\mathcal{B}(\mathcal{N} \times X)$ that is universal for $\mathcal{B}(X)$. Also show that no two of the sets $\mathcal{G}_0(X)$, $\mathcal{F}_0(X)$, $\ldots, \mathcal{G}_\alpha(X), \mathcal{F}_\alpha(X), \ldots, \mathcal{B}(X)$ are equal.

10. Let X be a Polish space, and let Y be a metrizable space. Show that if $A \in \mathcal{B}(X)$ and if $f: A \to Y$ is Borel measurable, then $f(A)$ is separable. (Hint: Let d be a metric for Y, and suppose that $f(A)$ is not separable. Choose a positive number ε and an uncountable subset C of $f(A)$ such that $d(x,y) \geq \varepsilon$ holds for each pair x, y of points in C; then choose a function $g: C \to A$ such that $y = f(g(y))$ holds for each y in C (check that C and g exist). Show that each subset of $g(C)$ is analytic, and then use Exercises 1 and 2 to derive a contradiction.)

8.3   The Separation Theorem and Its Consequences

This section is devoted to a fundamental technical fact about analytic sets (Theorem 8.3.1) and to some of its applications. The reader should take particular note of Theorems 8.3.6 and 8.3.7.

Let X be a Polish space, and let A_1 and A_2 be disjoint subsets of X. Then A_1 and A_2 can be *separated by Borel sets* if there are disjoint Borel subsets B_1 and B_2 of X such that $A_1 \subseteq B_1$ and $A_2 \subseteq B_2$.

Theorem 8.3.1. *Let X be a Polish space, and let A_1 and A_2 be disjoint analytic subsets of X. Then A_1 and A_2 can be separated by Borel sets.*

Proof. Let us begin by showing that

(a) if $C_1, C_2, \ldots$, and D are subsets of X such that for each n the sets C_n and D can be separated by Borel sets, then $\cup_n C_n$ and D can be separated by Borel sets, and

(b) if $E_1, E_2, \ldots$, and $F_1, F_2, \ldots$ are subsets of X such that for each m and n the sets E_m and F_n can be separated by Borel sets, then $\cup_m E_m$ and $\cup_n F_n$ can be separated by Borel sets.

First consider assertion (a). For each n choose disjoint Borel sets G_n and H_n such that $C_n \subseteq G_n$ and $D \subseteq H_n$. Then $\cup_n G_n$ and $\cap_n H_n$ are disjoint Borel sets that include $\cup_n C_n$ and D, respectively. Hence assertion (a) is proved.

Next consider assertion (b). Assertion (a) implies that for each m the sets E_m and $\cup_n F_n$ can be separated by Borel sets. Another application of assertion (a) now implies that $\cup_m E_m$ and $\cup_n F_n$ can be separated by Borel sets.

We turn to the proof of the theorem itself. So suppose that A_1 and A_2 are disjoint analytic subsets of X. Since the empty set can clearly be separated from an arbitrary subset of X by Borel sets, we can assume that neither A_1 nor A_2 is empty. Thus there are continuous functions $f, g: \mathcal{N} \to X$ such that $f(\mathcal{N}) = A_1$ and $g(\mathcal{N}) = A_2$ (Corollary 8.2.8). Suppose that A_1 and A_2 cannot be separated by Borel sets; we will derive a contradiction.

Recall (see Example 8.1.6(b)) that for positive integers k and $n_1, \ldots, n_k$ the set $\mathscr{N}(n_1, \ldots, n_k)$ is defined by

$$\mathscr{N}(n_1, \ldots, n_k) = \{\mathbf{m} \in \mathscr{N} : m_i = n_i \text{ for } i = 1, \ldots, k\}.$$

Since $A_1 = \cup_{m_1} f(\mathscr{N}(m_1))$ and $A_2 = \cup_{n_1} g(\mathscr{N}(n_1))$, assertion (b) above implies that there are positive integers m_1 and n_1 such that $f(\mathscr{N}(m_1))$ and $g(\mathscr{N}(n_1))$ cannot be separated by Borel sets. Likewise, since $f(\mathscr{N}(m_1)) = \cup_{m_2} f(\mathscr{N}(m_1, m_2))$ and $g(\mathscr{N}(n_1)) = \cup_{n_2} g(\mathscr{N}(n_1, n_2))$, there are positive integers m_2 and n_2 such that $f(\mathscr{N}(m_1, m_2))$ and $g(\mathscr{N}(n_1, n_2))$ cannot be separated by Borel sets. By continuing in this manner we can construct sequences $\mathbf{m} = \{m_i\}$ and $\mathbf{n} = \{n_i\}$ such that for each k the sets $f(\mathscr{N}(m_1, \ldots, m_k))$ and $g(\mathscr{N}(n_1, \ldots, n_k))$ cannot be separated by Borel sets. The points $f(\mathbf{m})$ and $g(\mathbf{n})$ must be equal, since otherwise they could be separated with open sets, which, by the continuity of f and g, would separate $f(\mathscr{N}(m_1, \ldots, m_k))$ and $g(\mathscr{N}(n_1, \ldots, n_k))$ for all large k. However, since $f(\mathbf{m}) \in A_1$ and $g(\mathbf{n}) \in A_2$, the equality of $f(\mathbf{m})$ and $g(\mathbf{n})$ contradicts the disjointness of A_1 and A_2. So we must conclude that A_1 and A_2 can be separated with Borel sets, and with this the proof is complete. $\square$

Corollary 8.3.2. *Let X be a Polish space, and let A_1, A_2, $\ldots$ be disjoint analytic subsets of X. Then there are disjoint Borel subsets B_1, B_2, $\ldots$ of X such that $A_n \subseteq B_n$ holds for each n.*

Proof. For each positive integer n the set $\cup_{m \neq n} A_m$ is analytic, and so we can use Theorem 8.3.1 and the disjointness of A_n and $\cup_{m \neq n} A_m$ to choose a Borel set C_n such that $A_n \subseteq C_n$ and $\cup_{m \neq n} A_m \subseteq C_n^c$. Now define the Borel sets B_1, B_2, $\ldots$ by letting B_n be equal to $C_n - (\cup_{m \neq n} C_m)$. $\square$

Corollary 8.3.3. *Let X be a Polish space, and let A be a subset of X. If both A and A^c are analytic, then A is Borel.*

Proof. According to Theorem 8.3.1 there are disjoint Borel subsets B_1 and B_2 of X such that $A \subseteq B_1$ and $A^c \subseteq B_2$. It follows immediately that $A = B_1$ and $A^c = B_2$, and hence that A is Borel. $\square$

Proposition 8.3.4. *Let X and Y be Polish spaces, let A be a Borel subset of X, and let f be a function from A to Y. Then f is Borel measurable if and only if its graph is a Borel subset of $X \times Y$.*

Proof. Proposition 8.1.8 implies that if f is Borel measurable, then $\mathrm{gr}(f)$ is a Borel subset of $A \times Y$ and hence (Lemma 7.2.2) of $X \times Y$. Now consider the converse. Suppose that $\mathrm{gr}(f)$ is a Borel subset of $X \times Y$ and that B is a Borel subset of Y. Then $\mathrm{gr}(f) \cap (X \times B)$ and $\mathrm{gr}(f) \cap (X \times B^c)$ are Borel, and hence analytic, subsets of $X \times Y$. Thus the projections of these sets on X are analytic. But these projections are $f^{-1}(B)$ and $f^{-1}(B^c)$, respectively. Furthermore the sets $f^{-1}(B)$ and $f^{-1}(B^c)$ are disjoint, and so, by Theorem 8.3.1, there are Borel sets B_1 and B_2 that separate them. It is easy to check that $f^{-1}(B)$ is equal to $A \cap B_1$ and so is a Borel set. Since B was an arbitrary Borel subset of Y, the measurability of f follows. $\square$

Proposition 8.3.5. *Let X and Y be Polish spaces, let A be a Borel subset of X, let $f: A \to Y$ be Borel measurable, and let $B = f(A)$. If f is injective and if[4] $B \in \mathscr{B}(Y)$, then f^{-1} is Borel measurable.*

Proof. Note that $\mathrm{gr}(f^{-1})$ is the image of $\mathrm{gr}(f)$ under the homeomorphism $(x, y) \mapsto (y, x)$ of $X \times Y$ onto $Y \times X$; hence $\mathrm{gr}(f^{-1})$ is a Borel subset of $Y \times X$ if and only if $\mathrm{gr}(f)$ is a Borel subset of $X \times Y$. Now apply Proposition 8.3.4 twice, once to conclude that $\mathrm{gr}(f)$ is a Borel subset of $X \times Y$ and once to conclude that f^{-1} is Borel measurable.   □

Let $(X, \mathscr{A})$ and $(Y, \mathscr{B})$ be measurable spaces. A bijection $f: X \to Y$ is an *isomorphism* if f is measurable with respect to $\mathscr{A}$ and $\mathscr{B}$ and f^{-1} is measurable with respect to $\mathscr{B}$ and $\mathscr{A}$. Equivalently, the bijection f is an isomorphism if the subsets A of X that belong to $\mathscr{A}$ are exactly those for which $f(A)$ belongs to $\mathscr{B}$. The spaces $(X, \mathscr{A})$ and $(Y, \mathscr{B})$ are *isomorphic* if there exists such an isomorphism. We will also call subsets X_0 and Y_0 of X and Y *isomorphic* if the spaces[5] $(X_0, \mathscr{A}_{X_0})$ and $(Y_0, \mathscr{B}_{Y_0})$ are isomorphic. In case $(X, \mathscr{B}(X))$ and $(Y, \mathscr{B}(Y))$ are Polish spaces, together with their Borel σ-algebras, we will often use the term *Borel isomorphism* instead of isomorphism.

The concept of a Borel isomorphism is a natural one; it is especially important because of the following easy-to-state but nontrivial result.[6]

Theorem 8.3.6. *Let A and B be Borel subsets of Polish spaces. Then A and B are Borel isomorphic if and only if they have the same cardinality. Furthermore, the cardinality of each uncountable Borel subset of a Polish space is that of the continuum.*

Proof. If A and B are isomorphic, then they certainly have the same cardinality. We turn to the converse.

Suppose that A and B have the same cardinality. If these sets are finite or countably infinite, then each of their subsets is a Borel set, and each bijection between them is an isomorphism; hence A and B are isomorphic.

Now suppose that A and B are uncountable. Note that we are simply assuming that A and B are uncountable; we are not *assuming* that they have the same cardinality. Proposition 8.2.13 says that there are continuous injective maps $f: \mathscr{N} \to A$ and $g: \mathscr{N} \to B$ such that $A - f(\mathscr{N})$ and $B - g(\mathscr{N})$ are at most countably infinite. Since they are countable, the sets $A - f(\mathscr{N})$ and $B - g(\mathscr{N})$ are Borel sets; thus $f(\mathscr{N})$ and $g(\mathscr{N})$ are also Borel sets, and (see Proposition 8.3.5) f and g are Borel isomorphisms of $\mathscr{N}$ onto $f(\mathscr{N})$ and $g(\mathscr{N})$, respectively. Thus $g \circ f^{-1}$ is a Borel isomorphism of $f(\mathscr{N})$ onto $g(\mathscr{N})$. Now let I be a countably infinite subset of $f(\mathscr{N})$, and let h be a bijection of the countably infinite set $I \cup (A - f(\mathscr{N}))$ onto

---

[4] We will see (Theorem 8.3.7) that the injectivity and measurability of f imply that $B \in \mathscr{B}(Y)$.

[5] Of course $\mathscr{A}_{X_0}$ and $\mathscr{B}_{Y_0}$ are the traces of $\mathscr{A}$ and $\mathscr{B}$ on X_0 and Y_0 (see Exercise 1.5.11).

[6] See Exercise 5 for a proof of Theorem 8.3.6 that does not depend on Proposition 8.2.13 or 8.3.5.

the countably infinite set $g(f^{-1}(I)) \cup (B - g(\mathcal{N}))$. It is easy to check that the map that agrees with $g \circ f^{-1}$ on $f(\mathcal{N}) - I$ and with h on $I \cup (A - f(\mathcal{N}))$ is a Borel isomorphism of A onto B.

In particular, each uncountable Borel subset of a Polish space is Borel isomorphic to $\mathbb{R}$, and so has the cardinality of the continuum. $\square$

It follows from Theorem 8.3.6 that a Borel subset of a Polish space is Borel isomorphic to $\mathbb{R}$, to the set $\mathbb{N}$ of all positive integers, to the set $\{1, 2, \ldots, n\}$ for some positive integer n, or to $\varnothing$.

We now show that the hypothesis that $f(A)$ belongs to $\mathscr{B}(Y)$ can be removed from Proposition 8.3.5.

Theorem 8.3.7. *Let X and Y be Polish spaces, let A be a Borel subset of X, and let $f : A \to Y$ be Borel measurable and injective. Then $f(A)$ is a Borel subset of Y.*

The proof of this result will depend on the following lemma.

Lemma 8.3.8. *Let X and Y be Polish spaces, let A be a nonempty Borel subset of X, and let $f : A \to Y$ be Borel measurable and injective. Then there is a Borel measurable function $g : Y \to X$ such that $g(Y) \subseteq A$ and such that $g(f(x)) = x$ holds at each x in A.*

Proof. Let d be a metric for X, and let $\bar{x}$ be an element of A (we will hold $\bar{x}$ fixed throughout this proof). For each positive integer n we define a function $g_n : Y \to X$ as follows. Choose a finite or countably infinite partition $\{A_{n,k}\}_k$ of A into nonempty Borel subsets of diameter at most $1/n$, and in each $A_{n,k}$ choose a point $x_{n,k}$. The sets $f(A_{n,k})$, $k = 1, 2, \ldots$, are disjoint and analytic (Proposition 8.2.6), and so we can choose disjoint Borel sets $B_{n,k}$, $k = 1, 2, \ldots$, such that $f(A_{n,k}) \subseteq B_{n,k}$ holds for each k (Corollary 8.3.2). Now define $g_n : Y \to X$ by letting $g_n(y) = x_{n,k}$ if $y \in B_{n,k}$ and letting $g_n(y) = \bar{x}$ if $y \notin (\cup_k B_{n,k})$. It is easy to check that each g_n is Borel measurable. Define $g : Y \to A$ by letting $g(y) = \lim_n g_n(y)$ if the limit exists and belongs to A and letting $g(y) = \bar{x}$ otherwise. Proposition 8.1.11 implies that g is Borel measurable. If $x \in A$, then $d(x, g_n(f(x))) \leq 1/n$ holds for each n, and so $g(f(x)) = x$. Thus g is the required function. $\square$

Proof of Theorem 8.3.7. We can certainly assume that A is not empty. According to Lemma 8.3.8 there is a Borel measurable function $g : Y \to X$ such that $g(Y) \subseteq A$ and such that $g(f(x)) = x$ holds at each x in A. It is easy to check that

$$f(A) = \{y \in Y : f(g(y)) = y\}.$$

Thus Exercise 8.1.3, applied to the functions $y \mapsto f(g(y))$ and $y \mapsto y$, implies that $f(A) \in \mathscr{B}(Y)$. $\square$

Exercises

1. Let X and Y be Polish spaces, and let $f: X \to Y$ be a function whose graph is an analytic subset of $X \times Y$. Show that f is Borel measurable.

2. Let X and Y be uncountable Polish spaces. Show that the cardinality of the collection of Borel measurable functions from X to Y is that of the continuum.

3. Show that there is a Lebesgue measurable function $f: \mathbb{R} \to \mathbb{R}$ such that no real-valued (as opposed to $[-\infty, +\infty]$-valued) Borel measurable function f_1 satisfies $f(x) \le f_1(x)$ at each x in $\mathbb{R}$. Thus the $[-\infty, +\infty]$-valued functions f_0 and f_1 in Proposition 2.2.5 cannot necessarily be replaced with real-valued functions, even if the function f is real-valued. (Hint: Let K be the Cantor set. According to the preceding exercise, we can choose a bijection $x \mapsto g_x$ of K onto the set of real-valued Borel functions on K. Define $f: \mathbb{R} \to \mathbb{R}$ by

$$f(x) = \begin{cases} g_x(x) + 1 & \text{if } x \in K, \\ 0 & \text{otherwise,} \end{cases}$$

and check that f meets the requirements above.)

4. Let X be a Polish space, let μ be a Borel measure on X such that $\mu(X) = 1$, and let λ be Lebesgue measure on the Borel subsets of $[0, 1]$. Show that there is a Borel measurable function $f: [0, 1] \to X$ such that $\mu = \lambda f^{-1}$. (Hint: This is easy if X is finite or countably infinite. Otherwise use Theorem 8.3.6, together with either Exercise 2.6.6 or Proposition 10.1.15.)

5. Give an alternate proof of the isomorphism theorem for Borel sets (Theorem 8.3.6) by supplying the details missing from the following outline. (This proof depends neither on the separation theorem and its consequences nor on Proposition 8.2.13.)

   (a) Show that every Borel subset of a Polish space is Borel isomorphic to a Borel subset of $\{0, 1\}^{\mathbb{N}}$. (Hint: Begin by showing that the interval $[0, 1]$ is Borel isomorphic to a Borel subset of $\{0, 1\}^{\mathbb{N}}$ (consider binary expansions). From this conclude that $[0, 1]^{\mathbb{N}}$ is Borel isomorphic to a Borel subset of $(\{0, 1\}^{\mathbb{N}})^{\mathbb{N}}$ and hence to a Borel subset of $\{0, 1\}^{\mathbb{N}}$. Finally, use Exercise 8.1.12.)

   (b) Show that each uncountable Borel subset of a Polish space has a Borel subset that is Borel isomorphic to $\{0, 1\}^{\mathbb{N}}$. (Hint: Use Corollary 8.2.14 or, to avoid Proposition 8.2.13, Exercise 8.2.1.)

   (c) (A version of the Schröder–Bernstein theorem for Borel sets—see item A.7 in Appendix A.) Suppose that X and Y are Polish spaces, that A and B are Borel subsets of X and Y, respectively, that A is Borel isomorphic to a Borel subset of B, and that B is Borel isomorphic to a Borel subset of A. Show that A and B are Borel isomorphic to one another. (Hint: Let f and g be Borel isomorphisms of A and B onto Borel subsets of B and A, respectively. Define sequences $\{A_n\}_{n=0}^{\infty}$ and $\{B_n\}_{n=0}^{\infty}$ inductively by $A_0 = A$, $B_0 = B$,

$A_{n+1} = g(B_n)$, and $B_{n+1} = f(A_n)$. Show that

$$h(x) = \begin{cases} f(x) & \text{if } x \in \cap_0^\infty A_n \text{ or } x \in \cup_0^\infty (A_{2n} - A_{2n+1}), \\ g^{-1}(x) & \text{if } x \in \cup_0^\infty (A_{2n+1} - A_{2n+2}) \end{cases}$$

gives a Borel isomorphism $h \colon A \to B$. See the proof of Proposition G.2 in Appendix G for a more detailed description of the construction of the function h.)

8.4   The Measurability of Analytic Sets

Let $(X, \mathscr{A})$ be a measurable space, and let μ be a measure on $(X, \mathscr{A})$. Recall that in Sect. 1.5 we defined the completion of $\mathscr{A}$ under μ to be the collection $\mathscr{A}_\mu$ of subsets A of X for which there are sets E and F that belong to $\mathscr{A}$ and satisfy the relations $E \subseteq A \subseteq F$ and $\mu(F - E) = 0$. The sets in $\mathscr{A}_\mu$ are often called μ-*measurable*.

We also defined the outer measure $\mu^*(A)$ and the inner measure $\mu_*(A)$ of an arbitrary subset A of X by

$$\mu^*(A) = \inf\{\mu(B) : A \subseteq B \text{ and } B \in \mathscr{A}\} \tag{1}$$

and

$$\mu_*(A) = \sup\{\mu(B) : B \subseteq A \text{ and } B \in \mathscr{A}\}. \tag{2}$$

We saw that a set A such that $\mu^*(A) < +\infty$ belongs to $\mathscr{A}_\mu$ if and only if $\mu_*(A) = \mu^*(A)$, that $\mathscr{A}_\mu$ is a σ-algebra on X, and that the restriction of μ^* (or of μ_*) to $\mathscr{A}_\mu$ is a measure on $\mathscr{A}_\mu$, which is called the completion of μ and is denoted by $\overline{\mu}$. It is easy to see that $\overline{\mu}$ is the only measure on $\mathscr{A}_\mu$ that agrees on $\mathscr{A}$ with μ.

We can now state the main result of this section.

Theorem 8.4.1. *Let X be a Polish space, and let μ be a finite Borel measure on X. Then every analytic subset of X is μ-measurable.*

For the proof we need the following lemma.

Lemma 8.4.2. *Let $(X, \mathscr{A})$ be a measurable space, let μ be a finite measure on $(X, \mathscr{A})$, and let μ^* be defined by Eq. (1). If $\{A_n\}$ is an increasing sequence of subsets of X, then*

$$\mu^*\left(\bigcup_n A_n\right) = \lim_n \mu^*(A_n).$$

Proof. The monotonicity of μ^* implies that the limit $\lim_n \mu^*(A_n)$ exists and satisfies $\lim_n \mu^*(A_n) \le \mu^*(\cup_n A_n)$. We need to verify the reverse inequality. Let ε be a positive number, and for each positive integer n use (1) to choose a set B_n that belongs to $\mathscr{A}$, includes A_n, and satisfies $\mu(B_n) \le \mu^*(A_n) + \varepsilon$. By replacing B_n

with $\cap_{j=n}^{\infty} B_j$, we can assume that the sequence $\{B_n\}$ is increasing. Proposition 1.2.5 implies that $\mu(\cup_n B_n) = \lim_n \mu(B_n)$, and so we have

$$\mu^*\left(\bigcup_n A_n\right) \leq \mu\left(\bigcup_n B_n\right) = \lim_n \mu(B_n) \leq \lim_n \mu^*(A_n) + \varepsilon.$$

Since ε is arbitrary, the proof of the lemma is complete.   □

Proof of Theorem 8.4.1. Let A be an analytic subset of X. We will show that A is μ-measurable by showing that $\mu_*(A) = \mu^*(A)$, and we will do this by producing, for an arbitrary positive ε, a compact subset K of A such that $\mu(K) \geq \mu^*(A) - \varepsilon$.

We can certainly assume that A is nonempty. Thus we can choose a continuous function $f \colon \mathcal{N} \to X$ such that $f(\mathcal{N}) = A$ (Corollary 8.2.8). We need some notation. For positive integers k and $n_1, \ldots, n_k$ let $\mathscr{L}(n_1, \ldots, n_k)$ be the set of those elements $\mathbf{m}$ of $\mathcal{N}$ that satisfy $m_i \leq n_i$ for $i = 1, \ldots, k$. We will construct an element $\mathbf{n} = \{n_i\}$ of $\mathcal{N}$ such that

$$\mu^*(f(\mathscr{L}(n_1, \ldots, n_k))) > \mu^*(A) - \varepsilon \tag{3}$$

holds for each k. We begin by choosing the first term n_1 of the sequence $\mathbf{n}$. Note that $\{\mathscr{L}(n_1)\}_{n_1=1}^{\infty}$ is an increasing sequence of sets whose union is $\mathcal{N}$, and so $\{f(\mathscr{L}(n_1))\}_{n_1=1}^{\infty}$ is an increasing sequence of sets whose union is A. Thus $\mu^*(A) = \lim_{n_1} \mu^*(f(\mathscr{L}(n_1)))$ (Lemma 8.4.2), and so we can pick a positive integer n_1 such that $\mu^*(f(\mathscr{L}(n_1))) > \mu^*(A) - \varepsilon$. Since $\mathscr{L}(n_1) = \cup_{n_2} \mathscr{L}(n_1, n_2)$, a similar argument produces a positive integer n_2 such that $\mu^*(f(\mathscr{L}(n_1, n_2))) > \mu^*(A) - \varepsilon$. Continuing in this way we obtain a sequence $\mathbf{n} = \{n_k\}$ of positive integers such that (3) holds for each k. Now let $L = \cap_k \mathscr{L}(n_1, \ldots, n_k)$. Then L is equal to

$$\{\mathbf{m} \in \mathcal{N} : m_i \leq n_i \text{ for each } i\}$$

and so is compact (see D.20 or D.42); it follows that the set K defined by $K = f(L)$ is a compact subset of A. We will show that $\mu(K) \geq \mu^*(A) - \varepsilon$.

Let us begin by showing that

$$K = \bigcap_k f(\mathscr{L}_k)^-, \tag{4}$$

where for each k we have abbreviated $\mathscr{L}(n_1, \ldots, n_k)$ by $\mathscr{L}_k$. Since it is clear that $K \subseteq \cap_k f(\mathscr{L}_k)^-$, we turn to the reverse inclusion. Let d be a metric for the topology of X. Suppose that x is a member of $\cap_k f(\mathscr{L}_k)^-$. For each k we can choose an element $\mathbf{m}_k$ of $\mathscr{L}_k$ such that $d(f(\mathbf{m}_k), x) \leq 1/k$. Note that for each i the ith components of the terms of $\{\mathbf{m}_k\}$ form a bounded subset of $\mathbb{N}$; hence the terms of $\{\mathbf{m}_k\}$ form a relatively compact[7] subset of $\mathcal{N}$, and we can choose a convergent subsequence of $\{\mathbf{m}_k\}$. Let $\mathbf{m}$ be the limit of this subsequence. It is easy to check that $\mathbf{m} \in \cap_k \mathscr{L}_k$ and that $f(\mathbf{m}) = x$. Hence $\cap_k f(\mathscr{L}_k)^- \subseteq K$, and (4) is proved.

---

[7] A subset of a Hausdorff space is *relatively compact* if its closure is compact.

For each k the set $f(\mathscr{L}_k)^-$ is closed and includes $f(\mathscr{L}_k)$; hence (see (3))

$$\mu(f(\mathscr{L}_k)^-) \geq \mu^*(f(\mathscr{L}_k)) > \mu^*(A) - \varepsilon. \tag{5}$$

Furthermore the sequence $\{f(\mathscr{L}_k)^-\}$ is decreasing, and so (4), (5), and Proposition 1.2.5 imply that

$$\mu(K) = \lim_k \mu(f(\mathscr{L}_k)^-) \geq \mu^*(A) - \varepsilon. \tag{6}$$

Thus $\mu_*(A) \geq \mu^*(A) - \varepsilon$ and so, since ε was arbitrary, the proof is complete. □

Let $(X, \mathscr{A})$ be a measurable space. A subset of X is *universally measurable* (with respect to $(X, \mathscr{A})$) if it is μ-measurable for every finite measure μ on $(X, \mathscr{A})$. Let $\mathscr{A}_*$ be the family of all universally measurable subsets of X. Then $\mathscr{A}_* = \cap_\mu \mathscr{A}_\mu$, where μ ranges over the family of finite measures on $(X, \mathscr{A})$; hence $\mathscr{A}_*$ is a σ-algebra. It is easy to check that for each finite measure μ on $(X, \mathscr{A})$ there is a unique measure on $(X, \mathscr{A}_*)$ that agrees on $\mathscr{A}$ with μ.

Now assume that X is a Polish space. The *universally measurable* subsets of X are those that are universally measurable with respect to $(X, \mathscr{B}(X))$.

Theorem 8.4.1 can now be reformulated as follows.

Corollary 8.4.3. *Every analytic subset of a Polish space is universally measurable.*

Proof. This corollary is simply a restatement of Theorem 8.4.1. □

Let X be an uncountable Polish space. Corollary 8.4.3 implies that the σ-algebra of universally measurable subsets of X includes the σ-algebra generated by the analytic subsets of X. These σ-algebras contain the complements of the analytic sets, and so contain some nonanalytic sets; thus the collection of universally measurable subsets of X is larger than the collection of analytic subsets of X, which in turn is larger than $\mathscr{B}(X)$.

Suppose that X and Y are Polish spaces. Note that if C is a Borel (or even analytic) subset of $X \times Y$, then the projection of C on X is analytic and so is universally measurable. This fact has the following useful generalization, in which the space X is not required to be Polish.

Proposition 8.4.4. *Let $(X, \mathscr{A})$ be a measurable space, let Y be a Polish space, and let C be a subset of $X \times Y$ that belongs to the product σ-algebra $\mathscr{A} \times \mathscr{B}(Y)$. Then the projection of C on X is universally measurable with respect to $(X, \mathscr{A})$.*

The proof depends on the following two lemmas; they will allow us to replace X with a suitable Polish space.

Lemma 8.4.5. *Let $(X, \mathscr{A})$, Y, and C be as in Proposition 8.4.4, and let $Z = \{0,1\}^{\mathbb{N}}$. Then there exist a function $h \colon X \to Z$ and a subset D of $Z \times Y$ such that*

(a) *h is measurable with respect to $\mathscr{A}$ and $\mathscr{B}(Z)$,*
(b) *$D \in \mathscr{B}(Z \times Y)$, and*
(c) *$C = H^{-1}(D)$, where $H \colon X \times Y \to Z \times Y$ is the map that takes (x,y) to $(h(x),y)$.*

Proof. Recall that $\mathscr{A} \times \mathscr{B}(Y)$ is generated by the family of all rectangles $A \times B$ such that $A \in \mathscr{A}$ and $B \in \mathscr{B}(Y)$. It follows from Exercise 1.1.7 that this family of rectangles has a countable subfamily $\mathscr{C}$ such that $C \in \sigma(\mathscr{C})$. Let $A_1 \times B_1, A_2 \times B_2,$ ... be the rectangles belonging to $\mathscr{C}$, and define $h \colon X \to Z$ by letting $h(x)$ be the sequence $\{\chi_{A_n}(x)\}$. Since the subsets $E_1, E_2, \ldots$ of Z defined by

$$E_k = \{\{n_i\} \in Z : n_k = 1\}$$

generate $\mathscr{B}(Z)$ (see Proposition 8.1.7) and since $h^{-1}(E_k) = A_k$ holds for each k, Proposition 2.6.2 implies that h is measurable with respect to $\mathscr{A}$ and $\mathscr{B}(Z)$. Define $H \colon X \times Y \to Z \times Y$ by $H(x,y) = (h(x), y)$, and let

$$\mathscr{F} = \{H^{-1}(D) : D \in \mathscr{B}(Z \times Y)\}.$$

Then $\mathscr{F}$ is a σ-algebra on $X \times Y$ that contains each $A_i \times B_i$. Hence $\sigma(\mathscr{C}) \subseteq \mathscr{F}$, and so $C \in \mathscr{F}$. With this the lemma is proved. □

Lemma 8.4.6. *Let $(X, \mathscr{A})$ and $(Y, \mathscr{B})$ be measurable spaces, and let $f \colon X \to Y$ be measurable with respect to $\mathscr{A}$ and $\mathscr{B}$. Then f is measurable with respect to the σ-algebras $\mathscr{A}_*$ and $\mathscr{B}_*$ of universally measurable sets.*

Proof. Suppose that $B_* \in \mathscr{B}_*$. We need to show that $f^{-1}(B_*) \in \mathscr{A}_*$. Let μ be a finite measure on $\mathscr{A}$. Recall that μf^{-1} is the measure on $\mathscr{B}$ defined by $\mu f^{-1}(B) = \mu(f^{-1}(B))$. Since B_* belongs to $\mathscr{B}_*$, it belongs to $\mathscr{B}_{\mu f^{-1}}$, and so there are sets B_0 and B_1 in $\mathscr{B}$ that satisfy $B_0 \subseteq B_* \subseteq B_1$ and $\mu f^{-1}(B_1 - B_0) = 0$. Then the sets $f^{-1}(B_0)$ and $f^{-1}(B_1)$ belong to $\mathscr{A}$ and satisfy $f^{-1}(B_0) \subseteq f^{-1}(B_*) \subseteq f^{-1}(B_1)$ and $\mu(f^{-1}(B_1) - f^{-1}(B_0)) = 0$. Hence $f^{-1}(B_*) \in \mathscr{A}_\mu$. Since μ was arbitrary, we conclude that $f^{-1}(B_*)$ belongs to $\mathscr{A}_*$, and the proof is complete. □

Proof of Proposition 8.4.4. Let $(X, \mathscr{A})$, Y, and C be as in the statement of Proposition 8.4.4, and construct h, H, and D as in Lemma 8.4.5. Let π_X be the projection of $X \times Y$ onto X, and let π_Z be the projection of $Z \times Y$ onto Z. Corollary 8.4.3 implies that $\pi_Z(D)$ is a universally measurable subset of Z, and so Lemma 8.4.6 implies that $h^{-1}(\pi_Z(D))$ is a universally measurable subset of X. Thus, in view of the easily verified relation

$$\pi_X(C) = \pi_X(H^{-1}(D)) = h^{-1}(\pi_Z(D)),$$

$\pi_X(C)$ is universally measurable. □

Exercises

1. Let $(X, \mathscr{A})$ be a measurable space.

   (a) Show that a function $f \colon X \to [-\infty, +\infty]$ is $\mathscr{A}_*$-measurable if and only if for each finite measure μ on $(X, \mathscr{A})$ there are $\mathscr{A}$-measurable functions

$f_0, f_1 : X \to [-\infty, +\infty]$ that satisfy $f_0 \le f \le f_1$ everywhere on X and are equal to one another μ-almost everywhere on X. (Hint: See Proposition 2.2.5.)

(b) Show that if $f : X \to [-\infty, +\infty]$ is $\mathscr{A}_*$-measurable and if the functions f_0 and f_1 in part (a) can be chosen independently of μ, then f is $\mathscr{A}$-measurable.

2. Let $(X, \mathscr{A})$ be a measurable space.

   (a) Show that $(\mathscr{A}_*)_* = \mathscr{A}_*$.
   (b) Show that if μ is a finite measure on $(X, \mathscr{A})$, then $(\mathscr{A}_\mu)_* = \mathscr{A}_\mu$.

3. Show that there is a Lebesgue measurable subset of $\mathbb{R}$ that is not universally measurable.

4. Show that each uncountable Polish space has a subset that is not universally measurable. (Hint: Use Theorem 8.3.6.)

5. Show by example that Lemma 8.4.2 would not be valid if μ^* were allowed to be an arbitrary outer measure on X.

6. Let $(X, \mathscr{A})$ and $(Y, \mathscr{B})$ be measurable spaces, and let K be a kernel from $(X, \mathscr{A})$ to $(Y, \mathscr{B})$ such that $\sup\{K(x, Y) : x \in X\}$ is finite (see Exercise 2.4.7). For each x in X let $B \mapsto \overline{K}(x, B)$ be the restriction to $\mathscr{B}_*$ of the completion of the measure $B \mapsto K(x, B)$. Finally, for each finite measure μ on $(X, \mathscr{A})$ let μK be the measure on $(Y, \mathscr{B})$ defined by $(\mu K)(B) = \int K(x, B) \mu(dx)$ (see part (a) of Exercise 2.4.7).

   (a) Show that $(x, B) \mapsto \overline{K}(x, B)$ is a kernel from $(X, \mathscr{A}_*)$ to $(Y, \mathscr{B}_*)$. (Hint: Use Exercise 1. Let B belong to $\mathscr{B}_*$, and let μ be a finite measure on $(X, \mathscr{A})$. Choose sets B_0 and B_1 that belong to $\mathscr{B}$ and satisfy the conditions $B_0 \subseteq B \subseteq B_1$ and $(\mu K)(B_1 - B_0) = 0$; then consider the functions $x \mapsto K(x, B_0)$ and $x \mapsto K(x, B_1)$.)
   (b) Suppose that μ is a finite measure on $(X, \mathscr{A})$ and that $\overline{\mu}$ and $\overline{\mu K}$ are the restrictions to $\mathscr{A}_*$ and $\mathscr{B}_*$ of the completions of μ and μK. Show that $\overline{\mu K} = \overline{\mu} \overline{K}$ (that is, show that

$$\overline{\mu K}(B) = \int \overline{K}(x, B) \overline{\mu}(dx)$$

holds for each B in $\mathscr{B}_*$.)

7. Let X be a Hausdorff space. A *capacity* on X is a function $I : \mathscr{P}(X) \to [-\infty, +\infty]$ such that

   (i) if $A \subseteq B \subseteq X$, then $I(A) \le I(B)$,
   (ii) each increasing sequence $\{A_n\}$ of subsets of X satisfies $I(\cup_n A_n) = \lim_n I(A_n)$, and
   (iii) each decreasing sequence $\{K_n\}$ of compact subsets of X satisfies $I(\cap_n K_n) = \lim_n I(K_n)$.

   A subset A of X is *I-capacitable* if

$$I(A) = \sup\{I(K) : K \subseteq A \text{ and } K \text{ is compact}\}.$$

Show that if the Hausdorff space X is Polish and if I is a capacity on X, then every relatively compact analytic subset of X is I-capacitable. (Hint: Modify the proof of Theorem 8.4.1.)

8.(a) Show that the space $\mathscr{I}$ of irrational numbers in the interval $(0,1)$ is not σ-compact.

(b) Let X be a Polish space that is not σ-compact, and define $I \colon \mathscr{P}(X) \to [-\infty, +\infty]$ by letting $I(A)$ be 0 if A is included in some σ-compact set and letting $I(A)$ be 1 otherwise. Show that

  (i) I is a capacity on X, and
  (ii) there is an analytic subset of X that is not I-capacitable.

8.5   Cross Sections

Let X and Y be Polish spaces, let A be a Borel or analytic subset of $X \times Y$, and let A_0 be the projection of A on X. It is sometimes useful to have a measurable function from A_0 to Y whose graph is a subset of A. Of course, the axiom of choice guarantees the existence of a function from A_0 to Y whose graph is a subset of A, but it asserts nothing about the measurability of that function. We will see below, however, that the theory of analytic sets allows one to construct such a function in a way that makes it measurable with respect to the σ-algebra of universally measurable subsets of X.

One should note that this construction does not always produce a Borel measurable function. In fact, there is a Borel subset A of $[0, 1] \times [0, 1]$ such that

(a) the image of A under the projection $(x, y) \mapsto x$ is all of $[0, 1]$, and
(b) there is no Borel function from $[0, 1]$ to $[0, 1]$ whose graph is a subset of A

(see Blackwell [10] or Novikoff [94]).

We will need a few more facts about $\mathscr{N}$ for our proof of Theorem 8.5.3. Let $\leq$ be lexicographic order on $\mathscr{N}$. In other words, we define a relation $<$ on $\mathscr{N}$ by declaring that $\mathbf{m} < \mathbf{n}$ holds if

(a) $\mathbf{m} \neq \mathbf{n}$ and
(b) $m_{i_0} < n_{i_0}$, where i_0 is the smallest of those positive integers i for which $m_i \neq n_i$;

then we declare that $\mathbf{m} \leq \mathbf{n}$ means that $\mathbf{m} < \mathbf{n}$ or $\mathbf{m} = \mathbf{n}$. It is easy to check that $\leq$ is a linear order on $\mathscr{N}$.

Recall also (see Example 8.1.6(b)) that $\mathscr{N}(n_1, \ldots, n_k)$ is the set of all elements of $\mathscr{N}$ whose first k elements are $n_1, \ldots, n_k$.

Lemma 8.5.1. *Each nonempty closed subset of $\mathscr{N}$ has a smallest element.*

Proof. Let C be a nonempty closed subset of $\mathscr{N}$. We define a sequence $\{n_j\}$ of positive integers as follows. Let

$$n_1 = \inf\{k \in \mathbb{N} : k = m_1 \text{ for some } \mathbf{m} \text{ in } C\}.$$

Next suppose that $n_1, \ldots, n_j$ have been chosen, and let

$$n_{j+1} = \inf\{k \in \mathbb{N} : k = m_{j+1} \text{ for some } \mathbf{m} \text{ in } C \cap \mathcal{N}(n_1, \ldots, n_j)\}.$$

It is easy to check that the sequence $\mathbf{n} = \{n_j\}$ produced by continuing in this way is the required element of C.   □

Lemma 8.5.2. *Each subset of $\mathcal{N}$ that has the form*

$$\{\mathbf{m} \in \mathcal{N} : \mathbf{m} < \mathbf{n}\} \tag{1}$$

for some $\mathbf{n}$ in $\mathcal{N}$ is open. The collection of all subsets of $\mathcal{N}$ of the form (1) generates $\mathcal{B}(\mathcal{N})$.

Proof. Note that $\{\mathbf{m} \in \mathcal{N} : \mathbf{m} < \mathbf{n}\}$ is equal to $\bigcup_{k=1}^{\infty} \bigcup_{j<n_k} \mathcal{N}(n_1, \ldots, n_{k-1}, j)$, and so, as the union of a collection of open sets, is open.

Let $\mathcal{F}$ be the σ-algebra generated by the sets of the form (1). Since each set of the form (1) is open, $\mathcal{F}$ is included in $\mathcal{B}(\mathcal{N})$. On the other hand, it is easy to check that for each k and each $n_1, \ldots, n_k$ the set $\mathcal{N}(n_1, \ldots, n_k)$ is the intersection of

$$\{\mathbf{m} \in \mathcal{N} : \mathbf{m} < (n_1, n_2, \ldots, n_{k-1}, n_k + 1, 1, 1, \ldots)\}$$

with the complement of

$$\{\mathbf{m} \in \mathcal{N} : \mathbf{m} < (n_1, n_2, \ldots, n_{k-1}, n_k, 1, 1, \ldots)\}$$

and so belongs to $\mathcal{F}$. Since the sets $\mathcal{N}(n_1, \ldots, n_k)$ form a countable base for $\mathcal{N}$ (see Example 8.1.6(b)), they generate $\mathcal{B}(\mathcal{N})$, and it follows that $\mathcal{B}(\mathcal{N}) \subseteq \mathcal{F}$. Thus $\mathcal{B}(\mathcal{N}) = \mathcal{F}$.   □

For the following theorem we will, as usual, let $\mathcal{B}(X)_*$ denote the σ-algebra of universally measurable subsets of the Polish space X; we will also let $\mathscr{A}(X)$ denote the σ-algebra generated by the analytic subsets of X.

Theorem 8.5.3. *Let X and Y be Polish spaces, let A be an analytic subset of $X \times Y$, and let A_0 be the projection of A on X. Then there is a function $f \colon A_0 \to Y$ such that*

(a) *the graph of f is a subset of A, and*
(b) *f is measurable with respect to $\mathscr{A}(X)$ and $\mathcal{B}(Y)$ and with respect to $\mathcal{B}(X)_*$ and $\mathcal{B}(Y)$.*

Proof. We can assume that A is not empty, and so we can choose a continuous function $g \colon \mathcal{N} \to X \times Y$ such that $g(\mathcal{N}) = A$ (Corollary 8.2.8). Let π_X and π_Y be the projections of $X \times Y$ onto X and Y, respectively. Then $\pi_X \circ g \colon \mathcal{N} \to X$ is continuous, and $(\pi_X \circ g)(\mathcal{N}) = \pi_X(A) = A_0$. Hence if $x \in A_0$, then $(\pi_X \circ g)^{-1}(\{x\})$ is a nonempty closed subset of $\mathcal{N}$, and so has a smallest member (Lemma 8.5.1). Define $h \colon A_0 \to \mathcal{N}$ by letting $h(x)$ be this smallest member of $(\pi_X \circ g)^{-1}(\{x\})$. Let $f = \pi_Y \circ g \circ h$. It is easy to check that f is a function from A_0 to Y whose graph is included in A. Since g and π_Y are continuous and since $\mathscr{A}(X) \subseteq \mathscr{B}(X)_*$ (Corollary 8.4.3), the measurability of f will follow if we prove that h is measurable with respect to $\mathscr{A}(X)$ and $\mathscr{B}(\mathscr{N})$.

Note that if for each $\mathbf{n}$ in $\mathscr{N}$ we let

$$U_{\mathbf{n}} = \{\mathbf{m} \in \mathscr{N} : \mathbf{m} < \mathbf{n}\},$$

then $h^{-1}(U_{\mathbf{n}})$ is equal to $(\pi_X \circ g)(U_{\mathbf{n}})$, and so, as the image of the open set $U_{\mathbf{n}}$ under the continuous map $\pi_X \circ g$, is analytic. Since the sets $U_{\mathbf{n}}$ generate $\mathscr{B}(\mathscr{N})$ (Lemma 8.5.2), the measurability of h with respect to $\mathscr{A}(X)$ and $\mathscr{B}(\mathscr{N})$ follows (Proposition 2.6.2). Thus f is measurable, and the proof is complete. $\square$

Theorem 8.5.3 implies the following result, in which X is no longer required to be Polish. Recall that if $(X, \mathscr{A})$ is an arbitrary measurable space, then $\mathscr{A}_*$ is the σ-algebra of sets that are universally measurable with respect to $(X, \mathscr{A})$.

Corollary 8.5.4. *Let $(X, \mathscr{A})$ be a measurable space, let Y be a Polish space, let C be a subset of $X \times Y$ that belongs to the σ-algebra $\mathscr{A} \times \mathscr{B}(Y)$, and let C_0 be the projection of C on X. Then there is a function $f : C_0 \to Y$ such that*

(a) *the graph of f is a subset of C, and*
(b) *f is measurable with respect to $\mathscr{A}_*$ and $\mathscr{B}(Y)$.*

Proof. Let Z, h, H, and D be as in Lemma 8.4.5, and let D_0 be the projection of D on Z. Note that $C_0 = h^{-1}(D_0)$. According to Theorem 8.5.3 there is a function $f_0 : D_0 \to Y$ that is measurable with respect to $\mathscr{B}(Z)_*$ and $\mathscr{B}(Y)$ and whose graph is a subset of D. Define $f : C_0 \to Y$ by $f(x) = f_0(h(x))$. The fact that $C = H^{-1}(D)$ implies that the graph of f is included in C, and Lemma 8.4.6 implies that h is measurable with respect to $\mathscr{A}_*$ and $\mathscr{B}(Z)_*$ and hence that f is measurable with respect to $\mathscr{A}_*$ and $\mathscr{B}(Y)$. $\square$

Exercises

1. Show by example that the Polish space Y in Proposition 8.4.4 and Corollary 8.5.4 cannot be replaced with an arbitrary measurable space $(Y, \mathscr{B})$. (Hint: Let $(X, \mathscr{A})$ be $(\mathbb{R}, \mathscr{B}(\mathbb{R}))$, let Y be a subset of $\mathbb{R}$ that is not Lebesgue measurable, and let $\mathscr{B}$ be the trace of $\mathscr{B}(\mathbb{R})$ on Y. For Proposition 8.4.4 consider the subset $\{(x, y) : x = y\}$ of $X \times Y$.)

2. Let $(X, \mathscr{A})$ be a measurable space, let Y be a Polish space, and let C be a subset of $X \times Y$ such that

   (i) for each x in X the section C_x is closed and nonempty, and
   (ii) for each open subset U of Y the set $\{x \in X : C_x \cap U \neq \varnothing\}$ belongs to $\mathscr{A}$.

   Show that there is a function $f : X \to Y$ such that
   (a) f is measurable with respect to $\mathscr{A}$ and $\mathscr{B}(Y)$, and

(b) the graph of f is included in C.

(Hint: Let d be a complete metric for Y, and let D be a countable dense subset of Y. Choose a sequence $\{f_n\}$ of $\mathscr{A}$-measurable functions from X to D such that $d(f_n(x), C_x) < 1/2^n$ and $d(f_n(x), f_{n+1}(x)) < 1/2^n$ hold for all n and x; then define f by $f(x) = \lim_n f_n(x)$.)

3. Let $(X, \mathscr{A})$, Y, and C be as in Exercise 2. Show that there is a sequence $\{f_n\}$ of functions from X to Y such that
   (a) each f_n is measurable with respect to $\mathscr{A}$ and $\mathscr{B}(Y)$, and
   (b) for each x in X the section C_x is the closure of the set $\{f_n(x) : n \in \mathbb{N}\}$.
   (Hint: Let $\{U_n\}$ be an enumeration of the nonempty sets in some countable base for Y. Define sets X_1, X_2, ... by letting X_n be the set of x's for which $C_x \cap U_n$ is not empty; then use Exercise 2 to construct for each n a measurable function $g_n : X_n \to U_n$ whose graph is included in $C \cap (X_n \times U_n)$. Construct the f_n's by extending the g_n's to X in a suitable way.)

4. Let X be a Polish space, let $(Y, \mathscr{A})$ be a measurable space, and let $f : X \to Y$ be a function such that

   (i) if $y \in Y$, then $f^{-1}(\{y\})$ is a nonempty closed subset of X, and
   (ii) if U is an open subset of X, then $f(U)$ belongs to $\mathscr{A}$.

   Use Exercise 2 to show that there is a function $g : Y \to X$ that is measurable with respect to $\mathscr{A}$ and $\mathscr{B}(X)$ and that satisfies $y = f(g(y))$ for each y in Y.

5. Use Exercise 8.2.4, together with ideas from the proof of Theorem 8.5.3, to give an alternate construction of the function g in Exercise 4.

8.6   Standard, Analytic, Lusin, and Souslin Spaces

A measurable space $(X, \mathscr{A})$ is *standard* if there is a Polish space Z such that $(X, \mathscr{A})$ is isomorphic to $(Z, \mathscr{B}(Z))$, and is *analytic* if there is a Polish space Z and an analytic subset A of Z such that $(X, \mathscr{A})$ is isomorphic to $(A, \mathscr{B}(A))$ (recall that $\mathscr{B}(A)$ is the Borel σ-algebra of the subspace A, and so, according to Lemma 7.2.2, is the family of subsets of A that have the form $A \cap B$ for some Borel subset B of Z).

Of course, the earlier sections of this chapter contain a number of properties of standard and analytic measurable spaces. (For example, Theorem 8.3.6 implies that if $(X, \mathscr{A})$ is a standard measurable space, then either X is countable and $\mathscr{A}$ contains all the subsets of X or else $(X, \mathscr{A})$ is isomorphic to $(\mathbb{R}, \mathscr{B}(\mathbb{R}))$.) This section contains a few more such properties, plus some techniques for verifying that a measurable space is standard or analytic.

We need to define a few more terms. Let $(X, \mathscr{A})$ be a measurable space. A subfamily $\mathscr{C}$ of $\mathscr{A}$ *generates* $\mathscr{A}$ if $\sigma(\mathscr{C}) = \mathscr{A}$. The σ-algebra $\mathscr{A}$, or the measurable space $(X, \mathscr{A})$, is *countably generated* if $\mathscr{A}$ has a countable subfamily that generates it. A family $\mathscr{C}$ of subsets of X *separates the points* of X if for each pair x, y of distinct points in X there is a member of $\mathscr{C}$ that contains exactly one of x and y. The space $(X, \mathscr{A})$, or the σ-algebra $\mathscr{A}$, is *separated* if $\mathscr{A}$ separates the points of X, and is *countably separated* if $\mathscr{A}$ has a countable subfamily that separates the points of X.

See Exercises 1, 2, and 4 for some information about the relationships among the concepts just defined.

Let us begin with a couple of general facts about analytic measurable spaces (Lemma 8.6.1 and Proposition 8.6.2), and then turn to ways of recognizing the analytic and standard spaces among the countably generated spaces (Propositions 8.6.5 and 8.6.6).

Lemma 8.6.1. *Let $(X, \mathscr{A})$ be an analytic measurable space, let Y be a Polish space, and let $f\colon X \to Y$ be measurable with respect to $\mathscr{A}$ and $\mathscr{B}(Y)$. Then the images under f of the sets in $\mathscr{A}$ are analytic.*

Proof. Since $(X, \mathscr{A})$ is an analytic measurable space, we can choose a Polish space Z, an analytic subset A_0 of Z, and an isomorphism g of $(A_0, \mathscr{B}(A_0))$ onto $(X, \mathscr{A})$. Suppose that $A \in \mathscr{A}$. Then $g^{-1}(A) \in \mathscr{B}(A_0)$, and so there is a set B in $\mathscr{B}(Z)$ such that $g^{-1}(A) = B \cap A_0$ (Lemma 7.2.2). Consequently $f(A)$, as the image of $g^{-1}(A)$ under the measurable map $f \circ g$, is analytic (Proposition 8.2.6).   □

Proposition 8.6.2. *Each bijective measurable map between analytic measurable spaces is an isomorphism.*

Proof. Suppose that $(X, \mathscr{A})$ and $(Y, \mathscr{B})$ are analytic measurable spaces and that $f\colon X \to Y$ is a measurable bijection. We need to show that if $A \in \mathscr{A}$, then $f(A) \in \mathscr{B}$. Since $(Y, \mathscr{B})$ is analytic, there is a Polish space Z, an analytic subset A_0 of Z, and an isomorphism g of $(Y, \mathscr{B})$ onto $(A_0, \mathscr{B}(A_0))$. Of course g is measurable with respect to $\mathscr{B}$ and $\mathscr{B}(Z)$ (Lemma 7.2.2). Now suppose that $A \in \mathscr{A}$. The measurability of $g \circ f$ with respect to $\mathscr{A}$ and $\mathscr{B}(Z)$ implies that $g(f(A))$ and $g(f(A^c))$ are analytic subsets of Z (Lemma 8.6.1), while the injectivity of $g \circ f$ implies that $g(f(A))$ and $g(f(A^c))$ are disjoint; hence the separation theorem for analytic sets (Theorem 8.3.1) provides a Borel subset B of Z such that $g(f(A)) \subseteq B$ and $g(f(A^c)) \subseteq B^c$. It is easy to check that $f(A)$ is equal to $g^{-1}(B)$, and so belongs to $\mathscr{B}$. Since A was an arbitrary set in $\mathscr{A}$, the measurability of f^{-1} follows.   □

We need the following elementary construction for our proof of Proposition 8.6.5.

Lemma 8.6.3. *Let $(X, \mathscr{A})$ be a countably generated measurable space, and suppose that the sets $A_1, A_2, \ldots$ generate $\mathscr{A}$. Define $F\colon X \to \{0, 1\}^{\mathbb{N}}$ by letting F take x to the sequence $\{\chi_{A_n}(x)\}$. Then*

$$\mathscr{A} = \{B \subseteq X : B = F^{-1}(C) \text{ for some } C \text{ in } \mathscr{B}(\{0, 1\}^{\mathbb{N}})\}. \tag{1}$$

Proof. Let us denote the set on the right-hand side of (1) by $\mathscr{A}_F$. Since the sets E_1, $E_2, \ldots$ defined by

$$E_k = \{\{n_i\} \in \{0, 1\}^{\mathbb{N}} : n_k = 1\}$$

generate $\mathscr{B}(\{0, 1\}^{\mathbb{N}})$ (Proposition 8.1.7) and since $A_k = F^{-1}(E_k)$ holds for each k, Proposition 2.6.2 implies that F is measurable with respect to $\mathscr{A}$ and $\mathscr{B}(\{0, 1\}^{\mathbb{N}})$, and hence that $\mathscr{A}_F \subseteq \mathscr{A}$. On the other hand, $\mathscr{A}_F$ is a σ-algebra on X that contains each A_k, and hence includes the σ-algebra these sets generate, namely $\mathscr{A}$. Thus $\mathscr{A} = \mathscr{A}_F$. $\square$

Corollary 8.6.4. *Let $(X, \mathscr{A})$ be a separated and countably generated measurable space. Then there is a subset A of $\{0,1\}^{\mathbb{N}}$ such that $(X, \mathscr{A})$ is isomorphic to $(A, \mathscr{B}(A))$.*

Proof. Use Lemma 8.6.3 to construct a map $F : X \to \{0,1\}^{\mathbb{N}}$ such that

$$\mathscr{A} = \{B \subseteq X : B = F^{-1}(C) \text{ for some } C \text{ in } \mathscr{B}(\{0,1\}^{\mathbb{N}})\}. \tag{2}$$

Let $A = F(X)$. Since $\mathscr{A}$ was assumed to separate the points of X, (2) implies first that F is injective and then that F is an isomorphism between $(X, \mathscr{A})$ and $(A, \mathscr{B}(A))$ (note that if $B = F^{-1}(C)$, then $F(B) = C \cap A$; also see Lemma 7.2.2). $\square$

Proposition 8.6.5. *Let $(X, \mathscr{A})$ be an analytic measurable space, let $(Y, \mathscr{B})$ be a separated and countably generated measurable space, and let $f : X \to Y$ be surjective and measurable. Then $(Y, \mathscr{B})$ is analytic.*

Proof. Use Corollary 8.6.4 to construct a function $F : Y \to \{0,1\}^{\mathbb{N}}$ that induces an isomorphism of $(Y, \mathscr{B})$ onto $(F(Y), \mathscr{B}(F(Y)))$. Lemma 8.6.1, applied to the map $F \circ f$, implies that $F(Y)$ is an analytic subset of $\{0,1\}^{\mathbb{N}}$. Thus $(Y, \mathscr{B})$, since it is isomorphic to $(F(Y), \mathscr{B}(F(Y)))$, is an analytic space. $\square$

Proposition 8.6.6. *Let $(X, \mathscr{A})$ be a standard measurable space, let $(Y, \mathscr{B})$ be a separated and countably generated measurable space, and let $f : X \to Y$ be bijective and measurable. Then $(Y, \mathscr{B})$ is standard.*

Proof. Proposition 8.6.5 implies that $(Y, \mathscr{B})$ is analytic, and Proposition 8.6.2 then implies that $(Y, \mathscr{B})$ is isomorphic to $(X, \mathscr{A})$. Since $(X, \mathscr{A})$ is standard, $(Y, \mathscr{B})$ must also be standard. $\square$

We turn to an important result due to Blackwell and to some of its consequences. For this we need to define the atoms of a σ-algebra. Let $(X, \mathscr{A})$ be a measurable space, and let x be an element of X. The *atom* of $\mathscr{A}$ determined by x is the intersection of those sets that belong to $\mathscr{A}$ and contain x. Note that a point y belongs to the atom determined by x if and only if x and y belong to exactly the same sets in $\mathscr{A}$. It is easy to check that the atoms of $\mathscr{A}$ form a partition of X, that an atom of $\mathscr{A}$ does not necessarily belong to $\mathscr{A}$ (see Exercise 5), and that an atom of $\mathscr{A}$ can contain more than one point (see Exercise 3).

Theorem 8.6.7 (Blackwell). *Let $(X, \mathscr{A})$ be an analytic measurable space, and let $\mathscr{A}_0$ be a countably generated sub-σ-algebra of $\mathscr{A}$. Then a subset of X belongs to $\mathscr{A}_0$ if and only if it belongs to $\mathscr{A}$ and is the union of a family of atoms of $\mathscr{A}_0$.*

Proof. Certainly every set that belongs to $\mathscr{A}_0$ also belongs to $\mathscr{A}$ and is the union of a family of atoms of $\mathscr{A}_0$. We need to prove the converse.

Use Lemma 8.6.3 to choose a function $F : X \to \{0,1\}^{\mathbb{N}}$ such that

$$\mathscr{A}_0 = \{B \subseteq X : B = F^{-1}(C) \text{ for some } C \text{ in } \mathscr{B}(\{0,1\}^{\mathbb{N}})\}. \tag{3}$$

Note that F is measurable with respect to $\mathscr{A}_0$ and $\mathscr{B}(\{0,1\}^{\mathbb{N}})$ and so with respect to $\mathscr{A}$ and $\mathscr{B}(\{0,1\}^{\mathbb{N}})$, and that the atoms of $\mathscr{A}_0$ are the nonempty subsets of X that are inverse images under F of one-point subsets of $\{0,1\}^{\mathbb{N}}$. Now suppose that A belongs to $\mathscr{A}$ and is the union of a family of atoms of $\mathscr{A}_0$. Then $F(A)$ and $F(A^c)$ are disjoint analytic subsets of $\{0,1\}^{\mathbb{N}}$ (use Lemma 8.6.1 and the assumption that A is the union of a collection of atoms of $\mathscr{A}_0$). Hence the separation theorem for analytic sets provides a Borel subset C of $\{0,1\}^{\mathbb{N}}$ such that $F(A) \subseteq C$ and $F(A^c) \subseteq C^c$. Then A is equal to $F^{-1}(C)$ and so in view of (3) belongs to $\mathscr{A}_0$. With this the proof is complete.   □

Corollary 8.6.8. *Let $(X, \mathscr{A})$ be an analytic measurable space, and let $\mathscr{A}_0$ be a separated and countably generated sub-σ-algebra of $\mathscr{A}$. Then $\mathscr{A}_0 = \mathscr{A}$.*

Proof. Since $\mathscr{A}_0$ is separated, each of its atoms contains only one point, and so each subset of X is the union of a family of atoms of $\mathscr{A}_0$. Thus Theorem 8.6.7 implies that a subset of X belongs to $\mathscr{A}_0$ if and only if it belongs to $\mathscr{A}$.   □

The following strengthened versions of Propositions 8.6.5 and 8.6.6 now follow. They will be useful for the study of Lusin and Souslin spaces later in this section.

Corollary 8.6.9. *Let $(X, \mathscr{A})$ be an analytic measurable space, let $(Y, \mathscr{B})$ be a countably separated measurable space, and let $f: X \to Y$ be surjective and measurable. Then $(Y, \mathscr{B})$ is analytic.*

Proof. We begin by showing that $\mathscr{B}$ is countably generated. Choose a countable subfamily $\mathscr{C}$ of $\mathscr{B}$ that separates the points of Y. We will show that $\mathscr{B}$ is equal to the countably generated σ-algebra $\sigma(\mathscr{C})$. Let B be an arbitrary element of $\mathscr{B}$, and let $\mathscr{B}_0 = \sigma(\mathscr{C} \cup \{B\})$. Then $\mathscr{B}_0$ is separated and countably generated, and f is measurable with respect to $\mathscr{A}$ and $\mathscr{B}_0$; hence $(Y, \mathscr{B}_0)$ is analytic (Proposition 8.6.5). Furthermore, $\sigma(\mathscr{C})$ is a separated and countably generated sub-σ-algebra of $\mathscr{B}_0$, and so Corollary 8.6.8 implies that $\sigma(\mathscr{C}) = \mathscr{B}_0$. Thus $B \in \sigma(\mathscr{C})$. Since B was an arbitrary member of $\mathscr{B}$, it follows that $\mathscr{B} = \sigma(\mathscr{C})$.

Now that we have proved that $\mathscr{B}$ is countably generated, we can use Proposition 8.6.5 to conclude that $(Y, \mathscr{B})$ is analytic.   □

Corollary 8.6.10. *Let $(X, \mathscr{A})$ be a standard measurable space, let $(Y, \mathscr{B})$ be a countably separated measurable space, and let $f: X \to Y$ be bijective and measurable. Then $(Y, \mathscr{B})$ is standard.*

Proof. This follows from Corollary 8.6.9 in the same way that Proposition 8.6.6 follows from Proposition 8.6.5.   □

Let us turn to the study of some not necessarily metrizable topological spaces that are closely related to the Polish spaces. A *Lusin space* is a Hausdorff space that is the image of a Polish space under a continuous bijection, and a *Souslin space* is a Hausdorff space that is the image of a Polish space under a continuous surjection. Of course, every Lusin space is a Souslin space.

Examples 8.6.11.

(a) It is clear that the Souslin subspaces of a Polish space X are exactly the analytic subsets of X. Proposition 8.2.10 (or Exercise 8.2.5) and Theorem 8.3.7 imply that the Lusin subspaces of a Polish space X are exactly the Borel subsets of X.

(b) Now suppose that X is a Polish space, and let X_0 be constructed by replacing the topology of X with a weaker Hausdorff topology. The function $f: X \to X_0$ defined by $f(x) = x$ is continuous, and so X_0 is a Lusin space. In particular, if X is a separable Banach space, then X with its weak topology[8] is a Lusin space. Likewise, if the dual X^* of the Banach space X is separable, then X^* with its weak* topology is a Lusin space. Furthermore, if the Banach space X is infinite dimensional, then the weak topology on X and the weak* topology on X^* are not metrizable.[9] Thus non-metrizable Lusin spaces arise in a natural way. □

The rest of this section is devoted to some basic measure-theoretic facts about Lusin and Souslin spaces. We will prove that if X is a Lusin space, then $(X, \mathscr{B}(X))$ is a standard measurable space, that if X is a Souslin space, then $(X, \mathscr{B}(X))$ is an analytic measurable space, and that if X is a Souslin space, then every finite Borel measure on X is regular.

Lemma 8.6.12. *If X is a Souslin space, then $\mathscr{B}(X)$ is countably separated.*

Proof. Choose a Polish space Z and a continuous surjection $f: Z \to X$. Define $F: Z \times Z \to X \times X$ by $F(z_1, z_2) = (f(z_1), f(z_2))$. Let Δ be the subset of $X \times X$ defined by

$$\Delta = \{(x_1, x_2) \in X \times X : x_1 = x_2\},$$

and let $\mathscr{U}$ be the collection of those open rectangles in $X \times X$ that are included in the complement of Δ. Then F is continuous, Δ is closed, and $\Delta^c = \cup \mathscr{U}$. Hence

$$F^{-1}(\Delta^c) = \bigcup \{F^{-1}(U) : U \in \mathscr{U}\},$$

---

[8]This example assumes more Banach space theory than is developed in this book.

[9]Suppose that X is an infinite-dimensional Banach space. If the weak topology on X is metrizable, then there is an infinite sequence $\{f_i\}$ in X^* such that each f in X^* is a linear combination of f_1, $\ldots, f_n$ for some n (choose $\{f_i\}$ so that for each weakly open neighborhood U of 0 there is a positive integer n and a positive number ε such that $x \in U$ holds whenever x satisfies $|f_i(x)| < \varepsilon$ for $i = 1$, $\ldots, n$; then use Lemma V.3.10 in [42]). Thus X^* is the union of a sequence of finite-dimensional subspaces of X^*. Since this is impossible (use Exercise 3.5.6, Corollary IV.3.2 in [42], and the Baire category theorem), we have a contradiction, and the weak topology on X is not metrizable. A similar argument shows that the weak* topology on X^* is not metrizable.

and D.11, applied to $\{F^{-1}(U) : U \in \mathcal{U}\}$, implies that there is a countable subcollection $\mathcal{U}_0$ of $\mathcal{U}$ such that

$$F^{-1}(\Delta^c) = \bigcup \{F^{-1}(U) : U \in \mathcal{U}_0\}.$$

This and the surjectivity of F imply that

$$\Delta^c = \bigcup \mathcal{U}_0.$$

Thus for each pair x, y of distinct points in X there is a set $V \times W$ in $\mathcal{U}_0$ such that $(x,y) \in V \times W$ and hence (recall that $(V \times W) \cap \Delta = \varnothing$) such that $x \in V$, $y \in W$, and $V \cap W = \varnothing$. Consequently the sides of the rectangles in $\mathcal{U}_0$ form a countable subfamily of $\mathcal{B}(X)$ that separates the points of X.   □

Proposition 8.6.13. *If X is a Souslin space, then $(X, \mathcal{B}(X))$ is an analytic measurable space, while if X is a Lusin space, then $(X, \mathcal{B}(X))$ is a standard measurable space.*

Proof. Let X be a Souslin space, and choose a Polish space Z and a continuous surjection $f \colon Z \to X$. Since f is Borel measurable and $\mathcal{B}(X)$ is countably separated (Lemma 8.6.12), Corollary 8.6.9 implies that $(X, \mathcal{B}(X))$ is analytic. A similar argument, based on Lemma 8.6.12 and Corollary 8.6.10, shows that if X is a Lusin space, then $(X, \mathcal{B}(X))$ is standard.   □

Theorem 8.6.14. *Every finite Borel measure on a Souslin space is regular.*

Proof. Let X be a Souslin space, and let μ be a finite Borel measure on X. We will show that

$$\mu(B) = \sup\{\mu(K) : K \subseteq B \text{ and } K \text{ is compact}\} \tag{4}$$

holds for each B in $\mathcal{B}(X)$. This gives the inner regularity of μ. It also implies the outer regularity of μ, since for each B in $\mathcal{B}(X)$ we can use (4), applied to B^c, to approximate B^c from below by compact sets and hence to approximate B from above by open sets.

So suppose that B belongs to $\mathcal{B}(X)$. We can assume that B is not empty. Let us begin by producing a continuous function $f \colon \mathcal{N} \to X$ such that $f(\mathcal{N}) = B$. For this choose a Polish space Z and a continuous surjection $g \colon Z \to X$, note that $g^{-1}(B)$ is a Borel and hence analytic subset of Z, choose a continuous function $h \colon \mathcal{N} \to Z$ such that $h(\mathcal{N}) = g^{-1}(B)$, and let $f = g \circ h$.

For each positive number ε we can use the constructions in the proof of Theorem 8.4.1 to choose sets $\mathcal{L}(n_1, \ldots, n_k)$ (to be abbreviated as $\mathcal{L}_k$) such that $\mu^*(f(\mathcal{L}_k)) > \mu(B) - \varepsilon$ holds for each k. Arguments used in the proof of Theorem 8.4.1 show that the sets L and K defined by $L = \cap_k \mathcal{L}_k$ and $K = f(L)$ are compact, that $\mu(\cap_k f(\mathcal{L}_k)^-) \geq \mu(B) - \varepsilon$, and that $K \subseteq \cap_k f(\mathcal{L}_k)^-$.

Our earlier proof of the reverse inclusion works only if X is metrizable; hence it must be replaced. Suppose that $x \in \cap_k f(\mathcal{L}_k)^-$ and that U is an open neighborhood of x. For each k choose an element $\mathbf{m}_k$ of $\mathcal{L}_k$ such that $f(\mathbf{m}_k) \in U$. As before, the sequence $\{\mathbf{m}_k\}$ has a convergent subsequence, say with limit $\mathbf{m}$. Then $\mathbf{m} \in L$, and the continuity of f implies that $f(\mathbf{m}) \in U^-$ and hence that U^- meets K. Since this is valid for each open neighborhood U of x, it follows that $x \in K$ (otherwise, since K is compact, Proposition 7.1.2 would provide disjoint open sets U_0 and V_0 such that $x \in U_0$ and $K \subseteq V_0$, and U_0 would be an open neighborhood of x such that $\overline{U_0} \cap K = \varnothing$). Since x was an arbitrary element of $\cap_k f(\mathscr{L}_k)^-$, it follows that $\cap_k f(\mathscr{L}_k)^- \subseteq K$ and hence that $K = \cap_k f(\mathscr{L}_k)^-$. With this we have constructed a compact subset K of B such that $\mu(K) \geq \mu(B) - \varepsilon$, and (4) is proved.   □

Exercises

1. Let $(X, \mathscr{A})$ be a measurable space. Show that if $\mathscr{A}$ is separated and countably generated, then $\mathscr{A}$ is countably separated.
2. Give a σ-algebra on $\mathbb{R}$ that is included in $\mathscr{B}(\mathbb{R})$ and is separated but not countably separated.
3. Let $(X, \mathscr{A})$ be a measurable space. Show that each atom of $\mathscr{A}$ contains only one point if and only if $\mathscr{A}$ separates the points of X.
4. Give an example of a measurable space that is countably separated but not countably generated.
5. Let $X = \{0, 1\}^{\mathbb{R}}$ and let $\mathscr{A}$ be the smallest σ-algebra on X that makes each coordinate projection of X onto $\{0, 1\}$ measurable (of course, $\{0, 1\}$ is to have the σ-algebra consisting of all of its subsets).
   (a) Show that for each A in $\mathscr{A}$ there is a countable subset S of $\mathbb{R}$ such that if $x \in A$, if $y \in X$, and if $x(s) = y(s)$ holds at each s in S, then $y \in A$. (Hint: See Exercise 1.1.7.)
   (b) Show that the atoms of $\mathscr{A}$ do not belong to $\mathscr{A}$.
6. Show by example that the hypothesis that $\mathscr{A}_0$ is countably generated cannot be removed from Theorem 8.6.7.
7. Show by example that the hypothesis that $(X, \mathscr{A})$ is analytic cannot be removed from Theorem 8.6.7. (Hint: Let $X = \mathbb{R}$, let A be a subset of $\mathbb{R}$ that is not Borel, and let $\mathscr{A} = \sigma(\mathscr{B}(\mathbb{R}) \cup \{A\})$.)
8. Let X be a Souslin space. Show that if $\mathscr{U}$ is a collection of open subsets of X, then there is a countable subcollection $\mathscr{U}_0$ of $\mathscr{U}$ such that $\cup \mathscr{U}_0 = \cup \mathscr{U}$. (Hint: Study the proof of Lemma 8.6.12.)
9. Show that if X and Y are Souslin spaces, then $\mathscr{B}(X \times Y) = \mathscr{B}(X) \times \mathscr{B}(Y)$. (Hint: Apply Exercise 8 to the space $X \times Y$.)
10. Show that each compact Souslin space is metrizable.

Notes

The classical theory of analytic sets was developed by the Polish and Russian schools of mathematics between the First and Second World Wars. See, for example, Kuratowski [77]. In the mid-1950s Blackwell [11] noted that the theory of analytic sets can be applied profitably to probability theory, while Mackey [86] noted that it is useful for the study of group representations; their work has done much to stimulate interest in the subject. Rather thorough recent treatments of analytic sets have been given by Kechris [68] and Srivastava [112]. See also [29,62,83,101,107].

Analytic and Borel subsets of non-separable spaces have been studied by A.H. Stone and his students. See Stone [113] for a survey and for further references.

Exercise 8.1.14 is due to Dudley [39].

The reader who wants to see additional (and more explicit) examples of analytic sets that are not Borel should see Mazurkiewicz [88] and Dellacherie [35]. For example, Mazurkiewicz shows that the subset A of $C[0,1]$ consisting of the differentiable functions (that is, of the continuous functions on $[0,1]$ that are differentiable at each point in $[0,1]$) is the complement of an analytic set, but is not itself analytic (thus A^c is analytic but not Borel).

The proof of Theorem 8.3.6 given in the text was suggested by Kuratowski and Mostowski [78], while that in Exercise 8.3.5 is taken from Parthasarathy [96]. The proof given here for Theorem 8.3.7 is due to Dellacherie [36].

Theorems 8.4.1 and 8.5.3 are classical. That they imply Proposition 8.4.4 and Corollary 8.5.4 has been noticed (independently) by a number of people. See Castaing and Valadier [26] and Wagner [121] (and of course [62,68,101,112]) for further information and references. The concepts of capacity and capacitability are due to Choquet [28].

The results in the first part of Sect. 8.6 are due to Blackwell [11] and Mackey [86]. Bourbaki (see Chapter IX of [17]) introduced the terms Lusin space and Souslin space for metrizable spaces that are images of Polish spaces under continuous bijections and surjections; Cartier [25] noted that the assumption of metrizability is not needed.

Chapter 9
Haar Measure

We saw in Chap. 1 that Lebesgue measure on $\mathbb{R}^d$ is translation invariant, in the sense that $\lambda(A+x) = \lambda(A)$ holds for each A in $\mathscr{B}(\mathbb{R}^d)$ and each x in $\mathbb{R}^d$. Furthermore, we saw that Lebesgue measure is essentially the only such Borel measure on $\mathbb{R}^d$: if μ is a nonzero Borel measure on $\mathbb{R}^d$ that is finite on the compact subsets of $\mathbb{R}^d$ and satisfies $\mu(A+x) = \mu(A)$ for each A in $\mathscr{B}(\mathbb{R}^d)$ and each x in $\mathbb{R}^d$, then there is a positive number c such that $\mu(A) = c\lambda(A)$ holds for every Borel subset A of $\mathbb{R}^d$.

It turns out that very similar results hold for every locally compact group (see Sect. 9.1 for the definition of such groups); the role of Lebesgue measure is played by what is called Haar measure. This chapter is devoted to an introduction to Haar measure.

Section 9.1 contains some basic definitions and facts about topological groups. Section 9.2 contains a proof of the existence and uniqueness of Haar measure, and Sect. 9.3 contains additional basic properties of Haar measures. In Sect. 9.4 we construct two algebras, $L^1(G)$ and $M(G)$, which are fundamental for the study of harmonic analysis on a locally compact group G.

9.1 Topological Groups

A *topological group* is a set G that has the structure of a group (say with group operation $(x,y) \mapsto xy$) and of a topological space and is such that the operations $(x,y) \mapsto xy$ and $x \mapsto x^{-1}$ are continuous. Note that $(x,y) \mapsto xy$ is a function from the product space $G \times G$ to G and that we are requiring that it be continuous with respect to the product topology on $G \times G$; thus xy must be "jointly continuous" in x and y and not merely continuous in x with y held fixed and continuous in y with x held fixed (see Exercise 3). A *locally compact topological group*, or simply a *locally compact group*, is a topological group whose topology is locally compact and Hausdorff. A *compact group* is a topological group whose topology is compact and Hausdorff.

D.L. Cohn, *Measure Theory: Second Edition*, Birkhäuser Advanced
Texts Basler Lehrbücher, DOI 10.1007/978-1-4614-6956-8_9,
© Springer Science+Business Media, LLC 2013

Examples 9.1.1.

(a) The set $\mathbb{R}$, with its usual topology and with addition as the group operation, is a locally compact group.
(b) Likewise, $\mathbb{R}^d$, $\mathbb{Z}$, and $\mathbb{Z}^d$ are locally compact groups.
(c) The set $\mathbb{R}^*$ of nonzero real numbers, with the topology it inherits as a subspace of $\mathbb{R}$ and with multiplication as the group operation, is a locally compact group.
(d) Let $\mathbb{T}$ be the set consisting of those complex numbers z that satisfy $|z| = 1$. Then $\mathbb{T}$, with the topology it inherits as a subspace of $\mathbb{C}$ and with multiplication as the group operation, is a compact group.
(e) The set $\mathbb{Q}$ of rational numbers, with the topology it inherits as a subspace of $\mathbb{R}$ and with addition as the group operation, is a topological group; it is not locally compact.
(f) An arbitrary group G, with the topology that makes every subset of G open, is a locally compact group; it is compact if and only if G is finite. $\square$

See Exercises 9–11 for some additional examples.

Let X be a topological space, and let x belong to X. Recall that a family $\mathcal{U}$ of subsets of X is a *base for the family of neighborhoods of x* if

(a) each member of $\mathcal{U}$ is an open neighborhood of x, and
(b) for each open neighborhood V of x there is a set that belongs to $\mathcal{U}$ and is included in V.

Let G be a group. If a is an element of G and if B is a subset of G, then the products aB and Ba are defined by

$$aB = \{ab : b \in B\}$$

and

$$Ba = \{ba : b \in B\}.$$

Likewise, if B and C are subsets of G, then BC and B^{-1} are defined by

$$BC = \{bc : b \in B \text{ and } c \in C\},$$

and

$$B^{-1} = \{b^{-1} : b \in B\}.$$

The set B is *symmetric* if $B = B^{-1}$. Thus B is symmetric if and only if the condition $x \in B$ is equivalent to the condition $x^{-1} \in B$.

Proposition 9.1.2. *Let G be a topological group, let e be the identity element of G, and let a be an arbitrary element of G.*

(a) *The functions $x \mapsto ax$, $x \mapsto xa$, and $x \mapsto x^{-1}$ are homeomorphisms of G onto G.*
(b) *If $\mathcal{U}$ is a base for the family of neighborhoods of e, then $\{aU : U \in \mathcal{U}\}$ and $\{Ua : U \in \mathcal{U}\}$ are bases for the family of neighborhoods of a.*

(c) *If K and L are compact subsets of G, then aK, Ka, KL, and K^{-1} are compact subsets of G.*

Proof. The definition of a topological group, together with the continuity of the maps $x \mapsto (x,a)$ and $x \mapsto (a,x)$, implies the continuity of the functions in part (a). Since these functions have continuous inverses (namely the functions that take x to $a^{-1}x$, to xa^{-1}, and to x^{-1}), they are homeomorphisms of G onto G.

Part (b) is an immediate consequence of part (a).

Part (c) follows from the fact that the image of a compact set under a continuous map is compact (as usual, the compactness of the subset $K \times L$ of $G \times G$ is given by Tychonoff's theorem, Theorem D.20). $\qquad\Box$

Proposition 9.1.3. *Let G be a topological group, let e be the identity element of G, and let U be an open neighborhood of e.*

(a) *There is an open neighborhood V of e such that $VV \subseteq U$.*
(b) *There is a symmetric open neighborhood of e that is included in U.*

Proof. Since the map $(x,y) \mapsto xy$ is continuous, the set W defined by $W = \{(x,y) : xy \in U\}$ is an open neighborhood of (e,e) in $G \times G$, and so there are open neighborhoods V_1 and V_2 of e that satisfy $V_1 \times V_2 \subseteq W$. The set V defined by $V = V_1 \cap V_2$ is then an open neighborhood of e that satisfies $VV \subseteq U$.

We turn to part (b). The continuity of the map $x \mapsto x^{-1}$ implies that if U is an open neighborhood of e, then U^{-1} is also an open neighborhood of e. Thus $U \cap U^{-1}$ is a symmetric open neighborhood of e that is included in U. $\qquad\Box$

Proposition 9.1.4. *Let G be a topological group, let K be a compact subset of G, and let U be an open subset of G that includes K. Then there are open neighborhoods V_R and V_L of e such that $KV_R \subseteq U$ and $V_LK \subseteq U$.*

Proof. For each x in K choose open neighborhoods W_x and V_x of e such that $xW_x \subseteq U$ and $V_xV_x \subseteq W_x$ (see Propositions 9.1.2 and 9.1.3). Then $\{xV_x\}_{x \in K}$ is an open cover of the compact set K, and so there is a finite collection $x_1, \dots, x_n$ of points in K such that the sets $x_iV_{x_i}$, $i = 1, \dots, n$, cover K. Let $V_R = \cap_{i=1}^n V_{x_i}$. If $x \in K$, then there is an index i such that $x \in x_iV_{x_i}$, and so

$$xV_R \subseteq x_iV_{x_i}V_{x_i} \subseteq x_iW_{x_i} \subseteq U.$$

Since x was an arbitrary element of K, it follows that $KV_R \subseteq U$. The construction of V_L is similar. $\qquad\Box$

Let G be a topological group, and let f be a real- or complex-valued function on G. Then f is *left uniformly continuous* if for each positive number ε there is an open neighborhood U of e such that $|f(x) - f(y)| < \varepsilon$ holds whenever x and y belong to G and satisfy $y \in xU$. Likewise, f is *right uniformly continuous* if for each positive number ε there is an open neighborhood U of e such that $|f(x) - f(y)| < \varepsilon$ holds whenever x and y belong to G and satisfy $y \in Ux$. Note that we can replace the neighborhoods of e appearing in this definition with smaller symmetric neighborhoods of e (Proposition 9.1.3) and that for such symmetric neighborhoods U the condition $x \in yU$ is equivalent to the condition $y \in xU$ and the condition $x \in Uy$ is equivalent to the condition $y \in Ux$. Thus x and y do in fact enter our definition symmetrically.

Proposition 9.1.5. *Let G be a locally compact group. Then each function in $\mathscr{K}(G)$ is left uniformly continuous and right uniformly continuous.*

Proof. Let f belong to $\mathscr{K}(G)$, and let K be the support of f. Suppose that ε is a positive number. For each x in K choose first an open neighborhood U_x of e such that $|f(x) - f(y)| < \varepsilon/2$ holds whenever y belongs to xU_x and then an open neighborhood V_x of e such that $V_x V_x \subseteq U_x$ (see Propositions 9.1.2 and 9.1.3). The family $\{xV_x\}_{x \in K}$ is an open cover of the compact set K, and so there is a finite collection $x_1, \dots, x_n$ of points in K such that the sets $x_i V_{x_i}$, $i = 1, \dots, n$, cover K. Let V be a symmetric open neighborhood of e that is included in $\cap_{i=1}^{n} V_{x_i}$ (Proposition 9.1.3). We will show that if x and y belong to G and satisfy $y \in xV$, then $|f(x) - f(y)| < \varepsilon$.

This inequality certainly holds if neither x nor y belongs to K (for then $f(x) = f(y) = 0$). Now suppose that $x \in K$ and $y \in xV$. Then there is an index i such that $x \in x_i V_{x_i}$ and hence such that x and y belong to $x_i U_{x_i}$ (note that $x \in x_i V_{x_i} \subseteq x_i U_{x_i}$ and $y \in xV \subseteq x_i V_{x_i} V_{x_i} \subseteq x_i U_{x_i}$). It follows that $|f(x) - f(x_i)| < \varepsilon/2$ and $|f(y) - f(x_i)| < \varepsilon/2$ and hence that $|f(x) - f(y)| < \varepsilon$. The remaining case to deal with is where $y \in K$ and $y \in xV$. Since V is symmetric, this is exactly the case where $y \in K$ and $x \in yV$, and the details we just looked at (with x and y interchanged) handle this. The left uniform continuity of f follows. The right uniform continuity of f can be proved in a similar way.   □

Corollary 9.1.6. *Let G be a locally compact group, let μ be a regular Borel measure on G, and let f belong to $\mathscr{K}(G)$. Then the functions $x \mapsto \int f(xy)\,\mu(dy)$ and $x \mapsto \int f(yx)\,\mu(dy)$ are continuous.*

Proof. We will check the continuity of $x \mapsto \int f(yx)\,\mu(dy)$ at an arbitrary point x_0 in G; the proof for $x \mapsto \int f(xy)\,\mu(dy)$ is similar.

Let K be the support of f, and let W be an open neighborhood of x_0 whose closure is compact. It is easy to check that for each x in W the function $y \mapsto f(yx)$ is continuous and vanishes outside the compact set $K(W^-)^{-1}$. Suppose that ε is a positive number, choose a positive number ε' such that $\varepsilon'\mu(K(W^-)^{-1}) < \varepsilon$, and use the left uniform continuity of f (Proposition 9.1.5) to choose an open neighborhood V of e such that $|f(s) - f(t)| < \varepsilon'$ holds whenever s and t belong to G and satisfy $s \in tV$. Then for each x in $W \cap x_0 V$ and each y in G we have $yx \in yx_0 V$, and so

$$\left| \int f(yx)\,\mu(dy) - \int f(yx_0)\,\mu(dy) \right| \leq \int |f(yx) - f(yx_0)|\,\mu(dy)$$

$$\leq \varepsilon'\mu(K(W^-)^{-1}) \leq \varepsilon.$$

Since ε is arbitrary, the proof is complete.   □

The next two results will be used only in Sect. 9.4.

Proposition 9.1.7. *Let G be a topological group, and let H be an open subgroup of G. Then H is closed.*

Proof. The complement of H is the union of the left cosets xH, where x ranges through the complement of H. Proposition 9.1.2 implies that each of these cosets is open. It follows that the complement of H is open and hence that H itself is closed. $\square$

Proposition 9.1.8. *Let G be a locally compact group. Then there is a subgroup H of G that is open, closed, and σ-compact.*

Proof. Since G is locally compact, we can choose an open neighborhood U of e whose closure is compact. Use Proposition 9.1.3 to choose a symmetric open neighborhood V of e that is included in U. Of course V^- is compact. Define sets V^n, $n = 1, 2, \ldots$, inductively by means of the equations $V^1 = V$ and $V^n = V^{n-1}V$, and then define H by $H = \cup_n V^n$. If $x \in V^m$ and $y \in V^n$, then $xy \in V^{m+n}$ and $x^{-1} \in V^m$ (recall that V is symmetric); hence H is a subgroup of G. It is clear that H is open and so also closed (see Exercise 4 and Proposition 9.1.7). Since V^- is compact and H is closed, the closure of each V^n is compact and included in H; the σ-compactness of H follows. $\square$

Exercises

1. Suppose that G is a group and a topological space. Show that G is a topological group if and only if the map $(x,y) \mapsto xy^{-1}$ from $G \times G$ to G is continuous.
2. Let G be $\mathbb{R}$, with addition as the group operation and with the weakest topology that makes each interval of the form $(a,b]$ open. Show that $(x,y) \mapsto x+y$ is continuous, but that $x \mapsto -x$ is not continuous. Thus G is not a topological group.
3. Let G be $\mathbb{R}$, with addition as the group operation and with the topology for which the open sets are those that either are empty or have a countable complement (check that these sets do form a topology on G). Show that
   (a) $x \mapsto -x$ is continuous,
   (b) $(x,y) \mapsto x+y$ is continuous in x when y is held fixed and continuous in y when x is held fixed, and
   (c) $(x,y) \mapsto x+y$ is not continuous.
   Thus G is not a topological group.
4. Let G be a topological group, let U be an open subset of G, and let A be an arbitrary subset of G. Show that AU and UA are open subsets of G. (Hint: Note that $AU = \cup_{a \in A} aU$.)
5. Show that if G_1 and G_2 are topological groups, then $G_1 \times G_2$, with the product topology and with the operation defined by $(x_1,x_2)(y_1,y_2) = (x_1y_1,x_2y_2)$, is a topological group.

6. Let G be a topological group. Show that the following conditions are equivalent:

   (i) The topology of G is Hausdorff.
   (ii) For each a in G the set $\{a\}$ is closed.
   (iii) For some a in G the set $\{a\}$ is closed.

7. Find all closed subgroups of $\mathbb{R}$. In other words, find all subgroups of the additive group $\mathbb{R}$ that are closed in the usual topology for $\mathbb{R}$.

8. Let G be a Hausdorff topological group, and let E and F be subsets of G.

   (a) Show that if E is compact and F is closed, then EF is closed.
   (b) Show by example that if E and F are closed (but not compact), then EF can fail to be closed. (Hint: Such examples can be found in the case where $G = \mathbb{R}$.)

9. Let G consist of the 2 by 2 matrices of the form $\begin{pmatrix} a & b \\ 0 & 1 \end{pmatrix}$, where a is a positive real number and b is an arbitrary real number. Show that G, with the operation of matrix multiplication and with the topology it inherits as a subspace of $\mathbb{R}^4$, is a locally compact group.

10. Let $GL(d, \mathbb{R})$ be the collection of all invertible d by d matrices with real entries. Show that $GL(d, \mathbb{R})$, with the operation of matrix multiplication and with the topology it inherits as a subspace of $\mathbb{R}^{d^2}$, is a locally compact group (it is called the *general linear group*). (Hint: See Lemma 6.1.2, and recall how Cramer's rule for the solution of systems of linear equations gives an explicit formula for the inverse of a matrix.)

11. Let $O(d)$ be the collection of all orthogonal[1] d by d matrices. Show that $O(d)$, with the operation of matrix multiplication and with the topology it inherits as a subspace of $\mathbb{R}^{d^2}$, is a compact group (it is called the *orthogonal group*).

12. Let G be the locally compact group introduced in Exercise 9. Construct a real-valued function on G that is right uniformly continuous, but not left uniformly continuous. (Hint: Consider

$$\begin{pmatrix} a & b \\ 0 & 1 \end{pmatrix} \mapsto \varphi(b),$$

   where φ is a suitable function from $\mathbb{R}$ to $\mathbb{R}$.)

13. Derive Proposition 9.1.5 from Proposition 9.1.4. (Hint: Suppose that f belongs to $\mathscr{K}(G)$. Consider the group $G \times G$ and the sets K and U defined by $K = \{(x,x) : x \in \mathrm{supp}(f)\}$ and $U = \{(x,y) : |f(x) - f(y)| < \varepsilon\}$.)

---

[1] Recall that a square matrix with real entries is *orthogonal* if the product of it with its transpose is the identity matrix.

9.2   The Existence and Uniqueness of Haar Measure

Let G be a locally compact group, and let μ be a nonzero regular Borel measure on G. Then μ is a *left Haar measure* (or simply a *Haar measure*) if it is *invariant under left translations* (or simply *translation invariant*), in the sense that $\mu(xA) = \mu(A)$ holds for each x in G and each A in $\mathscr{B}(G)$. Likewise, μ is a *right Haar measure* if $\mu(Ax) = \mu(A)$ holds for each x in G and each A in $\mathscr{B}(G)$. (Lemma 7.2.1 and Proposition 9.1.2 imply that if $x \in G$ and if A is a Borel subset of G, then xA and Ax are Borel subsets of G; hence the expressions $\mu(xA)$ and $\mu(Ax)$ appearing in the preceding definition are meaningful.)

In this section we prove that there is a left Haar measure on each locally compact group and that it is unique up to multiplication by a constant. A few properties of Haar measures, plus the relationship between left and right Haar measures, will be dealt with in Sect. 9.3. In Sect. 9.4 we will use these results to discuss some measure-theoretic tools for harmonic analysis.

Examples 9.2.1.

(a) Lebesgue measure on $\mathbb{R}$ (or on $\mathbb{R}^d$) is a left and a right Haar measure; see Proposition 1.4.4.
(b) If G is a group with the discrete topology (that is, with the topology that makes every subset of G open), then counting measure on G is a left and a right Haar measure; in particular, counting measure on the group $\mathbb{Z}$ of integers is a Haar measure.
(c) Let $\mathbb{T}$ be the set of complex numbers z such that $|z| = 1$, made into a topological group as in Example 9.1.1(d) in the previous section. Then linear Lebesgue measure on $\mathbb{T}$ is a Haar measure. More precisely, if λ_0 is Lebesgue measure on $\mathbb{R}$, restricted to the Borel subsets of the interval $[0, 2\pi)$, and if $F: [0, 2\pi) \to \mathbb{T}$ is defined by $F(\theta) = e^{i\theta}$, then $\lambda_0 F^{-1}$ is a left and a right Haar measure on $\mathbb{T}$. $\square$

See Exercises 3 and 5 below and also Exercises 4 and 6 in Sect. 9.3, for additional examples of Haar measures.

We need a bit of notation. Let G be a group, let x be an element of G, and let f be a function on G. The *left translate of f by x*, written $_xf$, is defined by $_xf(t) = f(x^{-1}t)$, and the *right translate of f by x*, written f_x, is defined by $f_x(t) = f(tx^{-1})$. The function $\check{f}$ (or $f^\vee$) is defined by $\check{f}(t) = f(t^{-1})$. Note that if x, y, and t belong to G, then

$$_{xy}f(t) = f((xy)^{-1}t) = f(y^{-1}x^{-1}t) = {}_yf(x^{-1}t) = {}_x({}_yf)(t);$$

hence

$$_{xy}f = {}_x({}_yf).$$

A similar argument shows that

$$f_{xy} = (f_x)_y.$$

If A is a subset of G, then the characteristic functions of the sets A, xA, and Ax are related by the identities

$$(\chi_A)_x = \chi_{Ax}$$

and

$$_x(\chi_A) = \chi_{xA}.$$

This gives one reason for defining $_x f(t)$ and $f_x(t)$ to be $f(x^{-1}t)$ and $f(tx^{-1})$, rather than $f(xt)$ and $f(tx)$. (The definitions of $_x f$ and f_x are not entirely standard; some authors use $f(xt)$ and $f(tx)$ where we used $f(x^{-1}t)$ and $f(tx^{-1})$.)

If G is a locally compact group and if μ is a left Haar measure on G, then

$$\int {}_x f \, d\mu = \int f \, d\mu \tag{1}$$

holds for each Borel function f that is either nonnegative or μ-integrable (note that $\int {}_x f \, d\mu = \mu(xA) = \mu(A) = \int f \, d\mu$ holds if f is the characteristic function of the Borel set A, and then use the linearity of the integral and the monotone convergence theorem).

Theorem 9.2.2. *Let G be a locally compact group. Then there is a left Haar measure on G.*

Proof. Let K be a compact subset of G, and let V be a subset of G whose interior V^o is nonempty. Then $\{xV^o\}_{x \in G}$ is an open cover of the compact set K, and so there are finite sequences $\{x_i\}_{i=1}^n$ of elements of G such that $K \subseteq \cup_{i=1}^n x_i V$. Let $\#(K:V)$ be the smallest nonnegative integer n for which such a sequence $\{x_i\}_{i=1}^n$ exists. Of course, $\#(K:V) = 0$ if and only if $K = \varnothing$.

Let us choose a compact set K_0 whose interior is nonempty; it will serve as a standard for measuring the sizes of various subsets of G and will remain fixed throughout this proof. Roughly speaking, we will measure the size of an arbitrary compact subset K of G by computing the ratio $\#(K:U)/\#(K_0:U)$ for each open neighborhood U of e and then finding a sort of limit of this ratio as the neighborhood U becomes smaller. We will use this "limit" to construct an outer measure μ^* on G, and then we will show that the restriction of μ^* to $\mathscr{B}(G)$ is the required measure.

We turn to the details. Let $\mathscr{C}$ be the family of all compact subsets of G, and let $\mathscr{U}$ be the family of all open neighborhoods of e. For each U in $\mathscr{U}$ define $h_U : \mathscr{C} \to \mathbb{R}$ by $h_U(K) = \#(K:U)/\#(K_0:U)$.

Lemma 9.2.3. *The relations*

(a) $0 \leq h_U(K) \leq \#(K:K_0)$,
(b) $h_U(K_0) = 1$,
(c) $h_U(xK) = h_U(K)$,
(d) $h_U(K_1) \leq h_U(K_2)$ if $K_1 \subseteq K_2$,
(e) $h_U(K_1 \cup K_2) \leq h_U(K_1) + h_U(K_2)$, and
(f) $h_U(K_1 \cup K_2) = h_U(K_1) + h_U(K_2)$ if $K_1 U^{-1} \cap K_2 U^{-1} = \varnothing$

hold for all U, K, K_1, K_2, and x.

Proof. The relation

$$\#(K : U) \leq \#(K : K_0)\#(K_0 : U) \tag{2}$$

holds for all K and U, as we can see by noting that if $\{x_i\}_{i=1}^m$ and $\{y_j\}_{j=1}^n$ are sequences in G such that $K \subseteq \bigcup_{i=1}^m x_i K_0$ and $K_0 \subseteq \bigcup_{j=1}^n y_j U$, then $K \subseteq \bigcup_{i=1}^m \bigcup_{j=1}^n x_i y_j U$. Dividing both sides of (2) by $\#(K_0 : U)$ gives assertion (a). Assertions (b), (c), (d), and (e) are clear. In view of (e), we can prove (f) by checking that

$$\#(K_1 \cup K_2 : U) \geq \#(K_1 : U) + \#(K_2 : U) \tag{3}$$

holds whenever

$$K_1 U^{-1} \cap K_2 U^{-1} = \varnothing. \tag{4}$$

So suppose that (4) holds and that $\{x_i\}_{i=1}^n$ is a sequence of points such that $n = \#(K_1 \cup K_2 : U)$ and $K_1 \cup K_2 \subseteq \bigcup_{i=1}^n x_i U$. Then each set $x_i U$ meets at most one of K_1 and K_2 (for if $x_i U$ met both K_1 and K_2, then x_i would belong to $K_1 U^{-1} \cap K_2 U^{-1}$), and so we can partition the sequence $\{x_i\}_{i=1}^n$ into sequences $\{y_i\}_{i=1}^j$ and $\{z_i\}_{i=1}^k$ such that $K_1 \subseteq \bigcup_{i=1}^j y_i U$ and $K_2 \subseteq \bigcup_{i=1}^k z_i U$. Relation (3) and part (f) of the lemma follow. $\qquad\square$

We now turn to the "limit" of the ratios $\#(K : U)/\#(K_0 : U)$—that is, of the functions $\{h_U\}_{U \in \mathcal{U}}$. We will find this "limit" by constructing a certain product space that contains all the functions h_U and then using a compactness argument to produce the "limit" function.

For each K in $\mathscr{C}$ let I_K be the subinterval $[0, \#(K : K_0)]$ of $\mathbb{R}$. Let X be the product space $\prod_{K \in \mathscr{C}} I_K$, endowed with the product topology. Since each interval I_K is compact, Tychonoff's theorem (Theorem D.20) implies that X is compact. According to part (a) of Lemma 9.2.3, each function h_U belongs to X. For each open neighborhood V of e let $S(V)$ be the closure in X of the set $\{h_U : U \in \mathcal{U} \text{ and } U \subseteq V\}$. If $V_1, \ldots, V_n$ belong to $\mathcal{U}$ (that is, if they are open neighborhoods of e) and if V is defined by $V = \bigcap_{i=1}^n V_i$, then $h_V \in \bigcap_{i=1}^n S(V_i)$; since $V_1, \ldots, V_n$ were arbitrary, this implies that the closed sets $\{S(V)\}_{V \in \mathcal{U}}$ satisfy the finite intersection property. The compactness of X now implies that $\bigcap_{V \in \mathcal{U}} S(V)$ is nonempty. Let us choose, once and for all, an element $h_\bullet$ of $\bigcap_{V \in \mathcal{U}} S(V)$. This function $h_\bullet$ is our "limit" of the functions h_U.

Lemma 9.2.4. *The function $h_\bullet$ satisfies*

(a) $0 \leq h_\bullet(K)$,
(b) $h_\bullet(\varnothing) = 0$,
(c) $h_\bullet(K_0) = 1$,
(d) $h_\bullet(xK) = h_\bullet(K)$,
(e) $h_\bullet(K_1) \leq h_\bullet(K_2)$ *if* $K_1 \subseteq K_2$,
(f) $h_\bullet(K_1 \cup K_2) \leq h_\bullet(K_1) + h_\bullet(K_2)$, *and*
(g) $h_\bullet(K_1 \cup K_2) = h_\bullet(K_1) + h_\bullet(K_2)$ *if* $K_1 \cap K_2 = \varnothing$

for all x in G and all K, K_1, and K_2 in $\mathscr{C}$.

Proof. Let us begin with part (f). Recall that X, as the product space $\prod_{K \in \mathscr{C}} I_K$, is a certain set of functions on $\mathscr{C}$, with its topology defined so that for each compact subset K of G (i.e., for each element K of the index set $\mathscr{C}$) the projection from X to $\mathbb{R}$ defined by $h \mapsto h(K)$ is continuous. Hence for each choice of compact subsets K_1 and K_2 of G the map from X to $\mathbb{R}$ defined by

$$h \mapsto h(K_1) + h(K_2) - h(K_1 \cup K_2) \tag{5}$$

is continuous. Since this map is, in addition, nonnegative at each h_U (see part (e) of Lemma 9.2.3), it is nonnegative at each point in each set $S(V)$. In particular, it is nonnegative at $h_\bullet$, and so part (f) is proved.

Property (a) is clear, and properties (b) through (e) can be proved with arguments similar to the one given above for part (f). We turn to part (g). Suppose that K_1 and K_2 are disjoint compact subsets of G. According to Proposition 7.1.2 there are disjoint open sets U_1 and U_2 such that $K_1 \subseteq U_1$ and $K_2 \subseteq U_2$, and according to Proposition 9.1.4 there are open neighborhoods V_1 and V_2 of e such that $K_1 V_1 \subseteq U_1$ and $K_2 V_2 \subseteq U_2$. Let $V = V_1 \cap V_2$. Then $K_1 V$ and $K_2 V$ are disjoint, and so for each U that belongs to $\mathscr{U}$ and satisfies $U \subseteq V^{-1}$ we have

$$h_U(K_1 \cup K_2) = h_U(K_1) + h_U(K_2)$$

(see part (f) of Lemma 9.2.3). Consequently the map defined by (5) vanishes at each element of $S(V^{-1})$. Since $h_\bullet \in S(V^{-1})$, part (g) follows.   □

Let us return to the proof of Theorem 9.2.2. We are now in a position to construct the promised outer measure on G. Define μ^* on the collection of open subsets of G by

$$\mu^*(U) = \sup\{h_\bullet(K) : K \subseteq U \text{ and } K \in \mathscr{C}\}, \tag{6}$$

and extend it to the collection of all subsets of G by

$$\mu^*(A) = \inf\{\mu^*(U) : A \subseteq U \text{ and } U \text{ is open}\}. \tag{7}$$

It is clear that μ^* is nonnegative, that it is monotone, and that $\mu^*(\varnothing) = 0$.

In view of (7), we can verify the countable subadditivity of μ^* by checking that each sequence $\{U_i\}$ of open subsets of G satisfies

$$\mu^*\left(\bigcup_i U_i\right) \leq \sum_i \mu^*(U_i). \tag{8}$$

So suppose that $\{U_i\}$ is a sequence of open subsets of G. Let K be a compact subset of $\cup_i U_i$. Then there is a positive integer n such that $K \subseteq \cup_{i=1}^n U_i$, and there are compact subsets $K_1, \ldots, K_n$ of $U_1, \ldots, U_n$ such that $K = \cup_{i=1}^n K_i$ (use Lemma 7.1.10 and mathematical induction). It follows that

$$h_\bullet(K) \leq \sum_{i=1}^{n} h_\bullet(K_i) \leq \sum_{i=1}^{n} \mu^*(U_i) \leq \sum_{i=1}^{\infty} \mu^*(U_i)$$

(see Lemma 9.2.4 and Eq. (6)); since K was an arbitrary compact subset of $\cup_i U_i$, another application of (6) gives (8).

We can prove that each Borel subset of G is μ^*-measurable by verifying that if U and V are open subsets of G and if $\mu^*(V) < +\infty$, then

$$\mu^*(V) \geq \mu^*(V \cap U) + \mu^*(V \cap U^c) \tag{9}$$

(see the proof of Proposition 7.2.9). We proceed to check this inequality. Let ε be a positive number. Choose a compact subset K of $V \cap U$ such that

$$h_\bullet(K) > \mu^*(V \cap U) - \varepsilon, \tag{10}$$

and then choose a compact subset L of $V \cap K^c$ such that $h_\bullet(L) > \mu^*(V \cap K^c) - \varepsilon$. Then K and L are disjoint, and, since $V \cap U^c \subseteq V \cap K^c$, L satisfies

$$h_\bullet(L) > \mu^*(V \cap U^c) - \varepsilon. \tag{11}$$

It follows from these inequalities and Lemma 9.2.4 that

$$h_\bullet(K \cup L) = h_\bullet(K) + h_\bullet(L) \geq \mu^*(V \cap U) + \mu^*(V \cap U^c) - 2\varepsilon.$$

Since ε is arbitrary and $h_\bullet(K \cup L) \leq \mu^*(V)$, inequality (9) follows. Consequently $\mathscr{B}(G)$ is included in the σ-algebra of μ^*-measurable sets, and the restriction of μ^* to $\mathscr{B}(G)$ is a measure (Theorem 1.3.6). Call this measure μ.

We turn to the regularity of μ. Note that if K is compact, if U is open, and if $K \subseteq U$, then $h_\bullet(K) \leq \mu(U)$. It follows from this and (7) that

$$h_\bullet(K) \leq \mu(K). \tag{12}$$

Furthermore, if K is a compact set and U is an open set that includes K and has a compact closure (see Proposition 7.1.4), then

$$h_\bullet(L) \leq h_\bullet(U^-)$$

holds for each compact subset L of U, and so

$$\mu(K) \leq \mu(U) \leq h_\bullet(U^-).$$

It follows that μ is finite on the compact subsets of G. The outer regularity of μ follows from (7), and the inner regularity follows from (6) and (12).

It is easy to check that μ is nonzero and translation-invariant (use Lemma 9.2.4 and Eqs. (6) and (7)). Thus μ is the required measure.   □

The following lemma gives a fundamental elementary property of Haar measures; we will need it for our proof of Theorem 9.2.6.

Lemma 9.2.5. *Let G be a locally compact group, and let μ be a left Haar measure on G. Then each nonempty open subset U of G satisfies $\mu(U) > 0$, and each nonnegative function f that belongs to $\mathscr{K}(G)$ and is not identically zero satisfies $\int f\,d\mu > 0$.*

Proof. Since μ is regular and not the zero measure, we can choose a compact set K such that $\mu(K) > 0$. Let U be a nonempty open subset of G. Then $\{xU\}_{x\in G}$ is an open cover of the compact set K, and so there is a finite collection, say $x_1, \ldots, x_n$, of elements of G such that the sets x_iU, $i = 1, \ldots, n$, cover K. The relation $\mu(K) \le \sum_i \mu(x_iU)$ and the translation invariance of μ imply that $\mu(K) \le n\mu(U)$ and hence that $\mu(U) > 0$. Thus the first half of the lemma is proved.

Now suppose that f is a nonnegative function that belongs to $\mathscr{K}(G)$ and does not vanish identically. Then there is a positive number ε and a nonempty open set U such that $f \ge \varepsilon\chi_U$. It follows that $\int f\,d\mu \ge \varepsilon\mu(U) > 0$. $\qquad\qquad\square$

Theorem 9.2.6. *Let G be a locally compact group, and let μ and v be left Haar measures on G. Then there is a positive real number c such that $v = c\mu$.*

Proof. Let g be a nonnegative function that belongs to $\mathscr{K}(G)$ and does not vanish identically (g will be held fixed throughout this proof), and let f be an arbitrary function in $\mathscr{K}(G)$. Since $\int g\,d\mu \ne 0$ (Lemma 9.2.5), we can form the ratio $\int f\,d\mu / \int g\,d\mu$. We will show that this ratio depends only on the functions f and g and not on the particular Haar measure μ used in its computation. It follows that the Haar measure v satisfies

$$\frac{\int f\,dv}{\int g\,dv} = \frac{\int f\,d\mu}{\int g\,d\mu}$$

and hence satisfies $\int f\,dv = c\int f\,d\mu$, where c is defined by $c = \int g\,dv / \int g\,d\mu$. Since this holds for each f in $\mathscr{K}(G)$, Theorem 7.2.8 implies that $v = c\mu$.

We turn to the ratio $\int f\,d\mu / \int g\,d\mu$. If $h \in \mathscr{K}(G \times G)$, then Proposition 7.6.4 implies that the iterated integrals $\int\int h(x,y)\,\mu(dx)\,v(dy)$ and $\int\int h(x,y)\,v(dy)\,\mu(dx)$ exist and are equal. If in the second of these integrals we reverse the order of integration, use the translation invariance of the Haar measure μ to replace x with $y^{-1}x$ (see (1)), again reverse the order of integration, and finally replace y with xy, we find that

$$\int\int h(x,y)\,v(dy)\,\mu(dx) = \int\int h(y^{-1}x,y)\,\mu(dx)\,v(dy)$$

$$= \int\int h(y^{-1},xy)\,v(dy)\,\mu(dx). \qquad (13)$$

Let us apply this identity to the function h defined by

$$h(x,y) = \frac{f(x)g(yx)}{\int g(tx)\,v(dt)}.$$

(Note that h does belong to $\mathscr{K}(G \times G)$: Corollary 9.1.6 and Lemma 9.2.5 imply that $x \mapsto \int g(tx)\,v(dt)$ is continuous and never vanishes; furthermore, if $K = \mathrm{supp}(f)$ and $L = \mathrm{supp}(g)$, then $\mathrm{supp}(h) \subseteq K \times LK^{-1}$.) For this function h we have $h(y^{-1}, xy) = f(y^{-1})g(x)/\int g(ty^{-1})\,v(dt)$, and so Eq. (13) implies that

$$\int f(x)\,\mu(dx) = \int g(x)\,\mu(dx) \int \frac{f(y^{-1})}{\int g(ty^{-1})\,v(dt)}\,v(dy).$$

Thus the ratio of $\int f\,d\mu$ to $\int g\,d\mu$ depends on f and g, but not on μ, and the proof is complete.   □

The reader should note that if the locally compact group G is commutative (and if, for convenience, the group operation is written additively), then a simpler proof of Theorem 9.2.6 can be given. In fact, it is easy to check that if f and g belong to $\mathscr{K}(G)$, then

$$\int f\,d\mu \int g\,dv = \int \int f(x)g(y)\,\mu(dx)\,v(dy)$$

$$= \int \cdot \int f(x+y)g(y)\,\mu(dx)\,v(dy)$$

$$= \int \int f(y)g(y-x)\,v(dy)\,\mu(dx)$$

$$= \int \int f(y)g(-x)\,\mu(dx)\,v(dy)$$

$$= \int f\,dv \int \check{g}\,d\mu.$$

Thus if we let g be a nonnegative function that belongs to $\mathscr{K}(G)$ and does not vanish identically and if we define c by $c = \int g\,dv/\int \check{g}\,d\mu$, then $\int f\,dv = c\int f\,d\mu$ holds for each f in $\mathscr{K}(G)$. It follows that $v = c\mu$.

Exercises

1. Let G be the set of rational numbers, with addition as the group operation and with the topology inherited from $\mathbb{R}$. Show that there is no nonzero translation-invariant regular Borel measure on G.
2. Let G be a locally compact group, let μ be a left Haar measure on G, and let f and g be continuous real-valued functions on G. Show that if f and g are equal μ-almost everywhere, then they are equal everywhere.

3. Let G be the multiplicative group of positive real numbers, with the topology it inherits as a subspace of $\mathbb{R}$. Show that the formula

$$\mu(A) = \int_A \frac{1}{x} \lambda(dx)$$

defines a Haar measure on G.

4. Let G be a locally compact group that is homeomorphic to an open subset (say U) of $\mathbb{R}^d$, and let φ be a homeomorphism of G onto U.

   (a) Show that if for each a in G the function $u \mapsto \varphi(a\varphi^{-1}(u))$ is the restriction to U of an affine[2] map $L_a \colon \mathbb{R}^d \to \mathbb{R}^d$, then the formula

   $$\mu(A) = \int_{\varphi(A)} \frac{1}{|\det(L_{\varphi^{-1}(u)})|} \lambda(du)$$

   defines a left Haar measure on G.

   (b) Likewise, show that if for each a in G the function $u \mapsto \varphi(\varphi^{-1}(u)a)$ is the restriction to U of an affine map $R_a \colon \mathbb{R}^d \to \mathbb{R}^d$, then the formula

   $$\mu(A) = \int_{\varphi(A)} \frac{1}{|\det(R_{\varphi^{-1}(u)})|} \lambda(du)$$

   defines a right Haar measure on G.

5. Let G be the group defined in Exercise 9.1.9. Suppose that we identify G with the right half-plane in $\mathbb{R}^2$ by associating the point (a,b) with the matrix $\begin{pmatrix} a & b \\ 0 & 1 \end{pmatrix}$. Show that the formula

$$\mu(A) = \iint_A \frac{1}{a^2} \, da\, db$$

defines a left Haar measure on G and that the formula

$$\mu(A) = \iint_A \frac{1}{a} \, da\, db$$

defines a right Haar measure on G. (Hint: Use the preceding exercise.)

6. Let G be a locally compact group, and let μ be a left Haar measure on G. Show that the topology of G is discrete if and only if $\mu(\{x\}) \neq 0$ holds for some (and hence for each) x in G.

---

[2] A map $F \colon \mathbb{R}^d \to \mathbb{R}^d$ is *affine* if there exist a linear map $G \colon \mathbb{R}^d \to \mathbb{R}^d$ and an element b of $\mathbb{R}^d$ such that $F(x) = G(x) + b$ holds for each x in $\mathbb{R}^d$. If F is affine, then G and b are uniquely determined by F, and we will (for simplicity) denote by $\det(F)$ the determinant of the linear part G of F (see Sect. 6.1).

9.3   Properties of Haar Measure

Let G be a locally compact group, and let μ be a regular Borel measure on G. The map $x \mapsto x^{-1}$ is a homeomorphism of G onto itself (Proposition 9.1.2), and so the subsets A of G that belong to $\mathscr{B}(G)$ are exactly those for which A^{-1} belongs to $\mathscr{B}(G)$ (Lemma 7.2.1). Define a function $\check{\mu}$ on $\mathscr{B}(G)$ by $\check{\mu}(A) = \mu(A^{-1})$. It is easy to check that $\check{\mu}$ is a regular Borel measure on G. The relation

$$\int f \, d\check{\mu} = \int \check{f} \, d\mu \tag{1}$$

holds if f is a Borel function that is nonnegative or $\check{\mu}$-integrable; this is clear if f is a characteristic function and follows in general from the linearity of the integral and the monotone convergence theorem.

Proposition 9.3.1. *Let G be a locally compact group, and let μ be a regular Borel measure on G. Then μ is a left Haar measure if and only if $\check{\mu}$ is a right Haar measure, and is a right Haar measure if and only if $\check{\mu}$ is a left Haar measure.*

Proof. The identity $(Ax)^{-1} = x^{-1}A^{-1}$ implies that $\check{\mu}(Ax) = \check{\mu}(A)$ holds for each x in G and each A in $\mathscr{B}(G)$ if and only if $\mu(x^{-1}A^{-1}) = \mu(A^{-1})$ holds for each x in G and each A in $\mathscr{B}(G)$. The first half of the proposition follows. We can derive the second half from the first by replacing μ with $\check{\mu}$ and noting that $\check{\check{\mu}} = \mu$. □

Corollary 9.3.2. *Let G be a locally compact group. Then there is one and, up to a constant multiple, only one right Haar measure on G.*

Proof. In view of Proposition 9.3.1, this is an immediate consequence of Theorems 9.2.2 and 9.2.6. □

Proposition 9.3.3. *Let G be a locally compact group, and let μ be a left Haar measure on G. Then μ is finite if and only if G is compact.*

Proof. The regularity of μ implies that μ is finite if G is compact.

We turn to the converse. Suppose that μ is finite. Let K be a compact subset of G such that $\mu(K) > 0$ (for instance, K can be a compact set whose interior is nonempty; see Lemma 9.2.5). The finiteness of $\mu(G)$ implies that there is an upper bound, for instance $\mu(G)/\mu(K)$, for the lengths of those finite sequences $\{x_i\}_1^n$ for which the sets $x_i K$, $i = 1, \dots, n$, are disjoint. Thus we can choose a positive integer n and points $x_1, \dots, x_n$ such that the sets $x_i K$, $i = 1, \dots, n$, are disjoint, but such that for no choice of x_{n+1} are the sets $x_i K$, $i = 1, \dots, n+1$, disjoint. It follows that if $x \in G$, then xK meets $\cup_{i=1}^n x_i K$, and so x belongs to $(\cup_{i=1}^n x_i K)K^{-1}$; hence G is equal to the compact set $(\cup_{i=1}^n x_i K)K^{-1}$. □

It follows that each compact group G has a Haar measure μ such that $\mu(G) = 1$. In dealing with compact groups one often assumes that the corresponding Haar measures have been "normalized" in this way.

Let G be a locally compact group, and let μ be a left Haar measure on G. The maps $u \mapsto ux$ are homeomorphisms of G onto itself (Proposition 9.1.2), and so for each x in G the formula $\mu_x(A) = \mu(Ax)$ defines a regular Borel measure μ_x on G. The translation invariance of μ implies that μ_x satisfies $\mu_x(yA) = \mu(yAx) = \mu(Ax) = \mu_x(A)$ for each y in G and each A in $\mathscr{B}(G)$. Thus μ_x is a left Haar measure, and so Theorem 9.2.6 implies that for each x there is a positive number, say $\Delta(x)$, such that $\mu_x = \Delta(x)\mu$. The function $\Delta \colon G \to \mathbb{R}$ defined in this way is called the *modular function* of G. See Exercises 2 and 4 for some examples.

If ν is another left Haar measure on G, then there is a positive constant c such that $\nu = c\mu$, and so $\nu_x = c\mu_x = c\Delta(x)\mu = \Delta(x)\nu$ holds for each x in G. Thus the modular function Δ is determined by the group G and does not depend on the particular left Haar measure used in its definition.

Recall that $(\chi_A)_x = \chi_{Ax}$ holds for each member x and subset A of G. It follows that

$$\int f_x \, d\mu = \Delta(x) \int f \, d\mu \tag{2}$$

holds if f is the characteristic function of a Borel subset of G and hence if f is a Borel function that is nonnegative or μ-integrable.

Proposition 9.3.4. *Let G be a locally compact group, and let Δ be the modular function of G. Then*

(a) *Δ is continuous, and*
(b) *$\Delta(xy) = \Delta(x)\Delta(y)$ holds for each x and y in G.*

Proof. Let μ be a left Haar measure on G, and let f be a nonnegative function that belongs to $\mathscr{K}(G)$ and does not vanish identically. Then $\int f \, d\mu \neq 0$ (Lemma 9.2.5), and so Corollary 9.1.6 and Eq. (2) imply the continuity of Δ. The relation $\Delta(xy) = \Delta(x)\Delta(y)$ follows from the calculation

$$\Delta(xy)\mu(A) = \mu(Axy) = \Delta(y)\mu(Ax) = \Delta(y)\Delta(x)\mu(A).$$

$\square$

A locally compact group G is *unimodular* if its modular function satisfies $\Delta(x) = 1$ at each x in G. Thus a locally compact group G is unimodular if and only if each left Haar measure on G is a right Haar measure and so if and only if the collection of all left Haar measures on G coincides with the collection of all right Haar measures on G. Of course every commutative locally compact group is unimodular.

Proposition 9.3.5. *Every compact group is unimodular.*

Proof. Let G be a compact group, and let Δ be its modular function. The relation $\Delta(x^n) = (\Delta(x))^n$ holds for each positive integer n and each element x of G (Proposition 9.3.4); hence Δ is unbounded if there is an element x of G that satisfies

$\Delta(x) > 1$ or that satisfies $0 < \Delta(x) < 1$ (for then x^{-1} satisfies $\Delta(x^{-1}) > 1$). However the continuity of Δ and the compactness of G imply that Δ is bounded; thus $\Delta(x) = 1$ must hold at each x in G. $\qquad\square$

The remaining results in this section will be needed only for a few exercises and for the definition and study of $M_a(G)$ in Sect. 9.4; they can be omitted on a first reading.

Proposition 9.3.6. *Let G be a locally compact group, and let μ be a left Haar measure on G. Then each Borel subset A of G satisfies*

$$\check{\mu}(A) = \int_A \Delta(x^{-1})\,\mu(dx).$$

Proof. Define a measure v on $\mathscr{B}(G)$ by

$$v(A) = \int_A \Delta(x^{-1})\,\mu(dx).$$

We will show that v is regular, that v is a right Haar measure, and finally that $v = \check{\mu}$.

We begin with the regularity of v. For each positive integer n let G_n be the open subset of G defined by

$$G_n = \left\{ x \in G : \frac{1}{n} < \Delta(x^{-1}) < n \right\}.$$

Let U be an open subset of G. Since $v(U) = \lim_n v(U \cap G_n)$ (Proposition 1.2.5), we can show that

$$v(U) = \sup\{v(K) : K \subseteq U \text{ and } K \text{ is compact}\}$$

by checking that

$$v(U \cap G_n) = \sup\{v(K) : K \subseteq U \cap G_n \text{ and } K \text{ is compact}\}$$

holds for each n. However this last equation is an easy consequence of the regularity of μ and the fact that $1/n < \Delta(x^{-1}) < n$ holds at each x in G_n (consider the cases where $\mu(U \cap G_n) = +\infty$ and where $\mu(U \cap G_n) < +\infty$ separately). Now suppose that A is an arbitrary Borel subset of G. We need to show that

$$v(A) = \inf\{v(U) : A \subseteq U \text{ and } U \text{ is open}\}. \qquad (3)$$

We can certainly restrict our attention to the case where $v(A)$ is finite. Let ε be a positive number. Then for each n we can choose an open subset U_n of G_n that includes $A \cap G_n$ and satisfies $v(U_n) < v(A \cap G_n) + \varepsilon/2^n$ (use the regularity of μ and the fact that $1/n < \Delta(x^{-1}) < n$ holds at each x in G_n). The set U defined by $U = \cup_n U_n$ then includes A and satisfies $v(U) < v(A) + \varepsilon$; since ε is arbitrary, (3) is proved. It is easy to see that each compact subset K of G satisfies $v(K) < +\infty$ (note that $\mu(K)$ is finite and that the function $x \mapsto \Delta(x^{-1})$ is bounded on K). With this the proof of the regularity of v is complete.

Since v is regular and nonzero, the calculation

$$v(Ay) = \int \chi_{A_y}(x)\Delta(x^{-1})\mu(dx)$$

$$= \int \chi_{A_y}(x)\Delta(y^{-1})\Delta((xy^{-1})^{-1})\mu(dx)$$

$$= \Delta(y^{-1}) \int (\chi_A)_y(x)\Delta((xy^{-1})^{-1})\mu(dx)$$

$$= \Delta(y^{-1})\Delta(y) \int \chi_A(x)\Delta(x^{-1})\mu(dx)$$

$$= v(A)$$

(here we used (2) and part (b) of Proposition 9.3.4) implies that v is a right Haar measure.

Thus there is a positive number c such that $v = c\check{\mu}$ (see Proposition 9.3.1 and Corollary 9.3.2), and so

$$c = \frac{v(A)}{\check{\mu}(A)} = \frac{v(A)}{\mu(A^{-1})} = \frac{1}{\mu(A^{-1})} \int_A \Delta(x^{-1})\mu(dx)$$

holds whenever A is a Borel set that satisfies $0 < \check{\mu}(A) < +\infty$. Since Δ is continuous and has value 1 at e, we can make the right side of the equation arbitrarily close to 1 by letting A be a sufficiently small symmetric neighborhood of e. Thus $c = 1$, and so $v = \check{\mu}$. $\square$

Corollary 9.3.7. *Let G be a locally compact group, let μ be a left Haar measure on G, and let v be a right Haar measure on G. Then a Borel subset A of G satisfies $\mu(A) = 0$ if and only if it satisfies $v(A) = 0$.*

Proof. The formula $A \mapsto \int_A \Delta(t^{-1})\mu(dt)$ defines a right Haar measure on G (Proposition 9.3.6), and so there is a positive constant c such that for each A in $\mathscr{B}(G)$ we have $v(A) = c \int_A \Delta(t^{-1})\mu(dt)$. Since Δ is positive everywhere on G, it follows that A satisfies $v(A) = 0$ if and only if it satisfies $\mu(A) = 0$ (see Corollary 2.3.12). $\square$

Exercises

1. Let G be a locally compact group, and let μ be a right Haar measure on G. Show that $\mu(xA) = \Delta(x^{-1})\mu(A)$ holds for each x in G and each A in $\mathscr{B}(G)$.

2. Let G be the group considered in Exercises 9.1.9 and 9.2.5, and let Δ be the modular function of G. Show that $\Delta \begin{pmatrix} a & b \\ 0 & 1 \end{pmatrix} = 1/a$ holds for each $\begin{pmatrix} a & b \\ 0 & 1 \end{pmatrix}$ in G.

3. Let G be as in the preceding exercise. Find a Borel subset of G that has finite measure under the left Haar measures on G but infinite measure under the right Haar measures on G.

4. Show that the formula

$$\mu(A) = \int_A \frac{1}{|\det(u)|^d} \lambda(du),$$

where λ is Lebesgue measure on $\mathbb{R}^{d^2}$, defines a left and a right Haar measure on $GL(d,\mathbb{R})$. Hence $GL(d,\mathbb{R})$ is unimodular (note, however, that it is neither compact nor abelian). (Hint: See Exercise 9.2.4.)

5. Let G be a locally compact group and let μ be a left Haar measure on G. Show that G is unimodular if and only if $\mu = \check{\mu}$.

6. Let H be $\{0,1\}$, with the discrete topology and with addition modulo 2 as the group operation. Let G be $H^{\mathbb{N}}$, with the product topology and with the group operation defined component-by-component in terms of the operation on H.

   (a) Show that G is a compact group.

   (b) Let μ be the Haar measure on G for which $\mu(G) = 1$ (see Proposition 9.3.3 and the remark following it). Show that

   $$\mu(\{\{a_j\} \in G : a_{n_i} = b_i \text{ for } i = 1, \dots, k\}) = \frac{1}{2^k}$$

   holds for each sequence $n_1, \dots, n_k$ of distinct positive integers and each sequence $b_1, \dots, b_k$ of elements of $\{0,1\}$.

   (c) Show that there are compact subsets K and L of G such that $\mu(K) = \mu(L) = 0$, but $KL = G$.

   (d) Let $f: G \to [0,1]$ be the map that takes the sequence $\{a_i\}$ to the number $\sum_{i=1}^{\infty} a_i 2^{-i}$. Show that $\lambda(B) = \mu(f^{-1}(B))$ holds for each Borel subset B of $[0,1]$.

7. Let G be a locally compact group, and let μ be a left Haar measure on G. Show that μ is σ-finite if and only if G is σ-compact.

8. Let G be a locally compact group that is not unimodular, let μ be a left Haar measure on G, and let ν be a right Haar measure on G. Show that there is a Borel subset A of G such that $\mu(A) < +\infty$ and $\nu(A) = +\infty$. (Hint: See Proposition 9.3.6 or Exercise 9.3.1.)

9. Let G be a locally compact group, let μ be a left Haar measure on G, and let ν be a right Haar measure on G. Suppose that outer measures μ^* and ν^* and measures μ_1 and ν_1 are associated to μ and ν as in Sect. 7.5.

   (a) Show that $\mathscr{M}_{\mu^*} = \mathscr{M}_{\nu^*}$.

   (b) Show that a subset of G is locally μ_1-null if and only if it is locally ν_1-null.

9.4   The Algebras $L^1(G)$ and $M(G)$

Since most of the topics dealt with in this section involve measures and integrals on products of locally compact groups, we begin by recalling some of the necessary facts.

Suppose that X and Y are locally compact Hausdorff spaces and that μ and v are regular Borel measures on X and Y, respectively. If X and Y have countable bases for their topologies, then $\mathscr{B}(X \times Y)$ is equal to $\mathscr{B}(X) \times \mathscr{B}(Y)$, μ and v are σ-finite, and the product measure $\mu \times v$ (as defined in Sect. 5.1) is a regular Borel measure (see Proposition 7.6.2). Thus the theory of product measures contained in Chap. 5 is adequate for the study of products of regular Borel measures on second countable locally compact Hausdorff spaces.[3]

We dealt with products of arbitrary locally compact Hausdorff spaces in Sect. 7.6; there we showed that if μ and v are regular Borel measures on X and Y, then

$$\int \int f(x,y)\,\mu(dx)\,v(dy) = \int \int f(x,y)\,v(dy)\,\mu(dx)$$

holds for each f in $\mathscr{K}(X \times Y)$, and we used the Riesz representation theorem (applied to the functional $f \mapsto \int \int f(x,y)\,\mu(dx)\,v(dy)$) to construct a regular Borel measure $\mu \times v$ on $X \times Y$ such that

$$\int f\,d(\mu \times v) = \int \int f(x,y)\,\mu(dx)\,v(dy) = \int \int f(x,y)\,v(dy)\,\mu(dx) \qquad (1)$$

holds for each f in $\mathscr{K}(X \times Y)$. We proved that (1) also holds for many other functions on $X \times Y$ (see Theorem 7.6.7 and Exercises 7.6.3 and 7.6.4).

Now let G be an arbitrary locally compact group, let μ be a left Haar measure on G, and let f and g belong to $\mathscr{L}^1(G, \mathscr{B}(G), \mu)$. The *convolution* of f and g is the function $f * g$ from G to $\mathbb{R}$ (or to $\mathbb{C}$) defined by

$$(f * g)(t) = \begin{cases} \int f(s)g(s^{-1}t)\,\mu(ds) & \text{if } s \mapsto f(s)g(s^{-1}t) \text{ is integrable,} \\ 0 & \text{otherwise.} \end{cases}$$

Some basic properties of convolutions are given by the following propositions.

Proposition 9.4.1. *Let G be a locally compact group, let μ be a left Haar measure on G, and let f and g belong to $\mathscr{L}^1(G, \mathscr{B}(G), \mu)$.*

(a) *The function $s \mapsto f(s)g(s^{-1}t)$ belongs to $\mathscr{L}^1(G, \mathscr{B}(G), \mu)$ for μ-almost every t in G.*

(b) *The convolution $f * g$ of f and g belongs to $\mathscr{L}^1(G, \mathscr{B}(G), \mu)$ and satisfies $\|f * g\|_1 \le \|f\|_1 \|g\|_1$.*

---

[3]In particular, the reader who is interested only in second countable locally compact groups can ignore the references to Sect. 7.6 in what follows.

We need the following two lemmas for the proof of Proposition 9.4.1.

Lemma 9.4.2. *Let G be a locally compact group, let μ be a left Haar measure on G, and let f belong to $\mathscr{L}^1(G, \mathscr{B}(G), \mu)$. Then there is a sequence $\{K_n\}$ of compact subsets of G such that f vanishes outside $\bigcup_n K_n$.*

Proof. We can use Corollary 2.3.11 and the regularity of μ to produce a sequence $\{U_n\}$ of open subsets of G that have finite measure under μ and are such that f vanishes outside $\bigcup_n U_n$. Let H be a subgroup of G that is open and σ-compact (see Proposition 9.1.8). Since each nonempty open subset of G has nonzero measure under μ (Lemma 9.2.5), it follows that each U_n meets at most countably many left cosets of H and hence that $\bigcup_n U_n$ is included in the union of a countable collection of left cosets of H. Since H, along with each of its cosets, is σ-compact, the lemma follows. $\square$

Lemma 9.4.3. *Let G be a locally compact group, let μ be a left Haar measure on G, and let $F : G \times G \to G \times G$ be defined by $F(s,t) = (s, s^{-1}t)$. Then F is a measure-preserving homeomorphism of $G \times G$ onto itself. That is, F is a homeomorphism such that each Borel subset A of $G \times G$ satisfies $(\mu \times \mu)(A) = (\mu \times \mu)(F^{-1}(A))$.*

Proof. The inverse of F is given by $F^{-1}(s,t) = (s, st)$; thus F and F^{-1} are both continuous, and F is a homeomorphism. The regularity of the measure $(\mu \times \mu)F^{-1}$ follows. Now suppose that U is an open subset of $G \times G$. For each s in G the sections U_s and $(F^{-1}(U))_s$ are related by $(F^{-1}(U))_s = sU_s$, and so Proposition 7.6.5 and the translation invariance of μ imply that $(\mu \times \mu)(U) = (\mu \times \mu)(F^{-1}(U))$. It follows from this and the regularity of the measures $\mu \times \mu$ and $(\mu \times \mu)F^{-1}$ that $(\mu \times \mu)(A) = (\mu \times \mu)(F^{-1}(A))$ holds for each A in $\mathscr{B}(G \times G)$. $\square$

Proof of Proposition 9.4.1. It follows from Exercise 7.6.4 that the function $(s,t) \mapsto f(s)g(t)$ belongs to $\mathscr{L}^1(G \times G, \mathscr{B}(G \times G), \mu \times \mu)$ and then from Lemma 9.4.3 that the function $(s,t) \mapsto f(s)g(s^{-1}t)$ belongs to $\mathscr{L}^1(G \times G, \mathscr{B}(G \times G), \mu \times \mu)$ (see Sect. 2.6). Since in addition $(s,t) \mapsto f(s)g(s^{-1}t)$ vanishes outside a σ-compact set (apply Lemma 9.4.2 to f and g, and then use Lemma 9.4.3), Theorem 7.6.7 implies part (a) and the first half of part (b). The second half of part (b) follows from the calculation

$$\int |(f * g)(t)| \mu(dt) \leq \int \int |f(s)g(s^{-1}t)| \mu(ds) \mu(dt)$$

$$= \int \int |f(s)g(t)| \mu(dt) \mu(ds) = \|f\|_1 \|g\|_1.$$

$\square$

Note that if f_1, f_2, g_1, and g_2 belong to $\mathscr{L}^1(G, \mathscr{B}(G), \mu)$, if $f_1 = f_2$ μ-a.e., and if $g_1 = g_2$ μ-a.e., then $f_1 * g_1 = f_2 * g_2$ μ-a.e.; this follows, for example, from the calculation

$$\|f_1 * g_1 - f_2 * g_2\|_1 \leq \|f_1 * (g_1 - g_2)\|_1 + \|(f_1 - f_2) * g_2\|_1$$

$$\leq \|f_1\|_1 \|g_1 - g_2\|_1 + \|f_1 - f_2\|_1 \|g_2\|_1 = 0$$

(see also Exercise 4). Thus convolution on $\mathscr{L}^1(G,\mathscr{B}(G),\mu)$ induces an operation on $L^1(G,\mathscr{B}(G),\mu)$; this operation is also denoted by $*$ and called *convolution*.

We will show that $L^1(G,\mathscr{B}(G),\mu)$, with convolution as multiplication, is a Banach algebra. (This Banach algebra is often denoted by $L^1(G)$.) Recall that an *algebra* is a vector space A on which there is defined an operation $\cdot$ (called multiplication) for which the identities

$$u\cdot(v\cdot w)=(u\cdot v)\cdot w,$$

$$u\cdot(v+w)=u\cdot v+u\cdot w,$$

$$(u+v)\cdot w=u\cdot w+v\cdot w,\text{ and}$$

$$\alpha(u\cdot v)=(\alpha u)\cdot v=u\cdot(\alpha v)$$

hold for all u, v, and w in A and all scalars α. A *Banach algebra* is an algebra for which

(a) the underlying vector space has the structure of a Banach space, say with norm $\|\cdot\|$, and
(b) the relation $\|u\cdot v\|\le\|u\|\|v\|$ holds for all u and v in A.

Proposition 9.4.4. *Let G be a locally compact group, and let μ be a left Haar measure on G. Then $L^1(G,\mathscr{B}(G),\mu)$, with convolution as multiplication, is a Banach algebra.*

Proof. With the exception of the associative law for convolutions, the conditions that define a Banach algebra are either immediate or given by Theorem 3.4.1 and Proposition 9.4.1.

We turn to the associative law. Suppose that f, g, and h belong to $\mathscr{K}(G)$ (or to $\mathscr{K}^{\mathbb{C}}(G)$) and that x belongs to G. Then the functions involved in computing $f*(g*h)$ and $(f*g)*h$ are all integrable, and these convolutions are given by

$$(f*(g*h))(x)=\int f(s)(g*h)(s^{-1}x)\,\mu(ds)$$

$$=\int\int f(s)g(t)h(t^{-1}s^{-1}x)\,\mu(dt)\,\mu(ds)$$

and

$$((f*g)*h)(x)=\int(f*g)(t)h(t^{-1}x)\,\mu(dt)$$

$$=\int\int f(s)g(s^{-1}t)h(t^{-1}x)\,\mu(ds)\,\mu(dt).$$

Consider the last of these integrals; in it reverse the order of integration and use the translation invariance of μ to replace t with st. It follows that $(f*(g*h))(x)=((f*g)*h)(x)$. Thus the associative law holds for those elements of $L^1(G,\mathscr{B}(G),\mu)$ that are determined by functions in $\mathscr{K}(G)$ (or in $\mathscr{K}^{\mathbb{C}}(G)$); since these elements are dense in $L^1(G, \mathscr{B}(G), \mu)$ (Proposition 7.4.3), the associative law follows (see Exercise 2). □

Let us turn to the convolution of measures. We begin with the following lemma.

Lemma 9.4.5. *Let G be a locally compact group. If μ and ν are finite positive regular Borel measures on G and if $\mu \times \nu$ is the regular Borel product of μ and ν, then the formula*

$$(\mu * \nu)(A) = (\mu \times \nu)(\{(x,y) \in G \times G : xy \in A\})$$

defines a regular Borel measure on G. Furthermore,

$$(\mu * \nu)(A) = \int \nu(x^{-1}A)\,\mu(dx) = \int \mu(Ay^{-1})\,\nu(dy) \qquad (2)$$

holds for each A in $\mathscr{B}(G)$.

Note that Corollary 7.6.6 implies that the functions appearing on the right side of (2) are Borel measurable.

Proof. Let $F \colon G \times G \to G$ be the group operation (in other words, define F by $F(x,y) = xy$). Then $\mu * \nu$ is given by the equation $(\mu * \nu)(A) = (\mu \times \nu)(F^{-1}(A))$, and so is a measure on $\mathscr{B}(G)$ (see Sect. 2.6). Corollary 7.6.6 implies that each A in $\mathscr{B}(G)$ satisfies (2). We need to check the regularity of $\mu * \nu$.

We begin by checking that an arbitrary Borel subset A of G satisfies

$$(\mu * \nu)(A) = \sup\{(\mu * \nu)(K) : K \subseteq A \text{ and } K \text{ is compact}\}. \qquad (3)$$

Suppose that ε is a positive number, that K_0 is a compact subset of $F^{-1}(A)$ such that $(\mu \times \nu)(K_0) > (\mu \times \nu)(F^{-1}(A)) - \varepsilon$ (see Proposition 7.2.6), and that $K = F(K_0)$. Then K is a compact subset of A such that $F^{-1}(K) \supseteq K_0$ and hence such that $(\mu * \nu)(K) > (\mu * \nu)(A) - \varepsilon$. Since ε is arbitrary, (3) follows. In particular, $\mu * \nu$ is inner regular. Since for each A in $\mathscr{B}(G)$ we can use (3), applied to A^c, to approximate A^c from below by compact sets and hence to approximate A from above by open sets, the outer regularity of $\mu * \nu$ follows. □

Recall that $M_r(G, \mathbb{R})$ is the Banach space of all finite signed regular Borel measures on G (the norm of μ is the total variation of μ). Likewise, $M_r(G, \mathbb{C})$ is the Banach space of all complex regular Borel measures on G. Here we will denote each of those spaces by $M(G)$.

Let μ and ν belong to $M(G)$. We define their *convolution* $\mu * \nu$ by

$$(\mu * \nu)(A) = \int \nu(x^{-1}A)\,\mu(dx) = \int \mu(Ay^{-1})\,\nu(dy). \qquad (4)$$

It follows from the preceding lemma and the Jordan decomposition theorem that the two integrals appearing in (4) exist and are equal, and that $\mu * \nu$ is regular. Thus $\mu * \nu \in M(G)$.

It is easy to check that if μ and v belong to $M(G)$ and if f is a bounded Borel function on G, then

$$\int f\,d(\mu * v) = \int\int f(xy)\,\mu(dx)\,v(dy) = \int\int f(xy)\,v(dy)\,\mu(dx) \qquad (5)$$

(first check (5) for characteristic functions, and then use the linearity of the integral and the dominated convergence theorem).

Proposition 9.4.6. *Let G be a locally compact group. Then M(G), with convolution as multiplication, is a Banach algebra.*

Proof. Let v_1, v_2, and v_3 belong to $M(G)$. Then each Borel subset A of G satisfies

$$(v_1 * (v_2 * v_3))(A) = \int (v_2 * v_3)(x^{-1}A)\,v_1(dx)$$

$$= \int\int v_3(y^{-1}x^{-1}A)\,v_2(dy)\,v_1(dx)$$

and

$$((v_1 * v_2) * v_3)(A) = \int v_3(u^{-1}A)\,(v_1 * v_2)(du)$$

$$= \int\int v_3((xy)^{-1}A)\,v_2(dy)\,v_1(dx)$$

(in the last step of this calculation we used (5)). The associativity of convolution follows.

We turn to the inequality $\|\mu * v\| \le \|\mu\|\,\|v\|$. Let $\{A_i\}_1^n$ be a finite partition of G into Borel sets. Then Exercise 4.2.8 implies that

$$\sum_i |(\mu * v)(A_i)| = \sum_i \left| \int \mu(A_i y^{-1})\,v(dy) \right|$$

$$\le \int \sum_i |\mu(A_i y^{-1})|\,|v|(dy) \le \int \|\mu\|\,d|v| = \|\mu\|\,\|v\|.$$

Since the partition $\{A_i\}$ was arbitrary, the inequality $\|\mu * v\| \le \|\mu\|\,\|v\|$ follows. The remaining conditions in the definition of a Banach algebra are clearly satisfied. $\square$

Let us consider the relationship between the convolution of functions and the convolution of measures. Corollary 9.3.7 implies that an element of $M(G)$ is absolutely continuous with respect to the left Haar measures on G if and only if it is absolutely continuous with respect to the right Haar measures on G. Thus we can define $M_a(G)$ to be the collection of elements of $M(G)$ that are absolutely continuous with respect to some (and hence every) Haar measure on G.

Recall that an *ideal* in an algebra A is a linear subspace I of A such that $u \cdot v$ and $v \cdot u$ belong to I whenever u belongs to I and v belongs to A.

Proposition 9.4.7. *Let G be a locally compact group. Then*

(a) $M_a(G)$ *is an ideal in the algebra* $M(G)$,
(b) *if* μ *is a left Haar measure on G, then the map* $f \mapsto \nu_f$ *(where* ν_f *is defined by* $\nu_f(A) = \int_A f \, d\mu$*) induces a norm-preserving algebra homomorphism of* $L^1(G, \mathscr{B}(G), \mu)$ *into* $M(G)$, *and*
(c) *the image of* $L^1(G, \mathscr{B}(G), \mu)$ *under this homomorphism is* $M_a(G)$.

Proof. It is clear that $M_a(G)$ is a linear subspace of $M(G)$. Suppose that μ is a left Haar measure on G, that $\nu_1 \in M(G)$, and that $\nu_2 \in M_a(G)$. Let A be a Borel subset of G that satisfies $\mu(A) = 0$. The translation invariance of μ implies that $\mu(x^{-1}A) = 0$ holds for each x in G; since $\nu_2 \ll \mu$, the relation $\nu_2(x^{-1}A) = 0$ also holds for each x in G. The definition of $\nu_1 * \nu_2$ now implies that $(\nu_1 * \nu_2)(A) = 0$. Hence $\nu_1 * \nu_2 \in M_a(G)$. The proof that $\nu_2 * \nu_1 \in M_a(G)$ is similar (use Corollary 9.3.7 to conclude that if $\mu(A) = 0$, then $\mu(Ay^{-1}) = 0$ holds for each y in G). Thus $M_a(G)$ is an ideal in $M(G)$.

We already know that the map $f \mapsto \nu_f$ induces a norm-preserving linear map whose image is $M_a(G)$ (Proposition 7.3.10). The calculation

$$\nu_{f*g}(A) = \int \chi_A(t) \int f(s) g(s^{-1}t) \mu(ds) \mu(dt)$$

$$= \int \int \chi_A(st) f(s) g(t) \mu(dt) \mu(ds)$$

$$= \int \int \chi_A(st) \nu_g(dt) \nu_f(ds)$$

$$= (\nu_f * \nu_g)(A)$$

shows that it preserves convolutions. $\square$

Proposition 9.4.7 provides a "coordinate-free" description of $L^1(G, \mathscr{B}(G), \mu)$: it is isomorphic to the algebra $M_a(G)$, whose definition depends only on the existence of Haar measures and not on the choice of a particular left or right Haar measure.

Let us close this section by returning to the map T constructed in Sect. 3.5 (see also Theorem 4.5.1, Example 4.5.2, Theorem 7.5.4, and the remarks following the proof of Theorem 7.5.4).

Theorem 9.4.8. *Let G be a locally compact group, and let* μ *be a regular Borel measure on G. Then the map T constructed in Sect. 3.5 is an isometric isomorphism of* $L^\infty(G, \mathscr{B}(G), \mu)$ *onto the dual of* $L^1(G, \mathscr{B}(G), \mu)$.

Proof. According to Proposition 3.5.5 we need only show that T is surjective. So suppose that F belongs to $(L^1(G, \mathscr{B}(G), \mu))^*$.

Let H be a subgroup of G that is open and σ-compact (see Proposition 9.1.8), and let $\mathscr{H}$ be the family of left cosets of H. For each C in $\mathscr{H}$ let $\mathscr{B}(C)$ be the σ-algebra of Borel subsets of C, let μ_C be the restriction of μ to $\mathscr{B}(C)$, and let F_C be the linear functional on $L^1(C, \mathscr{B}(C), \mu_C)$ defined by $F_C(\langle f \rangle) = F(\langle f' \rangle)$ (here f' is the function on G that agrees with f on C and vanishes on C^c). Since μ_C is

σ-finite (recall that C, as a coset of H, is σ-compact), we can choose a bounded Borel measurable function g_C on C such that $F_C(\langle f \rangle) = \int f g_C \, d\mu_C$ holds for each f in $\mathcal{L}^1(C, \mathcal{B}(C), \mu_C)$ (see Theorem 4.5.1). By modifying g_C on a μ_C-null set if necessary, we can assume that $|g_C(x)| \leq \|F_C\| \leq \|F\|$ holds at each x in C. Now choose a sequence $\{g_n\}$ of continuous functions on G such that

(a) $|g_n(x)| \leq \|F\|$ holds at each x in G, and
(b) for each C in $\mathcal{H}$ the sequence $\{g_n\}$ converges to g_C μ-almost everywhere on C

(construct the functions g_n on each C separately, using Lusin's theorem (Theorem 7.4.4) and the σ-compactness of the sets in $\mathcal{H}$; see also D.6). Finally, define[4] g by $g = \limsup_n g_n$ (in case we are dealing with complex-valued functions, define the real and imaginary parts of g separately). Then g is a bounded Borel function, and the relation $F(\langle f \rangle) = \int f g \, d\mu$ holds for each f in $\mathcal{L}^1(G, \mathcal{B}(G), \mu)$. Thus T is surjective, and the proof is complete. $\square$

Exercises

Note: In the following exercises G is a locally compact group with identity element e, and μ is a left Haar measure on G.

1. Show that if f and g belong to $\mathcal{K}(G)$, then $f * g$ belongs to $\mathcal{K}(G)$.
2. Show that if f and g belong to $\mathcal{L}^1(G, \mathcal{B}(G), \mu)$ and if $\{f_n\}$ and $\{g_n\}$ are sequences in $\mathcal{L}^1(G, \mathcal{B}(G), \mu)$ such that $\lim_n \|f_n - f\|_1 = 0$ and $\lim_n \|g_n - g\|_1 = 0$, then $\lim_n \|f_n * g_n - f * g\|_1 = 0$.
3. Suppose that f and g belong to $\mathcal{L}^1(G, \mathcal{B}(G), \mu)$. Show that in the definition of $f * g$ the expression $f(s)g(s^{-1}t)$ can be replaced
   (a) with $f(ts)g(s^{-1})$,
   (b) with $f(s^{-1})g(st)\Delta(s^{-1})$, and
   (c) with $f(ts^{-1})g(s)\Delta(s^{-1})$.
4. Show that if f_1, f_2, g_1, and g_2 belong to $\mathcal{L}^1(G, \mathcal{B}(G), \mu)$, if $f_1 = f_2$ μ-almost everywhere, and if $g_1 = g_2$ μ-almost everywhere, then $f_1 * g_1 = f_2 * g_2$ *everywhere*.
5. Show that G is commutative if and only if convolution is a commutative operation on $L^1(G)$. (Hint: To show that the commutativity of $L^1(G)$ implies that of G, consider $f * g$ and $g * f$ for suitable nonnegative functions f and g in $\mathcal{K}(G)$.)
6. (a) Suppose that the locally compact group G has a countable base for its topology. Show that there is a sequence $\{\varphi_n\}$ of nonnegative functions in $\mathcal{L}^1(G, \mathcal{B}(G), \mu)$ (or even in $\mathcal{K}(G)$) such that $\int \varphi_n \, d\mu = 1$ holds for each n and such that

---

[4]The function g cannot be defined simply be requiring that its restriction to each C in $\mathcal{H}$ be g_C; see Exercise 12.

$$\lim_n \|f * \varphi_n - f\|_1 = \lim_n \|\varphi_n * f - f\|_1 = 0 \qquad (6)$$

holds for each f in $\mathscr{L}^1(G, \mathscr{B}(G), \mu)$. Such a sequence is called an *approximate identity*. (Hint: Let $\{U_n\}$ be a decreasing sequence of open neighborhoods of e such that each open neighborhood of e includes some U_n. For each n let φ_n be a nonnegative function that belongs to $\mathscr{K}(X)$, vanishes outside U_n, and satisfies the relations $\varphi_n = (\varphi_n)^\vee$ and $\int \varphi_n \, d\mu = 1$. In verifying (6) it might be convenient to begin with the case where $f \in \mathscr{K}(G)$.)

(b) Now omit the assumption that G has a countable base for its topology. Show that there is a net[5] $\{\varphi_\alpha\}_{\alpha \in A}$ of nonnegative functions in $\mathscr{K}(G)$ such that $\int \varphi_\alpha \, d\mu = 1$ holds for each α and such that $\lim_\alpha \|f * \varphi_\alpha - f\|_1 = \lim_\alpha \|\varphi_\alpha * f - f\|_1 = 0$ holds for each f in $\mathscr{L}^1(G, \mathscr{B}(G), \mu)$. (Hint: Let the directed set A be the collection of all open neighborhoods of e, and declare that $U \le V$ holds if and only if $V \subseteq U$.)

7. Show that δ_e, the point mass concentrated at e, is an identity for the algebra $M(G)$.

8. Show that G is commutative if and only if convolution is a commutative operation on $M(G)$.

9. Suppose that $\nu \in M(G)$, that $f \in \mathscr{L}^1(G, \mathscr{B}(G), \mu)$, and that μ_f is the finite signed or complex regular Borel measure defined by $\mu_f(A) = \int_A f \, d\mu$ (see Proposition 7.3.8). Define functions g and h on G by

$$g(t) = \begin{cases} \int f(s^{-1}t) \, \nu(ds) & \text{if } s \mapsto f(s^{-1}t) \text{ is } |\nu|\text{-integrable}, \\ 0 & \text{otherwise}, \end{cases}$$

and

$$h(s) = \begin{cases} \int f(st^{-1})\Delta(t^{-1}) \, \nu(dt) & \text{if } t \mapsto f(st^{-1})\Delta(t^{-1}) \text{ is } |\nu|\text{-integrable}, \\ 0 & \text{otherwise}. \end{cases}$$

Show that g and h belong to $\mathscr{L}^1(G, \mathscr{B}(G), \mu)$ and that $(\nu * \mu_f)(A) = \int_A g \, d\mu$ and $(\mu_f * \nu)(A) = \int_A h \, d\mu$ hold for each A in $\mathscr{B}(G)$.

10. Let ν, f, and μ_f be as in Exercise 9. Show that $\nu * \mu_f = 0$ holds for each f in $\mathscr{L}^1(G, \mathscr{B}(G), \mu)$ if and only if $\nu = 0$. (Hint: Use Exercise 9 and Corollary 9.1.6

---

[5]Recall that a *directed set* is a partially ordered set A (say ordered by $\le$) such that for each α and β in A, there is an element γ of A that satisfies $\alpha \le \gamma$ and $\beta \le \gamma$. A *net* is a family indexed by a directed set. A net $\{x_\alpha\}_{\alpha \in A}$ in a topological space X is said to *converge* to a point x of X if for each open neighborhood U of x there is an element α_0 of A such that $x_\alpha \in U$ holds whenever α satisfies $\alpha \ge \alpha_0$. Thus $\lim_\alpha \|f * \varphi_\alpha - f\|_1 = 0$ holds if and only if for each positive ε there is an element α_0 of A such that $\|f * \varphi_\alpha - f\|_1 < \varepsilon$ holds whenever α satisfies $\alpha \ge \alpha_0$. See Kelley [69] for an extended treatment of nets.

to show that if f belongs to $\mathscr{K}(G)$ and satisfies $\nu * \mu_f = 0$, then $\int \check{f} d\nu = 0$; then use Theorem 7.3.6.)

11. Show that $L^1(G)$ has an identity if and only if the topology of G is discrete. (Hint: Use Exercise 6 in Sect. 9.2 and Exercises 7 and 10.)

12. Let G be $\mathbb{R}^2$, with the usual group operation but with the topology defined in Exercise 7.2.4. Show that

    (a) G is a locally compact group,

    (b) $\{0\} \times \mathbb{R}$ is an open, closed, and σ-compact subgroup of G, and

    (c) there is a function $f \colon G \to \mathbb{R}$ that is not Borel measurable, but for which each section f_x is Borel measurable. (Hint: See Exercises 8.2.7 and 8.2.9.)

    This explains the footnote in the proof of Theorem 9.4.8.

Notes

The history of Haar measure is summarized in the notes at the ends of Sections 15 and 16 of Hewitt and Ross [58].

The reader can find a more extensive introduction to topological groups in Pontryagin [98] or in Hewitt and Ross [58].

The proof given here for the existence of Haar measure (which is a modification of Halmos's modification of Weil's [126] proof) depends on the axiom of choice. Proofs that do not depend on this axiom have been given by Cartan [24] and Bredon [19]. Cartan's proof is given by Hewitt and Ross [58] and by Nachbin [93]. Hewitt and Ross and Nachbin also give calculations of Haar measure for a number of groups.

Chapter 10
Probability

This chapter is devoted to an introduction to probability theory. It contains some of the fundamental results of probability theory—the strong law of large numbers, the central limit theorem, the martingale convergence theorem, the construction of Brownian motion processes, and Kolmogorov's consistency theorem.

One purpose of this chapter is to give the reader a chance to work through some applications of measure theory and thereby to get some practice with the techniques presented earlier. Another, perhaps more significant, goal is to give the reader a broader picture of how σ-algebras, measures, measurable functions, and integrals arise.

10.1 Basics

In probability theory one describes and analyzes random situations, often called *experiments*. Let us look at how such situations can be modeled using measure theory. We begin with some terminology.

A *probability space* is a measure space $(\Omega, \mathscr{A}, P)$ such that $P(\Omega) = 1$. The elements of Ω are called the *elementary outcomes* or the *sample points* of our experiment, and the members of $\mathscr{A}$ are called *events*. If $A \in \mathscr{A}$, then $P(A)$ is the *probability* of the event A.

Example 10.1.1. We illustrate these concepts with a very simple example. Suppose we toss a fair coin (one for which a head has probability $1/2$) twice. There are four possible outcomes: we get two heads, we get a head and then a tail, we get a tail and then a head, or we get two tails. So we can let our set Ω of elementary outcomes be $\{HH, HT, TH, TT\}$. It is natural in this case to let $\mathscr{A}$ contain all the subsets of Ω. For example, $\{HT, TH\}$ is one of the subsets of Ω; it corresponds to the real-world event in which we get a head on exactly one of the tosses. Finally, in this situation each elementary outcome has probability $1/4$ of occurring, and so we let the probability of an event A be $1/4$ times the number of elements of A. □

D.L. Cohn, *Measure Theory: Second Edition*, Birkhäuser Advanced
Texts Basler Lehrbücher, DOI 10.1007/978-1-4614-6956-8_10,
© Springer Science+Business Media, LLC 2013

A *real-valued random variable* on a probability space $(\Omega, \mathscr{A}, P)$ is an $\mathscr{A}$-measurable function from Ω to $\mathbb{R}$. Such a variable represents a numerical observation or measurement whose value depends on the outcome of the random experiment represented by $(\Omega, \mathscr{A}, P)$. More generally, a *random variable* with values in a measurable space $(S, \mathscr{B})$ is a measurable function from $(\Omega, \mathscr{A}, P)$ to $(S, \mathscr{B})$. Let X be a random variable with values in $(S, \mathscr{B})$. The *distribution* of X is the measure PX^{-1} defined on $(S, \mathscr{B})$ by $(PX^{-1})(A) = P(X^{-1}(A))$ (see Sect. 2.6). We will often write P_X for the distribution of a random variable X. If $X_1, \ldots, X_d$ are $(S, \mathscr{B})$-valued random variables on $(\Omega, \mathscr{A}, P)$, then the formula $X(\omega) = (X_1(\omega), \ldots, X_d(\omega))$ defines an S^d-valued random variable X; the distribution of X is called the *joint distribution* of $X_1, \ldots, X_d$.

Example 10.1.2. Let us continue with our coin-tossing example. The number of heads that appear when our two coins are tossed can be represented with the random variable X defined by

$$X(\omega) = \begin{cases} 0 & \text{if } \omega = TT, \\ 1 & \text{if } \omega = HT \text{ or } \omega = TH, \text{ and} \\ 2 & \text{if } \omega = HH. \end{cases}$$

The distribution P_X of X is given by $P_X = \frac{1}{4}\delta_0 + \frac{1}{2}\delta_1 + \frac{1}{4}\delta_2$. $\square$

An abbreviated notation for events is common in probability. We introduce it with a couple of examples. Suppose that $(\Omega, \mathscr{A}, P)$ is a probability space and that X and X_n, $n = 1, 2, \ldots$, are real-valued random variables on Ω. Then the events

$$\{\omega \in \Omega : X(\omega) \geq 0\},$$

$$\{\omega \in \Omega : X(\omega) = \lim_n X_n(\omega)\},$$

and

$$\{\omega \in \Omega : \lim_n X_n(\omega) \text{ exists}\}$$

are often abbreviated as $\{X \geq 0\}$, $\{X = \lim_n X_n\}$, and $\{\lim_n X_n \text{ exists}\}$. Sometimes one goes a bit further and simply writes $P(X \geq 0)$ instead of $P(\{X \geq 0\})$ or $P(\{\omega \in \Omega : X(\omega) \geq 0\})$.

If a real-valued random variable X on a probability space $(\Omega, \mathscr{A}, P)$ is integrable with respect to P, then its *expected value*, or *expectation*, written $E(X)$, is the integral of X with respect to P. That is, $E(X) = \int X \, dP$. If X is integrable, one also says that X *has a finite expected value* or that X *has an expected value*. Note that Proposition 2.6.8 gives a way to compute the expected value of a real-valued random variable in terms of its distribution, namely $E(X) = \int_{\mathbb{R}} x P_X(dx)$. That proposition in fact gives the more general formula $E(f \circ X) = \int_{\mathbb{R}} f \, dP_X$, by which we can compute the expected value of a Borel function f of a random variable X in terms of the distribution of X.

We often have use for the expected value of the square of a real-valued random variable X, or the *second moment* of X. If X has a finite second moment, then it follows from the inequality $|X| \leq X^2 + 1$ that X has a finite expectation. In this case, one calls the expected value of $(X - E(X))^2$ the *variance* of X; it gives a measure of the amount by which the values of X differ from the expected value of X. The nonnegative square root of the variance of X is called the *standard deviation* of X. One often denotes the expected value of a random variable X with μ_X or simply μ, the variance with $\mathrm{var}(X)$ or σ_X^2, and the standard deviation with σ_X.

Lemma 10.1.3. *Let X be a random variable with a finite second moment, and let a and b be real numbers. Then*

(a) $\mathrm{var}(X) = E(X^2) - (E(X))^2$, *and*
(b) $\mathrm{var}(aX + b) = a^2 \mathrm{var}(X)$.

Proof. The lemma follows from basic algebra and the linearity of the integral. $\square$

Suppose that X is a real-valued random variable with a discrete distribution—that is, suppose that there is a countable subset C of $\mathbb{R}$ such that $P(X \in C) = 1$. Then X has a finite expected value if and only if $\sum_{x \in C} |x| P(X = x) < +\infty$, and in that case $E(X) = \sum_{x \in C} x P(X = x)$. Likewise, if the distribution P_X of X is absolutely continuous with respect to Lebesgue measure and if f_X is the Radon–Nikodym derivative of P_X with respect to Lebesgue measure, then X has a finite expected value if and only if $\int_{\mathbb{R}} |x| f_X(x) \, dx < +\infty$, and in that case $E(X) = \int_{\mathbb{R}} x f_X(x) \, dx$. As these remarks may suggest, it turns out that discrete and continuous random variables,[1] which receive separate treatments in elementary discussions of probability theory, can be given a fairly uniform treatment in terms of measure theory.

We have seen (in Propositions 1.3.9 and 1.3.10) that there is a correspondence between finite Borel measures on $\mathbb{R}$ and bounded nondecreasing right-continuous functions $F: \mathbb{R} \to \mathbb{R}$ for which $\lim_{x \to -\infty} F(x) = 0$. In the present context, this means that the distribution P_X of a real-valued random variable X is determined by the function $F_X: \mathbb{R} \to \mathbb{R}$ defined by

$$F_X(x) = P_X((-\infty, x]) = P(X \leq x).$$

The function F_X is called the *cumulative distribution function*, or just the *distribution function*, of X.

Let $\{X_i\}_{i \in I}$ be an indexed family of random variables on a probability space $(\Omega, \mathscr{A}, P)$. Then $\sigma(X_i, i \in I)$ is the smallest σ-algebra on Ω that makes all these variables measurable. Likewise, if $\{X_n\}$ is a sequence of random variables on $(\Omega, \mathscr{A}, P)$, then one often writes $\sigma(X_1, X_2, \ldots)$ for the smallest σ-algebra on Ω that makes each X_n measurable.

---

[1] A real-valued random variable is *discrete* if its distribution is discrete and is *continuous* if its distribution is absolutely continuous with respect to Lebesgue measure.

Examples 10.1.4.

(a) We begin by returning to coin tossing. Suppose that now our experiment is to toss a fair coin repeatedly, until we first get a head, and then to stop. It seems reasonable to define Ω by

$$\Omega = \{H, TH, TTH, \ldots, TTTTTTTH, \ldots\}$$

and to let $\mathscr{A}$ consist of all subsets of Ω. We will (by countable additivity) determine the probability of all the events in $\mathscr{A}$ if we specify the probabilities of the one-point subsets of Ω. It seems reasonable to let $P(\{H\}) = 1/2$, $P(\{TH\}) = 1/4$, $P(\{TTH\}) = 1/8$, ... (the reader should think through this assignment of probabilities again, after reading the discussion of independence that occurs later in this section). Note that the sum of the geometric series $\sum_{n=1}^{\infty}(1/2)^n$ is 1, and so this assignment of probabilities does give a probability measure.

(b) Now suppose that we choose a real number from the interval $[a,b]$ in such a way that the probability that the number chosen lies in a subinterval I of $[a,b]$ is proportional to the length of I. We can describe this situation with the probability space $([a,b], \mathscr{B}([a,b]), P)$, where the measure P is given by $P(A) = \lambda(A)/(b-a)$. In this case one has a *uniform distribution* on $[a,b]$. Of course, if the interval $[a,b]$ is the unit interval $[0,1]$, then the measure P is just the restriction of Lebesgue measure to the Borel subsets of $[0,1]$.

(c) Now suppose that f is a nonnegative Borel measurable function on $\mathbb{R}$ such that $\int f \, d\lambda = 1$. Then the formula $P(A) = \int_A f \, d\lambda$ defines a probability measure on the measurable space $(\mathbb{R}, \mathscr{B}(\mathbb{R}))$. The function f is called the *density* of P (or of a random variable having distribution P). Note that the measures in part (b) above can be viewed as special cases of the situation here, with the uniform distribution on $[a,b]$ given by the density function that has value $1/(b-a)$ on $[a,b]$ and 0 elsewhere.

(d) In a similar way, a nonnegative Borel measurable function on $\mathbb{R}^2$ such that $\int \int f(x,y) \lambda(dx)\lambda(dy) = 1$ defines a probability measure on the measurable space $(\mathbb{R}^2, \mathscr{B}(\mathbb{R}^2))$.

(e) Let us now look at *normal*, or *Gaussian*, distributions, which are given by the familiar bell-shaped curves. We begin by evaluating the integral $\int_{\mathbb{R}} e^{-x^2/2} \, dx$. Let us denote the value of this integral by A for a moment. If we interpret A^2 as an integral over $\mathbb{R}^2$ and evaluate the integral using polar coordinates, we find

$$A^2 = \int_{-\infty}^{\infty} \int_{-\infty}^{\infty} e^{-(x^2+y^2)/2} \, dx\, dy = \int_0^{2\pi} \int_0^{\infty} r e^{-r^2/2} \, dr\, d\theta = 2\pi.$$

Thus $A = \sqrt{2\pi}$, and so the function $x \mapsto \frac{1}{\sqrt{2\pi}} e^{-x^2/2}$ is a probability density function on $\mathbb{R}$ (that is, it is nonnegative and its integral over $\mathbb{R}$ is 1).

Now suppose that X is a random variable whose distribution has density $x \mapsto \frac{1}{\sqrt{2\pi}}e^{-x^2/2}$. It is easy to check that

$$\frac{1}{\sqrt{2\pi}}\int_{\mathbb{R}} xe^{-x^2/2}\,dx = 0$$

and hence that $E(X) = 0$. If in the following calculation we use integration by parts to convert the first integral into the second, whose value we know, we find that

$$\frac{1}{\sqrt{2\pi}}\int_{\mathbb{R}} x^2 e^{-x^2/2}\,dx = \frac{1}{\sqrt{2\pi}}\int_{\mathbb{R}} e^{-x^2/2}\,dx = 1$$

and hence that $E(X^2) = 1$. Thus X has expected value 0 and variance 1.

It is easy to check that if X is as above and if μ and σ are constants, with $\sigma > 0$, then the random variable $\sigma X + \mu$ has mean μ and variance σ^2 (see Lemma 10.1.3). Furthermore, according to Lemma 10.1.5, $\sigma X + \mu$ has density g_{μ,σ^2} given by

$$g_{\mu,\sigma^2}(x) = \frac{1}{\sqrt{2\pi}\sigma}e^{-(x-\mu)^2/2\sigma^2}.$$

With this we have the densities of the *normal* or *Gaussian* random variables with mean μ and variance σ^2.

One often writes $N(0,1)$ for the distribution of a normal random variable with mean 0 and variance 1 and $N(\mu,\sigma^2)$ for the distribution of a normal random variable with mean μ and variance σ^2. Thus $N(0,1)$ is the measure on $(\mathbb{R},\mathscr{B}(\mathbb{R}))$ with density $x \mapsto \frac{1}{\sqrt{2\pi}}e^{-x^2/2}$, and $N(\mu,\sigma^2)$ is the measure on $(\mathbb{R},\mathscr{B}(\mathbb{R}))$ with density g_{μ,σ^2}. □

Lemma 10.1.5. *Let X be a real-valued random variable with density f_X, let a and b be real constants with $a > 0$, and let $Y = aX + b$. Then Y is a continuous random variable whose density f_Y is given by*

$$f_Y(t) = \frac{1}{a}f_X\left(\frac{t-b}{a}\right).$$

Proof. Define a function $T: \mathbb{R} \to \mathbb{R}$ by $T(t) = at + b$. Then $\lambda(T(A)) = a\lambda(A)$ holds for each subinterval A of $\mathbb{R}$ and consequently for each Borel subset A of $\mathbb{R}$. Thus

$$a\int h\,d\lambda = \int h \circ T^{-1}\,d\lambda$$

holds for each nonnegative measurable h (check this first in the case where h is the indicator function of a Borel set), and so we have

$$P_Y(A) = P_X(T^{-1}(A)) = \int_{T^{-1}(A)} f_X \, d\lambda$$

$$= (1/a) \int (\chi_{T^{-1}(A)} \circ T^{-1})(f_X \circ T^{-1}) \, d\lambda$$

$$= (1/a) \int_A f_X\left(\frac{t-b}{a}\right) \lambda(dt).$$

Thus P_Y can be calculated by integrating the function $t \mapsto \frac{1}{a} f_X(\frac{t-b}{a})$, and the proof is complete.   □

We will need the following fact about normal distributions.

Lemma 10.1.6. *Let Z be a normal random variable with mean 0 and variance 1. Then*

$$P(Z \geq A) \leq \frac{1}{\sqrt{2\pi}A} e^{-A^2/2}$$

holds for each positive real number A.

Proof. We have

$$P(Z \geq A) = \frac{1}{\sqrt{2\pi}} \int_A^\infty e^{-x^2/2} \, dx \leq \frac{1}{\sqrt{2\pi}} \int_A^\infty \frac{x}{A} e^{-x^2/2} \, dx = \frac{1}{\sqrt{2\pi}A} e^{-A^2/2}.$$

□

Let us turn to a few definitions and results involving independence.

Let $(\Omega, \mathscr{A}, P)$ be a probability space, and let $\{A_i\}_{i \in I}$ be an indexed family of events. The events[2] A_i, $i \in I$, are called *independent* if for each finite subset I_0 of I we have

$$P(\cap_{i \in I_0} A_i) = \prod_{i \in I_0} P(A_i).$$

Let $\{X_i\}_{i \in I}$ be an indexed family of random variables, defined on $(\Omega, \mathscr{A}, P)$ and with values in the measurable space $(S, \mathscr{B})$. The random variables X_i, $i \in I$, are called *independent* if for each choice of sets A_i in $\mathscr{B}$, $i \in I$, the events $X_i^{-1}(A_i)$ are independent.

Finally, let $(\Omega, \mathscr{A}, P)$ be a probability space and let $\{\mathscr{A}_i\}_{i \in I}$ be an indexed family of sub-σ-algebras of $\mathscr{A}$. The σ-algebras $\mathscr{A}_i$, $i \in I$, are *independent* if for each choice of sets A_i in $\mathscr{A}_i$, $i \in I$, the events A_i are independent.

Note that if $\{X_i\}_{i \in I}$ is an indexed family of random variables on a probability space $(\Omega, \mathscr{A}, P)$, then the random variables X_i, $i \in I$, are independent if and only if the σ-algebras $\sigma(X_i)$, $i \in I$, are independent.

---

[2] Although the independence of A_i, $i \in I$, depends on the relationship between the events A_i, rather than on the events individually, it is standard to call the events, rather than the indexed family, independent.

Proposition 10.1.7. *Let $(\Omega, \mathscr{A}, P)$ be a probability space, let $\{\mathscr{A}_i\}_{i \in I}$ be an indexed family of independent sub-σ-algebras of $\mathscr{A}$, let $\{S_j\}_{j \in J}$ be a partition of I, and for each j in J let $\mathscr{B}_j = \sigma(\cup_{i \in S_j} \mathscr{A}_i)$. Then the σ-algebras $\mathscr{B}_j$ are independent.*

Proof. For each j in J let $\mathscr{P}_j$ consist of all finite intersections of sets in $\cup_{i \in S_j} \mathscr{A}_i$. Note that each $\mathscr{P}_j$ is a π-system such that $\mathscr{B}_j = \sigma(\mathscr{P}_j)$. Let J_0 be a nonempty finite subset of J, and for each j in J_0 let A_j be a member of $\mathscr{P}_j$. The relation

$$P(\cap_{j \in J_0} A_j) = \prod_{j \in J_0} P(A_j) \tag{1}$$

follows from the independence of the $\mathscr{A}_i$'s. Now suppose that the elements of J_0 are $j_1, j_2, \ldots, j_n$, and let $\mathscr{D}$ be the class of all A in $\mathscr{B}_{j_n}$ such that

$$P(A_{j_1} \cap \cdots \cap A_{j_{n-1}} \cap A) = P(A_{j_1}) \cdots P(A_{j_{n-1}}) P(A)$$

holds for all A_{j_i} in $\mathscr{P}_{j_i}$, $i = 1, \ldots, n-1$. Then $\mathscr{D}$ is a Dynkin class (i.e., a d-system) that includes $\mathscr{P}_{j_n}$, and so Theorem 1.6.2 implies that $\mathscr{D} = \mathscr{B}_{j_n}$. Similar arguments, $n-1$ of them, show that (1) holds for all A_j in $\mathscr{B}_j$, $j \in J_0$. Since the independence of the $\mathscr{B}_j$, $j \in J$ depends only on the independence of finite subfamilies, the proof is complete. □

Example 10.1.8. Proposition 10.1.7 may look overly abstract, but it allows simple proofs of some results for which a rigorous proof might otherwise be awkward. For example, suppose that $\{X_n\}_{n=1}^{\infty}$ is a sequence of independent random variables on a probability space $(\Omega, \mathscr{A}, P)$. Then it is an immediate consequence of Proposition 10.1.7 that the random variables $X_{2i-1} + X_{2i}$, $i = 1, 2, \ldots$ are independent. Proving this independence in other ways would probably take more work. □

Proposition 10.1.9. *Let $(\Omega, \mathscr{A}, P)$ be a probability space, let $(S, \mathscr{B})$ be a measurable space, let $X_1, X_2, \ldots, X_d$ be S-valued random variables on Ω, and let X be the S^d-valued random variable with components $X_1, X_2, \ldots, X_d$. Let $P_{X_1}, P_{X_2}, \ldots, P_{X_d}$, and P_X be the distributions of $X_1, X_2, \ldots, X_d$, and X, respectively. Then $X_1, X_2, \ldots, X_d$ are independent if and only if the joint distribution P_X is equal to the product measure $P_{X_1} \times P_{X_2} \times \cdots \times P_{X_d}$.*

Proof. If we rewrite the definition of independence, we find that $X_1, \ldots, X_d$ are independent if and only if

$$P_X(A_1 \times \cdots \times A_d) = \prod_i P_{X_i}(A_i)$$

holds for each choice of sets A_i in $\mathscr{B}$, $i = 1, \ldots, d$. Thus if P_X is equal to the product of the measures P_{X_i}, then $X_1, X_2, \ldots, X_d$ are independent. The converse follows from the uniqueness of product measures (see Theorem 5.1.4 and the discussion at the end of Sect. 5.2). □

Proposition 10.1.10. *Let* $(\Omega, \mathscr{A}, P)$ *be a probability space and let* $X_1, X_2, \ldots, X_n$ *be independent real-valued random variables on* $(\Omega, \mathscr{A}, P)$, *each of which has a finite expectation. Then the product* $\prod_i X_i$ *has a finite expectation, and* $E(\prod_i X_i) = \prod_i E(X_i)$.

Proof. Let X be the $\mathbb{R}^n$-valued random variable with components $X_1, \ldots, X_n$, and let P_X and $P_{X_1}, \ldots, P_{X_n}$ be the distributions of X and $X_1, \ldots, X_n$. We will use these distributions for the calculation of $E(\prod_i X_i)$ and $\prod_i E(X_i)$. Since the random variables $X_i, \ldots, X_n$ are independent, P_X is the product of the measures $P_{X_1}, \ldots, P_{X_n}$ (Proposition 10.1.9). Thus we can use Proposition 5.2.1 and Theorem 5.2.2, together with the finiteness of the expectations $E(X_i)$ and the remarks at the end of Sect. 5.2, to conclude that $\prod_i X_i$ has a finite expectation and that $E(\prod_i X_i) = \prod_i E(X_i)$. $\qquad\square$

Corollary 10.1.11. *Let* $X_1, X_2, \ldots, X_n$ *be independent real-valued random variables with finite second moments, and let* $S = X_1 + \cdots + X_n$. *Then* $\mathrm{var}(S) = \sum_i \mathrm{var}(X_i)$.

Proof. By the independence of X_i and X_j (where $i \neq j$), the expectation of the product $(X_i - E(X_i))(X_j - E(X_j))$ is the product of the expectations of $X_i - E(X_i)$ and $X_j - E(X_j)$, namely 0. Thus

$$\mathrm{var}(S) = E\left(\left(\sum_i (X_i - E(X_i))\right)^2\right) = \sum_i \sum_j E\left((X_i - E(X_i))(X_j - E(X_j))\right)$$

$$= \sum_i E\left((X_i - E(X_i))^2\right) = \sum_i \mathrm{var}(X_i). \qquad\square$$

Now suppose that X_1 and X_2 are independent real-valued (or $\mathbb{R}^d$-valued) random variables with distributions P_{X_1} and P_{X_2}. In view of Proposition 10.1.9, we can use the product measure $P_{X_1} \times P_{X_2}$ to compute the distribution $P_{X_1 + X_2}$ of $X_1 + X_2$:

$$P_{X_1 + X_2}(A) = (P_{X_1} \times P_{X_2})(\{(x_1, x_2) : x_1 + x_2 \in A\}). \tag{2}$$

One defines the *convolution* $\nu_1 * \nu_2$ of finite measures ν_1 and ν_2 on $(\mathbb{R}^d, \mathscr{B}(\mathbb{R}^d))$ by

$$(\nu_1 * \nu_2)(A) = (\nu_1 \times \nu_2)(\{(x_1, x_2) : x_1 + x_2 \in A\});$$

thus (2) says that the distribution of the sum of two independent random variables is the convolution of their distributions: $P_{X_1 + X_2} = P_{X_1} * P_{X_2}$.

Note that convolution satisfies the associative law $\nu_1 * (\nu_2 * \nu_3) = (\nu_1 * \nu_2) * \nu_3$, since if X_1, X_2, and X_3 are independent random variables with distributions ν_1, ν_2, and ν_3, then both $\nu_1 * (\nu_2 * \nu_3)$ and $(\nu_1 * \nu_2) * \nu_3$ give the distribution of $X_1 + X_2 + X_3$. More generally, the convolution of the distributions of n independent random variables gives the distribution of their sum.

We can compute convolutions as follows.

Proposition 10.1.12. *Let v_1 and v_2 be probability measures on $(\mathbb{R}^d, \mathscr{B}(\mathbb{R}^d))$.*

(a) *The convolution $v_1 * v_2$ satisfies*

$$(v_1 * v_2)(A) = \int v_1(A - y)\, dv_2(y) = \int v_2(A - x)\, dv_1(x)$$

for each A in $\mathscr{B}(\mathbb{R}^d)$.

(b) *If v_1 is absolutely continuous (with respect to Lebesgue measure), with density f, then $v_1 * v_2$ is absolutely continuous, with density $x \mapsto \int f(x - y)\, v_2(dy)$.*

(c) *If v_1 and v_2 are absolutely continuous, with densities f and g, then $v_1 * v_2$ is absolutely continuous, with density $x \mapsto \int f(x - y)g(y)\lambda(dy)$.*

Proof. Since the sections of the set $\{(x, y) : x + y \in A\}$ are equal to $A - x$ and $A - y$, part (a) is an immediate consequence of Theorem 5.1.4. Part (b) follows from the calculation

$$v_1 * v_2(A) = \int\int \chi_A(x + y)f(x)\lambda(dx)\, v_2(dy)$$

$$= \int\int \chi_A(x)f(x - y)\lambda(dx)\, v_2(dy)$$

$$= \int \chi_A(x) \int f(x - y)\, v_2(dy)\lambda(dx).$$

(The finiteness of $\int f(x - y)\, v_2(dy)$ for almost every x follows from this calculation, applied in the case where $A = \mathbb{R}$.) Part (c) follows from part (b), since in this case we have $\int f(x - y)\, v_2(dy) = \int f(x - y)g(y)\lambda(dy)$ (recall Exercise 4.2.3). $\qquad\square$

In the remainder of this section we look at some random variables that arise when we consider the binary expansions of the values of certain uniformly distributed random variables. The techniques discussed here will give us a way to construct arbitrary sequences of independent (real-valued) random variables.

It will be convenient to have a bit of standard terminology. A random variable X is said to have a *Bernoulli distribution with parameter p* if the possible values[3] of X are 0 and 1, with 1 having probability p and 0 having probability $1 - p$.

So let us suppose that $(\Omega, \mathscr{A}, P)$ is a probability space and that X is a random variable on $(\Omega, \mathscr{A}, P)$ that is uniformly distributed on $[0, 1]$. By redefining X on a null set, if necessary, we can assume that *every* value of X belongs to $[0, 1)$. Define a sequence $\{Y_n\}$ on $(\Omega, \mathscr{A}, P)$ by letting $Y_n(\omega)$ be the nth bit in the binary expansion[4]

---

[3] Actually, we are only assuming that $P(X \in \{0, 1\}) = 1$ and not that $X(\omega) \in \{0, 1\}$ for *every* ω in Ω.

[4] In case the value $X(\omega)$ has two binary expansions, take the one that ends in an infinite sequence of 0's. See B.9 in Appendix B.

of $X(\omega)$. Then $Y_1(\omega)$ is 0 if $X(\omega)$ belongs to the interval $[0, 1/2)$ and is 1 if $X(\omega)$ belongs to $[1/2, 1)$. Likewise $Y_2(\omega)$ is 0 if $X(\omega)$ belongs to $[0, 1/4) \cup [1/2, 3/4)$ and is 1 if $X(\omega)$ belongs to $[1/4, 1/2) \cup [3/4, 1)$. In general, $Y_n(\omega)$ is 0 if $X(\omega)$ satisfies $2i/2^n \le X(\omega) < (2i+1)/2^n$ for some i and is 1 otherwise; from that it is not difficult to check that the variables $\{Y_n\}$ are measurable and independent, with each having a Bernoulli distribution with parameter $1/2$.

Proposition 10.1.13. *Let $(\Omega, \mathscr{A}, P)$ be a probability space.*

(a) *Suppose that X is a random variable on $(\Omega, \mathscr{A}, P)$ that is uniformly distributed on $[0, 1]$, and define a sequence $\{Y_n\}$ on $(\Omega, \mathscr{A}, P)$ by letting $\{Y_n(\omega)\}$ be the sequence of 0's and 1's in the binary expansion of $X(\omega)$. Then $\{Y_n\}$ is a sequence of independent random variables, each of which has a Bernoulli distribution with parameter $1/2$.*

(b) *Conversely, suppose that $\{Y_n\}$ is a sequence of independent random variables on $(\Omega, \mathscr{A}, P)$, each of which has a Bernoulli distribution with parameter $1/2$. Then the random variable X defined by $X = \sum_n Y_n/2^n$ is uniformly distributed on the interval $[0, 1]$.*

Proof. A proof for part (a) was given just before the statement of the proposition. We turn to part (b). By modifying the variables Y_n on a null set if necessary, we can assume that for every ω the sequence $\{Y_n(\omega)\}$ contains only 0's and 1's and does not end with an infinite string of 1's. Consider the dyadic rational $i/2^n$, where i satisfies $0 \le i < 2^n$. Then $i/2^n$ has an n-bit binary expansion, say $0.b_1 b_2 \dots b_n$, and $X(\omega)$ belongs to the interval $[i/2^n, (i+1)/2^n)$ if and only if $Y_j(\omega) = b_j$ holds for $j = 1, \dots, n$. Thus $P_X(I) = \lambda(I)$ holds for intervals I of the form $[i/2^n, (i+1)/2^n)$ and hence (see Lemma 1.4.2) for all open subsets I of $(0, 1)$. In view of the regularity of P_X and λ (Proposition 1.5.6), the proof is complete. $\qquad\square$

Corollary 10.1.14. *There is an infinite sequence of independent random variables, each of which is uniformly distributed on $[0, 1]$. Such a sequence can be constructed on the probability space $([0, 1], \mathscr{B}([0, 1]), \lambda)$.*

Proof. Let X be a random variable that is uniformly distributed on $[0, 1]$; such a random variable can of course be defined on the probability space $([0, 1], \mathscr{B}([0, 1]), \lambda)$. Let $\{Y_n\}$ be the sequence of random variables constructed in part (a) of Proposition 10.1.13. Since the set $\mathbb{N}$ of positive integers has the same cardinality as the set $\mathbb{N} \times \mathbb{N}$ of pairs of positive integers, we can reindex the sequence $\{Y_n\}$, obtaining a doubly indexed sequence $\{Y'_{m,n}\}$. For each n define a random variable Z_n by $Z_n = \sum_m Y'_{m,n}/2^m$. Then part (b) Proposition 10.1.13 implies that each Z_n is uniformly distributed on $[0, 1]$, while Proposition 10.1.7 implies that the variables $\{Z_n\}$ are independent. $\qquad\square$

It is possible to use uniformly distributed random variables to construct random variables having arbitrary distributions on $(\mathbb{R}, \mathscr{B}(\mathbb{R}))$. This can be done as follows:

Proposition 10.1.15. *Let μ be a probability measure on $(\mathbb{R}, \mathscr{B}(\mathbb{R}))$ with cumulative distribution function F, and let X be a random variable that is uniformly distributed on the interval $(0, 1)$. Then the function $F^{-1}: (0, 1) \to \mathbb{R}$ defined by*

$$F^{-1}(t) = \inf\{x \in \mathbb{R} : t \le F(x)\}$$

is Borel measurable, and $F^{-1} \circ X$ has distribution μ.

Proof. The function F satisfies $\lim_{x \to -\infty} F(x) = 0$ and $\lim_{x \to +\infty} F(x) = 1$, from which it follows that for each t in $(0,1)$ the set $\{x \in \mathbb{R} : t \le F(x)\}$ is nonempty and bounded below and hence that each $F^{-1}(t)$ is finite. If $t_1 < t_2$, then

$$\{x \in \mathbb{R} : t_2 \le F(x)\} \subseteq \{x \in \mathbb{R} : t_1 \le F(x)\},$$

and taking the infima of these sets gives $F^{-1}(t_1) \le F^{-1}(t_2)$. In other words, F^{-1} is nondecreasing, and so it is Borel measurable.

Let us check that

$$F^{-1}(t) \le x \tag{3}$$

holds if and only if

$$t \le F(x). \tag{4}$$

It is immediate from the definition of F^{-1} that (4) implies (3). On the other hand, if (3) holds, then there is a sequence $\{x_n\}$ that decreases to x and is such that $t \le F(x_n)$ holds for each n. Since F is right continuous, (4) follows and the proof of the equivalence of (3) and (4) is complete.

Finally, the equivalence of (3) and (4) implies that for each x in $\mathbb{R}$ we have

$$P(F^{-1} \circ X \le x) = P(X \le F(x)) = F(x);$$

thus $F^{-1} \circ X$ has distribution function F and distribution μ. ∎

Corollary 10.1.16. *Let μ be a probability distribution on $(\mathbb{R}, \mathscr{B}(\mathbb{R}))$. Then there is an infinite sequence of independent random variables, each of which has distribution μ. Such a sequence of random variables can be constructed on the probability space $([0,1], \mathscr{B}([0,1]), \lambda)$.*

Proof. This is an immediate consequence of Corollary 10.1.14 and Proposition 10.1.15. ∎

Given a source of independent and uniformly distributed random numbers (for instance, a table of random numbers or a random number generator on a computer), one can use the techniques of Proposition 10.1.15 and Corollary 10.1.16 to simulate a sequence of observations from an arbitrary distribution.

Exercises

1. Let $(\Omega, \mathscr{A}, P)$ be a probability space, and let $A_1, A_2, \ldots, A_n$ be a finite indexed family of events in $\mathscr{A}$. Show that the conditions

   (i) the events $A_1, A_2, \ldots, A_n$ are independent,
   (ii) the equation

$$P(B_1 \cap B_2 \cap \cdots \cap B_n) = P(B_1)P(B_2) \cdots P(B_n)$$

   holds for every choice of $B_1, B_2, \ldots, B_n$, where for each i the event B_i is either A_i or A_i^c,
   (iii) the events $A_1^c, A_2^c, \ldots, A_n^c$ are independent, and
   (iv) the random variables $\chi_{A_1}, \chi_{A_2}, \ldots, \chi_{A_n}$ are independent

are equivalent.
2. Let $(\Omega, \mathscr{A}, P)$ be a probability space, let $X_1, \ldots, X_d$ be real-valued random variables on Ω, and let X be the $\mathbb{R}^d$-valued random vector whose components are $X_1, \ldots, X_d$. Suppose that $F_{X_1}, \ldots, F_{X_d}$ are the cumulative distribution functions of $X_1, \ldots, X_d$ and that F_X is the cumulative distribution function of X, defined by

$$F_X(t_1, \ldots, t_d) = P(X_i \le t_i \text{ for all } i).$$

Show that $X_1, \ldots, X_d$ are independent if and only if

$$F_X(t_1, \ldots, t_d) = F_{X_1}(t_1) \cdots F_{X_d}(t_d)$$

holds for all $(t_1, \ldots, t_d)$ in $\mathbb{R}^d$. (Hint: Use Theorem 1.6.2.)
3. Let $(\Omega, \mathscr{A}, P)$ be a probability space, let $X_1, \ldots, X_d$ be real-valued random variables on Ω, and let X be the $\mathbb{R}^d$-valued random vector whose components are $X_1, \ldots, X_d$. Let $\mu_1, \ldots, \mu_d$ be the distributions of $X_1, \ldots, X_d$, and let μ be the distribution of X.

   (a) Show that if μ is absolutely continuous (with respect to Lebesgue measure), then $\mu_1, \ldots, \mu_d$ are absolutely continuous.
   (b) Show by example that the absolute continuity of μ does not follow from the absolute continuity of $\mu_1, \ldots, \mu_d$.

4. Let $(\Omega, \mathscr{A}, P)$ be a probability space, let $X_1, \ldots, X_d$ be real-valued random variables on Ω, and let X be the $\mathbb{R}^d$-valued random vector whose components are $X_1, \ldots, X_d$. Suppose that the distributions of $X_1, \ldots, X_d$ are absolutely continuous, with densities $f_1, \ldots, f_d$. Show that $X_1, \ldots, X_d$ are independent if and only if the random vector X is an absolutely continuous random variable whose density is given by $(t_1, \ldots, t_d) \mapsto f_1(t_1) \ldots f_d(t_d)$.
5. Let $X_1, X_2, \ldots, X_n$ be independent random variables, each of which has a Bernoulli distribution with parameter p, and let $S = X_1 + X_2 + \cdots + X_n$.

   (a) Show that S has a *binomial distribution with parameters n and p*—that is, that it is concentrated on the set $\{0, 1, \ldots, n\}$, with $P(S = k)$ being given by $\binom{n}{k} p^k (1-p)^{n-k}$ for each k in $\{0, 1, \ldots, n\}$.
   (b) Show that $E(S) = np$ and $\text{var}(S) = np(1-p)$.

6. A real-valued random variable has a *Poisson distribution with parameter* λ if its values are nonnegative integers, with $P(X = k) = \lambda^k e^{-\lambda}/k!$ for each nonnegative integer k.

   (a) Check that the formula above indeed defines a probability measure on $(\mathbb{R}, \mathscr{B}(\mathbb{R}))$.
   (b) Verify that if the random variable X has a Poisson distribution with parameter λ, then $E(X) = \lambda$ and $\text{var}(X) = \lambda$.
   (c) Show that if X_1 and X_2 are independent random variables that have Poisson distributions with parameters λ_1 and λ_2, respectively, then $X_1 + X_2$ has a Poisson distribution with parameter $\lambda_1 + \lambda_2$.

7. Let X_1 and X_2 be independent random variables, each of which is uniformly distributed on the interval $[0, 1]$. Find the density function of $X_1 + X_2$.

8. Let X and Y be independent normal random variables with mean 0 and variance 1, and let R and Θ be random variables with values in $[0, +\infty)$ and $[0, 2\pi)$ that correspond to writing (X, Y) in polar coordinates.

   (a) Show that R and Θ are independent, that R has distribution function given by $t \mapsto 1 - e^{-t^2/2}$ for nonnegative t, and that Θ has a uniform distribution.
   (b) Derive from this a way to use Proposition 10.1.15 to simulate values for normally distributed random variables by using easily available functions, rather than by using the inverse of the distribution function of a normal distribution.

10.2 Laws of Large Numbers

This section contains an introduction to the laws of large numbers.

Let X and X_1, X_2, ... be random variables on the probability space $(\Omega, \mathscr{A}, P)$. Then $\{X_n\}$ is said to *converge in probability* to X if

$$\lim_n P(|X_n - X| > \varepsilon) = 0$$

holds for each positive number ε and to *converge almost surely* to X (or to *converge a.s. to X*) if

$$P(X = \lim_n X_n) = 1.$$

In other words, $\{X_n\}$ converges to X in probability if it converges to X in measure, and $\{X_n\}$ converges to X almost surely if it converges to X almost everywhere.[5] Thus a number of relationships between convergence in probability and almost sure convergence can be found in Chap. 3.

---

[5]More generally, an arbitrary (probabilistic) assertion holds *almost surely* if it holds almost everywhere.

Random variables X_i, $i \in I$, are said to be *identically distributed* if they all have the same distribution—that is, if $P_{X_i} = P_{X_j}$ for all i, j in I. Sequences $\{X_n\}$ of random variables that are independent and identically distributed occur frequently, and one often abbreviates a little and calls such sequences *i.i.d.*

Theorem 10.2.1 (Weak Law of Large Numbers). *Let $\{X_n\}$ be a sequence of independent identically distributed real-valued random variables with finite second moments. For each n let $S_n = X_1 + \cdots + X_n$. Then S_n/n converges to $E(X_1)$ in probability.*

Proof. Let ε be a positive number. Since $\mathrm{var}(S_n/n) = (1/n)\,\mathrm{var}(X_1)$ (see Corollary 10.1.11 and Lemma 10.1.3), Proposition 2.3.10 implies that

$$P\left(\left|\frac{S_n}{n} - E(X_1)\right| > \varepsilon\right) = P\left(\left|\frac{S_n - E(S_n)}{n}\right|^2 > \varepsilon^2\right)$$

$$\leq \frac{1}{\varepsilon^2}\,\mathrm{var}(S_n/n) = \frac{\mathrm{var}(X_1)}{n\varepsilon^2}.$$

Thus $\lim_n P(|\frac{S_n}{n} - E(X_1)| > \varepsilon) = 0$, and so S_n/n converges to $E(X_1)$ in probability. $\square$

Suppose that $(\Omega, \mathscr{A}, P)$ is a probability space and that $\{A_n\}$ is a sequence of events in $\mathscr{A}$. Then

$$\{\omega \in \Omega : \omega \in A_n \text{ for infinitely many } n\}$$

is equal to $\bigcap_{m=1}^{\infty} \bigcup_{n=m}^{\infty} A_n$; it is the event that infinitely many of the events A_n occur, and it is often written as $\{A_n \text{ i.o.}\}$ ("i.o." is an abbreviation for "infinitely often"). For example, if we are dealing with an infinite sequence of tosses of a coin, and if for each n we let A_n be the event that a head appears on the nth toss, then $\{A_n \text{ i.o.}\}$ is the event that a head appears on infinitely many of the tosses.

Proposition 10.2.2 (Borel–Cantelli Lemmas). *Let $(\Omega, \mathscr{A}, P)$ be a probability space, and let $\{A_n\}$ be a sequence of events in $\mathscr{A}$.*

(a) *If $\sum_n P(A_n) < +\infty$, then $P(\{A_n \text{ i.o.}\}) = 0$.*
(b) *If the events A_n, $n = 1, 2, \ldots$, are independent and if $\sum_n P(A_n) = +\infty$, then $P(\{A_n \text{ i.o.}\}) = 1$.*

Note that part (b) of Proposition 10.2.2 implies that if the events $\{A_n\}$ are independent and satisfy $P(\{A_n \text{ i.o.}\}) = 0$, then $\sum_n P(A_n) < +\infty$. Combining this with part (a) of the proposition, we see that for independent events the conditions $P(\{A_n \text{ i.o.}\}) = 0$ and $\sum_n P(A_n) < +\infty$ are equivalent.

Proof. Since $\{A_n \text{ i.o.}\} = \bigcap_{m=1}^{\infty} \bigcup_{n=m}^{\infty} A_n$, we have

$$P(\{A_n \text{ i.o.}\}) \leq P(\cup_{n=m}^{\infty} A_n) \leq \sum_{n=m}^{\infty} P(A_n)$$

for each m. Thus if $\sum_n P(A_n) < +\infty$, then $P(\{A_n \text{ i.o.}\}) \leq \lim_m \sum_{n=m}^{\infty} P(A_n) = 0$ and so $P(\{A_n \text{ i.o.}\}) = 0$; with this, part (a) is proved.

To prove part (b), let us look at the complement of $\{A_n \text{ i.o.}\}$. We have

$$\{A_n \text{ i.o.}\}^c = \cup_{m=1}^{\infty} \cap_{n=m}^{\infty} A_n^c,$$

and so we can prove that $P(\{A_n \text{ i.o.}\}) = 1$ by checking that $P(\cap_{n=m}^{\infty} A_n^c) = 0$ holds for each m. Since the events $A_n^c, A_{n+1}^c, \ldots$ are independent (see Exercise 10.1.1), we have

$$P(\cap_{n=m}^{\infty} A_n^c) = \prod_{n=m}^{\infty} (1 - P(A_n)).$$

We can now derive the relation

$$\prod_{n=m}^{\infty} (1 - P(A_n)) = 0 \tag{1}$$

from the hypothesis that $\sum_n P(A_n) = +\infty$: If $P(A_n) = 1$ for some n that is greater than or equal to m, or if there is a positive ε such that $P(A_n) \geq \varepsilon$ holds for infinitely many n, then (1) certainly holds. Otherwise, $\log(1 - P(A_n))$ is asymptotic to $-P(A_n)$, and so $\sum_{n=m}^{\infty} \log(1 - P(A_n)) = -\infty$, from which (1) follows. $\qquad \square$

Proposition 10.2.3 (Kolmogorov's Zero–One Law). *Suppose that $\{X_n\}$ is a sequence of independent random variables. Then each event that belongs to the σ-algebra $\cap_n \sigma(X_n, X_{n+1}, \ldots)$ has probability 0 or 1.*

The intersection of the σ-algebras $\sigma(X_n, X_{n+1}, \ldots)$ is, of course, a σ-algebra. It is called the *tail σ-algebra* of the sequence $\{X_n\}$, and its members are called *tail events*. Thus Kolmogorov's zero–one law can be rephrased so as to say that each tail event of a sequence of independent random variables has probability 0 or 1.

Proof. Let $\mathscr{T}$ be the tail σ-algebra for the sequence $\{X_n\}$. Proposition 10.1.7 implies that for each n the σ-algebras $\sigma(X_1), \ldots, \sigma(X_{n-1})$, and $\sigma(X_n, X_{n+1}, \ldots)$ are independent and hence that $\sigma(X_1), \ldots, \sigma(X_{n-1})$, and $\mathscr{T}$ are independent. Since this is true for every n, it follows that the collection consisting of $\sigma(X_n)$, $n = 1$, 2, $\ldots$, together with $\mathscr{T}$, is independent. Applying Proposition 10.1.7 once more shows that $\sigma(X_1, X_2, \ldots)$ and $\mathscr{T}$ are independent. Since $\mathscr{T}$ is a sub-σ-algebra of $\sigma(X_1, X_2, \ldots)$, $\mathscr{T}$ must be independent of $\mathscr{T}$. Thus each A in $\mathscr{T}$ satisfies $P(A) = P(A \cap A) = P(A)P(A)$, from which it follows that $P(A) = 0$ or $P(A) = 1$. $\qquad \square$

Example 10.2.4. Suppose that $\{X_n\}$ is a sequence of independent random variables, and for each n let $S_n = X_1 + \cdots + X_n$. For each k the convergence or divergence of the sequence $\{S_n(\omega)\}$ does not depend on the values $X_1(\omega), \ldots, X_k(\omega)$ but only on the later terms in the sequence $\{X_n(\omega)\}$. Thus the event $\{\lim_n S_n \text{ exists}\}$ is a tail event and so by Kolmogorov's zero–one law has probability 0 or 1. A similar argument shows that the event $\{\lim_n S_n/n \text{ exists}\}$ has probability 0 or 1. $\qquad \square$

Theorem 10.2.5 (Strong Law of Large Numbers). *Let $\{X_n\}$ be a sequence of independent identically distributed random variables with finite expected values. For each n let $S_n = X_1 + \cdots + X_n$. Then $\{S_n/n\}$ converges to $E(X_1)$ almost surely.*

We will need the following two results for the proof of the strong law of large numbers.

Proposition 10.2.6 (Kolmogorov's Inequality). *Let $X_1, X_2, \ldots, X_n$ be independent random variables, each of which has mean 0 and a finite second moment, and for each i let $S_i = X_1 + \cdots + X_i$. Then*

$$P(\max_{1 \leq i \leq n} |S_i| > \varepsilon) \leq (1/\varepsilon^2) \sum_{i=1}^{n} E(X_i^2)$$

holds for each positive ε.

Proof. Define events A and $A_1, \ldots, A_n$ by $A = \{\max_i |S_i| > \varepsilon\}$ and

$$A_i = \{|S_i| > \varepsilon \text{ and } |S_j| \leq \varepsilon \text{ for } j = 1, 2, \ldots, i-1\}.$$

Let us check that for each i we have

$$\int_{A_i} S_i^2 \, dP \leq \int_{A_i} S_n^2 \, dP. \tag{2}$$

To see this, note that the random variables $\chi_{A_i} S_i$ and $S_n - S_i$ are independent, while $E(S_n - S_i) = 0$, and so Proposition 10.1.10 implies that $\int_{A_i} S_i(S_n - S_i) = 0$. Hence, if we write S_n^2 as $(S_i + (S_n - S_i))^2$ and expand, we find

$$\int_{A_i} S_n^2 \, dP = \int_{A_i} S_i^2 \, dP + 2 \int_{A_i} S_i(S_n - S_i) \, dP + \int_{A_i} (S_n - S_i)^2 \, dP$$

$$= \int_{A_i} S_i^2 \, dP + \int_{A_i} (S_n - S_i)^2 \, dP$$

$$\geq \int_{A_i} S_i^2 \, dP,$$

and (2) follows. Using Proposition 2.3.10 and relation (2), we find

$$\varepsilon^2 P(A) = \sum_i \varepsilon^2 P(A_i) \leq \sum_i \int_{A_i} S_i^2 \leq \sum_i \int_{A_i} S_n^2 \leq \int S_n^2;$$

since the variables X_i are independent and have mean 0, we have $E(S_n^2) = \sum E(X_i^2)$, and the proof is complete.   □

Proposition 10.2.7. *Let $\{X_n\}$ be a sequence of independent random variables that have mean 0 and satisfy $\sum_n E(X_n^2) < +\infty$. Then $\sum_n X_n$ converges almost surely.*

Proof. For each n define S_n by $S_n = X_1 + X_2 + \cdots + X_n$. If for each m and n such that $m > n$ we apply Kolmogorov's inequality (Proposition 10.2.6) to the sequence $X_{n+1}, \ldots, X_m$ and then let m approach infinity, we find

$$P(\{\sup_{i>n} |S_i - S_n| > \varepsilon\}) \leq \frac{1}{\varepsilon^2} \sum_{i=n+1}^{\infty} E(X_i^2).$$

Choose a sequence $\{\varepsilon_k\}$ of positive numbers that decreases to 0, and for each k choose a positive integer n_k such that $\sum_{i=n_k+1}^{\infty} E(X_i^2) < \varepsilon_k^2/2^k$. For each k define A_k by $A_k = \{\sup_{i>n_k} |S_i - S_{n_k}| > \varepsilon_k\}$. Then

$$\sum_k P(A_k) < \sum_k \frac{1}{\varepsilon_k^2} \frac{\varepsilon_k^2}{2^k} = \sum_k 1/2^k < +\infty,$$

and so $P(\{A_k \text{ i.o.}\}) = 0$. However, for each ω outside $\{A_k \text{ i.o.}\}$ the sequence $\{S_n(\omega)\}$ is a Cauchy sequence, and so $\{S_n\}$ converges almost surely. $\qquad\square$

Proof of Strong Law of Large Numbers. For each i let Y_i be the truncated version of X_i defined by

$$Y_i(\omega) = \begin{cases} X_i(\omega) & \text{if } |X_i(\omega)| \leq i, \text{ and} \\ 0 & \text{otherwise.} \end{cases}$$

Of course, the variables $\{Y_i\}$ are independent and have finite expected values.

Claim. The series $\sum_i \frac{Y_i - E(Y_i)}{i}$ converges almost surely.

Since $E((Y_i - E(Y_i))^2) \leq E(Y_i^2)$, the claim will follow from Proposition 10.2.7 if we verify that $\sum_i E(Y_i^2/i^2) < +\infty$. Let μ be the common distribution of the X_i's, and for each positive integer j define I_j by $I_j = \{x \in \mathbb{R} : j-1 < |x| \leq j\}$. There is a constant C such that $\sum_{i=j}^{\infty} 1/i^2 \leq C/j$ holds for each j (use basic calculus), and so

$$\sum_i E(Y_i^2/i^2) = \sum_i \frac{1}{i^2} \int_{[-i,i]} x^2 \mu(dx)$$

$$= \sum_i \sum_{j \leq i} \frac{1}{i^2} \int_{I_j} x^2 \mu(dx) = \sum_j \sum_{i \geq j} \frac{1}{i^2} \int_{I_j} x^2 \mu(dx)$$

$$\leq \sum_j C \int_{I_j} \frac{x^2}{j} \mu(dx) \leq C \int_{\mathbb{R}} |x| \mu(dx) = CE(|X_1|) < +\infty.$$

With this the claim is proved.

For each n let T_n be $\sum_{i=1}^{n} \frac{Y_i - E(Y_i)}{i}$, the nth partial sum of $\sum_i \frac{Y_i - E(Y_i)}{i}$. The plan is to relate the partial sums of $\sum_i (Y_i - E(Y_i))$ to the T_n's and to $\{S_n/n\}$; this will give us the information that we need about the sequence $\{S_n/n\}$. We begin by noting that

$$\sum_{i=1}^{n}(Y_i - E(Y_i)) = \sum_{i=1}^{n} i(T_i - T_{i-1})$$

$$= nT_n - \sum_{i=1}^{n-1} T_i.$$

Since (by the claim above) $\lim_n T_n$ exists almost surely, if we divide both sides of the preceding equation by n and use item B.7 in Appendix B, we find

$$\lim_n \frac{1}{n}\sum_{i=1}^{n}(Y_i - E(Y_i)) = \lim_n \left(T_n - \frac{1}{n}\sum_{i=1}^{n-1} T_i\right) = 0 \quad \text{a.s.} \tag{3}$$

As preparation for the final step we check that

$$\lim_n \frac{1}{n}\sum_{i=1}^{n}(X_i - Y_i) = 0 \quad \text{a.s.} \tag{4}$$

and that

$$\lim_n \frac{1}{n}\sum_{i=1}^{n} E(Y_i) = E(X_1). \tag{5}$$

Let us begin with Eq. (4). Note that the finiteness of $E(|X_1|)$ and Exercise 2.4.6 imply that $\sum_i P(\{X_i \neq Y_i\}) = \sum_i P(|X_i| > i) < +\infty$; from this and the Borel–Cantelli lemma, we conclude that $P(\{X_i \neq Y_i \text{ i.o.}\}) = 0$ and hence that (4) holds. Equation (5) follows from the fact that $\lim_i E(Y_i) = E(X_1)$, plus another use of B.7. Finally, Eqs. (3) and (5) imply that

$$\lim_n \frac{1}{n}\sum_{i=1}^{n} Y_i = E(X_1)$$

holds almost surely, and from this, together with (4), we conclude that $\lim_n S_n/n = E(X_1)$ holds almost surely. With this the proof of the strong law is complete. $\square$

Theorem 10.2.8 (Converse to the Strong Law of Large Numbers). *Let $\{X_n\}$ be a sequence of independent identically distributed random variables that do not have finite expected values. For each n let $S_n = X_1 + \cdots + X_n$. Then $\limsup_n |S_n/n| = +\infty$ almost surely.*

Proof. Let K be a positive integer, fixed for a moment, and for each n let A_n be the event $\{|X_n| \geq Kn\}$. Since the variables $\{X_i\}$ have a common distribution, but do not have a finite expected value, it follows from Exercise 2.4.6 that $\sum_n P(A_n) = +\infty$. The second part of the Borel–Cantelli lemmas implies that $P(\{A_n \text{ i.o.}\}) = 1$ and hence that

$$P\left(\limsup_n \frac{|X_n|}{n} \geq K\right) = 1.$$

This is true for each positive integer K, and so it follows that $\limsup_n |X_n/n| = +\infty$ almost surely. However,

$$\frac{X_n}{n} = \frac{S_n}{n} - \frac{n-1}{n}\frac{S_{n-1}}{n-1},$$

from which it follows that $\limsup_n |X_n/n| \leq 2\limsup_n |S_n/n|$; thus $\limsup |S_n/n|$ is also almost surely infinite.   □

Exercises

1. The Weierstrass approximation theorem says that every continuous function on a closed bounded subinterval of $\mathbb{R}$ can be uniformly approximated by polynomials. This exercise is devoted to a derivation of the Weierstrass approximation theorem for functions on $[0, 1]$ from the weak law of large numbers.

   Let f be a continuous real-valued function on $[0, 1]$, let $\{X_n\}$ be a sequence of independent random variables, each of which has a Bernoulli distribution with parameter p, and for each n let $S_n = X_1 + \cdots + X_n$ and $Y_n = S_n/n$. For each p in $[0, 1]$ let $g_n(p)$ be $E_p(f \circ Y_n)$, the expected value of $f \circ Y_n$ in the case where the underlying Bernoulli distribution has parameter p. Then (see Exercise 10.1.5)

$$g_n(p) = \sum_{k=0}^{n} f(k/n) \binom{n}{k} p^k (1-p)^{n-k},$$

   and so g_n is a polynomial in p. Show that the sequence $\{g_n\}$ converges uniformly to f. (Hint: The weak law of large numbers says that for each ε we have $\lim_n P(|S_n/n - p| > \varepsilon) = 0$; check that this convergence is uniform in p. Use this and the uniform continuity of f to conclude that the convergence of $E_p(f \circ Y_n)$ to $f(p)$ is uniform in p.)

2. Suppose that $\{X_n\}$ is a sequence of independent random variables and that $\mathscr{T}$ is the σ-algebra of tail events of $\{X_n\}$. Show that every $[-\infty, +\infty]$-valued random variable that is $\mathscr{T}$-measurable is almost surely constant.

3. Let b be an integer such that $b \geq 2$. The digits that can occur in a base b expansion of a number are, of course, $0, 1, \ldots, b-1$. A number x in $[0, 1]$ is *normal to base b* if each value in $\{0, 1, \ldots, b-1\}$ occurs the expected fraction (namely $1/b$) of the time in the base b expansion of x—that is, if

$$\lim_n \frac{\text{number of times } k \text{ occurs among the first } n \text{ digits of } x}{n} = \frac{1}{b}$$

   holds for $k = 0, 1, \ldots, b-1$. The value x is *normal* if it is normal to base b for every b.

   (a) For a given b, show that almost every number in $[0,1]$ is normal to base b. (Hint: Modify part (a) of Proposition 10.1.13 and use the strong law of large numbers.)

   (b) Conclude that almost every number in $[0,1]$ is normal.

4. (The Glivenko–Cantelli Theorem) Let $(\Omega,\mathscr{A},P)$ be a probability space, let μ be a probability distribution on $(\mathbb{R},\mathscr{B}(\mathbb{R}))$, let F be its distribution function, and let $\{X_n\}$ be a sequence of independent random variables on $(\Omega,\mathscr{A},P)$, each of which has distribution μ. For each ω in Ω, $\{X_n(\omega)\}$ is a sequence of real numbers, and we can define a sequence $\{\mu_n^{\omega}\}_{n=1}^{\infty}$ of measures on $(\mathbb{R},\mathscr{B}(\mathbb{R}))$ by letting $\mu_n^{\omega} = (1/n)\sum_{k=1}^{n}\delta_{X_k(\omega)}$. Also, let F_n^{ω} be the distribution function of the measure μ_n^{ω}; thus,

$$F_n^{\omega}(x) = (1/n)\sum_{1}^{n}\chi_{(-\infty,x]}\circ X_k(\omega)$$

$$= \frac{\text{number of } k \text{ in } \{1,2,\ldots,n\} \text{ for which } X_k(\omega)\leq x}{n}$$

holds for all n, ω, and x. (Such functions F_n^{ω} are called *empirical distribution functions*.) Since μ describes the distribution of values of the X_n's, it seems plausible that for a typical ω, the measures μ_n^{ω} might approach μ as n becomes large. This is in fact true, and the Glivenko–Cantelli theorem makes a rather strong version of this precise, namely that for all ω outside some set of probability zero, the sequence $\{F_n^{\omega}(x)\}_{n=1}^{\infty}$ converges to $F(x)$, *uniformly in x*.

   (a) As a first step, show that if $x\in\mathbb{R}$, then $F(x) = \lim_n F_n^{\omega}(x)$ and $F(x-) = \lim_n F_n^{\omega}(x-)$ hold for almost every ω in Ω.

   (b) Show that if ε is a positive number, if $x_1, x_2, \ldots, x_k$ are real numbers such that $x_1 < x_2 < \cdots < x_k$ and such that the intervals $(-\infty,x_1), (x_1,x_2), \ldots, (x_k,+\infty)$ all have measure less than ε under μ, and if ω is such that $\lim_n F_n^{\omega}(x_i) = F(x_i)$ and $\lim_n F_n^{\omega}(x_i-) = F(x_i-)$ hold for $i = 1, 2, \ldots, k$, then $\sup_x |F_n^{\omega}(x) - F(x)| \leq \varepsilon$ holds for all large n.

   (c) Use parts (a) and (b) to prove the Glivenko–Cantelli theorem.

5. Let $\{X_n\}$ be a sequence of independent identically distributed random variables that are nonnegative and satisfy $E(X_n) = +\infty$ for each n. Show that $\lim_n \frac{S_n}{n} = +\infty$ almost surely.

6. (a) Let $X_1, X_2, \ldots, X_n$ be independent random variables on $(\Omega,\mathscr{A},P)$, each of which has mean 0, for each i let $S_i = X_1 + X_2 + \cdots + X_i$, let c be a positive constant such that $|X_i| \leq c$ holds almost surely for each i, and for each i let σ_i^2 be the variance of X_i. Show that for each positive number a,

$$P(\max_i |S_i| > a) \geq 1 - \frac{(a+c)^2}{\sum_i \sigma_i^2}.$$

   (Hint: Start by using ideas from the proof of Kolmogorov's inequality to show that

$$E(S_n^2) \leq a^2(1 - P(A)) + (a+c)^2 P(A) + \sum_i (\sigma_{i+1}^2 + \cdots \sigma_n^2) P(A_i),$$

where $A_1, \ldots, A_n$ are given by

$$A_i = \{|S_i| > a \text{ and } |S_j| \leq a \text{ for } j = 1, 2, \ldots, i-1\}$$

and $A = \cup_i A_i$.)

(b) Let $X_1, X_2, \ldots$ be independent random variables on $(\Omega, \mathscr{A}, P)$, each of which has mean 0, and for each i let σ_i^2 be the variance of X_i. Show that if there is a constant c such that $|X_i| \leq c$ holds almost surely for each i and if the series $\sum_i X_i$ is almost surely convergent, then $\sum_i \sigma_i^2 < +\infty$.

(c) Show that part (b) remains true if the assumption that each X_i has mean 0 is omitted. (Hint: Define random variables $Y_1, Y_2, \ldots$ on the product of $(\Omega, \mathscr{A}, P)$ with itself by letting $Y_i(\omega_1, \omega_2) = X_i(\omega_1) - X_i(\omega_2)$, and apply part (b) to the series $\sum_i Y_i$.)

7. Let $\{X_n\}$ be a sequence of independent random variables such that $P(X_n = 1) = P(X_n = -1) = \frac{1}{2}$ holds for each n, and let $\{a_n\}$ be a sequence of real numbers. Show that the series $\sum_n a_n X_n$ converges almost surely if and only if $\{a_n\} \in \ell^2$. (Hint: See Exercise 6.)

8. Let $X_1, X_2, \ldots$ be independent random variables on $(\Omega, \mathscr{A}, P)$, let c be a positive constant, and for each i define a new random variable, the truncation $X_i^{(c)}$ of X_i by c, as follows:

$$X_i^{(c)}(\omega) = \begin{cases} X_i(\omega) & \text{if } |X_i(\omega)| \leq c, \text{ and} \\ 0 & \text{otherwise.} \end{cases}$$

The *three series theorem* says that the series $\sum_i X_i$ converges almost surely if and only if the series

(i) $\sum_i P(|X_i| > c)$,

(ii) $\sum_i E(X_i^{(c)})$, and

(iii) $\sum_i \text{var}(X_i^{(c)})$

all converge. Prove the three series theorem. (Hint: Use the Borel–Cantelli lemma, Proposition 10.2.7, and Exercise 6.)

10.3 Convergence in Distribution and the Central Limit Theorem

In this section we look at circumstances under which probability distributions on $(\mathbb{R}, \mathscr{B}(\mathbb{R}))$, or on $(\mathbb{R}^d, \mathscr{B}(\mathbb{R}^d))$, give good approximations to one another. As a rather trivial example, if n is large, then the point mass $\delta_{1/n}$ concentrated at $1/n$ should be considered to be close to the point mass δ_0 concentrated at 0. As a somewhat less trivial example, for large values of n the measure $(1/n)\sum_{i=1}^{n}\delta_{i/n}$ would seem to give a reasonable approximation to the uniform distribution on $[0,1]$. More significantly, we will see in Theorem 10.3.16 (the central limit theorem) that the distributions of certain normalized sums of random variables are well approximated by Gaussian distributions.

We should note that for our current purposes the total variation norm (defined in Sect. 4.1) does not lead to a reasonable criterion for closeness. For example, the total variation distance between $\delta_{1/n}$ and δ_0 is 2, however large n is. We need a definition that, for large n, will classify these measures as close.

We will deal with such questions in terms of convergence of sequences of probability measures (for a bit about an approach using distances, see Exercise 12 and the notes at the end of the chapter). Let μ and μ_1, μ_2, ... be probability measures on $(\mathbb{R}^d, \mathscr{B}(\mathbb{R}^d))$. The sequence $\{\mu_n\}$ is said to *converge in distribution*, or to *converge weakly*, to μ if

$$\int f\,d\mu = \lim_n \int f\,d\mu_n$$

holds for each bounded continuous f on $\mathbb{R}^d$.

Before doing anything else, we should verify that limits in distribution of sequences of probability measures are unique. In other words, we should check that if the sequence $\{\mu_n\}$ converges in distribution to μ and to ν, then $\mu = \nu$. This, however, is an immediate consequence of the following lemma.

Lemma 10.3.1. *Let μ and ν be probability measures on $(\mathbb{R}^d, \mathscr{B}(\mathbb{R}^d))$. If $\int f\,d\mu = \int f\,d\nu$ holds for each bounded continuous f on $\mathbb{R}^d$, then $\mu = \nu$.*

Lemma 10.3.1 is an immediate consequence of the Riesz representation theorem (Theorem 7.2.8). The following proof, however, does not depend on the Riesz representation theorem and so avoids unnecessary dependence on Chap. 7.

Proof. Since μ and ν are regular (see Proposition 1.5.6), it is enough to prove that each compact subset K of $\mathbb{R}^d$ satisfies $\mu(K) = \nu(K)$. So let K be a nonempty compact subset of $\mathbb{R}^d$. Recall that the distance $d(x, K)$ between the point x and the set K is continuous as a function of x (see D.27) and is equal to 0 exactly when $x \in K$. For each k define a function $f_k \colon \mathbb{R}^d \to \mathbb{R}$ by $f_k(x) = \max(0, 1 - kd(x, K))$. These functions are bounded (by 0 and 1) and continuous, and they form a sequence that decreases to the indicator function χ_K of K. Furthermore $\int f_k\,d\mu = \int f_k\,d\nu$ holds for each k, and so we can use the dominated convergence theorem (or the monotone convergence theorem) to conclude that

$$\mu(K) = \lim_k \int f_k\,d\mu = \lim_k \int f_k\,d\nu = \nu(K).$$

With this the proof of the lemma is complete.   □

Proposition 10.3.2. *Suppose that μ and μ_n, $n = 1, 2, \ldots$, are probability measures on $(\mathbb{R}^d, \mathscr{B}(\mathbb{R}^d))$. Then the conditions*

(a) *the sequence $\{\mu_n\}$ converges in distribution to μ,*
(b) *each bounded uniformly continuous f on $\mathbb{R}^d$ satisfies $\int f \, d\mu = \lim_n \int f \, d\mu_n$,*
(c) *each closed subset F of $\mathbb{R}^d$ satisfies $\limsup_n \mu_n(F) \le \mu(F)$,*
(d) *each open subset U of $\mathbb{R}^d$ satisfies $\mu(U) \le \liminf_n \mu_n(U)$, and*
(e) *each Borel subset B of $\mathbb{R}^d$ whose boundary has measure 0 under μ satisfies*
$$\mu(B) = \lim_n \mu_n(B)$$

are equivalent.

Proof. Since every uniformly continuous function is continuous, condition (b) is an immediate consequence of condition (a). Now assume that condition (b) holds. If F is a nonempty closed subset of $\mathbb{R}^d$, then the functions $f_k \colon \mathbb{R}^d \to \mathbb{R}$ defined by $f_k(x) = \max(0, 1 - kd(x, F))$ are bounded (by 0 and 1) and uniformly continuous (again see D.27). Since these functions decrease to the indicator function of F, it follows that $\mu(F) = \lim_k \int f_k \, d\mu$. Now suppose that ε is a positive constant, and choose k such that $\int f_k \, d\mu < \mu(F) + \varepsilon$. Then, since $\mu_n(F) \le \int f_k \, d\mu_n$ holds for each n, we have

$$\limsup_n \mu_n(F) \le \lim_n \int f_k \, d\mu_n = \int f_k \, d\mu < \mu(F) + \varepsilon,$$

and condition (c) follows. It is easy to check that condition (d) is equivalent to condition (c). Now suppose that conditions (c) and (d) hold, and let B be a Borel set whose boundary has μ-measure 0. Let F and U be the closure and interior of B. Then $F - U$ is the boundary of B, and so $\mu(F) = \mu(U) = \mu(B)$, from which it follows that

$$\mu(B) = \mu(U) \le \liminf_n \mu_n(U)$$

$$\le \liminf_n \mu_n(B) \le \limsup_n \mu_n(B)$$

$$\le \limsup_n \mu_n(F) \le \mu(F) = \mu(B).$$

Thus, condition (e) follows from conditions (c) and (d).

Finally, we derive condition (a) from condition (e). So suppose that condition (e) holds, and let f be a bounded continuous function on $\mathbb{R}^d$. Suppose that ε is a positive number. Let B be a positive number such that $-B \le f(x) < B$ holds for all x, and let $c_0, c_1, \ldots, c_k$ be numbers such that

$$-B = c_0 < c_1 < \cdots < c_k = B$$

(we still need to look at the details of how the c_i's are to be chosen). For $i = 1, \ldots, k$ let $C_k = \{x \in \mathbb{R}^d : c_{k-1} \leq f(x) < c_k\}$. The continuity of f implies that the boundary of C_k is included in the set of points x such that $f(x)$ is equal to c_{k-1} or c_k. Since the sets $\{x \in \mathbb{R}^d : f(x) = c\}$, where c ranges over $\mathbb{R}$, are disjoint and Borel, at most a countable number of them can have positive measure under μ. It follows that we can choose our points c_i so that the boundaries of the sets C_i have μ-measure 0 and so that each interval $[c_{i-1}, c_i)$ has length less than ε. If we define g by $g = \sum_{i=1}^{k} c_i \chi_{C_i}$, then $f \leq g \leq f + \varepsilon$, and so, if we apply condition (e) to the sets C_i, we find

$$\limsup_n \int f \, d\mu_n \leq \lim_n \int g \, d\mu_n = \int g \, d\mu \leq \int f \, d\mu + \varepsilon.$$

A similar calculation shows that $\int f \, d\mu - \varepsilon \leq \liminf_n \int f \, d\mu_n$. Since ε is arbitrary, condition (a) follows, and with that the proof of the proposition is complete. $\qquad \square$

As we have seen, probability measures on $(\mathbb{R}, \mathscr{B}(\mathbb{R}))$ can be identified with distribution functions. Here is a characterization of convergence in distribution on $\mathbb{R}$ in terms of distribution functions (in fact, convergence in distribution seems to have first been defined in terms of distribution functions).

Proposition 10.3.3. *Suppose that μ and μ_n, $n = 1, 2, \ldots$, are probability measures on $(\mathbb{R}, \mathscr{B}(\mathbb{R}))$, with distribution functions F and F_n, $n = 1, 2, \ldots$. Then the conditions*

(a) *$\{\mu_n\}$ converges in distribution to μ,*
(b) *$F(t) = \lim_n F_n(t)$ holds at each t at which F is continuous, and*
(c) *$F(t) = \lim_n F_n(t)$ holds at each t in some dense subset of $\mathbb{R}$*

are equivalent.

Proof. It follows from Proposition 10.3.2 that condition (a) implies condition (b) and from the fact that a monotone function has at most countably many discontinuities (see Lemma 6.3.2) that condition (b) implies condition (c). To show that condition (c) implies condition (a), we will assume that condition (c) holds and prove that each open subset U of $\mathbb{R}$ satisfies $\mu(U) \leq \liminf_n \mu_n(U)$ (see Proposition 10.3.2). So suppose that U is a nonempty open subset of $\mathbb{R}$. Let ε be a positive number. According to Proposition C.4, there is a sequence $\{U_i\}$ of disjoint open intervals whose union is U. We can choose an integer k such that $\mu(U) - \varepsilon < \mu(\cup_{i=1}^{k} U_i)$. Next we approximate the sets U_i, $i = 1, \ldots, k$, with subintervals C_i such that $\sum_{i=1}^{k} \mu(U_i) - \varepsilon < \sum_{i=1}^{k} \mu(C_i)$ and such that each C_i is of the form $(c_i, d_i]$, where c_i and d_i belong to the dense set given by condition (c). Then each C_i satisfies $\mu(C_i) = \lim_n \mu_n(C_i)$, and it follows that

$$\mu(U) - 2\varepsilon < \sum_i \mu(C_i) = \lim_n \sum_i \mu_n(C_i) \leq \liminf_n \mu_n(U).$$

Since ε is arbitrary, we have $\mu(U) \leq \liminf_n \mu_n(U)$, and the proof is complete. $\qquad \square$

Next we introduce the Fourier transform of a probability measure. For that we need to know a bit about the integration of complex-valued functions; see Sect. 2.6. We will also be using complex-valued exponential functions; see item B.10 in Appendix B for the facts we need.

In addition, we need the following basic result:

Lemma 10.3.4. *Let z and $\{z_n\}$, $n = 1, 2, \ldots$, be complex numbers such that $z = \lim_n z_n$. Then $\lim_n (1 + z_n/n)^n = e^z$.*

Proof. Choose a positive constant M that is larger than the absolute values of z and of every z_n. For each k the term in the binomial expansion of $(1 + z_n/n)^n$ that involves the kth power of z_n is

$$\binom{n}{k} \frac{z_n^k}{n^k}.$$

As n approaches infinity, this term approaches the term $z^k/k!$ from the series expansion of e^z. Let us check that the sum of the terms of the binomial expansion of $(1 + z_n/n)^n$ approaches the sum of the terms of the series for e^z. The issue here is the interchange of sums and limits, and this interchange can be justified with the dominated convergence theorem, if we apply that theorem to integrals (i.e., sums) on the space of nonnegative integers together with counting measure and if we note that the functions involved here are dominated by the terms in the series expansion of e^M). Thus $\lim_n (1 + z_n/n)^n = e^z$, and the proof is complete. $\qquad\square$

Now suppose that μ is a probability measure on $(\mathbb{R}^d, \mathscr{B}(\mathbb{R}^d))$. The *characteristic function*,[6] or *Fourier transform*, of μ is the function $\phi_\mu \colon \mathbb{R}^d \to \mathbb{C}$ defined[7] by $\phi_\mu(t) = \int e^{i(t,x)} \mu(dx)$. (The integrand here is bounded and measurable, and so the definition of ϕ_μ makes sense.) If X is an $\mathbb{R}^d$-valued random variable, then the characteristic function of X, written ϕ_X, is defined to be the characteristic function of the distribution P_X of X, and so $\phi_X(t) = \phi_{P_X}(t) = E(e^{i(t,X)})$.

Proposition 10.3.5. *Let μ be a probability measure on $(\mathbb{R}^d, \mathscr{B}(\mathbb{R}^d))$. Then*

(a) $\phi_\mu(0) = 1$,
(b) $|\phi_\mu(t)| \leq 1$ *holds for each t in $\mathbb{R}^d$, and*
(c) ϕ_μ *is continuous on $\mathbb{R}^d$.*

Proof. Part (a) is immediate, and part (b) follows from Proposition 2.6.7. For part (c), let t be an arbitrary element of $\mathbb{R}^d$, and suppose that $\{t_n\}$ is a sequence of elements of $\mathbb{R}^d$ such that $t = \lim_n t_n$. Then the dominated convergence theorem

---

[6] The phrase "characteristic function" is ambiguous; it can mean either "Fourier transform" or "indicator function" (see item A.3 in Appendix A). In this chapter we follow the usage of probabilists and use characteristic function to mean Fourier transform; in the rest of the book we use characteristic function to mean indicator function.

[7] Here (t, x) is the inner product of t and x, defined by $(t, x) = \sum_{j=1}^{d} t_j x_j$. In case we are dealing with measures on $\mathbb{R}$, rather than on $\mathbb{R}^d$, we write e^{itx}, rather than $e^{i(t,x)}$.

implies that

$$\lim_n \int e^{i(t_n, x)} \mu(dx) = \int e^{i(t, x)} \mu(dx)$$

and hence that $\lim_n \phi_\mu(t_n) = \phi_\mu(t)$. Since this holds for every sequence $\{t_n\}$ that converges to t, the continuity of ϕ_μ follows (see D.31 in Appendix D). $\quad\square$

Lemma 10.3.6. *Let X be a real-valued random variable, let a and b be real constants, and define a random variable Y by $Y = aX + b$. Then $\phi_Y(t) = e^{itb} \phi_X(at)$ holds for all real t.*

Proof. This follows from the calculation $\phi_Y(t) = E(e^{it(aX+b)}) = e^{itb} E(e^{iatX}) = e^{itb} \phi_X(at)$. $\quad\square$

Proposition 10.3.7. *Let μ be a probability measure on $(\mathbb{R}, \mathscr{B}(\mathbb{R}))$, and let n be a positive integer such that μ has a finite nth moment—that is, such that $\int |x|^n \mu(dx)$ is finite. Then ϕ_μ has n continuous derivatives, which are given by*

$$\phi_\mu^{(k)}(t) = i^k \int x^k e^{itx} \mu(dx)$$

for $k = 1, 2, \ldots, n$.

Proof. Note[8] that $|e^{iu} - 1| \leq |u|$ holds for all real u and that $\lim_{u \to 0} (e^{iu} - 1)/u = i$. We will use those facts in the calculations below.

We verify the formula for $\phi_\mu^{(k)}$ by using mathematical induction. Suppose that we have already verified that $\phi_\mu^{(k)}$ has the required form (certainly $\phi_\mu^{(0)}$ is ϕ_μ and has the required form). Then

$$\frac{\phi_\mu^{(k)}(t+h) - \phi_\mu^{(k)}(t)}{h} = i^k \int x^k \frac{e^{i(t+h)x} - e^{itx}}{h} \mu(dx)$$

$$= i^k \int x^k e^{itx} \frac{e^{ihx} - 1}{h} \mu(dx).$$

The integrand in the second integral above approaches $ix^{k+1} e^{itx}$ as h approaches 0, and it is dominated by $|x^{k+1}|$. It follows from the dominated convergence theorem that if $0 \leq k < n$ and if $\phi_\mu^{(k)}$ has the form given in the proposition, then $\phi_\mu^{(k+1)}$ has the analogous form with k replaced by $k+1$. (Note that, as in the proof of Proposition 10.3.5, we are actually taking limits as h approaches 0 along sequences.) The continuity of $\phi_\mu^{(k+1)}$ follows from another application of the dominated convergence theorem. $\quad\square$

---

[8] A geometric justification for the inequality $|e^{iu} - 1| \leq |u|$ comes from the fact that $|e^{iu} - 1|$ is the straight-line distance between the points $(\cos u, \sin u)$ and $(1, 0)$, while $|u|$ gives the length of a path that connects these points and lies on the unit circle. Alternatively, we can give this inequality and also the limit $\lim_{u \to 0} (e^{iu} - 1)/u = i$ non-geometric proofs if we rewrite the exponentials in terms of sines and cosines.

Proposition 10.3.8. *Suppose that P is the normal distribution on* $(\mathbb{R}, \mathscr{B}(\mathbb{R}))$ *with mean* μ *and variance* σ^2. *Then* $\phi_P(t) = e^{it\mu}e^{-\sigma^2t^2/2}$.

Proof. Let us begin with the special case where P is the standard normal distribution (i.e., the normal distribution with mean 0 and variance 1). Then the Fourier transform ϕ_P of P is given by

$$\phi_P(t) = \frac{1}{\sqrt{2\pi}}\int_{\mathbb{R}} e^{ixt}e^{-x^2/2}\,dx.$$

It is easy to check that P has a finite first moment (in fact, finite moments of all orders), and so it follows from Proposition 10.3.7 that

$$\phi'_P(t) = \frac{1}{\sqrt{2\pi}}\int_{\mathbb{R}} ixe^{ixt}e^{-x^2/2}\,dx.$$

If we integrate by parts (view the integrand above as the product of ie^{ixt} and the derivative of $-e^{-x^2/2}$), we find that $\phi'_P(t) = -t\phi_P(t)$. It follows that the derivative of $t \mapsto e^{t^2/2}\phi_P(t)$ is identically zero and so, since $\phi_P(0) = 1$, that $\phi_P(t) = e^{-t^2/2}$. The general case now follows from Lemma 10.3.6.   □

Proposition 10.3.9. *Let* v_1 *and* v_2 *be probability measures on* $(\mathbb{R}^d, \mathscr{B}(\mathbb{R}^d))$, *and let* v *be their convolution. Then* $\phi_v(t) = \phi_{v_1}(t)\phi_{v_2}(t)$ *holds at each t in* $\mathbb{R}^d$.

Proof. Let X_1 and X_2 be independent random variables with distributions v_1 and v_2. Then $X_1 + X_2$ has distribution v, and so Proposition 10.1.10 implies that

$$\phi_v(t) = E(e^{it(X_1+X_2)}) = E(e^{itX_1})E(e^{itX_2}) = \phi_{v_1}(t)\phi_{v_2}(t).$$ □

Example 10.3.10. Let us now try to invert the Fourier transform—to go from the Fourier transform of a probability measure back to the measure. We start with the Gaussian distributions and look at $t \mapsto e^{-\sigma^2t^2/2}$, the Fourier transform of the Gaussian distribution with mean 0 and variance σ^2. If we multiply this function by e^{-ixt}, integrate, and use Proposition 10.3.8 at the last step, we find

$$\int_{\mathbb{R}} e^{-ixt}e^{-\sigma^2t^2/2}\,dt = \int_{\mathbb{R}} e^{ixt}e^{-\sigma^2t^2/2}\,dt$$

$$= \frac{\sqrt{2\pi}}{\sigma}\frac{1}{\sqrt{2\pi\frac{1}{\sigma}}}\int_{\mathbb{R}} e^{ixt}e^{-\frac{t^2}{2\frac{1}{\sigma^2}}}\,dt$$

$$= \frac{\sqrt{2\pi}}{\sigma}e^{-x^2/2\sigma^2}.$$

It follows that

$$\frac{1}{2\pi}\int_{\mathbb{R}} e^{-ixt}e^{-\sigma^2t^2/2}\,dt = \frac{1}{\sqrt{2\pi}\sigma}e^{-x^2/2\sigma^2}.$$ □

In particular, we can go from the Fourier transform ϕ of the Gaussian distribution with mean 0 and variance σ^2 back to its density, say g, by using the *Fourier inversion formula*

$$\frac{1}{2\pi} \int_{\mathbb{R}} e^{-ixt} \phi(t) dt = g(x), \tag{1}$$

which says that the *inverse Fourier transform* of ϕ is equal to g. The Fourier inversion formula works for many distributions, but not all (see Exercise 13). However, we now have enough information to prove the following uniqueness theorem.

Proposition 10.3.11. *Let μ and ν be probability measures on $(\mathbb{R}^d, \mathscr{B}(\mathbb{R}^d))$. Then $\mu = \nu$ if and only if $\phi_\mu = \phi_\nu$.*

Proof. The following is a proof for measures on $\mathbb{R}$, rather than on $\mathbb{R}^d$. We can convert it to a proof for measures on $\mathbb{R}^d$ by changing the constant $1/2\pi$ in the Fourier inversion formula to $1/(2\pi)^d$, replacing e^{-ixt} with $e^{-i\langle x,t\rangle}$, and checking that the Fourier inversion formula works for probabilities on $\mathbb{R}^d$ that are products of d Gaussian distributions, each with mean 0 and variance σ^2.

So let us turn to the proof when $d = 1$. It is certainly true that if $\mu = \nu$, then $\phi_\mu = \phi_\nu$, and so we need only check that if $\phi_\mu = \phi_\nu$, then $\mu = \nu$. So let μ and ν be probability measures on $(\mathbb{R}, \mathscr{B}(\mathbb{R}))$ such that $\phi_\mu = \phi_\nu$. In addition, let γ_σ be the Gaussian distribution on $\mathbb{R}$ with mean 0 and variance σ^2; let ϕ_{γ_σ} and g_σ be its Fourier transform and density function. Let us calculate the inverse Fourier transform of $\phi_{\gamma_\sigma * \mu}$, or equivalently of $\phi_{\gamma_\sigma} \phi_\mu$ (Proposition 10.3.9), using the fact that we know from Example 10.3.10 that the Fourier inversion formula works in the Gaussian case:

$$\frac{1}{2\pi} \int_{\mathbb{R}} e^{-ixt} \phi_{\gamma_\sigma}(t) \phi_\mu(t) dt = \frac{1}{2\pi} \int_{\mathbb{R}} e^{-ixt} \phi_{\gamma_\sigma}(t) \int_{\mathbb{R}} e^{its} \mu(ds) dt$$

$$= \frac{1}{2\pi} \int_{\mathbb{R}} \int_{\mathbb{R}} e^{-it(x-s)} \phi_{\gamma_\sigma}(t) dt \, \mu(ds)$$

$$= \int_{\mathbb{R}} g_\sigma(x-s) \mu(ds)$$

(we were able to apply Fubini's theorem because μ is finite and ϕ_{γ_σ} is integrable with respect to Lebesgue measure). Note that the result of this calculation is the density of $\gamma_\sigma * \mu$ (see Proposition 10.1.12). In other words, the inverse Fourier transform of $\phi_{\gamma_\sigma} \phi_\mu$ is the density of $\gamma_\sigma * \mu$. A similar calculation can be applied to ν. Since μ and ν are such that $\phi_\mu = \phi_\nu$, we can conclude from these calculations that $\gamma_\sigma * \mu = \gamma_\sigma * \nu$. Finally, $\gamma_\sigma * \mu$ and $\gamma_\sigma * \nu$ converge in distribution to μ and ν as σ approaches 0 (check this; you might use Exercise 7), and it follows that $\mu = \nu$. $\square$

Corollary 10.3.12. *Let $X_1, \ldots, X_d$ be real random variables, all defined on the same probability space, and let X be the $\mathbb{R}^d$-valued random variable whose* *components are $X_1, \ldots, X_d$. Then the random variables $X_1, \ldots, X_d$ are independent if and only if $\phi_X(t) = \prod_k \phi_{X_k}(t_k)$ holds for each vector $t = (t_1, \ldots, t_d)$ in $\mathbb{R}^d$.*

Proof. If the random variables $X_1, \ldots, X_d$ are independent, then the relation $\phi_X(t) = \prod_k \phi_{X_k}(t_k)$ follows from Proposition 10.1.10, which can easily be extended to apply to complex-valued functions.

We turn to the converse. Let μ_X and $\mu_{X_1}, \ldots, \mu_{X_d}$ be the distributions of X and $X_1, \ldots, X_d$. Since the characteristic function (call it ϕ_{prod}) of the product measure $\mu_{X_1} \times \cdots \times \mu_{X_d}$ is given by $\phi_{\text{prod}}(t) = \prod_k \phi_{X_k}(t_k)$, it follows from Proposition 10.3.11 that the relation $\phi_X(t) = \prod_k \phi_{X_k}(t_k)$ implies that μ is equal to the product measure $\mu_{X_1} \times \cdots \times \mu_{X_d}$ and then from Proposition 10.1.9 that the random variables $X_1, \ldots, X_d$ are independent. $\square$

Our goal for the rest of this section is to prove the central limit theorem (Theorem 10.3.16). The main tool for this will be Proposition 10.3.15.

Suppose that $\{\mu_n\}$ is a sequence of probability measures on $(\mathbb{R}, \mathscr{B}(\mathbb{R}))$. Let us look at the relationship between convergence in distribution of the sequence $\{\mu_n\}$ and pointwise convergence of the corresponding sequence $\{\phi_{\mu_n}\}$ of characteristic functions. For this we need a concept related to regularity. We know (see Proposition 1.5.6) that if μ is a probability measure on $(\mathbb{R}^d, \mathscr{B}(\mathbb{R}^d))$, then

$$\sup\{\mu(K) : K \text{ is compact}\} = 1.$$

Measures satisfying this condition are sometimes called *tight*. A collection $\mathscr{C}$ of probability measures on $(\mathbb{R}^d, \mathscr{B}(\mathbb{R}^d))$ is called *uniformly tight* if for every positive ε there is a compact set K such that

$$\mu(K) > 1 - \varepsilon$$

holds for each μ in $\mathscr{C}$.

The following result is sometimes useful for establishing the uniform tightness of a family of probability measures on $(\mathbb{R}, \mathscr{B}(\mathbb{R}))$. See, for example, the proof of Proposition 10.3.15.

Proposition 10.3.13. *Suppose that μ is a probability measure on $(\mathbb{R}, \mathscr{B}(\mathbb{R}))$ and that ϕ is its characteristic function. Then for each positive ε we have*

$$\mu\left(\left\{x \in \mathbb{R} : |x| \geq \frac{2}{\varepsilon}\right\}\right) \leq \frac{1}{\varepsilon} \int_{-\varepsilon}^{\varepsilon} (1 - \phi(t)) \, dt.$$

Since characteristic functions are complex-valued functions, it's conceivable that the integral on the right-hand side of the inequality above could have a non-real value, in which case the inequality would be meaningless. We'll see in the proof below that this difficulty does not occur.

Proof. Using Fubini's theorem and basic calculus, we find

$$\int_{-\varepsilon}^{\varepsilon} \phi(t)\,dt = \int_{-\varepsilon}^{\varepsilon}\int_{\mathbb{R}} e^{itx}\,\mu(dx)\,dt$$

$$= \int_{\mathbb{R}}\int_{-\varepsilon}^{\varepsilon} (\cos tx + i\sin tx)\,dt\,\mu(dx) = \int_{\mathbb{R}} \frac{2\sin\varepsilon x}{x}\,\mu(dx).$$

Since $(1 - \frac{\sin\varepsilon x}{\varepsilon x}) \geq \frac{1}{2}$ if $|\varepsilon x| \geq 2$, we have

$$\frac{1}{2\varepsilon}\int_{-\varepsilon}^{\varepsilon}(1 - \phi(t))\,dt = \int_{\mathbb{R}}\left(1 - \frac{\sin\varepsilon x}{\varepsilon x}\right)\mu(dx) \geq \frac{1}{2}\mu\left(\left\{x\in\mathbb{R}: |x| \geq \frac{2}{\varepsilon}\right\}\right)$$

and the proposition follows.   □

Proposition 10.3.14. *Let $\{\mu_n\}$ be a uniformly tight sequence of probability measures on $(\mathbb{R}, \mathcal{B}(\mathbb{R}))$. Then $\{\mu_n\}$ has a subsequence that converges in distribution to some probability measure on $(\mathbb{R}, \mathcal{B}(\mathbb{R}))$.*

Proof. Suppose that $\{F_n\}$ is the sequence of distribution functions corresponding to $\{\mu_n\}$ and that $\{x_k\}$ is an enumeration of some countable dense subset D of $\mathbb{R}$. We will use a diagonal argument to choose a convergent subsequence of $\{\mu_n\}$. To begin, choose a subsequence $\{F_{1,n}\}_n$ of $\{F_n\}_n$ such that $\{F_{1,n}(x_1)\}_n$ is convergent, and then continue inductively, for each k choosing a subsequence $\{F_{k+1,n}\}_n$ of $\{F_{k,n}\}_n$ such that $\{F_{k+1,n}(x_{k+1})\}_n$ is convergent. Now take the diagonal subsequence $\{F_{j,j}\}$ of $\{F_n\}$, and let $\{\mu_{n_j}\}$ be the corresponding subsequence of $\{\mu_n\}$. We will show that $\{\mu_{n_j}\}$ converges in distribution to some probability measure μ.

We can define a function G_0 on the countable dense set D by letting $G_0(x) = \lim_j F_{j,j}(x)$ hold for each x in D. Then G_0 is a nondecreasing function and, since the sequence $\{\mu_n\}$ is uniformly tight, G_0 satisfies $\lim_{x\to-\infty} G_0(x) = 0$ and $\lim_{x\to+\infty} G_0(x) = 1$. Next, define $G: \mathbb{R} \to \mathbb{R}$ by

$$G(x) = \inf\{G_0(t) : t \in D \text{ and } t > x\}.$$

Then G is nondecreasing, it has limits of 0 and 1 at $-\infty$ and $+\infty$, and it is right continuous; let μ be the corresponding probability measure (recall Proposition 1.3.10). We show that the sequence $\{\mu_{n_j}\}$ converges in distribution to μ by checking that $G(x) = \lim_j F_{j,j}(x)$ holds at each x at which G is continuous. To do this, suppose that G is continuous at x, let ε be a positive number, and choose values t_0 and t_1 in D such that $t_0 < x < t_1$, $G(x) - \varepsilon < G_0(t_0)$, and $G_0(t_1) < G(x) + \varepsilon$. Note that if j is large enough that $|F_{j,j}(t_1) - G_0(t_1)| < \varepsilon$, then

$$F_{j,j}(x) \leq F_{j,j}(t_1) < G_0(t_1) + \varepsilon < G(x) + 2\varepsilon.$$

A similar calculation gives a lower bound of $G(x) - 2\varepsilon$ for $F_{j,j}(x)$, and so we can conclude that $|G(x) - F_{j,j}(x)| < 2\varepsilon$ holds for all large j. Thus $G(x) = \lim_j F_{j,j}(x)$, and Proposition 10.3.3 implies that $\{\mu_{n_j}\}$ converges in distribution to μ. □

Proposition 10.3.15. *Let μ and μ_1, μ_2, ... be probability measures on $(\mathbb{R}, \mathscr{B}(\mathbb{R}))$. Then the sequence $\{\mu_n\}$ converges in distribution to μ if and only if the sequence $\{\phi_{\mu_n}\}$ converges pointwise to ϕ_μ.*

Proof. For each t the function $x \mapsto e^{itx}$ is bounded and continuous. Thus if $\{\mu_n\}$ converges in distribution to μ, then $\int e^{itx} \mu(dx) = \lim_n \int e^{itx} \mu_n(dx)$ holds for each t, and $\{\phi_{\mu_n}\}$ converges pointwise to ϕ_μ.

Let us turn to the converse and assume that $\{\phi_{\mu_n}\}$ converges pointwise to ϕ_μ. We begin by showing that the sequence $\{\mu_n\}$ is uniformly tight. Choose a positive number ε, and then use the continuity of ϕ_μ at 0 (and the fact that $\phi_\mu(0) = 1$) to choose δ such that $\frac{1}{\delta} \int_{-\delta}^{\delta} (1 - \phi_\mu(t)) \, dt < \varepsilon$. Since $\{\phi_{\mu_n}\}$ converges pointwise to ϕ_μ, we can use the dominated convergence theorem to conclude that

$$\frac{1}{\delta} \int_{-\delta}^{\delta} (1 - \phi_{\mu_n}(t)) \, dt < \varepsilon$$

holds for all large n. Proposition 10.3.13 now implies that

$$\mu_n\left(\left[-\frac{2}{\delta}, \frac{2}{\delta}\right]\right) > 1 - \varepsilon \tag{2}$$

holds for all large n. By making δ smaller, if necessary, we can make (2) hold for all n. It follows that the sequence $\{\mu_n\}$ is uniformly tight.

We now check that $\{\mu_n\}$ converges in distribution to μ. Suppose it did not. Then there would be a bounded continuous function f on $\mathbb{R}$ such that $\{\int f \, d\mu_n\}$ does not converge to $\int f \, d\mu$. Choose a subsequence $\{\mu_{n_k}\}$ of $\{\mu_n\}$ such that $\{\int f \, d\mu_{n_k}\}$ converges to a value other that $\int f \, d\mu$. The uniform tightness of $\{\mu_n\}$, which we verified above, together with Proposition 10.3.14, lets us replace $\{\mu_{n_k}\}$ with a subsubsequence that converges to some probability measure ν. Then $\nu \neq \mu$, since $\int f \, d\nu \neq \int f \, d\mu$, yet $\phi_\nu = \phi_\mu$, since $\{\phi_{\mu_{n_k}}\}$ converges to both ϕ_ν and ϕ_μ. This is impossible, and so our hypothesis that $\{\mu_n\}$ does not converge to μ must be false. □

Let us make a last preparation for the proof of the central limit theorem. Suppose that X is a random variable with mean 0 and variance 1 and that ϕ is its characteristic function. Then $\phi(0) = 1$, $\phi'(0) = 0$, $\phi''(0) = -1$, and ϕ has at least two continuous derivatives (see Proposition 10.3.7). According to l'Hospital's rule, plus the facts in the previous sentence, we have

$$\lim_{x \to 0} \frac{\phi(x) - (1 - x^2/2)}{x^2} = 0.$$

and so ϕ can be written in terms of its second-degree Maclaurin polynomial $1 - x^2/2$ as $\phi(x) = 1 - x^2/2 + R(x)$, where $\lim_{x \to 0} R(x)/x^2 = 0$.

Theorem 10.3.16 (Central Limit Theorem). *Let X_1, X_2, ... be a sequence of independent identically distributed random variables, with common mean μ and variance σ^2, and for each n let $S_n = X_1 + \cdots + X_n$. Then the normalized sequence $\{(S_n - n\mu)/\sigma\sqrt{n}\}$ converges in distribution to a normal (i.e., Gaussian) distribution with mean 0 and variance 1.*

Proof. Each random variable $(X_i - \mu)/\sigma$ has mean 0 and variance 1 and hence has a characteristic function ϕ that is as described just before the statement of the theorem. Since the X_i's are identically distributed, the function ϕ does not depend on the index i. Note that

$$\frac{S_n - n\mu}{\sigma\sqrt{n}} = \frac{1}{\sqrt{n}} \sum_{i=1}^{n} \frac{X_i - \mu}{\sigma}. \tag{3}$$

If we use Eq. (3), the independence of the X_i's, Lemma 10.3.6, Proposition 10.3.9, and the fact that $\lim_{x \to 0} R(x)/x^2 = 0$ (where $R(x)$ is the remainder defined just before the statement of the theorem), we find that the characteristic function of $(S_n - n\mu)/\sigma\sqrt{n}$ is given by

$$\phi\left(\frac{t}{\sqrt{n}}\right)^n = \left(1 - \frac{t^2}{2n} + R(t/\sqrt{n})\right)^n = \left(1 - \frac{t^2/2 + \varepsilon_n}{n}\right)^n,$$

where $\varepsilon_n = -nR(t/\sqrt{n})$ and hence where $\lim_n \varepsilon_n = 0$. It follows (Lemma 10.3.4) that the characteristic functions of the normalized sums $(S_n - n\mu)/\sigma\sqrt{n}$ approach the function $t \mapsto e^{-t^2/2}$; since the limit is the characteristic function of the normal distribution with mean 0 and variance 1, the theorem follows (see Proposition 10.3.15).   □

Exercises

1. For each positive integer n define a probability measure μ_n on $(\mathbb{R}, \mathscr{B}(\mathbb{R}))$ by $\mu_n = (1/n)\sum_{i=1}^{n} \delta_{i/n}$. Show that the sequence $\{\mu_n\}$ converges in distribution to the uniform distribution on $[0, 1]$.
2. Suppose that μ and $\mu_1, \mu_2, \ldots$, are probability measures on $(\mathbb{R}, \mathscr{B}(\mathbb{R}))$, each of which is concentrated on the integers. Show that the sequence $\{\mu_n\}$ converges in distribution to μ if and only if $\mu(\{k\}) = \lim_n \mu_n(\{k\})$ holds for each k in $\mathbb{Z}$.
3. Show that
   (a) if μ is the point mass at a, then ϕ_μ is given by $\phi_\mu(t) = e^{iat}$,
   (b) if μ is the binomial distribution with parameters n and p, then ϕ_μ is given by $\phi_\mu(t) = (1 - p(1 - e^{it}))^n$,

(c) if μ is the Poisson distribution with parameter λ, then $\phi_\mu(t) =$ $e^{-\lambda(1-e^{it})}$, and

(d) if μ is the uniform distribution on the interval $[a,b]$, then ϕ_μ is given by $\phi_\mu(t) = \frac{e^{itb} - e^{ita}}{it(b-a)}$.

4. Show that if ϕ is the characteristic function of a probability measure on $(\mathbb{R}, \mathscr{B}(\mathbb{R}))$, then $\phi(-t) = \overline{\phi(t)}$.

5. Show that a probability measure μ on $(\mathbb{R}, \mathscr{B}(\mathbb{R}))$ is symmetric (i.e., $\mu(-A) = \mu(A)$ holds for each A in $\mathscr{B}(\mathbb{R})$) if and only if ϕ_μ is real-valued.

6. Show that if ϕ is the characteristic function of a probability measure on $(\mathbb{R}, \mathscr{B}(\mathbb{R}))$, then ϕ is uniformly continuous on $\mathbb{R}$.

7. Suppose that X and $X_1, X_2, \ldots$ are real-valued random variables and that μ and $\mu_1, \mu_2, \ldots$ are their distributions. Show that if $\{X_n\}$ converges in probability to X, then $\{\mu_n\}$ converges in distribution to μ.

8. Let μ be a probability distribution on $(\mathbb{R}, \mathscr{B}(\mathbb{R}))$. Show that $|\phi_\mu(t)| = 1$ for some nonzero number t if and only if there exist real numbers a and b such that μ is concentrated on the set $\{a + bn : n \in \mathbb{Z}\}$. (Such a distribution is called a *lattice distribution.*)

9. Show directly (i.e., using only the definition of convergence in distribution) that if a sequence $\{\mu_n\}$ of probability measures on $(\mathbb{R}^d, \mathscr{B}(\mathbb{R}^d))$ converges in distribution to some probability measure, then the sequence $\{\mu_n\}$ is uniformly tight.

10. Suppose that $\{\mu_n\}$ is a sequence of probability distributions on $(\mathbb{R}, \mathscr{B}(\mathbb{R}))$ whose characteristic functions $\{\phi_{\mu_n}\}$ converge pointwise to some function $\phi \colon \mathbb{R} \to \mathbb{C}$. Show that if ϕ is continuous at 0, then there is a probability distribution μ on $(\mathbb{R}, \mathscr{B}(\mathbb{R}))$ such that $\{\mu_n\}$ converges to μ in distribution.

11. For each n let μ_n be a binomial distribution with parameters n and p_n. Show that if $\{np_n\}$ is convergent, with $\lambda = \lim_n np_n$, then the sequence $\{\mu_n\}$ converges in distribution to the Poisson distribution with parameter λ. Do this

(a) by making a direct calculation of probabilities (see Exercise 2), and

(b) by using characteristic functions.

12. Suppose that for probability measures μ and ν on $(\mathbb{R}, \mathscr{B}(\mathbb{R}))$ we define $d(\mu, \nu)$ by

$$d(\mu, \nu) = \inf\{\varepsilon > 0 : F_\mu(t) \le F_\nu(t + \varepsilon) + \varepsilon \text{ and}$$

$$F_\nu(t) \le F_\mu(t + \varepsilon) + \varepsilon \text{ for all } t \text{ in } \mathbb{R}\}.$$

(The function d is known as *Lévy's metric.*)

(a) Show that d is a metric on the set of all probability measures on $(\mathbb{R}, \mathscr{B}(\mathbb{R}))$.

(b) Suppose that μ and $\mu_1, \mu_2, \ldots$ are probability measures on $(\mathbb{R}, \mathscr{B}(\mathbb{R}))$. Show that the sequence $\{\mu_n\}$ converges in distribution to μ if and only if $\lim_n d(\mu_n, \mu) = 0$.

13. Suppose that μ is a probability distribution on $\mathbb{R}$ such that ϕ_μ is integrable. (Note that for the inversion formula (1) to make sense with the integral interpreted as a Lebesgue integral, ϕ_μ must be integrable.)
    (a) Show that if μ is absolutely continuous with density function g and if the inversion formula (1) is valid for ϕ_μ and g, then g is bounded and continuous.
    (b) Show that if ϕ_μ is integrable, then μ is absolutely continuous and formula (1) works. (Hint: Use some ideas and calculations from Proposition 10.3.11. In particular, consider $\int_\mathbb{R} h(x)p(x)\,dx$, where h ranges over the continuous functions with compact support on $\mathbb{R}$ and p is the inverse Fourier transform of $\phi_{\gamma_\sigma * \mu}$.)

14. Show how to prove the central limit theorem without using Proposition 10.3.13. (Hint: For each n let μ_n be the distribution of $(S_n - n\mu)/\sigma\sqrt{n}$. Use Markov's inequality (that is, Proposition 2.3.10), rather than Proposition 10.3.13, to show that the sequence $\{\mu_n\}$ is tight.)

15. Let μ and $\mu_1, \mu_2, \ldots$ be probability measures on $(\mathbb{R}, \mathscr{B}(\mathbb{R}))$ such that the sequence $\{\mu_n\}$ converges in distribution to μ.
    (a) Suppose that X and $X_1, X_2, \ldots$ are random variables, all defined on the same probability space, whose distributions are μ and $\mu_1, \mu_2, \ldots$. Show (by giving a simple example) that it does not follow that $\{X_n\}$ converges almost surely to X.
    (b) On the other hand, show that there are random variables X and $X_1, X_2, \ldots$, all defined on the same probability space and with distributions μ and μ_1, $\mu_2, \ldots$, such that $\{X_n\}$ converges to X almost surely. (Hint: Let F and $F_1, F_2, \ldots$ be the distribution functions of μ and $\mu_1, \mu_2, \ldots$. Then the random variables F^{-1} and $F_1^{-1}, F_2^{-1}, \ldots$ constructed from F and F_1, F_2, $\ldots$ as in Proposition 10.1.15 do what is required. To verify the almost sure convergence, use the equivalence of inequalities (3) and (4) from Sect. 10.1 to verify that $\lim_n F_n^{-1}(t) = F^{-1}(t)$ holds at each t at which F^{-1} is continuous.)

10.4   Conditional Distributions and Martingales

Suppose that $(\Omega, \mathscr{A}, P)$ is a probability space, that A and B are events in $\mathscr{A}$, and that $P(B) \neq 0$. In elementary treatments of probability, the *conditional probability of A, given B*, written $P(A|B)$, is defined by

$$P(A|B) = \frac{P(A \cap B)}{P(B)}.$$

Example 10.4.1. Suppose that we select a number at random from the set $\{1,2,3,4,5,6\}$, with each number in that set having probability $1/6$ of being selected. Consider events E and F, where E is the event that the number selected is even and F is the event that the number selected is not equal to 6. Then we have

$$P(E|F) = \frac{P(E \cap F)}{P(F)} = \frac{2/6}{5/6} = 2/5$$

and

$$P(F|E) = \frac{P(F \cap E)}{P(E)} = \frac{2/6}{3/6} = 2/3,$$

which should agree with one's intuition.   □

Let us deal for a moment with a probability space $(\Omega, \mathscr{A}, P)$ such that Ω is finite and $\mathscr{A}$ contains all the subsets of Ω. Let X and Y be real-valued random variables on $(\Omega, \mathscr{A}, P)$ with values $x_1, \ldots, x_m$ and $y_1, \ldots, y_n$, and let us assume that $P(Y = y_j) \neq 0$ for each j. Then $E(X|Y = y_j)$, the *conditional expectation of X, given that* $Y = y_j$, is defined by

$$E(X|Y = y_j) = \sum_i x_i P(X = x_i | Y = y_j).$$

It follows that

$$E(X|Y = y_j) = \frac{\sum_i x_i P(X = x_i \text{ and } Y = y_j)}{P(Y = y_j)} = \frac{\int_{Y=y_j} X \, dP}{P(Y = y_j)}. \tag{1}$$

Of course, this defines a function $y_j \mapsto E(X|Y = y_j)$ on the set of values of Y. It is convenient to have a slightly different form of the conditional expectation, with the new form being defined on the probability space $(\Omega, \mathscr{A}, P)$. Let us define $E(X|Y) \colon \Omega \to \mathbb{R}$ by letting $E(X|Y)(\omega)$ be $E(X|Y = y_j)$ for those ω that satisfy $Y(\omega) = y_j$. In other words, $E(X|Y)$ is the composition of the functions $\omega \mapsto Y(\omega)$ and $y \mapsto E(X|Y = y)$. It follows from (1) that

$$\int_B E(X|Y) \, dP = \int_B X \, dP \tag{2}$$

holds for each B of the form $\{Y = y_j\}$. Since each B in the σ-algebra $\sigma(Y)$ generated by Y is a finite disjoint union of sets of the form $\{Y = y_j\}$, it follows that (2) holds for each B in $\sigma(Y)$. Furthermore, $E(X|Y)$ is $\sigma(Y)$-measurable (in this simple example, where Ω is finite, this just means that $E(X|Y)$ is constant on each set of the form $\{Y = y_j\}$).

We are now ready to look at how these ideas generalize to arbitrary probability spaces.

Let $(\Omega, \mathscr{A}, P)$ be a probability space and let $\mathscr{B}$ be a sub-σ-algebra of $\mathscr{A}$. Suppose that X is a real-valued random variable on $(\Omega, \mathscr{A}, P)$ that has a finite expected value. A *conditional expectation of X given* $\mathscr{B}$ is a random variable Y that is $\mathscr{B}$-measurable, is integrable (that is, has a finite expected value), and satisfies

$$\int_B Y \, dP = \int_B X \, dP$$

for each B in $\mathscr{B}$. One generally writes $E(X|\mathscr{B})$ for a conditional expectation of X given $\mathscr{B}$. When one needs to be more precise, one sometimes calls an integrable $\mathscr{B}$-measurable function Y that satisfies $\int_B Y\,dP = \int_B X\,dP$ for all B in $\mathscr{B}$ a *version* of the conditional expectation of X given $\mathscr{B}$ or a version of $E(X|\mathscr{B})$.

Proposition 10.4.2. *Let $(\Omega, \mathscr{A}, P)$ be a probability space, let X be a random variable on $(\Omega, \mathscr{A}, P)$ that has a finite expected value, and let $\mathscr{B}$ be a sub-σ-algebra of $\mathscr{A}$. Then*

(a) *X has a conditional expectation given $\mathscr{B}$, and*
(b) *the conditional expectation of X given $\mathscr{B}$ is unique, in the sense that if Y_1 and Y_2 are versions of $E(X|\mathscr{B})$, then $Y_1 = Y_2$ almost surely.*

Proof. The formula $\mu(B) = \int_B X\,dP$ defines a finite signed measure on $(\Omega, \mathscr{B})$; it is absolutely continuous with respect to the restriction of P to $\mathscr{B}$. Thus the Radon–Nikodym theorem (Theorem 4.2.4), applied to μ and the restriction of P to $\mathscr{B}$, gives a $\mathscr{B}$-measurable random variable Y such that

$$\int_B Y\,dP = \mu(B) = \int_B X\,dP$$

holds for each B in $\mathscr{B}$. Thus Y is a conditional expectation of X given B. The uniqueness assertion in the Radon–Nikodym theorem gives the uniqueness of the conditional expectation.   □

Proposition 10.4.3. *Let $(\Omega, \mathscr{A}, P)$ be a probability space, let $\mathscr{B}$ and $\mathscr{B}_0$ be sub-σ-algebras of $\mathscr{A}$, and let X and Y be random variables on $(\Omega, \mathscr{A}, P)$ that have finite expected values. Then*

(a) *if a and b are constants, then $E(aX + bY|\mathscr{B}) = aE(X|\mathscr{B}) + bE(Y|\mathscr{B})$ almost surely,[9]*
(b) *if $X \leq Y$, then $E(X|\mathscr{B}) \leq E(Y|\mathscr{B})$ almost surely,*
(c) *$\|E(X|\mathscr{B})\|_1 \leq \|X\|_1$,*
(d) *if X is $\mathscr{B}$-measurable, then $E(X|\mathscr{B}) = X$ almost surely (in particular, if c is a constant, then $E(c|\mathscr{B}) = c$ almost surely),*
(e) *if $\mathscr{B}_0 \subseteq \mathscr{B}$, then $E(X|\mathscr{B}_0) = E(E(X|\mathscr{B})|\mathscr{B}_0)$ almost surely,*
(f) *if $\mathscr{B}$ and X are independent (that is, if $\mathscr{B}$ and $\sigma(X)$ are independent), then $E(X|\mathscr{B})$ is almost surely equal to the constant $E(X)$, and*
(g) *if X is bounded and $\mathscr{B}$-measurable, then $E(XY|\mathscr{B}) = XE(Y|\mathscr{B})$ almost surely.*

Proof. Note that $aE(X|\mathscr{B}) + bE(Y|\mathscr{B})$ is a $\mathscr{B}$-measurable function that satisfies

---

[9] It is probably worth translating one of the parts of this proposition into more precise language. Part (a) says that if Z is a version of $E(aX + bY|\mathscr{B})$, if Z_1 is a version of $E(X|\mathscr{B})$, and if Z_2 is a version of $E(Y|\mathscr{B})$, then $Z = aZ_1 + bZ_2$ almost surely. Equivalently, part (a) can be viewed as saying that if Z_1 and Z_2 are versions of $E(X|\mathscr{B})$ and $E(Y|\mathscr{B})$, then $aZ_1 + bZ_2$ is a version of $E(aX + bY|\mathscr{B})$. Other assertions about conditional expectations can be made precise in similar ways.

$$\int_B (aE(X|\mathscr{B}) + bE(Y|\mathscr{B}))\, dP = \int_B (aX + bY)\, dP$$

for each B in $\mathscr{B}$ and hence is a conditional expectation of $aX + bY$ given $\mathscr{B}$. Part (a) then follows from the uniqueness of conditional expectations (part (b) of Proposition 10.4.2).

For part (b), note that

$$\int_B E(X|\mathscr{B})\, dP = \int_B X\, dP \le \int_B Y\, dP = \int_B E(Y|\mathscr{B})\, dP$$

holds for each B in $\mathscr{B}$. It now follows from Corollary 2.3.13 that $E(X|\mathscr{B}) \le E(Y|\mathscr{B})$ almost surely.

If we let A_+ and A_- be the sets $\{E(X|\mathscr{B}) \ge 0\}$ and $\{E(X|\mathscr{B}) < 0\}$, then part (c) follows from the calculation

$$\|E(X|\mathscr{B})\|_1 = \int_{A_+} E(X|\mathscr{B})\, dP - \int_{A_-} E(X|\mathscr{B})\, dP$$

$$= \int_{A_+} X\, dP - \int_{A_-} X\, dP \le \|X\|_1.$$

Part (d) is immediate, and part (e) follows from the calculation

$$\int_B E(E(X|\mathscr{B})|\mathscr{B}_0)\, dP = \int_B E(X|\mathscr{B})\, dP = \int_B X\, dP$$

which holds for every B in $\mathscr{B}_0$ (recall that $\mathscr{B}_0 \subseteq \mathscr{B}$).

We turn to part (f). If $\mathscr{B}$ and X are independent, then for each B in $\mathscr{B}$ the random variables χ_B and X are independent, and so Proposition 10.1.10 implies that

$$\int_B X\, dP = \int \chi_B X\, dP = \int \chi_B\, dP \int X\, dP$$

$$= P(B)E(X) = \int_B E(X)\, dP;$$

it follows that $E(X)$ is a version of $E(X|\mathscr{B})$.

Let us start our consideration of part (g) with the special case where $X = \chi_A$ for some A in $\mathscr{B}$. Then for each B in $\mathscr{B}$ we have

$$\int_B XY\, dP = \int_{B \cap A} Y\, dP = \int_{B \cap A} E(Y|\mathscr{B})\, dP$$

$$= \int_B \chi_A E(Y|\mathscr{B})\, dP = \int_B X E(Y|\mathscr{B})\, dP$$

and so

$$\int_B XY\, dP = \int_B X E(Y|\mathscr{B})\, dP. \tag{3}$$

Equation (3) now extends to the case where X is simple function and then (by the dominated convergence theorem) to the case where X is an arbitrary bounded $\mathscr{B}$-measurable function. Furthermore $XE(Y|\mathscr{B})$ is $\mathscr{B}$-measurable. Thus $XE(Y|\mathscr{B})$ is a version of $E(XY|\mathscr{B})$ and the proof is complete.   □

Proposition 10.4.4 (Monotone and Dominated Convergence Theorems for Conditional Expectations). *Let $(\Omega, \mathscr{A}, P)$ be a probability space, let $\mathscr{B}$ be a sub-σ-algebra of $\mathscr{A}$, and let $X_1, X_2, \ldots$ be random variables with finite expected values such that $\lim_n X_n$ exists almost surely. If*

(a) *$\{X_n\}$ is an increasing sequence such that $\lim_n E(X_n)$ is finite, or*
(b) *there exists a random variable Y with finite expected value such that each X_n satisfies $|X_n| \leq Y$ almost surely,*

then $\lim_n X_n$ has a finite expected value and $E(\lim_n X_n|\mathscr{B}) = \lim_n E(X_n|\mathscr{B})$ almost surely.

Proof. First suppose that condition (a) holds. Let us also temporarily assume that the random variables X_n are nonnegative. Since we are assuming that $\{X_n\}$ is an increasing sequence, it follows from part (b) of Proposition 10.4.3 that the sequence $\{E(X_n|\mathscr{B})\}$ is increasing almost surely and so has an almost sure limit, possibly with some of values of $\lim_n E(X_n|\mathscr{B})$ being infinite. The monotone convergence theorem implies that

$$\int \lim_n E(X_n|\mathscr{B})\,dP = \lim_n \int E(X_n|\mathscr{B})\,dP = \lim_n \int X_n\,dP < +\infty,$$

and so $\lim_n E(X_n|\mathscr{B})$ is finite almost everywhere. Applying the monotone convergence theorem twice more gives

$$\int_B \lim_n X_n\,dP = \lim_n \int_B X_n\,dP = \lim_n \int_B E(X_n|\mathscr{B})\,dP = \int_B \lim_n E(X_n|\mathscr{B})\,dP$$

for each B in $\mathscr{B}$; thus $\lim_n E(X_n|\mathscr{B})$ is a version of $E(\lim_n X_n|\mathscr{B})$ and the proof is complete in the case where condition (a) holds and the X_n's are nonnegative. We can complete the proof for the case where (a) holds by applying what we have just proved to the sequence $\{X_n - X_1\}$ and then using the linearity of conditional expectations.

Now suppose that condition (b) holds. Since we are assuming that $|X_n| \leq Y$ for each n, we have $|\lim_n X_n| \leq Y$ and so $\lim_n X_n$ has a finite expected value. For each n let $Y_n = \inf\{X_k : k \geq n\}$ and $Z_n = \sup\{X_k : k \geq n\}$. Then $\{Y_n\}$ is an increasing sequence that converges pointwise to $\liminf_n X_n$, and $\{Z_n\}$ is a decreasing sequence that converges pointwise to $\limsup_n X_n$; since $\lim X_n$ exists, both those sequences converge to it almost surely. If we apply the first half of the proposition to the sequence $\{Y_n\}$, we conclude that $\lim_n E(Y_n|\mathscr{B}) = E(\lim_n Y_n|\mathscr{B}) = E(\lim_n X_n|\mathscr{B})$ almost surely. A similar argument, applied to the sequence $\{Y - Z_n\}$, shows that $\lim_n E(Z_n|\mathscr{B}) = E(\lim_n X_n|\mathscr{B})$ almost surely. Finally, each variable $E(X_n|\mathscr{B})$ lies between the corresponding variables $E(Y_n|\mathscr{B})$ and $E(Z_n|\mathscr{B})$, and it follows that $\lim_n E(X_n|\mathscr{B}) = E(\lim_n X_n|\mathscr{B})$ almost surely. With this the proof is complete. □

In the remainder of this chapter we will be looking at *stochastic processes*. A rather abstract definition might say that a stochastic process is an indexed family $\{X_t\}_{t\in T}$ of random variables, where T is an arbitrary nonempty set and all the random variables are defined on the same probability space. However, one usually deals with more concrete situations, in which the index set T is a set of integers or else a nice set of real numbers (such as an interval), and the members of T are interpreted as times. For each t in T the random variable X_t is thought of as representing a quantity that can be observed at time t.

A *discrete-time* stochastic process is one for which T is a set of integers, and a *continuous-time* process is one for which T is an interval of real numbers. We will see a few discrete-time processes in this section, and we will see some continuous-time processes later in the chapter.

Let $(\Omega, \mathscr{A}, P)$ be a probability space. A *filtration*[10] is a sequence $\{\mathscr{F}_n\}_{n=0}^{\infty}$ of sub-σ-algebras of $\mathscr{A}$ that is increasing, in the sense that $\mathscr{F}_n \subseteq \mathscr{F}_{n+1}$ holds for each n. A discrete-time stochastic process (i.e., a sequence of random variables) $\{X_n\}_{n=0}^{\infty}$ is *adapted* to the filtration $\{\mathscr{F}_n\}_{n=0}^{\infty}$ if X_n is $\mathscr{F}_n$-measurable for each n. Note that the sequence $\{X_n\}_{n=0}^{\infty}$ is adapted to the filtration $\{\mathscr{F}_n\}_{n=0}^{\infty}$ if and only if $\sigma(X_0, \ldots, X_n) \subseteq \mathscr{F}_n$ holds for each n.

The intuition here is that the events in the σ-algebra $\mathscr{F}_n$ are those that could be known by time n. In one common situation, $\{X_n\}$ is an arbitrary sequence of random variables and for each n we let $\mathscr{F}_n$ be $\sigma(X_0, \ldots, X_n)$. In this case $\mathscr{F}_n$ contains exactly the events that are determined by the random variables $X_0, \ldots, X_n$.

Let $\{\mathscr{F}_n\}$ be a filtration on the probability space $(\Omega, \mathscr{A}, P)$. A *stopping time* or an *optional time* is a function $\tau: \Omega \to \mathbb{N}_0 \cup \{+\infty\}$ such that $\{\tau \leq n\} \in \mathscr{F}_n$ holds for each n in $\mathbb{N}_0$. It is easy to check that if τ is a stopping time, then τ is $\mathscr{A}$-measurable and that a function $\tau: \Omega \to \mathbb{N}_0 \cup \{+\infty\}$ is a stopping time if and only if $\{\tau = n\} \in \mathscr{F}_n$ holds for each n in $\mathbb{N}_0$.

One standard interpretation of a stopping time is the following: You are observing random variables X_0, X_1, ..., one after the other, and you may decide to stop observing at some random time τ. It is reasonable to decide whether or not to stop with the nth observation on the basis of the information that is available by time n, but it is not reasonable to use information about the future (e.g., the values of X_{n+1}, X_{n+2}, ...). In other words, $\{\tau = n\}$, the event that you stop just after observing X_n, should belong to $\mathscr{F}_n$.

---

[10]In this section we are dealing with discrete-time processes. On the other hand, a *filtration* $\{\mathscr{F}_t\}_{t\in T}$ in continuous time is defined by requiring that $\mathscr{F}_{t_1} \subseteq \mathscr{F}_{t_2}$ holds whenever t_1 and t_2 are elements of T such that $t_1 < t_2$. If $\{\mathscr{F}_t\}_{t\in T}$ is a filtration with $T = [0, +\infty)$, then a stopping time for it is a function $\tau: \Omega \to [0, +\infty]$ such that $\{\tau \leq t\} \in \mathscr{F}_t$ holds for all t in T. Except for a few exercises involving Brownian motion, we will not be dealing with filtrations in continuous time.

Example 10.4.5. Suppose that you take a random walk on the integers in the following way. You begin at 0, and every minute you toss a fair coin and move to the right by a distance of 1 if the coin yields a head and to the left by a distance of 1 if it yields a tail. To formalize this, we let $\{Y_i\}_{i=1}^{\infty}$ be a sequence of independent and identically distributed random variables such that

$$P(\{Y_i = -1\}) = P(\{Y_i = 1\}) = 1/2$$

holds for each i, and then we define $\{X_n\}_{n=0}^{\infty}$ by $X_0 = 0$ and $X_n = Y_1 + \cdots + Y_n$ if $n > 0$. Finally, we define the filtration $\{\mathscr{F}_n\}$ by letting $\mathscr{F}_n$ be $\sigma(X_0, \ldots, X_n)$ for each n.

Let us consider a rather simple stopping time for this process. The time you first reach 1 (if you ever reach it) is given by

$$\tau_{\{1\}}(\omega) = \inf\{n \in \mathbb{N}_0 : X_n(\omega) = 1\}. \tag{4}$$

Note that $\tau_{\{1\}}(\omega) = +\infty$ if the set on the right side of (4) is empty—in other words, if you never reach the point 1. Since

$$\{\tau_{\{1\}} \leq n\} = \bigcup_{i \leq n} \{X_i = 1\} \in \mathscr{F}_n,$$

the variable $\tau_{\{1\}}$ is in fact a stopping time. □

Example 10.4.6. Now suppose we have an arbitrary real-valued process $\{X_n\}_{n=0}^{\infty}$ that is adapted to some filtration $\{\mathscr{F}_n\}$ and we want to know the first time that X_n is in some Borel subset A of $\mathbb{R}$. The same reasoning as in Example 10.4.5 works if we replace (4) with

$$\tau_A(\omega) = \inf\{n \in \mathbb{N}_0 : X_n(\omega) \in A\}. \quad □$$

Let us now turn to martingales. Suppose that $(\Omega, \mathscr{A}, P)$ is a probability space, that $\{\mathscr{F}_n\}_{n=0}^{\infty}$ is a filtration on $(\Omega, \mathscr{A}, P)$, and that $\{X_n\}_{n=0}^{\infty}$ is a discrete-time process on $(\Omega, \mathscr{A}, P)$. Then $(\{X_n\}_{n=0}^{\infty}, \{\mathscr{F}_n\}_{n=0}^{\infty})$, or simply $\{X_n\}_{n=0}^{\infty}$, is a *martingale* if

(a) $\{X_n\}_{n=0}^{\infty}$ is adapted to $\{\mathscr{F}_n\}_{n=0}^{\infty}$,
(b) each X_n has a finite expected value, and
(c) for each n we have $X_n = E(X_{n+1}|\mathscr{F}_n)$ almost surely.

Sometimes we will say that $\{X_n\}$ is a martingale *relative to* $\{\mathscr{F}_n\}$. If condition (c) is replaced with

for each n we have $X_n \leq E(X_{n+1}|\mathscr{F}_n)$ almost surely

or with

for each n we have $X_n \geq E(X_{n+1}|\mathscr{F}_n)$ almost surely, then $(\{X_n\}_{n=0}^\infty, \{\mathscr{F}_n\}_{n=0}^\infty)$ or $\{X_n\}_{n=0}^\infty$ is a *submartingale* or a *supermartingale*. Note that we can verify condition (c) in the definition of a martingale by checking that $\int_B X_n\,dP = \int_B X_{n+1}\,dP$ holds for each n in $\mathbb{N}_0$ and each B in $\mathscr{F}_n$. Similar remarks apply to submartingales and supermartingales.

Examples 10.4.7.

(a) Let $(\Omega, \mathscr{A}, P)$ be a probability space, and let $\{Y_n\}_{n=1}^\infty$ be a sequence of independent (real-valued) random variables on Ω with finite expectations. Define $\{S_n\}_{n=0}^\infty$ by $S_0 = 0$ and $S_n = Y_1 + \cdots + Y_n$ if $n \geq 1$, and define a filtration $\{\mathscr{F}_n\}_{n=0}^\infty$ by $\mathscr{F}_n = \sigma(S_0, \ldots, S_n)$. If $E(Y_n) = 0$ for $n = 1, 2, \ldots$, then we can use parts (a), (d), and (f) of Proposition 10.4.3, together with the independence of the sequence $\{Y_n\}_{n=1}^\infty$, to show that

$$E(S_{n+1}|\mathscr{F}_n) = E(S_n + Y_{n+1}|\mathscr{F}_n) = S_n + E(Y_{n+1}|\mathscr{F}_n) = S_n$$

holds almost surely for each n, and hence that $\{S_n\}_{n=0}^\infty$ is a martingale. Similar calculations show that if $E(Y_n) \geq 0$ for $n = 1, 2, \ldots$ (or if $E(Y_n) \leq 0$ for $n = 1, 2, \ldots$), then $\{S_n\}_{n=0}^\infty$ is a submartingale (or a supermartingale).

(b) Suppose that you are gambling, making a sequence of wagers. Let $\{X_n\}_{n=0}^\infty$ be a sequence of random variables with finite expected values and defined on some probability space $(\Omega, \mathscr{A}, P)$, and suppose that X_0 represents your capital at the start and that X_n represents your capital after n wagers. Define a filtration by letting $\mathscr{F}_n = \sigma(X_0, \ldots, X_n)$ hold for each n. Then $\{X_n\}_{n=0}^\infty$ is a martingale if the wagers are fair (that is, if at each stage the conditional expectation of your gain from the next wager, namely $E(X_{n+1}|\mathscr{F}_n) - X_n$, is 0); it is a submartingale if the wagers favor you and is a supermartingale if they favor your opponent.

(c) Let $(\Omega, \mathscr{A}, P)$ be a probability space, let $\{\mathscr{F}_n\}_{n=0}^\infty$ be a filtration on $(\Omega, \mathscr{A}, P)$, and let X be an integrable $\mathscr{A}$-measurable function on Ω. For each n define X_n by $X_n = E(X|\mathscr{F}_n)$. Let us check that $\{X_n\}_{n=0}^\infty$ is a martingale. Condition (c) in the definition of martingales is the only thing to check, and that condition follows from the calculation

$$E(X_{n+1}|\mathscr{F}_n) = E(E(X|\mathscr{F}_{n+1})|\mathscr{F}_n) = E(X|\mathscr{F}_n) = X_n$$

(see part (e) of Proposition 10.4.3).

(d) We define a martingale on the probability space $((0,1], \mathscr{B}((0,1]), \lambda)$ as follows. Let $\mathscr{F}_0$ be the σ-algebra that contains only the sets $\varnothing$ and $(0,1]$. For positive n let $\mathscr{P}_n$ be the partition of $(0,1]$ that consists of the intervals $(i/2^n, (i+1)/2^n]$, $i = 0, \ldots, 2^n - 1$; then let $\mathscr{F}_n = \sigma(\mathscr{P}_n)$. Now suppose that μ is a finite Borel measure on $(0,1]$, and for each n define $X_n: (0,1] \to \mathbb{R}$ by $X_n(x) = \mu(I)/\lambda(I)$, where I is the interval in $\mathscr{P}_n$ that contains x. Then each interval I in $\mathscr{P}_n$ satisfies

$$\int_I X_n\,d\lambda = \mu(I) = \int_I X_{n+1}\,d\lambda.$$

It follows that the same equation holds if I is replaced with an arbitrary set in $\mathscr{F}_n$; hence $X_n = E(X_{n+1}|\mathscr{F}_n)$ and $\{X_n\}$ is a martingale. There are a couple of things to note here. First, if we consider the behavior of the sequence $\{X_n(x)\}$ as n goes to infinity, we seem to be dealing with some sort of derivative. We'll look harder at this later in this section. Second, we are dealing with pure analysis in this example; no probability seems to be involved.   □

The following is one of the major results of martingale theory.

Theorem 10.4.8 (Doob's Martingale Convergence Theorem). *Let* $(\Omega, \mathscr{A}, P)$ *be a probability space, and let* $(\{X_n\}_{n=0}^{\infty}, \{\mathscr{F}_n\}_{n=0}^{\infty})$ *be a submartingale on* $(\Omega, \mathscr{A}, P)$ *such that* $\sup_n E(X_n^+) < +\infty$. *Then the limit* $\lim_n X_n$ *exists almost surely, and* $E(|\lim_n X_n|) < +\infty$.

We need a few preliminary results before we prove the martingale convergence theorem.

Lemma 10.4.9. *Suppose that* $\{\mathscr{F}_n\}$ *is a filtration on the probability space* $(\Omega, \mathscr{A}, P)$ *and that* $\{X_n\}$ *and* $\{Y_n\}$ *are submartingales on* Ω *relative to* $\{\mathscr{F}_n\}$. *Then* $\{X_n \vee Y_n\}$ *is a submartingale relative to* $\{\mathscr{F}_n\}$.

Proof. It is clear that each $X_n \vee Y_n$ has a finite expectation and is $\mathscr{F}_n$-measurable. Define sets C_n, $n = 0, 1, \ldots$, by $C_n = \{X_n > Y_n\}$. Then each C_n belongs to the corresponding $\mathscr{F}_n$, and for each B in $\mathscr{F}_n$ we have

$$\int_B (X_n \vee Y_n)\, dP = \int_{B \cap C_n} X_n\, dP + \int_{B \cap C_n^c} Y_n\, dP$$

$$\leq \int_{B \cap C_n} X_{n+1}\, dP + \int_{B \cap C_n^c} Y_{n+1}\, dP \leq \int_B (X_{n+1} \vee Y_{n+1})\, dP.$$

Thus $\{X_n \vee Y_n\}$ is a submartingale relative to $\{\mathscr{F}_n\}$.   □

Let us for a moment view a martingale (or sub- or supermartingale) $\{X_n\}$ in terms of gambling, with X_n representing our capital after the nth of a sequence of games. It is sometimes useful to modify $\{X_n\}$ by allowing ourselves to skip certain of the games. More precisely, let $\{\varepsilon_n\}$ be a sequence of $\{0, 1\}$-valued random variables, with ε_n having value 1 if we participate in the nth game and having value 0 otherwise. Since $X_n - X_{n-1}$ would be our gain or loss from the nth game of the original sequence, $\varepsilon_n(X_n - X_{n-1})$ will be our gain or loss in the modified sequence. Thus we can describe our fortunes in the modified situation with a sequence $\{Y_n\}$, where $Y_0 = X_0$ and $Y_n = Y_{n-1} + \varepsilon_n(X_n - X_{n-1})$, or, equivalently, $Y_n = X_0 + \sum_{i=1}^{n} \varepsilon_i(X_i - X_{i-1})$. For this formalization to be reasonable, we must make our decisions about which games to play and which to skip using only information that is available at the time of the decision. Hence it is natural to assume that ε_n is $\mathscr{F}_{n-1}$-measurable.

We have the following proposition, which says that if we transform a submartingale $\{X_n\}$ as in the preceding paragraph, then the resulting sequence $\{Y_n\}$ is also a submartingale, with expected values no larger than those for the original sequence.

Proposition 10.4.10. *Suppose that $(\{X_n\}, \{\mathscr{F}_n\})$ is a submartingale on the probability space $(\Omega, \mathscr{A}, P)$ and that $\{\varepsilon_n\}_{n=1}^{\infty}$ is a sequence of $\{0,1\}$-valued random variables on Ω such that ε_n is $\mathscr{F}_{n-1}$-measurable for each n. Then the sequence $\{Y_n\}_{n=0}^{\infty}$ defined by $Y_0 = X_0$ and $Y_n = Y_{n-1} + \varepsilon_n(X_n - X_{n-1})$ for $n = 1, 2, \dots$ is a submartingale, and $E(Y_n) \leq E(X_n)$ holds for each n.*

Proof. It is clear that each Y_n is $\mathscr{F}_n$-measurable and has a finite expected value. Since $\{X_n\}$ is a submartingale,

$$E(X_n - X_{n-1}|\mathscr{F}_{n-1}) = E(X_n|\mathscr{F}_{n-1}) - X_{n-1} \geq 0$$

holds almost surely for $n = 1, 2, \dots$, and so (see Proposition 10.4.3)

$$E(Y_n|\mathscr{F}_{n-1}) = E(Y_{n-1}|\mathscr{F}_{n-1}) + E(\varepsilon_n(X_n - X_{n-1})|\mathscr{F}_{n-1})$$
$$= Y_{n-1} + \varepsilon_n E(X_n - X_{n-1}|\mathscr{F}_{n-1})$$
$$\geq Y_{n-1}$$

almost surely; thus $\{Y_n\}$ is a submartingale. We prove that $E(Y_n) \leq E(X_n)$ by induction. This inequality certainly holds when $n = 0$. For the induction step, note that, since $E(X_n - X_{n-1}|\mathscr{F}_{n-1}) \geq 0$, we have

$$E(Y_n) = E(Y_{n-1}) + E(\varepsilon_n(X_n - X_{n-1}))$$
$$= E(Y_{n-1}) + E(\varepsilon_n E(X_n - X_{n-1}|\mathscr{F}_{n-1}))$$
$$\leq E(X_{n-1}) + E(X_n - X_{n-1}) = E(X_n). \qquad \square$$

In order to prove the martingale convergence theorem, we will look a bit at how a sequence $\{x_n\}$ of real numbers might fail to converge. One way for this to happen is for $\liminf_n x_n$ to be less than $\limsup_n x_n$. In that case, there are real numbers a and b such that

$$\liminf_n x_n < a < b < \limsup_n x_n,$$

from which it follows that there is a subsequence $\{x_{n_k}\}$ of $\{x_n\}$ such that $x_{n_1} < a$, $x_{n_2} > b$, $x_{n_3} < a, \dots$. This suggests the following definition. A sequence $\{x_n\}$ is said to have an *upcrossing* of the interval $[a,b]$ as n increases from p to q if $x_p \leq a$, $x_n < b$ for n satisfying $p < n < q$, and $x_q \geq b$.

Now suppose that $(\Omega, \mathscr{A}, P)$ is a probability space, that $\{\mathscr{F}_n\}$ is a filtration on $(\Omega, \mathscr{A}, P)$, and that $\{X_n\}_{n=0}^{\infty}$ is a sequence of random variables adapted to $\{\mathscr{F}_n\}$. Let a and b be real numbers such that $a < b$. Our immediate goal is to count the upcrossings of the interval $[a,b]$ made by these random variables, and for this we use sequences $\{\sigma_n\}$ and $\{\tau_n\}$ of stopping times defined as follows. We define σ_1 by

$$\sigma_1(\omega) = \inf\{i \in \mathbb{N}_0 : X_i(\omega) \le a\},$$

and then we continue inductively, defining σ_n, $n \ge 2$, and τ_n, $n \ge 1$, by

$$\tau_n(\omega) = \inf\{i \in \mathbb{N}_0 : i > \sigma_n(\omega) \text{ and } X_i(\omega) \ge b\}$$

and

$$\sigma_n(\omega) = \inf\{i \in \mathbb{N}_0 : i > \tau_{n-1}(\omega) \text{ and } X_i(\omega) \le a\}$$

(recall that the infimum of the empty set is $+\infty$). We can check inductively that σ_n and τ_n are indeed stopping times by noting that

$$\{\sigma_1 \le k\} = \cup_{i=0}^{k}\{X_i \le a\} \in \mathscr{F}_k,$$

$$\{\sigma_n \le k\} = \cup_{i=1}^{k}\{\tau_{n-1} < i \text{ and } X_i \le a\} \in \mathscr{F}_k \quad \text{if } n \ge 2, \text{ and}$$

$$\{\tau_n \le k\} = \cup_{i=1}^{k}\{\sigma_n < i \text{ and } X_i \ge b\} \in \mathscr{F}_k.$$

The finite sequence $\{X_i(\omega)\}_{i=0}^{n}$ contains k or more upcrossings[11] of $[a,b]$ if and only if $\tau_k(\omega) \le n$. Thus, if we define functions $U_n^{[a,b]} : \Omega \to \mathbb{R}$ by letting $U_n^{[a,b]}(\omega)$ be the number of upcrossings of $[a,b]$ in the sequence $\{X_i(\omega)\}_{i=0}^{n}$, then $\{U_n^{[a,b]} \ge k\} = \{\tau_k \le n\}$; since each τ_k is a stopping time, it follows that $U_n^{[a,b]}$ is $\mathscr{F}_n$-measurable.

Proposition 10.4.11 (The upcrossing inequality). *Let $(\Omega, \mathscr{A}, P)$ be a probability space and let $(\{X_n\}, \{\mathscr{F}_n\})$ be a submartingale on $(\Omega, \mathscr{A}, P)$. If a and b are real numbers such that $a < b$, then for each n the number $U_n^{[a,b]}$ of upcrossings of $[a,b]$ by $\{X_i\}_{i=0}^{n}$ satisfies*

$$E(U_n^{[a,b]}) \le \frac{E((X_n - a)^+)}{b - a}.$$

Proof. Let us suppose that a and b are fixed. We can assume that each X_n satisfies $a \le X_n$, since replacing $\{X_n\}$ with $\{\max(X_n, a)\}$ gives a new sequence that is a submartingale (see Lemma 10.4.9), has the same number of upcrossings of $[a,b]$ as the original sequence, and is such that $E((X_n - a)^+)$ is the same for the old and new sequences. Let $\{\sigma_n\}$ and $\{\tau_n\}$ be the sequences of stopping times defined before the statement of the proposition, and define functions[12] $\varepsilon_n : \Omega \to \mathbb{R}$, $n = 1, 2, \ldots$, by

$$\varepsilon_n(\omega) = \begin{cases} 1 & \text{if there is an } i \text{ such that } \sigma_i(\omega) < n \le \tau_i(\omega), \text{ and} \\ 0 & \text{otherwise.} \end{cases}$$

Then

---

[11] Here we are, of course, counting non-overlapping upcrossings, where we call a sequence of upcrossings of $[a,b]$ non-overlapping if the sets of times (i.e., of subscripts) during which they occur are non-overlapping.

[12] The intuitive meaning of ε_n is that it tells whether X_n is part of an upcrossing.

$$\{\varepsilon_n = 1\} = \cup_i(\{\sigma_i \leq n-1\} \cap \{\tau_i \leq n-1\}^c) \in \mathscr{F}_{n-1},$$

and so ε_n is $\mathscr{F}_{n-1}$-measurable. Let $\{Y_n\}$ be the submartingale (see Proposition 10.4.10) defined by $Y_n = X_0 + \sum_{i=1}^n \varepsilon_i(X_i - X_{i-1})$. We will use $\{Y_n\}$ to bound the number of upcrossings of $[a,b]$ by $\{X_i\}_{i=0}^n$.

For an arbitrary element ω of Ω let us analyze the set of those k that satisfy $k \leq n$ and $\varepsilon_k(\omega) = 1$. Such values of k can arise in two ways. First, for each i such that $\tau_i(\omega) \leq n$ we have the set of k that satisfy $\sigma_i(\omega) < k \leq \tau_i(\omega)$. Those values correspond to the steps in the upcrossing of $[a,b]$ that begins at $\sigma_i(\omega)$ and ends at $\tau_i(\omega)$, and so we have

$$b - a \leq \sum_{k=\sigma_i(\omega)+1}^{\tau_i(\omega)} (X_k(\omega) - X_{k-1}(\omega)). \tag{5}$$

The other way that such k can arise is for there to be an i such that $\sigma_i(\omega) < k \leq n < \tau_i(\omega)$. These k correspond to a potential upcrossing that has started but has not finished by time n, and in this case we have

$$\sum_{k=\sigma_i(\omega)+1}^{n} (X_k(\omega) - X_{k-1}(\omega)) = X_n(\omega) - a \geq 0. \tag{6}$$

We are now ready to relate the number of upcrossings to the submartingale $\{Y_n\}$. In view of (5) and (6), we have

$$X_0 + (b-a)U_n^{[a,b]} \leq X_0 + \sum_{k=1}^{n} \varepsilon_k(X_k - X_{k-1}) = Y_n;$$

since $a \leq X_0$ and $E(Y_n) \leq E(X_n)$ (see Proposition 10.4.10), it follows that

$$a + (b-a)E(U_n^{[a,b]}) \leq E(Y_n) \leq E(X_n)$$

and hence that

$$(b-a)E(U_n^{[a,b]}) \leq E(X_n - a) \leq E((X_n - a)^+).$$

With this the proof of the upcrossing lemma is complete.   □

We are now in a position to prove the martingale convergence theorem.

Proof of the Martingale Convergence Theorem. As in the statement of the theorem, let $\{X_n\}_{n=0}^\infty$ be a submartingale such that $\sup_n E(X_n^+) < +\infty$. We begin by showing that $\liminf_n X_n = \limsup_n X_n$ almost surely, which we do by counting upcrossings.

For each pair a, b of real numbers such that $a < b$ we define $U^{[a,b]} \colon \Omega \to \mathbb{R}$ by letting $U^{[a,b]}(\omega)$ be the total number of upcrossings of $[a,b]$ in the sequence $\{X_n(\omega)\}_{n=0}^\infty$. (This differs from $U_n^{[a,b]}$, which only counts the upcrossings in the first

$n+1$ terms of $\{X_i(\omega)\}_{i=0}^{\infty}$). Note that the sequence $\{U_n^{[a,b]}\}_{n=1}^{\infty}$ is increasing and has $U^{[a,b]}$ as its limit, and also that $(X_n - a)^+ \le X_n^+ + |a|$. The monotone convergence theorem and the upcrossing inequality, together with assumption that $\sup_n E(X_n^+) < +\infty$, imply that

$$E(U^{[a,b]}) = \lim_n E(U_n^{[a,b]}) \le \sup_n \frac{E((X_n - a)^+)}{b-a} \le \frac{\sup_n EX_n^+ + |a|}{b-a} < +\infty.$$

It follows that $U^{[a,b]}$, the number of upcrossings of $[a,b]$, is almost surely finite. Since

$$\{\liminf_n X_n < \limsup_n X_n\} = \cup_{a,b}\{U^{[a,b]} = +\infty\},$$

where a and b range over all rational numbers such that $a < b$, we have $\liminf_n X_n = \limsup_n X_n$ almost surely. Thus $\lim_n X_n$ exists almost surely, as an element of $[-\infty, +\infty]$. We still need to show that $E(|\lim_n X_n|) < +\infty$ and hence that $\lim_n X_n$ is finite almost surely.

Note that $|X_n| = 2X_n^+ - X_n$, and so if we use Fatou's lemma (Theorem 2.4.4), plus the fact that $\{X_n\}$, as a submartingale, satisfies $E(X_0) \le E(X_n)$, we find

$$\int |\lim_n X_n| \, dP = \int \liminf_n |X_n| \, dP$$

$$\le \liminf_n \int |X_n| \, dP \le 2 \sup_n \int X_n^+ - \int X_0 \, dP < +\infty.$$

With this the proof of the martingale convergence theorem is complete. □

Let us return to a couple of the examples discussed above. We first look at Example 10.4.7(c), which we can extend as follows:

Proposition 10.4.12. *Let $(\Omega, \mathscr{A}, P)$ be a probability space, let X be an integrable random variable on Ω, let $\{\mathscr{F}_n\}$ be a filtration on $(\Omega, \mathscr{A}, P)$, and let $\mathscr{F}_\infty = \sigma(\cup_n \mathscr{F}_n)$. Then the martingale $\{X_n\}$ defined by $X_n = E(X|\mathscr{F}_n)$ converges almost surely and in mean (i.e., in the norm $\|\cdot\|_1$) to $E(X|\mathscr{F}_\infty)$.*

Proof. Since

$$E(X_n^+) = \int_{\{X_n \ge 0\}} X_n = \int_{\{X_n \ge 0\}} X \le \|X\|_1,$$

the martingale convergence theorem (Theorem 10.4.8) implies that the sequence $\{X_n\}$ converges almost surely, say to $X_{\lim}$.

Let $X_\infty = E(X|\mathscr{F}_\infty)$. Part (e) of Proposition 10.4.3 implies that $\{X_n\}$ is also given by $X_n = E(X_\infty|\mathscr{F}_n)$. Let us show that $\lim_n \|X_n - X_\infty\|_1 = 0$. Suppose that ε is a positive real number. It follows from Proposition 3.4.2 and Lemma 3.4.6 that there is a simple function X_ε of the form $\sum_i a_i \chi_{A_i}$, where each A_i belongs to $\cup_n \mathscr{F}_n$, such that $\|X_\varepsilon - X_\infty\|_1 < \varepsilon$. Since each A_i is in $\mathscr{F}_n$ for some n, there is a positive integer N such that X_ε is $\mathscr{F}_N$-measurable. It follows that if $n \ge N$, then $E(X_\varepsilon|\mathscr{F}_n) = X_\varepsilon$, and so (see also part (c) of Proposition 10.4.3)

$$\|X_n - X_\infty\|_1 = \|E(X_\infty|\mathscr{F}_n) - X_\infty\|_1$$
$$\leq \|E(X_\infty|\mathscr{F}_n) - E(X_\varepsilon|\mathscr{F}_n)\|_1 + \|E(X_\varepsilon|\mathscr{F}_n) - X_\varepsilon\|_1 + \|X_\varepsilon - X_\infty\|_1$$
$$\leq \|X_\varepsilon - X_\infty\|_1 + 0 + \|X_\varepsilon - X_\infty\|_1 \leq 2\varepsilon.$$

Since ε was arbitrary, we have $\lim_n \|X_n - X_\infty\|_1 = 0$.

We still need to show that $\{X_n\}$ converges to X_∞ almost surely. Since we have $\lim_n \|X_n - X_\infty\|_1 = 0$, there is a subsequence of $\{X_n\}$ that converges to X_∞ almost surely (see the discussion that follows the proof of Proposition 3.1.5). Since we already know that the sequence $\{X_n\}$ converges almost surely to $X_{\lim}$, we can conclude that $X_\infty = X_{\lim}$ and hence that $\{X_n\}$ converges to X_∞ both almost surely and with respect to $\|\cdot\|_1$. □

See Exercise 11 for another proof of Proposition 10.4.12.

Example 10.4.13. Let us now look at Example 10.4.7(d), which hinted at some relationships between martingales and derivatives. Let μ be the measure from that example, and define F by $F(x) = \mu((0,x])$. The martingale convergence theorem says that the limit

$$\lim_n \frac{F(b_n) - F(a_n)}{b_n - a_n}$$

exists for almost every x in $(0,1]$, where for each n we let $(a_n, b_n]$ be the interval in $\mathscr{P}_n$ that contains x. In case μ is absolutely continuous with respect to Lebesgue measure, Proposition 10.4.12 identifies this limit as the Radon–Nikodym derivative of μ with respect to Lebesgue measure. See Exercise 12 for the case of singular measures.

Note that the argument in the preceding paragraph is not a derivation of the almost everywhere differentiability of monotone functions from the martingale convergence theorem—there are uncountably many possible choices for the sequence $\{\mathscr{P}_n\}$ of partitions of $(0,1]$, and different sequences of partitions could give rise to different sets of values where the limit does not exist. Nevertheless, as noted by Doob [38, p. 347], these ideas can be made to work; see Chatterji [27] for the details. □

Exercises

1. Let $(\Omega, \mathscr{A}, P)$ be a probability space, let X and Y be random variables on $(\Omega, \mathscr{A}, P)$ such that the joint distribution of (X,Y) on $\mathbb{R}^2$ is absolutely continuous with respect to Lebesgue measure, and let $p: \mathbb{R}^2 \to \mathbb{R}$ be the density function for that joint distribution. Suppose that $F: \mathbb{R}^2 \to \mathbb{R}$ is a Borel measurable function such that $F \circ (X,Y)$ has a finite expected value. Define a

function $f: \mathbb{R} \to \mathbb{R}$ by letting

$$f(x) = \frac{\int F(x,y)p(x,y)dy}{\int p(x,y)dy}$$

for those x for which the expression above is defined and finite and by letting $f(x) = 0$ for other x. Show that $f \circ X$ is a version of the conditional expectation $E(F \circ (X,Y)|\sigma(X))$.

2. Suppose that $(\Omega, \mathscr{A}, P)$ is a probability space and that $\{\mathscr{F}_n\}$ is a filtration on $(\Omega, \mathscr{A}, P)$.

   (a) Show that if τ_1 and τ_2 are stopping times and n is a positive integer, then $\tau_1 + n$, $\tau_1 + \tau_2$, $\tau_1 \vee \tau_2$, and $\tau_1 \wedge \tau_2$ are stopping times.

   (b) Show that if $\{\tau_n\}$ is a sequence of stopping times, then $\inf_n \tau_n$, $\sup_n \tau_n$, $\liminf_n \tau_n$, and $\limsup_n \tau_n$ are stopping times.

3. Let $(\Omega, \mathscr{A}, P)$ be a probability space, let $\{\mathscr{F}_n\}_0^\infty$ be a filtration on $(\Omega, \mathscr{A}, P)$, and let τ be a stopping time. Define $\mathscr{F}_\tau$ to be the set of all sets A in $\sigma(\cup \mathscr{F}_n)$ such that $A \cap \{\tau \leq n\} \in \mathscr{F}_n$ holds for each nonnegative integer n.

   (a) Show that $\mathscr{F}_\tau$ is a sub-σ-algebra of $\mathscr{A}$.

   (b) Show that a set A belongs to $\mathscr{F}_\tau$ if and only if it satisfies $A \cap \{\tau = n\} \in \mathscr{F}_n$ for each nonnegative integer n, along with $A \cap \{\tau = +\infty\} \in \sigma(\cup \mathscr{F}_n)$.

4. Suppose that $\{X_n\}_1^\infty$ is a sequence of independent identically distributed random variables on $(\Omega, \mathscr{A}, P)$. Define a filtration $\{\mathscr{F}_n\}_0^\infty$ by $\mathscr{F}_0 = \{\varnothing, \Omega\}$ and $\mathscr{F}_n = \sigma(X_1, \ldots, X_n)$ for $n = 1, 2, \ldots$. Suppose that τ is a stopping time such that $P(\tau < +\infty) = 1$. Define a sequence $\{Y_n\}$ of random variables by

$$Y_n(\omega) = \begin{cases} X_{\tau+n}(\omega) & \text{if } \tau(\omega) < +\infty, \text{ and} \\ 0 & \text{otherwise.} \end{cases}$$

   (a) Show that the random variables $\{Y_n\}$ are independent and identically distributed, with the same distributions as the X_n's. (Hint: Consider the probabilities of events of the form $\{\tau = m\} \cap \{Y_1 \in A_1\} \cap \{Y_2 \in A_2\} \cap \cdots \cap \{Y_n \in A_n\}$.)

   (b) Show that the σ-algebra $\mathscr{F}_\tau$ and the process $\{Y_n\}$ are independent. That is, show that the σ-algebras $\mathscr{F}_\tau$ and $\sigma(Y_n, n = 1, 2, \ldots)$ are independent.

5. (Jensen's inequality for conditional expectations) Let $\varphi : \mathbb{R} \to \mathbb{R}$ be a convex function, let $(\Omega, \mathscr{A}, P)$ be a probability space, let $\mathscr{B}$ be a sub-σ-algebra of $\mathscr{A}$, and let X be a random variable on $(\Omega, \mathscr{A}, P)$ such that both X and $\varphi \circ X$ have finite expected values. Show that $\varphi \circ E(X|\mathscr{B}) \leq E(\varphi \circ X|\mathscr{B})$ holds almost surely. (Hint: Use ideas from Exercise 3.3.8 to show that there is a family $\mathscr{F}$ of functions, each of the form $x \mapsto ax + b$, such that

$$\varphi(x) = \sup\{f(x) : f \in \mathscr{F}\}$$

holds for each x in $\mathbb{R}$ and such that $f \circ E(X|\mathscr{B}) \leq E(\varphi \circ X|\mathscr{B})$ holds almost surely for each f in $\mathscr{F}$. To conclude that $\varphi \circ E(X|\mathscr{B}) \leq E(\varphi \circ X|\mathscr{B})$ holds almost surely, choose a countable subset $\mathscr{F}_0$ of $\mathscr{F}$ such that φ is the pointwise supremum of the functions in $\mathscr{F}_0$. (Why do we need $\mathscr{F}_0$ to be countable?) The existence of such a subset can be derived from item D.11 in the appendices.)

6. Show that if $\{X_n\}$ is a submartingale relative to $\{\mathscr{F}_n\}$, then it is a submartingale relative to $\{\sigma(X_0, X_1, \ldots, X_n)\}$.

7. Show that if $(\{X_n\}, \{\mathscr{F}_n\})$ is a submartingale and if τ is a stopping time, then $(\{X_{\tau \wedge n}\}, \{\mathscr{F}_n\})$ is a submartingale.

8. (This exercise has nothing to do with martingales or conditional expectations. It appears here as preparation for Exercise 10.) Suppose that $\{a_n\}$ is a sequence of real numbers such that the sequence $\{e^{it a_n}\}$ is convergent for all t in some Lebesgue measurable set of positive measure.
   (a) Show that $\{e^{it a_n}\}$ is convergent for all real t. (Hint: Use Proposition 1.4.10.)
   (b) Show that $\{a_n\}$ is convergent. (Hint: Choose an interval $[b, c]$ such that $\int_b^c \lim e^{it a_n}\, dt \neq 0$. Then consider the sequence $\{\int_b^c e^{it a_n}\, dt\}$.)

9. Suppose that $\{X_n\}$ is a sequence of independent random variables on some probability space. For each n define $\mathscr{F}_n$ and S_n by $\mathscr{F}_n = \sigma(X_1, X_2, \ldots, X_n)$ and $S_n = X_1 + X_2 + \cdots + X_n$. Suppose that t is a real number such that $\lim_n E(e^{it S_n})$ exists and is not equal to 0. Check that for such t we have $E(e^{it S_n}) \neq 0$ for all n. Let $Y_n = e^{it S_n}/E(e^{it S_n})$ for each n.
   (a) Verify that $(\{Y_n\}, \{\mathscr{F}_n\})$ is a martingale.
   (b) Conclude that the sequence $\{e^{it S_n}\}$ is almost surely convergent.

10. Let $\{X_n\}$ be a sequence of independent random variables, let $\sum_n X_n$ be the corresponding infinite series, let μ_n, $n = 1, 2, \ldots$ be the distributions of the partial sums of the series, and let ϕ_{μ_n}, $n = 1, 2, \ldots$ be the corresponding characteristic functions. Consider the following conditions:

   (i) The series $\sum_n X_n$ converges almost everywhere.
   (ii) The series $\sum_n X_n$ converges in probability.
   (iii) The series $\sum_n X_n$ converges in distribution (that is, the sequence $\{\mu_n\}$ converges in distribution to some probability measure).
   (iv) The sequence of characteristic functions $\{\phi_{\mu_n}\}$ has a nonzero pointwise limit on a set of positive measure. That is, $\lim_n \phi_{\mu_n}(t)$ exists and is nonzero for all t in some set of positive measure.

   We have seen that condition (i) implies condition (ii), condition (ii) implies condition (iii), and condition (iii) implies condition (iv) (see Proposition 3.1.2, Exercise 10.3.7, and Proposition 10.3.15). Now prove that condition (iv) implies condition (i). (Hint: Use Exercises 8 and 9.)

11. Let $(\{X_n\}, \{\mathscr{F}_n\})$ be a martingale on $(\Omega, \mathscr{A}, P)$ such that the sequence $\{X_n\}$ is uniformly integrable. (See Exercises 4.2.12–4.2.16.)
    (a) Show that $\{X_n\}$ converges almost surely and in mean to some random variable X.
    (b) Show that for each n the equality $X_n = E(X|\mathscr{F}_n)$ holds almost surely.

12. Suppose that μ is a finite measure on $((0,1],\mathscr{B}((0,1]))$ and that $\{X_n\}$ is the martingale defined in Example 10.4.7(d). Show that if μ is singular with respect to Lebesgue measure, then $\lim_n X_n = 0$ holds λ-almost everywhere on $(0,1]$.

13. Let $(\Omega,\mathscr{A},P)$ be a probability space. In this exercise we consider sequences $\{X_n\}$ and $\{\mathscr{F}_n\}$ that are indexed by the *negative* integers. The pair $(\{X_n\},\{\mathscr{F}_n\})$ is called a *reverse martingale* if

   (i)   each $\mathscr{F}_n$ is a sub-σ-algebra of $\mathscr{A}$,
   (ii) $\mathscr{F}_m \subseteq \mathscr{F}_n$ holds whenever $m \leq n$,
   (iii) each X_n is measurable with respect to the corresponding $\mathscr{F}_n$ and has a finite expected value, and
   (iv) $X_n = E(X_{n+1}|\mathscr{F}_n)$ holds for $n = -2, -3, \dots$.

   Prove the convergence theorem for reverse martingales: if $(\{X_n\},\{\mathscr{F}_n\})$ is a reverse martingale, then there is a function $X_{-\infty}$ such that $X_{-\infty} = \lim_{n\to-\infty} X_n$ holds almost surely and in mean. Furthermore, $X_{-\infty} = E(X_{-1}|\cap_n \mathscr{F}_n)$. (Hint: Use the upcrossing inequality, and verify and use the fact that the sequence $\{X_n\}$ is uniformly integrable. See Exercises 4.2.12–4.2.16

14. In this exercise we derive the strong law of large numbers from the convergence theorem for reverse martingales (see Exercise 13). Suppose that $(\Omega,\mathscr{A},P)$ is a probability space and that $\{X_i\}$ is a sequence of independent identically distributed random variables on $(\Omega,\mathscr{A},P)$ that have finite expected values. For each positive integer n let $S_n = X_1 + X_2 + \cdots + X_n$ and define the σ-algebra $\mathscr{F}_{-n}$ to be $\sigma(S_n,X_{n+1},X_{n+2},\dots)$.
    (a) Let $\mathscr{F} = \sigma(S_n)$. Show that $E(X_1|\mathscr{F}) = E(X_2|\mathscr{F}) = \cdots = E(X_n|\mathscr{F})$ and conclude that $E(X_1|\mathscr{F}) = S_n/n$. (Hint: Using the map

$$\omega \mapsto (X_1(\omega),X_2(\omega),\dots,X_n(\omega))$$

to convert this to a calculation on $\mathbb{R}^n$ might be useful.)
    (b) Show that $(\{S_n/n\},\{\mathscr{F}_n\})$ is a reverse martingale.
    (c) Use the convergence theorem for reverse martingales, together with Kolmogorov's zero–one law (see Exercise 10.2.2), to conclude that $\lim_n S_n/n = E(X_1)$ holds almost surely.

10.5   Brownian Motion

In this section we look at a continuous-time stochastic process that models Brownian motion, the random movement of a very small particle suspended in a fluid. Einstein seems to have been one of the first to study Brownian motion mathematically, and Norbert Wiener was the first to build a probability measure with which to describe Brownian motion. In fact, the basic probability measure defining a Brownian motion process is generally called a *Wiener measure*.

As we noted in Sect. 10.4, a continuous-time process is a stochastic process $\{X_t\}_{t \in T}$ for which the index set T is a reasonable subset of $\mathbb{R}$—typically an interval such as $[0,1]$ or $[0, +\infty)$. We will first construct a Brownian motion in which the index set is $[0,1]$ and then we'll note how to build one with index set $[0, +\infty)$.

Since one usually thinks of particles moving in three-dimensional space, it seems natural to construct a process $\{X_t\}_{t \in T}$ for which the variables X_t have values in $\mathbb{R}^3$. However, the trick of taking three independent one-dimensional process and using them to build a three-dimensional process works. More precisely, suppose that $\{X_t\}_{t \in T}$ is a one-dimensional Brownian motion on a probability space $(\Omega, \mathscr{A}, P)$. Then it turns out that the three-dimensional process $\{X_t'\}_{t \in T}$ that is defined on the product of three copies of $(\Omega, \mathscr{A}, P)$ by $X_t'((\omega_1, \omega_2, \omega_3)) = (X_t(\omega_1), X_t(\omega_2), X_t(\omega_3))$ has suitable properties. In any case, we will devote our attention to one-dimensional Brownian motion. We begin with a precise definition.

Suppose that $(\Omega, \mathscr{A}, P)$ is a probability space and that T is either $[0,1]$ or $[0, +\infty)$. A stochastic process $\{X_t\}_{t \in T}$ with values in $\mathbb{R}$ is a *Brownian motion*[13] if

(a) $X_0(\omega) = 0$ for all ω in Ω,
(b) for each choice of $t_0, t_1, \ldots, t_n$ in T such that $t_0 < t_1 < \cdots < t_n$ the increments $X_{t_i} - X_{t_{i-1}}$, $i = 1, \ldots, n$, are independent, with $X_{t_i} - X_{t_{i-1}}$ having distribution $N(0, t_i - t_{i-1})$, that is, a normal distribution with mean 0 and variance $t_i - t_{i-1}$, and
(c) for each ω in Ω the function $X_{\bullet}(\omega) \colon T \to \mathbb{R}$ defined by $t \mapsto X_t(\omega)$ is continuous.

Given a process $\{X_t\}_{t \in T}$, the functions $t \mapsto X_t(\omega)$ are called the *paths* of the process. Thus condition (c) says that we are requiring the paths of a Brownian motion process to be continuous.

Theorem 10.5.1. *Let $T = [0,1]$. Then a one-dimensional Brownian motion with parameter set T exists. That is, there exist a probability space $(\Omega, \mathscr{A}, P)$ and random variables X_t, $t \in T$, on Ω such that the stochastic process $\{X_t\}_{t \in T}$ is a Brownian motion.*

Proof. Let $(\Omega, \mathscr{A}, P)$ be a probability space on which there exists a sequence $\{Z_n\}_{n=0}^{\infty}$ of independent random variables, each of which has a normal distribution with mean 0 and variance 1. (Recall that according to Corollary 10.1.16, such a sequence can be constructed on the probability space $([0,1], \mathscr{B}([0,1]), \lambda)$.) We will use such a sequence $\{Z_n\}$ to build a sequence of piecewise linear approximations to a Brownian motion process. More precisely, we will construct processes $\{X_t^n\}_{t \in T}$, $n = 0, 1, \ldots$, such that

(a') for each n the paths of $\{X_t^n\}_{t \in T}$ satisfy $X_0^n(\omega) = 0$ for all ω and are piecewise linear, with the paths being linear on the intervals of the form $[(i-1)/2^n, i/2^n]$,

---

[13] Some authors only require conditions (a) and (c) in the definition of a Brownian motion to hold for all ω outside some P-null subset of Ω.

(b′) for each n the process $\{X_t^n\}_{t\in T}$, when restricted to the points $t_{i/2^n}$, $i = 0, \ldots,$ 2^n, looks like a Brownian motion (that is, it has independent increments whose distributions are normal and have the required means and variances),

(c′) for almost every ω the sequence of functions $\{t \mapsto X_t^n(\omega)\}_{n=1}^\infty$ converges uniformly on $[0, 1]$ as n approaches infinity, and

(d′) these processes satisfy $X_t^n(\omega) = X_t^{n+1}(\omega) = X_t^{n+2}(\omega) = \ldots$ for each n and ω and each t of the form $i/2^n$.

Now assume that we have constructed such a sequence of processes $\{X_t^n\}_{t\in T}$, and let A be an event of probability 1 such that if $\omega \in A$, then the sequence $\{t \mapsto X_t^n(\omega)\}_n$ converges uniformly on $\dot{T}$. Define a process $\{X_t\}_{t\in T}$ by

$$X_t(\omega) = \begin{cases} \lim_n X_t^n(\omega) & \text{if } t \in T \text{ and } \omega \in A, \text{ and} \\ 0 & \text{if } t \in T \text{ and } \omega \notin A. \end{cases}$$

Then, in view of the uniform convergence of the paths, condition (a′) implies that $X_0 = 0$ and that all the paths of $\{X_t\}_{t\in T}$ are continuous. Conditions (b′) and (d′) imply that if $t_0, t_1, \ldots, t_k$ are dyadic rationals such that $t_0 < t_1 < \cdots < t_k$, then the increments $X_{t_i} - X_{t_{i-1}}$, $i = 1, \ldots, k$, are independent, with $X_{t_i} - X_{t_{i-1}}$ having distribution $N(0, t_i - t_{i-1})$. We need to extend this to the case where the t_i are not necessarily dyadic rationals.

So suppose that t_i, $i = 0, \ldots, k$, are elements of $[0, 1]$ such that $t_0 < t_1 < \cdots < t_k$. Let us approximate these values by choosing sequences $\{t_{i,n}\}_n$, $i = 0, \ldots, k$, of dyadic rationals in $[0, 1]$ such that $t_i = \lim_n t_{i,n}$ holds for all i and $t_{i-1,n} < t_{i,n}$ holds for all i and n. Then for each n the increments $X_{t_{i,n}} - X_{t_{i-1,n}}$, $i = 1, \ldots, k$, are independent, with $X_{t_{i,n}} - X_{t_{i-1,n}}$ having distribution $N(0, t_{i,n} - t_{i-1,n})$. The increments $X_{t_{i,n}} - X_{t_{i-1,n}}$ converge pointwise (and so[14] in distribution) to the increments $X_{t_i} - X_{t_{i-1}}$, and so it follows that the increments $X_{t_i} - X_{t_{i-1}}$, $i = 1, \ldots, k$, are independent (see Corollary 10.3.12), with $X_{t_i} - X_{t_{i-1}}$ having distribution $N(0, t_i - t_{i-1})$. This will complete the proof, as soon as we construct the processes $\{X_t^n\}_{t\in T}$, $n = 0, \ldots$.

We turn to the construction of processes $\{X_t^n\}_{t\in T}$, $n = 0, \ldots$ satisfying conditions (a′)–(d′). Recall that we have a sequence $\{Z_n\}_{n=0}^\infty$ of independent normal random variables, each with mean 0 and variance 1. We define the process $\{X_t^0\}_{t\in T}$ by letting $X_t^0(\omega) = tZ_0(\omega)$ hold for each ω and each t. This process certainly satisfies conditions (a′) and (b′) above.

Given the process $\{X_t^{n-1}\}_{t\in T}$, we form the process $\{X_t^n\}_{t\in T}$ as follows. For each t of the form $i/2^{n-1}$ we let $X_t^n = X_t^{n-1}$. For each t of the form $(2i+1)/2^n$, $i = 0, \ldots,$ $2^{n-1} - 1$, we let

$$X_t^n = X_t^{n-1} + 2^{-(n+1)/2}Z_{2^{n-1}+i}.$$

---

[14] Use the definition of convergence in distribution, together with the dominated convergence theorem.

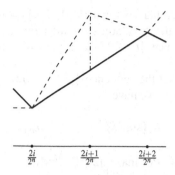

Fig. 10.1 Constructing X^n from X^{n-1}. Solid line: path of X^{n-1}. Dashed line: path of X^n. Vertical line: $2^{-(n+1)/2}Z_{2^{n-1}+i}$

Then we use straight line segments to interpolate between the points $(t, X_t^n(\omega))$, for which t has the form $i/2^n$ for some i. See Fig. 10.1. (The choice of $Z_{2^{n-1}+i}$ from the sequence of Z's is made so that the new Z's used in the construction of $\{X_t^n\}_{t \in T}$ are all distinct from those used earlier—that is, from those used in the construction of $\{X_t^k\}_{t \in T}$, where $k < n$. The coefficient of $Z_{2^{n-1}+i}$ will turn out to be what is needed to make the increments of $\{X_t^n\}_{t \in T}$ be independent and have the required distributions.) To simplify the notation a bit, let us denote $i/2^n$ by t_i, for $i = 0, \ldots,$ 2^n. Then the increment $X_{t_{2i+1}}^n - X_{t_{2i}}^n$ is given by

$$X_{t_{2i+1}}^n - X_{t_{2i}}^n = X_{t_{2i+1}}^{n-1} + 2^{-(n+1)/2}Z_{2^{n-1}+i} - X_{t_{2i}}^{n-1}$$

$$= (1/2)(X_{t_{2i}}^{n-1} + X_{t_{2i+2}}^{n-1}) + 2^{-(n+1)/2}Z_{2^{n-1}+i} - X_{t_{2i}}^{n-1}$$

$$= (1/2)(X_{t_{2i+2}}^{n-1} - X_{t_{2i}}^{n-1}) + 2^{-(n+1)/2}Z_{2^{n-1}+i}.$$

A similar calculation shows that

$$X_{t_{2i+2}}^n - X_{t_{2i+1}}^n = (1/2)(X_{t_{2i+2}}^{n-1} - X_{t_{2i}}^{n-1}) - 2^{-(n+1)/2}Z_{2^{n-1}+i}.$$

The variables $(1/2)(X_{t_{2i+2}}^{n-1} - X_{t_{2i}}^{n-1})$ and $2^{-(n+1)/2}Z_{2^{n-1}+i}$ are independent, with each having distribution $N(0, 1/2^{n+1})$, from which it follows that the increments $X_{t_{2i+1}}^n - X_{t_{2i}}^n$ and $X_{t_{2i+2}}^n - X_{t_{2i+1}}^n$ both have distribution $N(0, 1/2^n)$. Finally, if one calculates the characteristic function of the joint distribution of the increments $X_{t_{i+1}}^n - X_{t_i}^n$, one obtains the product of the characteristic functions of normal variables with mean 0 and variance $1/2^n$, and the independence of the increments follows. With this we have verified conditions (a'), (b'), and (d').

We turn to condition (c'), the almost sure uniform convergence of the sequence $\{X_t^n(\omega)\}$. Suppose that we can find a sequence $\{\varepsilon_n\}$ of positive numbers such that $\sum_n \varepsilon_n < +\infty$ and $\sum_n P(A_n) < +\infty$, where A_n is defined by

$$A_n = \{\sup_t |X_t^n - X_t^{n-1}| > \varepsilon_n\}.$$

Then the Borel–Cantelli lemma says that $P(\{A_n \text{ i.o.}\}) = 0$; since if $\omega \notin \{A_n \text{ i.o.}\}$, then $\sup_t |X_t^n(\omega) - X_t^{n-1}(\omega)| \le \varepsilon_n$ holds for all large n, the almost sure uniform convergence of the sequence $\{t \mapsto X_t^n(\omega)\}_{n=1}^{\infty}$ will follow from the condition $\sum_n \varepsilon_n < +\infty$.

We still need to construct the sequence $\{\varepsilon_n\}$. In view of the way $\{X_t^n\}_{t \in T}$ was constructed from $\{X_t^{n-1}\}_{t \in T}$, we have

$$P(A_n) = P(\{\sup_t |X_t^n - X_t^{n-1}| > \varepsilon_n\})$$

$$= P(\max_{0 \le i < 2^{n-1}} |2^{-(n+1)/2} Z_{2^{n-1}+i}| > \varepsilon_n)$$

$$\le \sum_{i=0}^{2^{n-1}-1} P(|2^{-(n+1)/2} Z_{2^{n-1}+i}| > \varepsilon_n)$$

$$= 2^{n-1} P(|Z_{2^{n-1}}| > 2^{(n+1)/2} \varepsilon_n).$$

Since $Z_{2^{n-1}}$ has a normal distribution with mean 0 and variance 1, it follows from Lemma 10.1.6 that

$$P(A_n) \le 2^{n-1} \frac{2}{\sqrt{2\pi} 2^{(n+1)/2} \varepsilon_n} e^{-(1/2)2^{n+1} \varepsilon_n^2} = \frac{2^{n/2}-1}{\sqrt{\pi} \varepsilon_n} e^{-2^n \varepsilon_n^2}.$$

If, for example, we let ε_n be $2^{-n/4}$, then $\sum_n \varepsilon_n < +\infty$ and $\sum_n P(A_n) < +\infty$, and the proof is complete.   □

Corollary 10.5.2. *A one-dimensional Brownian motion with parameter set $[0, +\infty)$ exists.*

Proof. We will use a sequence $\{X_t^{(n)}\}_{t \in [0,1]}$, $n = 1, 2, \ldots$, of independent Brownian motion processes, which we can construct as follows. According to Corollary 10.1.16 there exists a sequence $\{Z_n\}$ of independent normal random variables, each with mean 0 and variance 1. Using ideas from the proof of Corollary 10.1.14, we can divide the sequence $\{Z_n\}$ into a sequence of sequences $\{Z_{m,n}'\}_m$, $n = 1, 2, \ldots$. Finally, for each n, the construction in Theorem 10.5.1 can be applied to the sequence $\{Z_{m,n}'\}_m$ to produce the process $\{X_t^{(n)}\}_{t \in [0,1]}$; the independence of these processes follows from the independence of the sequences $\{Z_{m,n}'\}_m$, $n = 1, 2, \ldots$.

Next we define a process $\{X_t\}_{t \in [0,+\infty)}$ by splicing together the paths of the processes $\{X_t^{(n)}\}_{t \in [0,1]}$—that is, by letting $X_t(\omega) = X_t^{(1)}(\omega)$ if $t \le 1$, letting $X_t(\omega) = X_1^{(1)}(\omega) + X_{t-1}^{(2)}(\omega)$ if $1 < t \le 2, \ldots$. More precisely, we define X_t recursively by

$$X_t(\omega) = \begin{cases} X_t^{(1)}(\omega) & \text{if } 0 \le t \le 1, \text{ and} \\ X_{n-1}(\omega) + X_{t-(n-1)}^{(n)}(\omega) & \text{if } n > 1 \text{ and } n-1 < t \le n. \end{cases}$$

It is clear that the paths of $\{X_t\}_{t\in[0,+\infty)}$ are continuous and that $X_0 = 0$. Now suppose that we have a sequence $t_0, t_1, \ldots, t_m$ in $[0,+\infty)$ such that $t_{i-1} < t_i$, $i = 1, \ldots, m$. Add to this sequence those integers between t_0 and t_m that are not in the original sequence, forming a new sequence $s_0, s_1, \ldots, s_n$ such that $s_{i-1} < s_i$, $i = 1, \ldots, n$. Since $\{X_t^{(n)}\}_{t\in[0,1]}$, $n = 1, 2, \ldots$, is a collection of independent Brownian motions, the increments $X_{s_i} - X_{s_{i-1}}$, $i = 1, \ldots, n$, are independent normal variables with mean 0 and the appropriate variances. It follows that the increments $X_{t_i} - X_{t_{i-1}}$, $i = 1, \ldots,$ m, are independent and have the required distributions. □

Here is an interesting fact about the paths of a Brownian motion process.

Theorem 10.5.3. *Almost all the paths of a one-dimensional Brownian motion are nowhere differentiable. More precisely, let $T = [0,1]$ and let $\{X_t\}_{t\in T}$ be a one-dimensional Brownian motion on the probability space $(\Omega, \mathscr{A}, P)$. Then there is a set A in $\mathscr{A}$ such that $P(A) = 0$ and such that for each ω outside A the path $t \mapsto X_t(\omega)$ is nowhere differentiable.*

Proof. Let K be a positive integer, which we hold fixed for the moment. We will construct a sequence $\{B_n\}$ of $\mathscr{A}$-measurable subsets of Ω such that

(a) $\lim_n P(B_n) = 0$, and
(b) if ω is a element of Ω such that the path $t \mapsto X_t(\omega)$ is differentiable at some t_0 in $[0,1]$, with $|X'_{t_0}(\omega)| < K$, then ω belongs to B_n for all large n.

Suppose we have constructed such a sequence $\{B_n\}$. Let A_K be $\cup_m \cap_{n \geq m} B_n$, the set of points ω such that $\omega \in B_n$ holds for all large n. Then $P(\cap_{n\geq m}B_n) \leq \lim_n P(B_n) = 0$ holds for all m, and so $P(A_K) = 0$. Now suppose that we let K vary through the positive integers, and we define A by $A = \cup_{K=1}^{\infty} A_K$. Then A has P-measure 0, and it follows from condition (b) that A contains every ω for which the path $t \mapsto X_t(\omega)$ is differentiable at one or more points; in other words, A is as described in the statement of the theorem.

Now we turn to our remaining task, the construction of a sequence[15] $\{B_n\}$ of sets satisfying conditions (a) and (b) above. We once again consider K to be fixed; we do so through the end of the proof. For each n, where $n \geq 3$, we define sets $C_{n,k}$, $k = 1, \ldots, n$ by

$$C_{n,k} = \left\{ \omega : |X_{k/n}(\omega) - X_{(k-1)/n}(\omega)| < \frac{3K}{n} \right\},$$

and then we define sets $D_{n,k}$, $k = 2, \ldots, n-1$ by

$$D_{n,k} = C_{n,k-1} \cap C_{n,k} \cap C_{n,k+1}.$$

Finally, we define sets B_n by $B_n = C_{n,1} \cup C_{n,n} \cup (\cup_{k=2}^{n-1} D_{n,k})$. We will show that the sets B_n satisfy (a) and (b).

---

[15]Our sequence will start with n equal to 3, rather than equal to 0 or 1.

We begin our verification of condition (a) by estimating the probabilities of the sets $C_{n,k}$. Since the difference $X_{k/n} - X_{(k-1)/n}$ is normal with mean 0 and variance $1/n$, it has the same distribution as the variable $Z/\sqrt{n}$, where Z is a normal variable with mean 0 and variance 1. Thus

$$P(C_{n,k}) = P\left(|X_{k/n} - X_{(k-1)/n}| < \frac{3K}{n}\right) = P\left(|Z| < \frac{3K}{\sqrt{n}}\right)$$

$$= \frac{1}{\sqrt{2\pi}} \int_{-\frac{3K}{\sqrt{n}}}^{\frac{3K}{\sqrt{n}}} e^{-x^2/2} \, dx < \frac{K_1}{\sqrt{n}},$$

where K_1 is the constant $6K/\sqrt{2\pi}$. The independence of the events $C_{n,k}$, $k = 1, \ldots,$ n, implies that

$$P(D_{n,k}) = P(C_{n,k-1})P(C_{n,k})P(C_{n,k+1}) < K_1^3/n^{3/2}.$$

Since $B_n = C_{n,1} \cup C_{n,n} \cup (\cup_{k=2}^{n-1} D_{n,k})$, we have $P(B_n) < 2K_1/\sqrt{n} + (n-2)K_1^3/n^{3/2}$, and $\lim_n P(B_n) = 0$ follows. Thus condition (a) holds.

We turn to condition (b). Suppose that $t \mapsto X_t(\omega)$ is differentiable at the point t_0, and that $|X'_{t_0}(\omega)| < K$. Let n be large enough that

$$|X_t(\omega) - X_{t_0}(\omega)| < K|t - t_0| \tag{1}$$

holds when $|t - t_0| \le 2/n$. It follows that if $t_0 \in [\frac{k-1}{n}, \frac{k}{n}]$, then

$$|X_{k/n}(\omega) - X_{(k-1)/n}(\omega)| < K/n,$$

while if t_0 lies in an interval of length $1/n$ adjacent to the interval $[\frac{k-1}{n}, \frac{k}{n}]$, then

$$|X_{k/n}(\omega) - X_{(k-1)/n}(\omega)| \le |X_{k/n}(\omega) - X_{t_0}(\omega)| + |X_{t_0}(\omega) - X_{(k-1)/n}(\omega)|$$
$$< K/n + 2K/n = 3K/n.$$

Now suppose that k is such that $t_0 \in [\frac{k-1}{n}, \frac{k}{n}]$. The estimates we have just made show that $\omega \in C_{n,1} \cup C_{n,n}$ if k is 1 or n and that $\omega \in D_{n,k}$ otherwise. In any case, $\omega \in B_n$, and the verification of condition (b) is complete.   □

Exercises

1. Suppose that we have a stochastic process $\{X_t\}$ with index set $[0,1] \cap \mathbb{Q}$ that satisfies properties (a) and (b) in the definition of Brownian motion (where the values t_i are restricted to lie in $[0,1] \cap \mathbb{Q}$). In this exercise we prove that almost all the paths of this process are uniformly continuous on $[0,1] \cap \mathbb{Q}$. In the following

exercise we use this to give another construction of a Brownian motion process on $[0,1]$.

(a) Show that if a and b are rational numbers that satisfy $0 \le a < b \le 1$ and if C is a positive constant, then

$$P(\sup\{X_t - X_a : t \in [a,b] \cap \mathbb{Q}\} > C) \le 2P(X_b - X_a > C). \tag{2}$$

(Hint: First suppose that $a \le t_1 < t_2 < \cdots < t_k \le b$ and let A_i be the event that i is the smallest value of j for which $X_{t_j} - X_a > C$. Check that

$$P(A_i) = P(A_i \cap \{X_b - X_{t_i} \ge 0\}) + P(A_i \cap \{X_b - X_{t_i} < 0\})$$
$$\le 2P(A_i \cap \{X_b - X_a > C\}),$$

and then use this estimate to prove the analogue of (2) in which the supremum is taken as t ranges over $\{t_1, t_2, \ldots, t_n\}$. Finally, take limits as more and more points from $[a,b] \cap \mathbb{Q}$ are considered in the supremum.)

(b) For each positive δ define $v(\delta)$ by

$$v(\delta) = \sup\{|X_t - X_s| : s,t \in [0,1] \cap \mathbb{Q} \text{ and } |t - s| < \delta\}.$$

Use part (a), together with Lemma 10.1.6, to show that there exist sequences $\{\varepsilon_n\}$ and $\{\delta_n\}$ of positive numbers such that $\lim_n \varepsilon_n = \lim_n \delta_n = 0$ and

$$\sum_i P(v(\delta_n) > \varepsilon_n) < +\infty;$$

from this derive the almost sure uniform continuity of the paths.

2. In Exercise 10.6.4 we will construct a stochastic process $\{X_t\}$ with index set $[0,1] \cap \mathbb{Q}$ that satisfies properties (a) and (b) in the definition of Brownian motion. Given that result, use Exercise 1 to give a proof of the existence of Brownian motion on $[0,1]$ that is quite different from the proof in the text.

3. Let $T = [0, +\infty)$, let $(\Omega, \mathscr{A}, P)$ be a probability space, and let $\{X_t\}_{t \in T}$ be a Brownian motion process on $(\Omega, \mathscr{A}, P)$. Define a filtration $\{\mathscr{F}_t\}_{t \in T}$ by letting $\mathscr{F}_t = \sigma(\{X_s : s \le t\})$ hold for each t in T.

(a) Let a be a real number. Show that the function $\tau: \Omega \to [0, +\infty]$ defined by $\tau(\omega) = \inf\{t : X_t(\omega) = a\}$ is a stopping time.

(b) Suppose τ is a stopping time. Show that if n is a positive integer, then

$$\tau_n(\omega) = \inf\{i/2^n : \tau(\omega) \le i/2^n\}$$

defines a stopping time (of course, $\tau_n(\omega) = +\infty$ if $\tau(\omega) = +\infty$).

(c) Show that if τ is a stopping time, then X_τ is $\mathscr{F}_\tau$-measurable.

4. Let $T = [0, +\infty)$ and let $\{X_t\}_{t \in T}$ be a Brownian motion process.

(a) Fix a value t_0 such that $0 < t_0 < +\infty$ and define a process $\{Y_t\}_{t \in T}$ by $Y_t = X_{t+t_0} - X_{t_0}$ for t in T. Show that $\{Y_t\}_{t \in T}$ is a Brownian motion and that it is independent of $\mathscr{F}_{t_0}$ (in other words, the σ-algebras $\sigma(Y_t, t \in T)$ and $\mathscr{F}_t$ are independent).

(b) Suppose that τ is a stopping time that is finite almost surely, and define a process $\{Y_t\}_{t \in T}$ by

$$
Y_t(\omega) = \begin{cases} X_{t+\tau(\omega)}(\omega) - X_{\tau(\omega)}(\omega) & \text{if } \tau(\omega) < +\infty, \text{ and} \\ 0 & \text{otherwise.} \end{cases}
$$

Show that if the stopping time τ has only finitely many values, then $\{Y_t\}_{t \in T}$ is a Brownian motion that is independent of $\mathscr{F}_\tau$.

(c) Show that the assumption that τ has only finitely many values can be removed from part (b). (Hint: See Exercise 3.)

10.6 Construction of Probability Measures

This section contains two constructions of possibly infinite families of random variables with specified distributions. The first construction gives sequences of independent random variables, while the second gives families of not necessarily independent random variables.

Let us recall the methods we have been using to construct sequences of independent real-valued random variables. In simple cases, where we need only finitely many independent random variables, say with distributions $\mu_1, \mu_2, \ldots, \mu_d$, we saw that we can take the product measure $\mu_1 \times \cdots \times \mu_d$ on $\mathbb{R}^d$ and then let the random variables be the coordinate functions on $\mathbb{R}^d$. On the other hand, to construct an infinite sequence of independent real-valued random variables, we used a perhaps awkward-seeming ad hoc construction based on the binary expansion of numbers in the unit interval, together with a kind of inverse for distribution functions of probability measures (see the end of Sect. 10.1).

Here we will look at the use of product spaces to construct infinite families of random variables. Note that the random variables we construct do not need to be real valued—in our first construction, they can have values in arbitrary measurable spaces, while in our second construction, they can have values in rather general, but not arbitrary, spaces.

We begin by defining the measurable spaces on which we will construct families of random variables. Let I be an index set, and let $\{(\Omega_i, \mathscr{A}_i)\}_{i \in I}$ be an indexed family of measurable spaces. (In typical situations the measurable spaces $(\Omega_i, \mathscr{A}_i)$ will be equal to one another.) The *product* of these measurable spaces is the measurable space $(\Omega, \mathscr{A})$ defined as follows: The underlying set Ω is the product $\prod_i \Omega_i$ of the sets $\{\Omega_i\}_i$; that is, Ω is the set of all functions $\omega \colon I \to \cup_i \Omega_i$ such that $\omega(i) \in \Omega_i$ for each i in I. For each i we define the coordinate function $X_i \colon \Omega \to \Omega_i$ by

$X_i(\omega) = \omega(i)$. Finally, we let $\mathscr{A}$ be the smallest σ-algebra on Ω that makes each X_i measurable with respect to $\mathscr{A}$ and $\mathscr{A}_i$. Equivalently, we can let $\mathscr{A}$ be the σ-algebra on Ω generated by the collection of all sets that have the form

$$\{\omega \in \Omega : \omega(i) \in A_i \text{ holds for each } i \text{ in } I_0\}$$

for some finite subset I_0 of I and some sets A_i that satisfy $A_i \in \mathscr{A}_i$ for each i in I_0.

Let us turn to the construction of sequences of independent random variables.

Proposition 10.6.1. Let $\{(\Omega_i, \mathscr{A}_i, P_i)\}_{i \in \mathbb{N}}$ be a family of probability spaces indexed by the set $\mathbb{N}$ of positive integers, let $(\Omega, \mathscr{A})$ be the product of the measurable spaces $\{(\Omega_i, \mathscr{A}_i)\}_{i \in \mathbb{N}}$, and for each i in $\mathbb{N}$ let X_i be the coordinate projection from Ω to Ω_i. Then there is a unique probability measure P on $(\Omega, \mathscr{A})$ such that

(a) for each i the distribution of X_i is P_i, and
(b) the random variables $\{X_i\}_{i \in \mathbb{N}}$ are independent.

Proof. What we need here is a product measure with infinitely many factors. In particular, we need a measure P on $(\Omega, \mathscr{A})$ such that for each n and each choice of sets A_i in $\mathscr{A}_i$, $i = 1, \ldots, n$, we have

$$P(A) = P_1(A_1) P_2(A_2) \cdots P_n(A_n),$$

where A is the subset

$$A_1 \times \cdots \times A_n \times \Omega_{n+1} \times \cdots \tag{1}$$

of Ω—that is, where A consists of those sequences $\{x_i\}_1^\infty$ in Ω such that $x_i \in A_i$ holds for $i = 1, \ldots, n$.

The results in Chap. 5 give us a start on the construction of such measures. Namely for each n those results give us a product measure $P_1 \times \cdots \times P_n$ on the measurable space $(\prod_1^n \Omega_i, \prod_1^n \mathscr{A}_i)$. For each n let proj_n be the projection of the infinite product Ω onto $\prod_1^n \Omega_i$, that is, the function that takes an infinite sequence to the sequence of its first n components. Let $\mathscr{A}^{(1)}$ be the collection of subsets of Ω defined[16] by

$$\mathscr{A}^{(1)} = \bigcup_n \text{proj}_n^{-1}(\prod_1^n \mathscr{A}_i).$$

Since $\{\text{proj}_n^{-1}(\prod_1^n \mathscr{A}_i)\}_{n=1}^\infty$ is an increasing sequence of σ-algebras on Ω, it follows that $\mathscr{A}^{(1)}$ is an algebra of sets. Furthermore $\mathscr{A} = \sigma(\mathscr{A}^{(1)})$. We need to transfer our finite-dimensional product measures to $\mathscr{A}^{(1)}$. For that, define a function P on $\mathscr{A}^{(1)}$ by letting

$$P(\text{proj}_n^{-1}(A)) = (P_1 \times \cdots \times P_n)(A)$$

---

[16]Note that if X and Y are sets, if f is a function from X to Y, and if $\mathscr{C}$ is a family of subsets of Y, then $f^{-1}(\mathscr{C}) = \{f^{-1}(C) : C \in \mathscr{C}\}$.

hold for each n and each A in $\prod_1^n \mathscr{A}_i$ (the reader should check that P is well defined). Certainly P has the necessary value on each rectangular set of the form given in (1). Furthermore P is countably additive on each $\text{proj}_n^{-1}(\prod_1^n \mathscr{A}_i)$ and so is at least finitely additive on $\mathscr{A}^{(1)}$. If we show that P is countably additive on $\mathscr{A}^{(1)}$, then it will have a countably additive extension to $\mathscr{A}$ (see Exercise 1.3.5) and the proof of existence will be complete.

We need a bit of notation for the proof of countable additivity. For each n we want the analogue of Ω, $\mathscr{A}^{(1)}$, and P, but with the products starting with $(\Omega_n, \mathscr{A}_n)$ and P_n, rather than with $(\Omega_1, \mathscr{A}_1)$ and P_1. Let us use the notation $\Omega^{(n)}$, $\mathscr{A}^{(n)}$, and $P^{(n)}$ for such sets,[17] algebras, and finitely additive probabilities. Note that $\Omega^{(1)} = \Omega$, $P^{(1)} = P$, and $\mathscr{A}^{(1)}$ is the algebra discussed above. Note also that if A is a set in $\mathscr{A}^{(n)}$, then for each x in Ω_n the section A_x belongs to $\mathscr{A}^{(n+1)}$. Finally, let us introduce the following temporary notation for sections of sets. Instead of writing A_x we will write $A(x)$, and instead of writing $(A_{x_1})_{x_2}$ we will write $A(x_1, x_2)$. Continuing in this way gives a reasonable way to express the result of many iterations of the operation of taking a section of a set.

We prove the countable additivity of P by showing that if $\{A_j\}$ is a decreasing sequence of sets in $\mathscr{A}^{(1)}$ such that $\cap_j A_j = \varnothing$, then $\lim_j P(A_j) = 0$.[18] We do this by considering the contrapositive and showing that if $\{A_j\}$ is a decreasing sequence of sets in $\mathscr{A}^{(1)}$ such that $\lim_j P(A_j) > 0$, then $\cap_j A_j \neq \varnothing$. So let us fix a decreasing sequence $\{A_j\}$ and a positive number ε such that $P(A_j) \geq \varepsilon$ holds for all j. We will show that $\cap_j A_j \neq \varnothing$ by constructing an element of $\cap_j A_j$. Suppose that A_j is a member of the sequence $\{A_j\}$. Then there is a positive integer k and a set B_j in $\prod_1^k \mathscr{A}_i$ such that $A_j = \text{proj}_k^{-1}(B_j)$. We have (see Theorem 5.1.4)

$$(P_1 \times \cdots \times P_k)(B_j) = \int_{\Omega_1} (P_2 \times \cdots \times P_k)(B_j(x_1)) P_1(dx_1),$$

which translates into

$$P(A_j) = \int_{\Omega_1} P^{(2)}(A_j(x_1)) P_1(dx_1).$$

Since $\{A_j\}_j$ is a decreasing sequence of sets, $\{P^{(2)}(A_j(x_1))\}_j$ is (for each choice of x_1 in Ω_1) a decreasing sequence of numbers, and we can define a function $f_1: \Omega_1 \to \mathbb{R}$ by $f_1(x_1) = \lim_j P^{(2)}(A_j(x_1))$. The function f_1 is measurable, and it follows from the dominated convergence theorem that

$$\int_{\Omega_1} f_1(x_1) P_1(dx_1) = \lim_j \int_{\Omega_1} P^{(2)}(A_j(x_1)) P_1(dx_1) = \lim_j P(A_j) \geq \varepsilon.$$

---

[17] Be careful to note that $\Omega^{(n)}$ is a product space, while Ω_n is one of (in fact, the first of) its factors.

[18] See Proposition 1.2.6, whose proof can easily be modified so as to apply to finitely additive measures *on algebras*.

Since P_1 has total mass 1, there must be an element x_1 of Ω_1 such that $f_1(x_1) \geq \varepsilon$ and hence such that $P^{(2)}(A_j(x_1)) \geq \varepsilon$ holds for all j; fix such a value x_1. We can apply the same argument to the sequence $\{A_j(x_1)\}_j$, producing an element x_2 of Ω_2 such that $P^{(3)}(A_j(x_1,x_2)) \geq \varepsilon$ holds for each j. By repeating this argument over and over, we produce a sequence $\{x_n\}$ such that $P^{(n+1)}(A_j(x_1,\dots,x_n)) \geq \varepsilon$ holds for all j and n.

To complete our proof that P is countably additive on $\mathscr{A}^{(1)}$, we need to show that $\cap_j A_j \neq \varnothing$. We do this by verifying that the sequence $\{x_n\}$ constructed above belongs to $\cap_j A_j$. So fix a set A_j in $\{A_j\}$. Then there is a positive integer k and a set B_j in $\prod_1^k \mathscr{A}_i$ such that $A_j = \mathrm{proj}_k^{-1}(B_j)$. Note that, because of this representation of A_j, the section $A_j(x_1,\dots,x_k)$ is equal to either $\Omega^{(k+1)}$ or $\varnothing$, depending on whether $(x_1,\dots,x_k)$ belongs to B_j or not. However, we know that $A_j(x_1,\dots,x_k)$ is not empty (since $P^{(k+1)}(A_j(x_1,\dots,x_k)) \geq \varepsilon$). Thus, $A_j(x_1,\dots,x_k) = \Omega^{(k+1)}$, and *every* continuation of the finite sequence $x_1, \dots, x_k$ belongs to A_j; in particular $\{x_n\} \in A_j$. Since this argument works for every j, we have $\{x_n\} \in \cap A_j$, and the construction of our product measure is complete.

We turn to the uniqueness of P. The collection of sets of the form (1) (where $A_i \in \mathscr{A}_i$ holds for each i) is a π-system that generates $\mathscr{A}$, and so the uniqueness of P follows from Corollary 1.6.3. $\square$

See Exercise 2 for an extension of Proposition 10.6.1 to the case of uncountably many random variables.

Now we turn to the construction of families of random variables that are not necessarily independent. For the construction of such families we will once again build a suitable measure on an infinite product space. This time, however, the measure we construct will not be a product measure.

As before, let I be an index set and let $\{(\Omega_i, \mathscr{A}_i)\}_{i \in I}$ be an indexed family of measurable spaces. Let $(\Omega, \mathscr{A})$ and $\{X_i\}_{i \in I}$ be the measurable space and coordinate functions constructed at the beginning of this section. We need to look at how to describe the dependence between our random variables. To get an idea of what to do, let us temporarily assume that we already have a probability P on $(\Omega, \mathscr{A})$. We will get a consistency condition that the joint distributions of finite collections of the random variables $\{X_i\}$ must satisfy; then we will use this consistency condition as one of the hypotheses in our existence theorem (Theorem 10.6.2).

Let $\mathscr{I}$ be the collection of all nonempty finite subsets of I. For each I_0 in $\mathscr{I}$ consider the finite product $(\prod_{i \in I_0} \Omega_i, \prod_{i \in I_0} \mathscr{A}_i)$. Let us call this product $(\Omega_{I_0}, \mathscr{A}_{I_0})$. For each I_0 let $X_{I_0} : \Omega \to \Omega_{I_0}$ be the projection of Ω onto Ω_{I_0}. So in set-theoretic terms, $X_{I_0}(\omega)$ is the restriction of the function ω to the subset I_0 of its domain. It is easy to check that for each I_0 the function X_{I_0} is measurable with respect to $\mathscr{A}$ and $\mathscr{A}_{I_0}$. Let P_{I_0} be the distribution of X_{I_0} (in other words, let P_{I_0} be the joint distribution of the random variables $X_i, i \in I_0$); thus $P_{I_0}(A) = P(X_{I_0}^{-1}(A))$ holds for each A in $\mathscr{A}_{I_0}$.

We need to look at how these distributions on finite products are related to one another. So suppose that I_1 and I_2 belong to $\mathscr{I}$ and satisfy $I_2 \subseteq I_1$, and let $\mathrm{proj}_{I_2,I_1} : \Omega_{I_1} \to \Omega_{I_2}$ be the projection of Ω_{I_1} onto Ω_{I_2}. Certainly proj_{I_2,I_1} is measurable and $X_{I_2} = \text{proj}_{I_2, I_1} \circ X_{I_1}$; thus $P(X_{I_2}^{-1}(A)) = P(X_{I_1}^{-1}(\text{proj}_{I_2, I_1}^{-1}(A)))$ holds for each A in $\mathscr{A}_{I_2}$. That is, the distributions on the finite product spaces satisfy the condition

$$P_{I_2} = P_{I_1} \text{proj}_{I_2, I_1}^{-1} \text{ for all } I_1, I_2 \text{ in } \mathscr{I} \text{ such that } I_2 \subseteq I_1. \tag{2}$$

This is the consistency condition that will be one of the hypotheses in the following theorem.

The upcoming theorem would not hold if the spaces $(\Omega_i, \mathscr{A}_i)$ were allowed to be completely arbitrary (see Exercise 5). To get around that difficulty, we will assume that for each i there is a compact metric space K_i such that $(\Omega_i, \mathscr{A}_i)$ is Borel isomorphic to $(K_i, \mathscr{B}(K_i))$; in other words, there must be a bijection $f_i \colon \Omega_i \to K_i$ such that f_i and f_i^{-1} are both measurable. Such measurable spaces are called *standard*.[19] One can check (see Exercise 1) that $(\mathbb{R}, \mathscr{B}(\mathbb{R}))$ is isomorphic to $([0, 1], \mathscr{B}([0, 1])$ and hence that $(\mathbb{R}, \mathscr{B}(\mathbb{R}))$ is standard; from this one can conclude that $(\mathbb{R}^d, \mathscr{B}(\mathbb{R}^d))$ is also standard.

Theorem 10.6.2 (Kolmogorov Consistency Theorem). *Let I be a nonempty set, let $\{(\Omega_i, \mathscr{A}_i)\}_{i \in I}$ be an indexed family of measurable spaces, and let $\mathscr{I}$ be the collection of all nonempty finite subsets of I. As in the discussion above, define product measurable spaces $(\Omega, \mathscr{A})$ and $\{(\Omega_{I_0}, \mathscr{A}_{I_0})\}_{I_0 \in \mathscr{I}}$, plus projections $X_{I_0} \colon \Omega \to \Omega_{I_0}$ and $\text{proj}_{I_2, I_1} \colon \Omega_{I_1} \to \Omega_{I_2}$, where $I_0, I_1, I_2 \in \mathscr{I}$ and $I_2 \subseteq I_1$. Let $\{P_{I_0}\}_{I_0 \in \mathscr{I}}$ be an indexed family of probability measures on the spaces $\{(\Omega_{I_0}, \mathscr{A}_{I_0})\}_{I_0 \in \mathscr{I}}$. If*

(a) *the measurable spaces $\{(\Omega_i, \mathscr{A}_i)\}_{i \in I}$ are all standard, and*
(b) *the measures $\{P_{I_0}\}_{I_0 \in \mathscr{I}}$ are consistent, in the sense that they satisfy condition (2),*

then there is a unique probability measure P on $(\Omega, \mathscr{A})$ such that for each I_0 in $\mathscr{I}$ the distribution of X_{I_0} is P_{I_0}.

Proof. The hypothesis that the spaces $\{(\Omega_i, \mathscr{A}_i)\}_{i \in I}$ are standard implies that for each i there is a compact metrizable topology on Ω_i for which $\mathscr{B}(\Omega_i) = \mathscr{A}_i$. Fix such a topology for each i. It follows from Tychonoff's theorem (Theorem D.20) and Proposition 7.1.13 that the product topology on Ω is compact Hausdorff and that for each I_0 the product topology on Ω_{I_0} is compact and metrizable; furthermore, $\mathscr{B}(\Omega_{I_0}) = \mathscr{A}_{I_0}$ holds for each I_0 in $\mathscr{I}$ (see Proposition 7.6.2). We will construct a suitable positive linear functional L on the space $C(\Omega)$ of continuous real-valued functions on Ω. The Riesz representation theorem (Theorem 7.2.8) then gives a regular Borel measure μ on Ω such that $L(f) = \int f \, d\mu$ holds for each f in $C(\Omega)$. We will see that the restriction of μ to $\mathscr{A}$ is the measure we need.

We turn to the definition of the linear functional L. We begin by defining it on the algebra of functions on Ω generated by the functions that can be written in the form $g \circ X_i$ for some i in I and some g in $C(\Omega_i)$. Let us call this algebra $C_\bullet$. Since the functions h in $C_\bullet$ are finite sums of finite products of functions of the form $g \circ X_i$,

---

[19] See Chap. 8, and especially Sect. 8.6, for more information about standard spaces.

each can be written in the form $h_{I_0} \circ X_{I_0}$ for some I_0 in $\mathscr{I}$ and some h_{I_0} in $C(\Omega_{I_0})$. We want to define $L(h)$ for h in $C_\bullet$ by $L(h) = \int_{\Omega_{I_0}} h_{I_0}\, dP_{I_0}$, where h and h_{I_0} are related by $h = h_{I_0} \circ X_{I_0}$. The potential problem is that a function h can in general be written in many ways, say as $h_{I_1} \circ X_{I_1}$ and as $h_{I_2} \circ X_{I_2}$, and so we need to check that $L(h)$ does not depend on how h is written.[20] Suppose that I_1 and I_2 are as in the previous sentence, and let $I_3 = I_1 \cup I_2$. The relation $h_{I_1} \circ X_{I_1} = h = h_{I_2} \circ X_{I_2}$ implies that

$$h_{I_1} \circ \mathrm{proj}_{I_1, I_3} = h_{I_2} \circ \mathrm{proj}_{I_2, I_3}.$$

From this and the consistency condition (2), we find

$$\int_{\Omega_{I_1}} h_{I_1}\, dP_{I_1} = \int_{\Omega_{I_3}} h_{I_1} \circ \mathrm{proj}_{I_1, I_3}\, dP_{I_3}$$

$$= \int_{\Omega_{I_3}} h_{I_2} \circ \mathrm{proj}_{I_2, I_3}\, dP_{I_3} = \int_{\Omega_{I_2}} h_{I_2}\, dP_{I_2},$$

and it follows that L is well defined on $C_\bullet$. The Stone–Weierstrass theorem (Theorem D.22) implies that $C_\bullet$ is uniformly dense in $C(\Omega)$. Thus we can extend L from $C_\bullet$ to $C(\Omega)$. It is easy to check that the extended L is positive and linear. Thus the Riesz representation theorem gives a regular Borel measure μ on Ω such that $L(h) = \int h\, d\mu$ holds for each h in $C(\Omega)$. In particular, for each I_0 in $\mathscr{I}$ and each h_{I_0} in $C(\Omega_{I_0})$ we have

$$\int_{\Omega_{I_0}} h_{I_0}\, dP_{I_0} = L(h_{I_0} \circ X_{I_0}) = \int_{\Omega} h_{I_0} \circ X_{I_0}\, d\mu = \int_{\Omega_{I_0}} h_{I_0}\, d(\mu X_{I_0}^{-1}). \qquad (3)$$

Let P be the restriction of μ to $\mathscr{A}$. It follows from Eq. (3) that $P_{I_0} = P X_{I_0}^{-1}$. In other words, P_{I_0} is the distribution of X_{I_0} under P. Since this is true for each I_0 in $\mathscr{I}$, we have constructed the required measure on $(\Omega, \mathscr{A})$.

We turn to the uniqueness of P. Define $\mathscr{A}'$ by $\mathscr{A}' = \cup_{I_0 \in \mathscr{I}} X_{I_0}^{-1}(\mathscr{A}_{I_0})$. Then $\mathscr{A}'$ is a π-system on Ω and $\sigma(\mathscr{A}') = \mathscr{A}$. Suppose that P' and P'' are probabilities on $\mathscr{A}$ that satisfy $P_{I_0} = P' X_{I_0}^{-1} = P'' X_{I_0}^{-1}$ for each I_0 in $\mathscr{I}$. This means that P' and P'' agree on $\mathscr{A}'$, and it follows from Corollary 1.6.3 that $P' = P''$. With this the proof is complete. □

Exercises

1. Check that the measurable spaces $(\mathbb{R}, \mathscr{B}(\mathbb{R}))$ and $([0,1], \mathscr{B}([0,1]))$ are isomorphic. (Hint: This is an immediate consequence of some of the results in Chap. 8.

---

[20]This is where we use the consistency condition (2).

A more elementary proof is possible: start with a homeomorphism of $\mathbb{R}$ onto the open interval $(0,1)$, and modify it on a countable set so as to get a suitable bijection from $\mathbb{R}$ onto the closed interval $[0,1]$.)

2. Show that Proposition 10.6.1 also holds for uncountable families of independent random variables (i.e., for uncountable index sets). (Hint: Suppose that the index set I is uncountable. Combine the version of Proposition 10.6.1 for countable products with the fact that the product σ-algebra on $\prod_{i \in I} \Omega_i$ is the union of the inverse images (under projection) of the product σ-algebras on the countable products $\prod_{i \in I_0} \Omega_i$, where I_0 ranges over the countable subsets of I. See Exercise 1.1.7.)

3. Let $T = [0,1]$. For each t in T let $(\Omega_t, \mathscr{A}_t) = (\mathbb{R}, \mathscr{B}(\mathbb{R}))$, and let $(\Omega, \mathscr{A})$ be the product of these spaces. Show that the subset of Ω consisting of the continuous functions from T to $\mathbb{R}$ does not belong to $\mathscr{A}$.[21] (Hint: See Exercise 1.1.7.)

4. Use Theorem 10.6.2 to construct a stochastic process $\{X_t\}$ with index set $[0,1] \cap \mathbb{Q}$ that satisfies properties (a) and (b) in the definition of Brownian motion (where the values t_i are restricted to lie in $[0,1] \cap \mathbb{Q}$). (Given this result, Exercises 10.5.1 and 10.5.2 can be used to give a proof of Theorem 10.5.1 that is less technical than the one given in Sect. 10.5.)

5. Show that the conclusion of the Kolmogorov consistency theorem (Theorem 10.6.2) may fail if the assumption that the measurable spaces $(\Omega_i, \mathscr{A}_i)$ are standard is simply omitted. (Hint: Let $\{A_n\}$ be a decreasing sequence of subsets of $[0,1]$ such that $\lambda^*(A_n) = 1$ holds for each n, but for which $\cap_n A_n = \varnothing$. See Exercise 1.4.7. For each n let $\Omega_n = A_n$ and let $\mathscr{A}_n$ be the trace of $\mathscr{B}(\mathbb{R})$ on A_n. Finally, for index sets I_0 of the form $\{1,2,\ldots,n\}$ define P_{I_0} on $(\Omega_{I_0}, \mathscr{A}_{I_0})$ by letting it be the image of the trace of Lebesgue measure on A_n under the mapping $x \mapsto (x,x,\ldots,x)$.)

6. Assume that we modify the statement of the Kolmogorov consistency theorem (Theorem 10.6.2) by replacing the assumption that the spaces $(\Omega_i, \mathscr{A}_i)$ are standard with the assumption that each Ω_i is a universally measurable subset of some compact metric space K_i (and adding the assumption that $\mathscr{A}_i$ is the trace of $\mathscr{B}(K_i)$ on K_i). Prove that this modified version is true. (Hint: Don't work too hard—derive this modified version from the original version of Theorem 10.6.2.)

Notes

Kolmogorov was at the forefront of early work on measure-theoretic probability, as was Doob a few years later; see Kolmogorov's book on the foundations of

---

[21] Thus one often needs to say things like "There is a set A in $\mathscr{A}$ that has probability 1 and is such that $t \mapsto X_t(\omega)$ is continuous for each ω in A." rather than less pedantic things like "The set of all ω such that $t \mapsto X_t(\omega)$ is continuous has probability 1."

probability [72] and Doob's book on stochastic processes [38]. Dudley [40] gives detailed historical citations in his end-of-chapter notes.

See Billingsley [8], Dudley [40], Klenke [71], Lamperti [79], Walsh [124], and Williams [128] for introductions to probability that carry the ideas in this chapter much further and are at a level appropriate for people who have completed a course in measure theory.

Much more on dealing with convergence of probability measures using distances (see a remark near the start of Sect. 10.3, and see Exercise 10.3.12) can be found in Dudley [40] and Dudley [41].

Appendix A
Notation and Set Theory

See van Dalen et al. [118], Halmos [55], Hrbacek and Jech [63], or Moschovakis [90] for further information on the topics discussed in this appendix.

A.1. Let A and B be sets. We write $x \in A$, $x \notin A$, and $A \subseteq B$ to indicate that x is a member of A, that x is not a member of A, and that A is a subset of B, respectively. We will denote the union, intersection, and difference of A and B by $A \cup B$, $A \cap B$, and $A - B$, respectively (of course $A - B = \{x : x \in A \text{ and } x \notin B\}$). In case we are dealing with subsets of a fixed set X, the complement of A will be denoted by A^c; thus $A^c = X - A$.

The empty set will be denoted by $\varnothing$.

The symmetric difference of the sets A and B is defined by

$$A \triangle B = (A - B) \cup (B - A).$$

It is clear that $A \triangle A = \varnothing$ and that $A \triangle B = A^c \triangle B^c$. Furthermore, x belongs to $A \triangle (B \triangle C)$ if and only if it belongs either to exactly one, or else to all three, of A, B, and C; since a similar remark applies to $(A \triangle B) \triangle C$, we have

$$A \triangle (B \triangle C) = (A \triangle B) \triangle C.$$

Suppose that $A_1, \ldots, A_n$ is a finite sequence of sets. The union and intersection of these sets are defined by

$$\bigcup_{i=1}^{n} A_i = \{x : x \in A_i \text{ for some } i \text{ in the range } 1, \ldots, n\}$$

and

$$\bigcap_{i=1}^{n} A_i = \{x : x \in A_i \text{ for each } i \text{ in the range } 1, \ldots, n\}$$

D.L. Cohn, *Measure Theory: Second Edition*, Birkhäuser Advanced
Texts Basler Lehrbücher, DOI 10.1007/978-1-4614-6956-8,
© Springer Science+Business Media, LLC 2013

The union and intersection of an infinite sequence $\{A_i\}_{i=1}^{\infty}$ of sets, written $\cup_{i=1}^{\infty} A_i$ and $\cap_{i=1}^{\infty} A_i$ respectively, are defined in a similar way. (To simplify notation we will sometimes write $\cup_i A_i$ in place of $\cup_{i=1}^{n} A_i$ or $\cup_{i=1}^{\infty} A_i$, and $\cap_i A_i$ in place of $\cap_{i=1}^{n} A_i$ or $\cap_{i=1}^{\infty} A_i$.)

The union and intersection of an arbitrary family $\mathscr{S}$ of subsets of a set X are defined by

$$\cup \mathscr{S} = \{x \in X : x \in S \text{ for some } S \text{ in } \mathscr{S}\}$$

and

$$\cap \mathscr{S} = \{x \in X : x \in S \text{ for each } S \text{ in } \mathscr{S}\}.$$

De Morgan's laws hold: $(\cup \mathscr{S})^c = \cap \{S^c : S \in \mathscr{S}\}$ and $(\cap \mathscr{S})^c = \cup \{S^c : S \in \mathscr{S}\}$.

The set of all subsets of the set X is called the *power set of X*; we will denote it by $\mathscr{P}(X)$.

A.2. We will use $\mathbb{N}$, $\mathbb{N}_0$, $\mathbb{Z}$, $\mathbb{Q}$, $\mathbb{R}$, and $\mathbb{C}$ to denote the sets of positive integers, of nonnegative integers, of integers (positive, negative, or zero), of rational numbers, of real numbers, and of complex numbers, respectively. The subintervals $[a,b]$ and (a,b) of $\mathbb{R}$ are defined by

$$[a,b] = \{x \in \mathbb{R} : a \leq x \leq b\}$$

and

$$(a,b) = \{x \in \mathbb{R} : a < x < b\}.$$

Other types of intervals, such as $(a,b]$, $(-\infty,b)$, and $(-\infty,+\infty)$, are defined analogously.

A.3. We write $f : X \to Y$ to indicate that f is a function whose *domain* is X and whose values lie in Y (Y is then sometimes called the *codomain* of f); thus f associates a unique element $f(x)$ of Y to each element x of X. We will sometimes define a function $f : X \to Y$ by using the arrow $\mapsto$ to show the action of f on an element of X. For example, if we are dealing with real-valued functions on $\mathbb{R}$, it is often easier to say "the function $x \mapsto x+2$" than to say "the function $f : \mathbb{R} \to \mathbb{R}$ defined by $f(x) = x+2$." Be careful to distinguish between $\to$ and $\mapsto$: the arrow $\to$ is used to specify the domain and codomain of a function, while the arrow $\mapsto$ is used to describe the action of a function on an element of its domain.

Let X, Y, and Z be sets, and consider functions $g : X \to Y$ and $f : Y \to Z$. Then $f \circ g : X \to Z$ is the function defined by $(f \circ g)(x) = f(g(x))$; it is called the *composition* of f and g.

Suppose that f is a function from X to Y, that A is a subset of X, and that B is a subset of Y. The *image* of A under f, written $f(A)$, is defined by

$$f(A) = \{y \in Y : y = f(x) \text{ for some } x \text{ in } A\},$$

and the *inverse image* of B under f, written $f^{-1}(B)$, is defined by

$$f^{-1}(B) = \{x \in X : f(x) \in B\}.$$

If A is a subset of the domain of f, then the *restriction* of f to A is the function that agrees with f on A and is undefined elsewhere.

A function $f: X \to Y$ is *injective* (or *one-to-one*) if $f(x_1) \neq f(x_2)$ holds whenever x_1 and x_2 are distinct elements of X, and is *surjective* (or *onto*) if each element of Y is of the form $f(x)$ for some x in X. A function is *bijective* if it is both injective and surjective. A function that is injective (or surjective, or bijective) is sometimes called an *injection* (or a *surjection*, or a *bijection*).

If $f: X \to Y$ is bijective, then the *inverse* of f, written f^{-1}, is the function from Y to X that is defined by letting $f^{-1}(y)$ be the unique element of X whose image under f is y; thus $x = f^{-1}(y)$ holds if and only if $y = f(x)$.

Let A be a subset of the set X. The *characteristic function* (or *indicator function*) of A is the function $\chi_A: X \to \mathbb{R}$ defined by

$$\chi_A(x) = \begin{cases} 1 & \text{if } x \in A, \\ 0 & \text{if } x \notin A. \end{cases}$$

A.4. The *product* (or *Cartesian product*) of sets X and Y, written $X \times Y$, is the set of all ordered pairs (x,y) for which $x \in X$ and $y \in Y$.

A.5. Notation such as $\{A_i\}_{i \in I}$ or $\{A_i\}$ will be used for an indexed family of sets; here I is the *index set* and A_i is the set associated to the element i of I. An infinite sequence of sets is, of course, an indexed family of sets for which the index set is $\mathbb{N}$ (or perhaps $\mathbb{N}_0$). The *product* $\prod_i A_i$ of the indexed family of sets $\{A_i\}_{i \in I}$ is the set of all functions $a: I \to \cup \{A_i : i \in I\}$ such that $a(i) \in A_i$ holds for each i in I (here one usually writes a_i in place of $a(i)$). If each A_i is equal to the set A, we often write A^I instead of $\prod_i A_i$.

A.6. Sets X and Y have the *same cardinality* if there is a bijection of X onto Y. A set is *finite* if it is empty or has the same cardinality as $\{1, 2, \ldots, n\}$ for some positive integer n; it is *countably infinite* if it has the same cardinality as $\mathbb{N}$. An *enumeration* of a countably infinite set X is a bijection of $\mathbb{N}$ onto X. Thus an enumeration of X can be viewed as an infinite sequence $\{x_n\}$ such that

(a) each x_n belongs to X, and
(b) each element of X is of the form x_n for exactly one value of n.

A set is *countable* if it is finite or countably infinite.

It is easy to check that every subset of a countable set is countable. We should also note that if X and Y are countable, then

(a) $X \cup Y$ is countable, and
(b) $X \times Y$ is countable.

Let us check (b) in the case where X and Y are both countably infinite. Let f be an enumeration of X, and let g be an enumeration of Y. Then $(m,n) \mapsto (f(m), g(n))$ is a bijection of $\mathbb{N} \times \mathbb{N}$ onto $X \times Y$, and so we need only construct an enumeration of $\mathbb{N} \times \mathbb{N}$. This, however, can be done if we define $h: \mathbb{N} \to \mathbb{N} \times \mathbb{N}$ by following

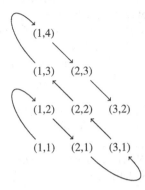

Fig. A.1 Enumeration of $\mathbb{N} \times \mathbb{N}$

the path indicated in Fig. A.1, letting $h(1) = (1,1)$, $h(2) = (1,2)$, $h(3) = (2,1)$, and so forth. (Alternatively, one can define an enumeration h of $\mathbb{N} \times \mathbb{N}$ by letting $h(n) = (r+1, s+1)$, where r and s are the nonnegative integers appearing in the factorization $n = 2^r(2s+1)$ of n into a product of a power of 2 and an odd integer.)

A similar argument can be used to show that the set $\mathbb{Q}$ of rational numbers is countable.

A.7. Suppose that A and B are sets. The Schröder–Bernstein theorem says that if A has the same cardinality as some subset of B and if B has the same cardinality as some subset of A, then A has the same cardinality as B; this can be proved with a version of the arguments used in Proposition G.2 and suggested in part (c) of Exercise 8.3.5 (alternatively, see Halmos [55, Section 22]).

A.8. The set $\mathbb{R}$ is not countable. To say that a set *has the cardinality of the continuum*, or *has cardinality c*, is to say that it has the same cardinality as $\mathbb{R}$. The product sets $\{0,1\}^{\mathbb{N}}$ and $\mathbb{R}^{\mathbb{N}}$ both have the cardinality of the continuum.

The *continuum hypothesis* says that if A is an infinite subset of $\mathbb{R}$, then either A is countably infinite or else A has the cardinality of the continuum. K. Gödel proved that the continuum hypothesis is consistent with the usual axioms for set theory, and P. J. Cohen proved that it is independent of these axioms (see [30, 50]).

A.9. A *relation* on a set X is a property that holds for some (perhaps none, perhaps all) of the ordered pairs in $X \times X$. For instance, $=$ and $\leq$ are relations on $\mathbb{R}$. If $\sim$ is a relation on X, we write $x \sim y$ to indicate that $\sim$ holds for the pair (x,y). A relation $\sim$ is usually represented by (or is considered to be) the set of ordered pairs (x,y) for which $x \sim y$ holds. Thus a relation on X is a subset of $X \times X$.

A.10. An *equivalence relation* on X is a relation $\sim$ that is reflexive ($x \sim x$ holds for each x in X), symmetric (if $x \sim y$, then $y \sim x$), and transitive (if $x \sim y$ and $y \sim z$, then

$x \sim z$). If $\sim$ is an equivalence relation on X, and if $x \in X$, then the *equivalence class* of x under $\sim$ is the set E_x defined by

$$E_x = \{y \in X : y \sim x\}.$$

Of course, x belongs to E_x. It is easy to check that if $x, y \in X$, then E_x and E_y are either equal or disjoint. Thus the equivalence classes under $\sim$ form a *partition* of X (i.e., a collection of nonempty disjoint sets whose union is X).

A.11. A *partial order* on a set X is a relation $\leq$ that is reflexive ($x \leq x$ holds for each x in X), antisymmetric (if $x \leq y$ and $y \leq x$, then $x = y$), and transitive (if $x \leq y$ and $y \leq z$, then $x \leq z$). A *partially ordered set* is a set, together with a partial order on it. A *linear order* on a set X is a partial order $\leq$ on X such that if $x, y \in X$, then either $x \leq y$ or $y \leq x$. The relation $\leq$ (with its usual meaning) is a linear order on $\mathbb{R}$. If X is a set with at least two elements, and if $\mathscr{P}(X)$ is the set of all subsets of X, then $\subseteq$ is a partial order, but not a linear order, on $\mathscr{P}(X)$.

If $\leq$ is a partial order on a set X, then $x < y$ means that x and y satisfy $x \leq y$ but are not equal.

Let X be a partially ordered set, with partial order $\leq$. A *chain* in X is a subset C of X such that if $x, y \in C$, then either $x \leq y$ or $y \leq x$. An element x of X is an *upper bound* for a subset A of X if $y \leq x$ holds for each y in A; a *lower bound* for A is defined analogously. An element x of X is *maximal* if $x \leq y$ can hold only if $y = x$ (in other words, x is maximal if there are no elements of X larger than x; there may be elements y of X for which neither $x \leq y$ nor $y \leq x$ holds).

A linear order on a set X is a *well ordering* if each nonempty subset of X has a smallest element (that is, if each nonempty subset A of X has a lower bound *that belongs to A*). A set X *can be well ordered* if there is a well ordering on X.

The set $\mathbb{N}$ of positive integers (with the usual order relation on it) is well ordered, but the set $\mathbb{Q}$ of rationals is not. However, one can easily define a well ordering on $\mathbb{Q}$, as follows: Let $f : \mathbb{N} \to \mathbb{Q}$ be a bijective function (that is, an enumeration of $\mathbb{Q}$), and let f^{-1} be its inverse. Define a binary relation $\prec$ on $\mathbb{Q}$ by declaring that $x \prec y$ holds if and only if $f^{-1}(x) < f^{-1}(y)$ (here $<$ is the usual less-than relation on $\mathbb{N}$). Since $<$ is a well ordering of $\mathbb{N}$, $\prec$ is a well ordering of $\mathbb{Q}$.

A.12. The *axiom of choice* says that if $\mathscr{S}$ is a set of disjoint nonempty sets, then there is a set that has exactly one element in common with each set in $\mathscr{S}$. Another (equivalent) formulation of the axiom of choice says that if $\{A_i\}_{i \in I}$ is an indexed family of nonempty sets, then $\prod_i A_i$ is nonempty.

A.13. (Theorem) *The following are equivalent:*

(a) *The axiom of choice holds.*
(b) *(Zorn's lemma) If X is a partially ordered set such that each chain in X has an upper bound in X, then X has a maximal element.*
(c) *(The well-ordering theorem) Every set can be well ordered.*

A proof of Theorem A.13 can be found in [55, 63, 90, 118]. (Gödel and Cohen showed that the axiom of choice is consistent with and independent of the remaining standard axioms for set theory; again see [30, 50].)

A.14. The reader will need some experience with ordinal numbers in order to do a few of the exercises in Chaps. 7 through 9. It is enough to know a bit about the countable ordinals and the first uncountable ordinal and to have some facility with transfinite recursion and induction. Once again, see [55, 63, 90, 118].

Appendix B
Algebra and Basic Facts About $\mathbb{R}$ and $\mathbb{C}$

B.1. A *field* is a set F, together with binary operations $+$ and $\cdot$ on F such that

(a) $(x+y)+z = x+(y+z)$ holds for all x, y, z in F,
(b) $x+y = y+x$ holds for all x, y in F,
(c) there is an element 0 of F such that $x+0 = x$ holds for all x in F,
(d) for each x in F there is an element $-x$ of F such that $x+(-x) = 0$,
(e) $(x \cdot y) \cdot z = x \cdot (y \cdot z)$ holds for all x, y, z in F,
(f) $x \cdot y = y \cdot x$ holds for all x, y in F,
(g) there is an element 1 of F, distinct from 0, such that $1 \cdot x = x$ holds for all x in F,
(h) for each nonzero x in F there is an element x^{-1} of F such that $x \cdot x^{-1} = 1$, and
(i) $x \cdot (y+z) = x \cdot y + x \cdot z$ holds for all x, y, z in F.

Of course, one usually writes xy in place of $x \cdot y$.

B.2. An *ordered field* is a field F, together with a linear order $\leq$ (see A.11) on F such that

(a) if x, y, and z belong to F and if $x \leq y$, then $x+z \leq y+z$, and
(b) if x and y belong to F and satisfy $x > 0$ and $y > 0$, then $x \cdot y > 0$.

Let F be an ordered field, and let A be a subset of F. An *upper bound* of A is an element x of F such that $a \leq x$ holds for each a in A; a *least upper bound* (or *supremum*) of A is an upper bound of A that is smaller than all other upper bounds of A. *Lower bounds* and *greatest lower bounds* (or *infima*) are defined analogously. An ordered field F is *complete* if each nonempty subset of F that has an upper bound in F has a least upper bound in F.

B.3. The field $\mathbb{R}$ of real numbers is a complete ordered field; it is essentially the only complete ordered field (see Birkhoff and MacLane [9, Chapter 4], Gleason [49, Chapters 8 and 9], or Spivak [111, Chapters 28 and 29] for a precise statement and proof of this assertion).

D.L. Cohn, *Measure Theory: Second Edition*, Birkhäuser Advanced
Texts Basler Lehrbücher, DOI 10.1007/978-1-4614-6956-8,
© Springer Science+Business Media, LLC 2013

B.4. The *extended real numbers* consist of the real numbers, together with $+\infty$ and $-\infty$. We will use $\overline{\mathbb{R}}$ or $[-\infty, +\infty]$ to denote the set of all extended real numbers. The relations $-\infty < x$ and $x < +\infty$ are declared to hold for each real number x (of course $-\infty < +\infty$). We define arithmetic operations on $\overline{\mathbb{R}}$ by declaring that

$$x + (+\infty) = (+\infty) + x = +\infty$$

and

$$x + (-\infty) = (-\infty) + x = -\infty$$

hold for each real x, that

$$x \cdot (+\infty) = (+\infty) \cdot x = +\infty$$

and

$$x \cdot (-\infty) = (-\infty) \cdot x = -\infty$$

hold for each positive real x, and that

$$x \cdot (+\infty) = (+\infty) \cdot x = -\infty$$

and

$$x \cdot (-\infty) = (-\infty) \cdot x = +\infty$$

hold for each negative real x; we also declare that

$$(+\infty) + (+\infty) = +\infty,$$

$$(-\infty) + (-\infty) = -\infty,$$

$$(+\infty) \cdot (+\infty) = (-\infty) \cdot (-\infty) = +\infty,$$

$$(+\infty) \cdot (-\infty) = (-\infty) \cdot (+\infty) = -\infty,$$

and
$$0 \cdot (+\infty) = (+\infty) \cdot 0 = 0 \cdot (-\infty) = (-\infty) \cdot 0 = 0.$$

The sums $(+\infty) + (-\infty)$ and $(-\infty) + (+\infty)$ are left undefined. (The products $0 \cdot (+\infty)$, $(+\infty) \cdot 0$, $(-\infty) \cdot 0$, and $0 \cdot (-\infty)$, even though left undefined in many other areas of mathematics, are defined to be 0 in the study of measure theory; this simplifies the definition of the Lebesgue integral.)

The absolute values of $+\infty$ and of $-\infty$ are defined by

$$|+\infty| = |-\infty| = +\infty.$$

The maximum and minimum of the extended real numbers x and y are often denoted by $x \vee y$ and $x \wedge y$.

B.5. Each subset of $\overline{\mathbb{R}}$ has a least upper bound, or supremum, and a greatest lower bound, or infimum, in $\overline{\mathbb{R}}$. The supremum and infimum of a subset A of $\overline{\mathbb{R}}$ are often denoted by $\sup(A)$ and $\inf(A)$. Note that the set under consideration here may be empty: each element of $\overline{\mathbb{R}}$ is an upper bound and a lower bound of $\varnothing$; hence $\sup(\varnothing) = -\infty$ and $\inf(\varnothing) = +\infty$. Note also that $\sup(A)$ is a real number (rather than $+\infty$ or $-\infty$) if and only if A is nonempty and bounded above; a similar remark applies to infima.

B.6. Let $\{x_n\}$ be a sequence of elements of $\overline{\mathbb{R}}$. The *limit superior* of $\{x_n\}$, written $\overline{\lim}_n x_n$ or $\limsup_n x_n$, is defined by

$$\overline{\lim_n} x_n = \inf_k \sup_{n \geq k} x_n.$$

Likewise, the *limit inferior* of $\{x_n\}$, written $\underline{\lim}_n x_n$ or $\liminf_n x_n$, is defined by

$$\underline{\lim_n} x_n = \sup_k \inf_{n \geq k} x_n.$$

The relation $\underline{\lim}_n x_n \leq \overline{\lim}_n x_n$ holds for each sequence $\{x_n\}$. The sequence $\{x_n\}$ has a *limit* (in $\overline{\mathbb{R}}$) if $\overline{\lim}_n x_n = \underline{\lim}_n x_n$; the limit of $\{x_n\}$ is then defined by

$$\lim_n x_n = \overline{\lim_n} x_n = \underline{\lim_n} x_n$$

(note that $\lim_n x_n$ can be $+\infty$ or $-\infty$).

In cases where each x_n, along with $\lim_n x_n$, is finite, the definition of limit given above is equivalent to the usual ε–δ (or ε–N) definition: $x = \lim_n x_n$ if and only if for every ε there is a positive integer N such that $|x_n - x| < \varepsilon$ holds for each n larger than N. (We need our definition of limits in $\overline{\mathbb{R}}$, involving lim sups and lim infs, because we need to handle infinite limits and sums, and sums some of whose terms may include $+\infty$ or $-\infty$.)

B.7. We will occasionally need the fact that if a and a_n, $n = 1, 2, \ldots$, are real (or complex) numbers such that $a = \lim_n a_n$, then $a = \lim_n (a_1 + \cdots + a_n)/n$. To verify this, note that if $1 \leq M < n$, then

$$\left| \frac{1}{n} \sum_{i=1}^n a_i - a \right| \leq \frac{1}{n} \sum_{i=1}^M |a_i - a| + \frac{1}{n} \sum_{i=M+1}^n |a_i - a|.$$

If we first make M so large that $|a_i - a| < \varepsilon$ if $i > M$ and then choose N so large that $(1/n) \sum_{i=1}^M |a_i - a|$ is less than ε if $n > N$, then $(1/n) \sum_{i=1}^n a_i$ is within 2ε of a if $n > \max(M, N)$.

B.8. Let $\sum_{k=1}^\infty x_k$ be an infinite series whose terms belong to $\overline{\mathbb{R}}$. This series *has a sum* if

(a) $+\infty$ and $-\infty$ do not both occur among the terms of $\sum_{k=1}^\infty x_k$, and
(b) the sequence $\{\sum_{k=1}^n x_k\}_{n=1}^\infty$ of partial sums of $\sum_{k=1}^\infty x_k$ has a limit in $\overline{\mathbb{R}}$.

The sum of the series $\sum_{k=1}^{\infty} x_k$ is then defined to be $\lim_n \sum_{k=1}^{n} x_k$ and is denoted by $\sum_{k=1}^{\infty} x_k$. (Note that condition (a) above is needed to guarantee that each of the partial sums $\sum_{k=1}^{n} x_k$ is defined.)

The reader can check that the sum of the series $\sum_{k=1}^{\infty} x_k$ exists *and belongs to $\mathbb{R}$* if and only if

(a) each term of $\sum_{k=1}^{\infty} x_k$ belongs to $\mathbb{R}$, and
(b) the series $\sum_{k=1}^{\infty} x_k$ is convergent (in the sense of elementary calculus).

Suppose that $\sum_{k=1}^{\infty} x_k$ is an infinite series whose terms belong to $[0, +\infty]$. It is easy to see that the sum of the series $\sum_{k=1}^{\infty} x_k$ exists and is the supremum of the set of sums $\sum_{k \in F} x_k$, where F ranges over the set of finite subsets of $\mathbb{N}$.

B.9. A *dyadic rational* is a number that can be written in the form $i/2^n$ for some integer i and some nonnegative integer n. If x is a dyadic rational that belongs to the interval $(0, 1)$, then x can be written in the form $i/2^n$, where n is a positive integer and i is an odd integer such that $0 < i < 2^n$. Such an x has a binary expansion $0.b_1 b_2 \ldots b_n$, where there are exactly n bits to the right of the binary point and where b_n, the rightmost of these bits, is equal to 1. Such an x also has an unending binary expansion, where $b_n = 0$ and all the later bits $(b_{n+1}, b_{n+2}, \ldots)$ are equal to 1. These dyadic rationals are the only values in the interval $(0, 1)$ that have more than one binary expansion; to see this, suppose that x has binary expansions $0.b_1 b_2 \ldots$ and $0.c_1 c_2 \ldots$, let n_0 be the smallest n such that $b_n \neq c_n$ (for definiteness, suppose that $b_{n_0} = 0$ and $c_{n_0} = 1$), and check that this can happen only if $b_{n_0+1} = b_{n_0+2} = \cdots = 1$ and $c_{n_0+1} = c_{n_0+2} = \cdots = 0$.

B.10. Roughly speaking, the *complex numbers* are those of the form $x + iy$, where x and y are real numbers and i satisfies $i^2 = -1$. They form a field. More precisely, the set $\mathbb{C}$ of complex numbers can be represented by the set of all ordered pairs (x, y) of real numbers; addition and multiplication are then defined on $\mathbb{C}$ by

$$(x, y) + (u, v) = (x + u, y + v)$$

and

$$(x, y) \cdot (u, v) = (xu - yv, xv + yu).$$

It is not hard to check that with these operations

(a) $\mathbb{C}$ is a field, and
(b) $(0, 1) \cdot (0, 1) = (-1, 0)$.

If we return to the usual informal notation and write $x + iy$ in place of (x, y), then assertions (a) and (b) above provide justification for the first two sentences of this paragraph.

If z is a complex number, then the real numbers x and y that satisfy $z = x + iy$ are called the *real* and *imaginary parts* of z; they are sometimes denoted by $\Re(z)$ and $\Im(z)$.

The *absolute value*, or *modulus*, of the complex number z (where $z = x + iy$) is defined by

$$|z| = \sqrt{x^2 + y^2}.$$

It is easy to check that $|z_1 z_2| = |z_1||z_2|$ and $|z_1 + z_2| \le |z_1| + |z_2|$ hold for all z_1, z_2 in $\mathbb{C}$.

Limits of sequences of complex numbers and sums of infinite series whose terms are complex are defined in the expected way. The exponential function is defined on $\mathbb{C}$ by the usual infinite series:

$$e^z = \sum_{n=0}^{\infty} z^n / n!.$$

With some elementary manipulations of this series, one can check that

(a) $e^0 = 1$,
(b) $e^{z_1 + z_2} = e^{z_1} e^{z_2}$ for all complex z_1 and z_2, and
(c) $e^{it} = \cos t + i \sin t$ for all real t.

B.11. Let F be a field (in this book it will generally be $\mathbb{R}$ or $\mathbb{C}$). A *vector space* over F is a set V, together with operations $(v_1, v_2) \mapsto v_1 + v_2$ from $V \times V$ to V and $(\alpha, v) \mapsto \alpha \cdot v$ from $F \times V$ to V such that

(a) $(x + y) + z = x + (y + z)$ holds for all x, y, z in V,
(b) $x + y = y + x$ holds for all x, y in V,
(c) there is an element 0 of V such that $x + 0 = x$ holds for all x in V,
(d) for each x in V there is an element $-x$ of V such that $x + (-x) = 0$,
(e) $1 \cdot x = x$ holds for all x in V,
(f) $(\alpha \beta) \cdot x = \alpha \cdot (\beta \cdot x)$ holds for all α, β in F and all x in V,
(g) $(\alpha + \beta) \cdot x = \alpha \cdot x + \beta \cdot x$ holds for all α, β in F and all x in V, and
(h) $\alpha \cdot (x + y) = \alpha \cdot x + \alpha \cdot y$ holds for all α in F and all x, y in V.

(We will, of course, usually write αx in place of $\alpha \cdot x$.)

Note that $\mathbb{R}^d$ is a vector space over $\mathbb{R}$ and that $\mathbb{C}^d$ is a vector space over $\mathbb{C}$ (it is also a vector space over $\mathbb{R}$). Note also that if F is a field, then F is a vector space over F.

A *subspace* (or a *linear subspace*) of a vector space V over F is a subset V_0 of V that is a vector space when the operations $+$ and $\cdot$ are restricted to $V_0 \times V_0$ and $F \times V_0$.

B.12. Let V_1 and V_2 be vector spaces over the same field F. A function $L \colon V_1 \to V_2$ is *linear* if

$$L(\alpha x + \beta y) = \alpha L(x) + \beta L(y)$$

holds for all α, β in F and all x, y in V_1. A bijective linear map is a *linear isomorphism*. It is easy to check that the inverse of a linear isomorphism is linear.

Let V be a vector space over the field F. A *linear functional* on V is a linear map from V to the field F.

B.13. Let V be a vector space over $\mathbb{R}$ or $\mathbb{C}$. For each pair x, y of elements of V, the *line segment* connecting x and y is the set of points that can be written in the form $tx + (1-t)y$ for some t in the interval $[0,1]$. A subset C of V is *convex* if for each pair x, y of points in C the line segment connecting x and y is included in C.

B.14. (We will need this and Sect. B.15 only for the discussion of the Banach–Tarski paradox in Appendix G.) Let V be a vector space over $\mathbb{R}$, and let $T : V \to V$ be a linear operator. If x is a nonzero vector and λ is a real number such that $T(x) = \lambda x$, then x is an *eigenvector* of T and λ is an *eigenvalue* of T.

Note that if λ is an eigenvalue of T and if x is a corresponding eigenvector, then $(T - \lambda I)(x) = 0$, and so $T - \lambda I$ is not invertible. If the vector space V is finite dimensional, the converse holds: λ is an eigenvalue of T if and only if the operator $T - \lambda I$ is not invertible.

Let T be a linear operator on the finite-dimensional vector space V, let $\{e_i\}$ be a basis for V, and let A be the matrix of T with respect to $\{e_i\}$. Define $p : \mathbb{R} \to \mathbb{R}$ by $p(\lambda) = \det(A - \lambda I)$. Then $p(\lambda)$ is a polynomial in λ, called the *characteristic polynomial* of A (or of T). The eigenvalues of T are exactly the roots of the polynomial $p(\lambda)$.

B.15. The *transpose* of a matrix A (with components a_{ij}) is the matrix A^t whose components are given by $a^t_{ij} = a_{ji}$. Note that if A is a d by d matrix, if $x, y \in \mathbb{R}^d$, with x and y viewed as column vectors, and if $(\cdot, \cdot)$ is the usual inner product function on $\mathbb{R}^d$, then $(Ax, y) = (x, A^t y)$.

B.16. A *group* is a set G, together with a binary operation $\cdot$ on G such that

(a) $(x \cdot y) \cdot z = x \cdot (y \cdot z)$ holds for all x, y, z in G,
(b) there is an element e of G such that $e \cdot x = x \cdot e = x$ holds for all x in G, and
(c) for each x in G there is an element x^{-1} of G such that $x \cdot x^{-1} = x^{-1} \cdot x = e$.

A group G is *commutative* (or *abelian*) if $x \cdot y = y \cdot x$ holds for all x, y in G. One often uses $+$, rather than $\cdot$, to denote the operation in a commutative group. A *subgroup* of the group G is a subset G_0 of G that is a group when the operation $\cdot$ is restricted to $G_0 \times G_0$.

B.17. Let G_1 and G_2 be groups. A function $f : G_1 \to G_2$ is a *homomorphism* if $f(x \cdot y) = f(x) \cdot f(y)$ holds for all x, y in G_1. A bijective function $f : G_1 \to G_2$ is an *isomorphism* if both f and f^{-1} are homomorphisms.

Appendix C
Calculus and Topology in $\mathbb{R}^d$

C.1. Recall that $\mathbb{R}^d$ is the set of all d-tuples of real numbers; it is a vector space over $\mathbb{R}$. (The d in $\mathbb{R}^d$ is for dimension; we write $\mathbb{R}^d$, rather than $\mathbb{R}^n$, in order to have n available for use as a subscript.) Let $x = (x_1, \ldots, x_d)$ and $y = (y_1, \ldots, y_d)$ be elements of $\mathbb{R}^d$. The *norm* of x is defined by

$$\|x\| = \left(\sum_{i=1}^{d} x_i^2 \right)^{1/2}$$

and the *distance* between x and y is defined to be $\|x - y\|$.

C.2. If $x \in \mathbb{R}^d$ and if r is a positive number, then the *open ball $B(x, r)$* with center x and radius r is defined by

$$B(x, r) = \{ y \in \mathbb{R}^d : \|y - x\| < r \}.$$

A subset U of $\mathbb{R}^d$ is *open* if for each x in U there is a positive number r such that $B(x, r) \subseteq U$. A subset of $\mathbb{R}^d$ is *closed* if its complement is open. A point x in $\mathbb{R}^d$ is a *limit point* of the subset A of $\mathbb{R}^d$ if for each positive r the open ball $B(x, r)$ contains infinitely many points of A (this is equivalent to requiring that for each positive r the ball $B(x, r)$ contain at least one point of A distinct from x). It is easy to check that a subset of $\mathbb{R}^d$ is closed if and only if it contains all of its limit points.

If A is a subset of $\mathbb{R}^d$, then the *closure* of A is the set $\overline{A}$ (or A^-) that consists of the points in A, together with the limit points of A; $\overline{A}$ is closed and is, in fact, the smallest closed subset of $\mathbb{R}^d$ that includes A.

C.3. A subset A of $\mathbb{R}^d$ is *bounded* if there is a real number M such that $\|x\| \leq M$ holds for each x in A.

C.4. (Proposition) *Let U be an open subset of $\mathbb{R}$. Then there is a countable collection $\mathscr{U}$ of disjoint open intervals such that $U = \cup \mathscr{U}$.*

D.L. Cohn, *Measure Theory: Second Edition*, Birkhäuser Advanced
Texts Basler Lehrbücher, DOI 10.1007/978-1-4614-6956-8,
© Springer Science+Business Media, LLC 2013

Proof. Let $\mathscr{U}$ consist of those open subintervals I of U that are maximal, in the sense that the only open interval J that satisfies $I \subseteq J \subseteq U$ is I itself. Of course $\cup \mathscr{U} \subseteq U$. One can verify the reverse inclusion by noting that if $x \in U$, then the union of those open intervals that contain x and are included in U is an open interval that contains x and belongs to $\mathscr{U}$. It is easy to check (do so) that the intervals in $\mathscr{U}$ are disjoint from one another. If for each I in $\mathscr{U}$ we choose a rational number x_I that belongs to I, then (since the sets in $\mathscr{U}$ are disjoint from one another) the map $I \mapsto x_I$ is an injection; thus $\mathscr{U}$ has the same cardinality as some subset of $\mathbb{Q}$, and so is countable (see item A.6 in Appendix A). $\qquad\square$

C.5. A sequence $\{x_n\}$ of elements of $\mathbb{R}^d$ *converges* to the element x of $\mathbb{R}^d$ if $\lim_n \|x_n - x\| = 0$; x is then called the *limit* of the sequence $\{x_n\}$ (note that here x and $x_1, x_2, \dots$ are elements of $\mathbb{R}^d$; in particular, $x_1, x_2, \dots$ are *not* the components of x). A sequence in $\mathbb{R}^d$ is *convergent* if it converges to some element of $\mathbb{R}^d$.

C.6. Let A be a subset of $\mathbb{R}^d$, and let x_0 belong to A. A function $f \colon A \to \mathbb{R}$ is *continuous at* x_0 if for each positive number ε there is a positive number δ such that $|f(x) - f(x_0)| < \varepsilon$ holds whenever x belongs to A and satisfies $\|x - x_0\| < \delta$; f is *continuous* if it is continuous at each point in A. The function $f \colon A \to \mathbb{R}$ is *uniformly continuous* if for each positive number ε there is a positive number δ such that $|f(x) - f(x')| < \varepsilon$ holds whenever x and x' belong to A and satisfy $\|x - x'\| < \delta$. A function $f \colon A \to \mathbb{R}$ is *continuous on* (or *uniformly continuous on*) the subset A_0 of A if the restriction of f to A_0 is continuous (or uniformly continuous).

C.7. Let A be a subset of $\mathbb{R}^d$, let f be a real- or complex-valued function on A, and let a be a limit point of A. Then $f(x)$ has *limit* L as x approaches a, written $\lim_{x \to a} f(x) = L$, if for every positive ε there is a positive δ such that $|f(x) - f(a)| < \varepsilon$ holds whenever x is a member of A that satisfies $0 < \|x - a\| < \delta$.

One can check that $\lim_{x \to a} f(x) = L$ if and only if $\lim_n f(x_n) = L$ for every sequence $\{x_n\}$ of elements of A, all different from a, such that $\lim_n x_n = a$. (Let us consider the more difficult half of that assertion, namely that if $\lim_n f(x_n) = L$ for every sequence $\{x_n\}$ of elements of A, all different from a, such that $\lim_n x_n = a$, then $\lim_{x \to a} f(x) = L$. We prove this by proving its contrapositive. So assume that $\lim_{x \to a} f(x) = L$ is not true. Then there exists a positive ε such that for every positive δ there is a value x in A such that $0 < \|x - a\| < \delta$ and $|f(x) - L| \geq \varepsilon$. If for each n we let $\delta = 1/n$ and choose an element x_n of A such that $0 < \|x_n - a\| < 1/n$ and $|f(x_n) - L| \geq \varepsilon$, we will have a sequence $\{x_n\}$ of elements of A, all different from a, that satisfy $\lim_n x_n = a$ but not $\lim_n f(x_n) = L$.)

C.8. Let A be a subset of $\mathbb{R}^d$. An *open cover* of A is a collection $\mathscr{S}$ of open subsets of $\mathbb{R}^d$ such that $A \subseteq \cup \mathscr{S}$. A *subcover* of the open cover $\mathscr{S}$ is a subfamily of $\mathscr{S}$ that is itself an open cover of A.

Proofs of the following results can be found in almost any text on advanced calculus or basic analysis (see, for example, Bartle [4], Hoffman [60], Rudin [104], or Thomson et al. [117]).

C.9. (Theorem) *Let A be a closed bounded subset of $\mathbb{R}^d$. Then every open cover of A has a finite subcover.*

Theorem C.9 is often called the *Heine–Borel* theorem.

C.10. (Theorem) *Let A be a closed bounded subset of $\mathbb{R}^d$. Then every sequence of elements of A has a subsequence that converges to an element of A.*

C.11. It is easy to check that the converses of Theorems C.9 and C.10 hold: if A satisfies the conclusion of Theorem C.9 or of Theorem C.10, then A is closed and bounded. The subsets of $\mathbb{R}^d$ that satisfy the conclusion of Theorem C.9 (hence the closed bounded subsets of $\mathbb{R}^d$) are often called *compact*. See also Appendix D.

C.12. (Theorem) *Let C be a nonempty closed bounded subset of $\mathbb{R}^d$, and let $f: C \to \mathbb{R}$ be continuous. Then*

(a) *f is uniformly continuous on C, and*
(b) *f is bounded on C. Moreover, there are elements x_0 and x_1 of C such that $f(x_0) \leq f(x) \leq f(x_1)$ holds at each x in C.*

C.13. (The Intermediate Value Theorem) *Let A be a subset of $\mathbb{R}$, and let $f: A \to \mathbb{R}$ be continuous. If the interval $[x_0, x_1]$ is included in A, then for each real number y between $f(x_0)$ and $f(x_1)$ there is an element x of $[x_0, x_1]$ such that $y = f(x)$.*

C.14. (The Mean Value Theorem) *Let a and b be real numbers such that $a < b$. If $f: [a, b] \to \mathbb{R}$ is continuous on the closed interval $[a, b]$ and differentiable at each point in the open interval (a, b), then there is a number c in (a, b) such that $f(b) - f(a) = f'(c)(b - a)$.*

Appendix D
Topological Spaces and Metric Spaces

A number of the results in this appendix are stated without proof. For additional details, the reader should consult a text on point-set topology (for example, Kelley [69], Munkres [91], or Simmons [109]).

D.1. Let X be a set. A *topology* on X is a family $\mathcal{O}$ of subsets of X such that

(a) $X \in \mathcal{O}$,
(b) $\varnothing \in \mathcal{O}$,
(c) if $\mathcal{S}$ is an arbitrary collection of sets that belong to $\mathcal{O}$, then $\cup \mathcal{S} \in \mathcal{O}$, and
(d) if $\mathcal{S}$ is a finite collection of sets that belong to $\mathcal{O}$, then $\cap \mathcal{S} \in \mathcal{O}$.

A *topological space* is a pair $(X, \mathcal{O})$, where X is a set and $\mathcal{O}$ is a topology on X (we will generally abbreviate the notation and simply call X a topological space). The *open* subsets of X are those that belong to $\mathcal{O}$. An *open neighborhood* of a point x in X is an open set that contains x.

The collection of all open subsets of $\mathbb{R}^d$ (as defined in Appendix C) is a topology on $\mathbb{R}^d$; it is sometimes called the *usual* topology on $\mathbb{R}^d$.

D.2. Let $(X, \mathcal{O})$ be a topological space. A subset F of X is *closed* if F^c is open. The union of a finite collection of closed sets is closed, as is the intersection of an arbitrary collection of closed sets (use De Morgan's laws and parts (c) and (d) of the definition of a topology). It follows that if $A \subseteq X$, then there is a smallest closed set that includes A, namely the intersection of all the closed subsets of X that include A; this set is called the *closure* of A and is denoted by $\overline{A}$ or by A^-. A point x in X is a *limit point* of A if each open neighborhood of x contains at least one point of A other than x (the point x itself may or may not belong to A). A set is closed if and only if it contains each of its limit points. The closure of the set A consists of the points in A, together with the limit points of A.

D.3. Let $(X, \mathcal{O})$ be a topological space, and let A be a subset of X. The *interior* of A, written A^o, is the union of the open subsets of X that are included in A; thus A^o is the largest open subset of A. It is easy to check that $A^o = ((A^c)^-)^c$.

D.L. Cohn, *Measure Theory: Second Edition*, Birkhäuser Advanced
Texts Basler Lehrbücher, DOI 10.1007/978-1-4614-6956-8,
© Springer Science+Business Media, LLC 2013

D.4. Let $(X, \mathscr{O})$ be a topological space, let Y be a subset of X, and let $\mathscr{O}_Y$ be the collection of all subsets of Y that have the form $Y \cap U$ for some U in $\mathscr{O}$. Then $\mathscr{O}_Y$ is a topology on Y; it is said to be *inherited from* X, or to be *induced* by $\mathscr{O}$. The space $(Y, \mathscr{O}_Y)$ (or simply Y) is called a *subspace* of $(X, \mathscr{O})$ (or of X).

Note that if Y is an open subset of X, then the members of $\mathscr{O}_Y$ are exactly the subsets of Y that are open as subsets of X. Likewise, if Y is a closed subset of X, then the closed subsets of the topological space $(Y, \mathscr{O}_Y)$ are exactly the subsets of Y that are closed as subsets of $(X, \mathscr{O}_X)$.

D.5. Let X and Y be topological spaces. A function $f : X \to Y$ is *continuous* if $f^{-1}(U)$ is an open subset of X whenever U is an open subset of Y. It is easy to check that f is continuous if and only if $f^{-1}(C)$ is closed whenever C is a closed subset of Y. A function $f : X \to Y$ is a *homeomorphism* if it is a bijection such that f and f^{-1} are both continuous. Equivalently, f is a homeomorphism if it is a bijection such that $f^{-1}(U)$ is open exactly when U is open. The spaces X and Y are *homeomorphic* if there is a homeomorphism of X onto Y.

D.6. We will on occasion need the following techniques for verifying the continuity of a function. Let X and Y be topological spaces, and let f be a function from X to Y. If $\mathscr{S}$ is a collection of open subsets of X such that $X = \cup \mathscr{S}$, and if for each U in $\mathscr{S}$ the restriction f_U of f to U is continuous (as a function from U to Y), then f is continuous (to prove this, note that if V is an open subset of Y, then $f^{-1}(V)$ is the union of the sets $f_U^{-1}(V)$, and so is open). Likewise, if $\mathscr{S}$ is a *finite* collection of closed sets such that $X = \cup \mathscr{S}$, and if for each C in $\mathscr{S}$ the restriction of f to C is continuous, then f is continuous.

D.7. If $\mathscr{O}_1$ and $\mathscr{O}_2$ are topologies on the set X, and if $\mathscr{O}_1 \subseteq \mathscr{O}_2$, then $\mathscr{O}_1$ is said to be *weaker* than $\mathscr{O}_2$.

Now suppose that $\mathscr{A}$ is an arbitrary collection of subsets of the set X. There exist topologies on X that include $\mathscr{A}$ (for instance, the collection of all subsets of X). The intersection of all such topologies on X is a topology; it is the weakest topology on X that includes $\mathscr{A}$ and is said to be *generated* by $\mathscr{A}$.

We also need to consider topologies generated by sets of functions. Suppose that X is a set and that $\{f_i\}$ is a collection of functions, where for each i the function f_i maps X to some topological space Y_i. A topology on X makes all these functions continuous if and only if $f_i^{-1}(U)$ is open (in X) for each index i and each open subset U of Y_i. The topology *generated* by the family $\{f_i\}$ is the weakest topology on X that makes each f_i continuous, or equivalently, the topology generated by the sets $f_i^{-1}(U)$.

D.8. A subset A of a topological space X is *dense* in X if $\overline{A} = X$. The space X is *separable* if it has a countable dense subset.

D.9. Let $(X, \mathscr{O})$ be a topological space. A collection $\mathscr{U}$ of open subsets of X is a *base* for $(X, \mathscr{O})$ if for each V in $\mathscr{O}$ and each x in V there is a set U that belongs to $\mathscr{U}$ and satisfies $x \in U \subseteq V$. Equivalently, $\mathscr{U}$ is a base for X if the open subsets of

X are exactly the unions of (possibly empty) collections of sets in $\mathscr{U}$. A topological space is said to be *second countable*, or to *have a countable base*, if it has a base that contains only countably many sets.

D.10. It is easy to see that if X is second countable, then X is separable (if $\mathscr{U}$ is a countable base for X, then we can form a countable dense subset of X by choosing one point from each nonempty set in $\mathscr{U}$). The converse is not true. (Construct a topological space $(X, \mathscr{O})$ by letting $X = \mathbb{R}$ and letting $\mathscr{O}$ consist of those subsets A of X such that either $A = \varnothing$ or $0 \in A$. Then $\{0\}$ is dense in X, and so X is separable; however, X is not second countable. Exercise 7.1.8 contains a more interesting example.)

D.11. If X is a second countable topological space, and if $\mathscr{V}$ is a collection of open subsets of X, then there is a countable subset $\mathscr{V}_0$ of $\mathscr{V}$ such that $\cup \mathscr{V}_0 = \cup \mathscr{V}$. (Let $\mathscr{U}$ be a countable base for X, and let $\mathscr{U}_0$ be the collection of those elements U of $\mathscr{U}$ for which there is a set in $\mathscr{V}$ that includes U. For each U in $\mathscr{U}_0$ choose an element of $\mathscr{V}$ that includes U. The collection of sets chosen is the required subset of $\mathscr{V}$.)

D.12. A topological space X is *Hausdorff* if for each pair x, y of distinct points in X there are open sets U, V such that $x \in U$, $y \in V$, and $U \cap V = \varnothing$.

D.13. Let A be a subset of the topological space X. An *open cover* of A is a collection $\mathscr{S}$ of open subsets of X such that $A \subseteq \cup \mathscr{S}$. A *subcover* of the open cover $\mathscr{S}$ is a subfamily of $\mathscr{S}$ that is itself an open cover of A. The set A is *compact* if each open cover of A has a finite subcover. A topological space X is *compact* if X, when viewed as a subset of the space X, is compact.

D.14. A collection $\mathscr{C}$ of subsets of a set X satisfies the *finite intersection property* if each finite subcollection of $\mathscr{C}$ has a nonempty intersection. It follows from De Morgan's laws that a topological space X is compact if and only if each collection of closed subsets of X that satisfies the finite intersection property has a nonempty intersection.

D.15. If X and Y are topological spaces, if $f \colon X \to Y$ is continuous, and if K is a compact subset of X, then $f(K)$ is a compact subset of Y.

D.16. Every closed subset of a compact set is compact. Conversely, every compact subset of a Hausdorff space is closed (this is a consequence of Proposition 7.1.2; in fact, the first half of the proof of that proposition is all that is needed in the current situation).

D.17. It follows from D.15 and D.16 that if X is a compact space, if Y is a Hausdorff space, and if $f \colon X \to Y$ is a continuous bijection, then f is a homeomorphism.

D.18. If X is a nonempty compact space, and if $f \colon X \to \mathbb{R}$ is continuous, then f is bounded and attains its supremum and infimum: there are points x_0 and x_1 in X such that $f(x_0) \le f(x) \le f(x_1)$ holds at each x in X.

D.19. Let $\{(X_\alpha, \mathcal{O}_\alpha)\}$ be an indexed family of topological spaces, and let $\prod_\alpha X_\alpha$ be the product of the corresponding indexed family of sets $\{X_\alpha\}$ (see A.5). The *product topology* on $\prod_\alpha X_\alpha$ is the weakest topology on $\prod_\alpha X_\alpha$ that makes each of the coordinate projections $\pi_\beta : \prod_\alpha X_\alpha \to X_\beta$ continuous (the projection π_β is defined by $\pi_\beta(x) = x_\beta$); see D.7. If $\mathcal{U}$ is the collection of sets that have the form $\prod_\alpha U_\alpha$ for some family $\{U_\alpha\}$ for which

(a) $U_\alpha \in \mathcal{O}_\alpha$ holds for each α and
(b) $U_\alpha = X_\alpha$ holds for all but finitely many values of α,

then $\mathcal{U}$ is a base for the product topology on $\prod_\alpha X_\alpha$.

D.20. (Tychonoff's Theorem) *Let $\{(X_\alpha, \mathcal{O}_\alpha)\}$ be an indexed collection of topological spaces. If each $(X_\alpha, \mathcal{O}_\alpha)$ is compact, then $\prod_\alpha X_\alpha$, with the product topology, is compact.*

D.21. Let X be a set. A collection $\mathcal{F}$ of functions on X *separates the points of X* if for each pair x, y of distinct points in X there is a function f in $\mathcal{F}$ such that $f(x) \neq f(y)$. A vector space $\mathcal{F}$ of real-valued functions on X is an *algebra* if fg belongs to $\mathcal{F}$ whenever f and g belong to $\mathcal{F}$ (here fg is the product of f and g, defined by $(fg)(x) = f(x)g(x)$). Now suppose that $\mathcal{F}$ is a vector space of bounded real-valued functions on X. A subset of $\mathcal{F}$ is *uniformly dense* in $\mathcal{F}$ if it is dense in $\mathcal{F}$ when $\mathcal{F}$ is given the topology induced by the uniform norm (see Example 3.2.1(f) in Sect. 3.2).

D.22. (Stone–Weierstrass Theorem) *Let X be a compact Hausdorff space. If A is an algebra of continuous real-valued functions on X that contains the constant functions and separates the points of X, then A is uniformly dense in the space $C(X)$ of continuous real-valued functions on X.*

D.23. (Stone–Weierstrass Theorem) *Let X be a locally compact[1] Hausdorff space, and let A be a subalgebra of $C_0(X)$ such that*

(a) *A separates the points of X, and*
(b) *for each x in X there is a function in A that does not vanish at x.*

Then A is uniformly dense in $C_0(X)$.

Theorem D.23 can be proved by applying Theorem D.22 to the one-point compactification of X.

D.24. Suppose that X is a set and that $\leq$ is a linear order on X. For each x in X define intervals $(-\infty, x)$ and $(x, +\infty)$ by

$$(-\infty, x) = \{z \in X : z < x\}$$

and

$$(x, +\infty) = \{z \in X : x < z\}.$$

---

[1]Locally compact spaces are defined in Sect. 7.1, and $C_0(X)$ is defined in Sect. 7.3.

The *order topology* on X is the weakest topology on X that contains all of these intervals. The set that consists of these intervals, the intervals of the form $\{z \in X : x < z < y\}$, and the set X, is a base for the order topology on X.

D.25. Let X be a set. A *metric* on X is a function $d : X \times X \to \mathbb{R}$ that satisfies

(a) $d(x,y) \geq 0$,
(b) $d(x,y) = 0$ if and only if $x = y$,
(c) $d(x,y) = d(y,x)$, and
(d) $d(x,z) \leq d(x,y) + d(y,z)$

for all x, y, and z in X. A *metric space* is a pair (X,d), where X is a set and d is a metric on X (of course, X itself is often called a metric space).

Let (X,d) be a metric space. If $x \in X$ and if r is a positive number, then the set $B(x,r)$ defined by

$$B(x,r) = \{y \in X : d(x,y) < r\}$$

is called the *open ball* with center x and radius r; the *closed ball* with center x and radius r is the set

$$\{y \in X : d(x,y) \leq r\}.$$

A subset U of X is *open* if for each x in U there is a positive number r such that $B(x,r) \subseteq U$. The collection of all open subsets of X is a topology on X; it is called the topology *induced* or *generated* by d.[2] The open balls form a base for this topology.

D.26. A topological space $(X, \mathcal{O})$ (or a topology $\mathcal{O}$) is *metrizable* if there is a metric d on X that generates the topology $\mathcal{O}$; the metric d is then said to *metrize* X (or $(X, \mathcal{O})$).

D.27. Let X be a metric space. The *diameter* of the subset A of X, written $\mathrm{diam}(A)$, is defined by

$$\mathrm{diam}(A) = \sup\{d(x,y) : x, y \in A\}.$$

The set A is *bounded* if $\mathrm{diam}(A)$ is not equal to $+\infty$. The *distance* between the point x and the nonempty subset A of X is defined by

$$d(x,A) = \inf\{d(x,y) : y \in A\}.$$

Note that if x_1 and x_2 are points in X, then

$$d(x_1,A) \leq d(x_1,x_2) + d(x_2,A).$$

Since we can interchange the points x_1 and x_2 in the formula above, we find that

$$|d(x_1,A) - d(x_2,A)| \leq d(x_1,x_2),$$

---

[2]When dealing with a metric space (X,d), we will often implicitly assume that X has been given the topology induced by d.

from which it follows that $x \mapsto d(x,A)$ is continuous (and, in fact, uniformly continuous).

D.28. Each closed subset of a metric space is a G_δ, and each open subset is an F_σ. To check the first of these claims, note that if C is a nonempty closed subset of the metric space X, then

$$C = \bigcap_n \left\{ x \in X : d(x,C) < \frac{1}{n} \right\},$$

and so C is the intersection of a sequence of open sets. Now use De Morgan's laws (see Sect. A.1) to check that each open set is an F_σ.

D.29. Let x and $x_1, x_2, \ldots$ belong to the metric space X. The sequence $\{x_n\}$ *converges* to x if $\lim_n d(x_n,x) = 0$; if $\{x_n\}$ converges to x, we say that x is the *limit* of $\{x_n\}$, and we write $x = \lim_n x_n$.

D.30. Let X be a metric space. It is easy to check that a point x in X belongs to the closure of the subset A of X if and only if there is a sequence in A that converges to x.

D.31. Let (X,d) and (Y,d') be metric spaces, and give X and Y the topologies induced by d and d' respectively. Then a function $f: X \to Y$ is continuous (in the sense of D.5) if and only if for each x_0 in X and each positive number ε there is a positive number δ such that $d'(f(x),f(x_0)) < \varepsilon$ holds whenever x belongs to X and satisfies $d(x,x_0) < \delta$. The observation at the end of C.7 generalizes to metric spaces, and a small modification of the argument given there yields the following characterization of continuity in terms of sequences: the function f is continuous if and only if $f(x) = \lim_n f(x_n)$ holds whenever x and $x_1, x_2, \ldots$ are points in X such that $x = \lim_n x_n$.

D.32. We noted in D.10 that every second countable topological space is separable. The converse holds for metrizable spaces: if d metrizes the topology of X, and if D is a countable dense subset of X, then the collection consisting of those open balls $B(x,r)$ for which $x \in D$ and r is rational is a countable base for X.

D.33. If X is a second countable topological space, and if Y is a subspace of X, then Y is second countable (if $\mathscr{U}$ is a countable base for X, then $\{U \cap Y : U \in \mathscr{U}\}$ is a countable base for Y). It follows from this, together with D.10 and D.32, that every subspace of a separable metrizable space is separable.

D.34. Let (X,d) be a metric space. A sequence $\{x_n\}$ of elements of X is a *Cauchy sequence* if for each positive number ε there is a positive integer N such that $d(x_m,x_n) < \varepsilon$ holds whenever $m \geq N$ and $n \geq N$. The metric space X is *complete* if every Cauchy sequence in X converges to an element of X.

D.35. (**Cantor's Nested Set Theorem**) *Let X be a complete metric space. If $\{A_n\}$ is a decreasing sequence of nonempty closed sets of X such that $\lim_n \mathrm{diam}(A_n) = 0$, then $\bigcap_{n=1}^\infty A_n$ contains exactly one point.*

Proof. For each positive integer n choose an element x_n of A_n. Then $\{x_n\}$ is a Cauchy sequence whose limit belongs to $\cap_{n=1}^{\infty} A_n$. Thus $\cap_{n=1}^{\infty} A_n$ is not empty. Since $\lim_n \text{diam}(A_n) = 0$, the set $\cap_{n=1}^{\infty} A_n$ cannot contain more than one point.   □

D.36. A subset A of a topological space X is *nowhere dense* if the interior of $\overline{A}$ is empty.

D.37. (Baire Category Theorem) *Let X be a nonempty complete metric space (or a nonempty topological space that can be metrized with a complete metric). Then X cannot be written as the union of a sequence of nowhere dense sets. Moreover, if $\{A_n\}$ is a sequence of nowhere dense subsets of X, then $(\cup_n A_n)^c$ is dense in X.*

D.38. The metric space (X,d) is *totally bounded* if for each positive ε there is a finite subset S of X such that

$$X = \bigcup\{B(x,\varepsilon) : x \in S\}.$$

D.39. (Theorem) *Let X be a metric space. Then the conditions*

(a) *the space X is compact,*
(b) *the space X is complete and totally bounded, and*
(c) *each sequence of elements of X has a subsequence that converges to an element of X*

are equivalent.

D.40. (Corollary) *Each compact metric space is separable.*

Proof. Let X be a compact metric space. Theorem D.39 implies that X is totally bounded, and so for each positive integer n we can choose a finite set S_n such that $X = \cup\{B(x, 1/n) : x \in S_n\}$. The set $\cup_n S_n$ is then a countable dense subset of X.   □

D.41. Note, however, that a compact Hausdorff space can fail to be second countable and can even fail to be separable (see Exercises 7.1.7, 7.1.8, and 7.1.10).

D.42. Let $\{X_n\}$ be a sequence of nonempty metrizable spaces, and for each n let d_n be a metric that metrizes X_n. Let $\mathbf{x}$ and $\mathbf{y}$ denote the points $\{x_n\}$ and $\{y_n\}$ of the product space $\prod_n X_n$. Then the formula

$$d(\mathbf{x},\mathbf{y}) = \sum_n \frac{1}{2^n} \min(1, d_n(x_n, y_n))$$

defines a metric on $\prod_n X_n$ that metrizes the product topology. This fact, together with Theorem D.39, can be used to give a fairly easy proof of Tychonoff's theorem for *countable* families of compact *metrizable* spaces.

Appendix E
The Bochner Integral

Let $(X, \mathscr{A})$ be a measurable space, let E be a real or complex Banach space (that is, a Banach space over $\mathbb{R}$ or $\mathbb{C}$), and let $\mathscr{B}(E)$ be the σ-algebra of Borel subsets of E (that is, let $\mathscr{B}(E)$ be the σ-algebra on E generated by the open subsets of E). We will sometimes denote the norm on E by $|\cdot|$, rather than by the more customary $\|\cdot\|$. This will allow us to use $\|\cdot\|$ for the norm of elements of certain spaces of E-valued functions; see, for example, formula (7) below. A function $f \colon X \to E$ is *Borel measurable* if it is measurable with respect to $\mathscr{A}$ and $\mathscr{B}(E)$, and is *strongly measurable* if it is Borel measurable and has a separable range (here by the range of f we mean the subset $f(X)$ of E). The function f is *simple* if it has only finitely many values. Of course, a simple function is strongly measurable if and only if it is Borel measurable.

It is easy to see that if f is Borel measurable, then $x \mapsto |f(x)|$ is $\mathscr{A}$-measurable (use Lemma 7.2.1 and Proposition 2.6.1).

Note that if E is separable, then every E-valued Borel measurable function is strongly measurable. On the other hand, if E is not separable and if $(X, \mathscr{A}) = (E, \mathscr{B}(E))$, then the identity map from X to E is Borel measurable, but is not strongly measurable.

E.1. (Proposition) *Let $(X, \mathscr{A})$ be a measurable space, and let E be a real or complex Banach space. Then*

(a) *the collection of Borel measurable functions from X to E is closed under the formation of pointwise limits, and*
(b) *the collection of strongly measurable functions from X to E is closed under the formation of pointwise limits.*

Proof. Part (a) is a special case of Proposition 8.1.10, and so we can turn to part (b).

Let $\{f_n\}$ be a sequence of strongly measurable functions from X to E, and suppose that $\{f_n\}$ converges pointwise to f. It follows from the separability of the sets $f_n(X)$, $n = 1, 2, \ldots$, that $\cup_n f_n(X)$ is separable, that the closure of $\cup_n f_n(X)$ is separable, and finally that $f(X)$ is separable (see D.33). Since f is Borel measurable (part (a)), the proof is complete. □

D.L. Cohn, *Measure Theory: Second Edition*, Birkhäuser Advanced Texts Basler Lehrbücher, DOI 10.1007/978-1-4614-6956-8,
© Springer Science+Business Media, LLC 2013

E.2. (Proposition) *Let $(X, \mathscr{A})$ be a measurable space, let E be a real or complex Banach space, and let $f: X \to E$ be strongly measurable. Then there is a sequence $\{f_n\}$ of strongly measurable simple functions such that*

$$f(x) = \lim_n f_n(x)$$

and

$$|f_n(x)| \leq |f(x)|, \text{ for } n = 1, 2, \ldots,$$

hold at each x in X.

Proof. We can certainly assume that $f(X)$ contains at least one nonzero element of E. Let C be a countable dense subset of $f(X)$, let $C^\sim$ be the set of rational multiples of elements of C, and let $\{y_n\}$ be an enumeration of $C^\sim$. We can assume that $y_1 = 0$. It is easy to check (do so) that

> for each y in $f(X)$ and each positive number ε there is a term
> y_m of $\{y_n\}$ that satisfies $|y_m| \leq |y|$ and $|y_m - y| < \varepsilon$.   (1)

For each x in X and each positive integer n define a subset $A_n(x)$ of E by

$$A_n(x) = \{y_j : j \leq n \text{ and } |y_j| \leq |f(x)|\}.$$

Since $y_1 = 0$, each $A_n(x)$ is nonempty.

We now construct the required sequence $\{f_n\}$ by letting $f_n(x)$ be the element of $A_n(x)$ that lies closest to $f(x)$ (in case

$$|f(x) - y_j| = \inf\{|f(x) - y_i| : y_i \in A_n(x)\}   (2)$$

holds for several elements y_j of $A_n(x)$, let $f_n(x)$ be y_{j_0}, where j_0 is the smallest value of j for which y_j belongs to $A_n(x)$ and satisfies (2)). It is clear that each f_n is a simple function and that $|f_n(x)| \leq |f(x)|$ holds for each n and x. Since the sets $\{x \in X : f_n(x) = y_j\}$ can be described by means of inequalities involving $|f(x)|$, $|y_i|$, $i = 1, \ldots, n$, and $|f(x) - y_i|$, $i = 1, \ldots, n$, each f_n is strongly measurable. Finally, observation (1) implies that $\{f_n\}$ converges pointwise to f (if y_m satisfies the inequalities $|y_m| \leq |f(x)|$ and $|y_m - f(x)| < \varepsilon$, then $|f_n(x) - f(x)| < \varepsilon$ holds whenever $n \geq m$).   □

Let us note two consequences of Propositions E.1 and E.2. The first is immediate: a function from X to E is strongly measurable if and only if it is the pointwise limit of a sequence of Borel (or strongly) measurable simple functions. The second is given by the following corollary (see, however, Exercise 2).

E.3. (Corollary) *Let $(X, \mathscr{A})$ be a measurable space, and let E be a real or complex Banach space. Then the set of all strongly measurable functions from X to E is a vector space.*

Proof. Suppose that f and g are strongly measurable and that a and b are real (or complex) numbers. Choose sequences $\{f_n\}$ and $\{g_n\}$ of strongly measurable simple functions that converge pointwise to f and g respectively (Proposition E.2). Since $\{af_n + bg_n\}$ converges pointwise to $af + bg$, and since each $af_n + bg_n$ is strongly measurable (it is simple and each of its values is attained on a measurable set), Proposition E.1 implies that $af + bg$ is strongly measurable.   □

We turn to the integration of functions with values in a Banach space. Let $(X, \mathscr{A}, \mu)$ be a measure space, and let E be a real or complex Banach space. A function $f : X \to E$ is *integrable* (or *strongly integrable*, or *Bochner integrable*) if it is strongly measurable and the function $x \to |f(x)|$ is integrable.[1]

The integral of such functions is defined as follows. First suppose that $f : X \to E$ is simple and integrable. Let $a_1, \ldots, a_n$ be the nonzero values of f, and suppose that these values are attained on the sets $A_1, \ldots, A_n$. Then Proposition 2.3.10, applied to the real-valued function $x \mapsto |f(x)|$, implies that each A_i has finite measure under μ. Thus the expression $\sum_{i=1}^{n} a_i \mu(A_i)$ makes sense; we define the *integral* of f, written $\int f \, d\mu$, to be this sum. It is easy to see that

$$\left| \int f \, d\mu \right| \leq \int |f| \, d\mu. \tag{3}$$

It is also easy to see that if f and g are simple integrable functions and a and b are real (or complex) numbers, then $af + bg$ is a simple integrable function, and

$$\int (af + bg) \, d\mu = a \int f \, d\mu + b \int g \, d\mu. \tag{4}$$

Now suppose that f is an arbitrary integrable function. Choose a sequence $\{f_n\}$ of simple integrable functions such that $f(x) = \lim_n f_n(x)$ holds at each x in X and such that the function $x \mapsto \sup_n |f_n(x)|$ is integrable (see Proposition E.2). The dominated convergence theorem for real-valued functions (Theorem 2.4.5) implies that $\lim_n \int |f_n - f| \, d\mu = 0$, and hence that $\lim_{m,n} \int |f_m - f_n| \, d\mu = 0$. Thus (see (3) and (4)) $\{\int f_n \, d\mu\}$ is a Cauchy sequence in E, and so is convergent. The *integral* (or *Bochner integral*) of f, written $\int f \, d\mu$, is defined to be the limit of the sequence $\{\int f_n \, d\mu\}$. (It is easy to check that the value of $\int f \, d\mu$ does not depend on the choice of the sequence $\{f_n\}$: if $\{g_n\}$ is another sequence having the properties required of $\{f_n\}$, then $\lim_n \int |f_n - g_n| \, d\mu = 0$, from which it follows that $\lim_n \int (f_n - g_n) \, d\mu = 0$ and hence that $\lim_n \int f_n \, d\mu = \lim_n \int g_n \, d\mu$.)

Let us note a few basic properties of the Bochner integral.

E.4. (Proposition) *Let $(X, \mathscr{A}, \mu)$ be a measure space, and let E be a real or complex Banach space. Suppose that $f, g : X \to E$ are integrable and that a and b are real (or complex) numbers. Then $af + bg$ is integrable, and*

---

[1] See Exercise 4 for an indication of another standard definition of Bochner integrability.

$$\int (af + bg)\, d\mu = a \int f\, d\mu + b \int g\, d\mu. \tag{5}$$

Proof. The integrability of $af + bg$ follows from Corollary E.3 and the inequality $|(af + bg)(x)| \leq |a|\,|f(x)| + |b|\,|g(x)|$. Let $\{f_n\}$ and $\{g_n\}$ be sequences of simple integrable functions that converge pointwise to f and g respectively and are such that $x \mapsto \sup_n |f_n(x)|$ and $x \mapsto \sup_n |g_n(x)|$ are integrable. Then the functions $af_n + bg_n$ are simple and integrable, and they satisfy

$$\int (af_n + bg_n)\, d\mu = a \int f_n\, d\mu + b \int g_n\, d\mu \tag{6}$$

(see (4)). Furthermore $x \mapsto \sup_n |(af_n + bg_n)(x)|$ is integrable, and so according to the definition of the integral, we can take limits in (6), obtaining (5).   □

E.5. (Proposition) *Let $(X, \mathscr{A}, \mu)$ be a measure space, and let E be a real or complex Banach space. If $f : X \to E$ is integrable, then $|\int f\, d\mu| \leq \int |f|\, d\mu$.*

Proof. Let f be an integrable function, and let $\{f_n\}$ be a sequence of simple integrable functions such that $\sup_n |f_n(x)| \leq |f(x)|$ and $f(x) = \lim_n f_n(x)$ hold at each x in X (Proposition E.2). Then

$$\left| \int f_n\, d\mu \right| \leq \int |f_n|\, d\mu \leq \int |f|\, d\mu$$

(see (3)); since $\int f\, d\mu = \lim_n \int f_n\, d\mu$, the proposition follows.   □

The dominated convergence theorem can be formulated as follows for E-valued functions.

E.6. (Theorem) *Let $(X, \mathscr{A}, \mu)$ be a measure space, let E be a real or complex Banach space, and let g be a $[0, +\infty]$-valued integrable function on X. Suppose that f and $f_1, f_2, \ldots$ are strongly measurable E-valued functions on X such that the relations*

$$f(x) = \lim_n f_n(x)$$

and

$$|f_n(x)| \leq g(x), \text{ for } n = 1, 2, \ldots,$$

hold at almost every x in X. Then f and $f_1, f_2, \ldots$ are integrable, and $\int f\, d\mu = \lim_n \int f_n\, d\mu$.

Proof. The integrability of f and $f_1, f_2, \ldots$ is immediate. Since $|f_n - f| \leq 2g$ holds almost everywhere, the dominated convergence theorem for real-valued functions (Theorem 2.4.5) implies that $\lim_n \int |f_n - f|\, d\mu = 0$. In view of Propositions E.4 and E.5, this implies that $\int f\, d\mu = \lim_n \int f_n\, d\mu$.   □

Let $\mathscr{L}^1(X, \mathscr{A}, \mu, E)$ be the set of all E-valued integrable functions on X. Then $\mathscr{L}^1(X, \mathscr{A}, \mu, E)$ is a vector space (see Proposition E.4). It is easy to check that the collection $L^1(X, \mathscr{A}, \mu, E)$ of equivalence classes (under almost everywhere equality) of elements of $\mathscr{L}^1(X, \mathscr{A}, \mu, E)$ can be made into a vector space in the natural way, and that the formula

$$\|f\|_1 = \int |f| \, d\mu \tag{7}$$

induces a norm on $L^1(X, \mathscr{A}, \mu, E)$ (and, of course, a seminorm on $\mathscr{L}^1(X, \mathscr{A}, \mu, E)$). The proof of Theorem 3.4.1 can be modified so as to show that $L^1(X, \mathscr{A}, \mu, E)$ is complete under $\| \cdot \|_1$.

One often finds it useful to be able to deal with vector-valued functions in terms of real- (or complex-) valued functions. For this we need to recall the Hahn–Banach theorem.

E.7. (Hahn–Banach Theorem) *Let E be a real or complex normed linear space, let F be a linear subspace of E, and let φ_0 be a continuous linear functional on F. Then there is a continuous linear functional φ on E such that $\|\varphi\| = \|\varphi_0\|$ and such that φ_0 is the restriction of φ to F. In other words, φ_0 can be extended to a continuous linear functional on all of E without increasing its norm.*

A proof of the Hahn–Banach theorem can be found in almost any basic text on functional analysis (see, for example, Conway [31], Kolmogorov and Fomin [73], Royden [102], or Simmons [109]).

We also need the following consequence of the Hahn–Banach theorem.

E.8. (Corollary) *Let E be a real or complex normed linear space that does not consist of 0 alone. Then for each y in E there is a continuous linear functional φ on E such that $\|\varphi\| = 1$ and $\varphi(y) = \|y\|$.*

Proof. Let y be a nonzero element of E, let F be the subspace of E consisting of all scalar multiples of y, and let φ_0 be the linear functional on F defined by $\varphi_0(ty) = t\|y\|$. Then φ_0 satisfies $\|\varphi_0\| = 1$ and $\varphi_0(y) = \|y\|$, and we can produce the required functional φ by applying Theorem E.7 to φ_0. (In case $y = 0$, let φ be an arbitrary linear functional on E that satisfies $\|\varphi\| = 1$.) $\square$

Let us now apply Theorem E.7 and Corollary E.8 to the study of vector-valued functions.

E.9. (Theorem) *Let $(X, \mathscr{A})$ be a measurable space, and let E be a real or complex Banach space. A function $f: X \to E$ is strongly measurable if and only if*

(a) *the image $f(X)$ of X under f is separable, and*
(b) *for each φ in E^* the function $\varphi \circ f$ is $\mathscr{A}$-measurable.*

We will use the following lemma in our proof of Theorem E.9.

E.10. (Lemma) *Let E be a separable normed linear space over $\mathbb{R}$ or $\mathbb{C}$. Then there is a sequence $\{\varphi_n\}$ of elements of E^* such that*

$$\|y\| = \sup\{|\varphi_n(y)| : n = 1, 2, \ldots\} \tag{8}$$

holds for each y in E.

Proof. We can assume that E does not consist of 0 alone. Choose a sequence $\{y_n\}$ whose terms form a dense subset of E. According to Corollary E.8, we can choose, for each n, an element φ_n of E^* that satisfies $\|\varphi_n\| = 1$ and $\varphi_n(y_n) = \|y_n\|$. Let us check that the sequence $\{\varphi_n\}$ meets the requirements of the lemma. Since each φ_n satisfies $\|\varphi_n\| = 1$, it follows that

$$\sup\{|\varphi_n(y)| : n = 1, 2, \dots\} \le \|y\|$$

holds for each y in E. For an arbitrary y in E we can find terms in the sequence $\{y_n\}$ that lie arbitrarily close to y, and so the calculations

$$\varphi_n(y) = \varphi_n(y - y_n) + \varphi_n(y_n) = \varphi_n(y - y_n) + \|y_n\|$$

and $|\varphi_n(y - y_n)| \le \|\varphi_n\| \|y - y_n\| = \|y - y_n\|$ imply that

$$\|y\| = \sup\{|\varphi_n(y)| : n = 1, 2, \dots\}.$$

Relation (8) follows.   □

Proof of Theorem E.9. Let us assume that we are dealing with Banach spaces over $\mathbb{R}$; the case of Banach spaces over $\mathbb{C}$ is similar.

If f is strongly measurable, then (a) is immediate and (b) follows from Lemma 7.2.1 and Proposition 2.6.1.

Now suppose that f satisfies (a) and (b). In view of (a), it suffices to show that f is Borel measurable. Let E_0 be the smallest closed linear subspace of E that includes $f(X)$. Then E_0 is separable (if C is a countable dense subset of $f(X)$, then E_0 is the closure of the set of finite sums of rational multiples of elements of C). We can show that f is Borel measurable (that is, measurable with respect to $\mathscr{A}$ and $\mathscr{B}(E)$) by showing that it is measurable with respect to $\mathscr{A}$ and $\mathscr{B}(E_0)$ (Lemma 7.2.2).

Let $\{\varphi_n\}$ be a sequence in $(E_0)^*$ such that

$$\|y\| = \sup\{|\varphi_n(y)| : n = 1, 2, \dots\} \tag{9}$$

holds for each y in E_0 (Lemma E.10). Since each continuous linear functional on E_0 is the restriction to E_0 of an element of E^* (Theorem E.7), condition (b) implies that for each n the function $\varphi_n \circ f$ is $\mathscr{A}$-measurable. If B is a closed ball in E_0, say with center y_0 and radius r, then $f^{-1}(B)$ is equal to

$$\bigcap_n \{x : |\varphi_n(f(x)) - \varphi_n(y_0)| \le r\},$$

and so belongs to $\mathscr{A}$. Since each open ball in E_0 is the union of a countable collection of closed balls, and since each open subset of E_0 is the union of a countable collection of open balls (recall that E_0 is separable), the collection of closed balls generates $\mathscr{B}(E_0)$. It now follows from Proposition 2.6.2 that f is measurable with respect to $\mathscr{A}$ and $\mathscr{B}(E_0)$ and the proof is complete.   □

E.11. (Proposition) *Let* $(X, \mathscr{A}, \mu)$ *be a measure space, let* E *be a real or complex Banach space, and let* $f: X \to E$ *be integrable. Then*

$$\int \varphi \circ f \, d\mu = \varphi \left(\int f \, d\mu \right) \tag{10}$$

holds for each φ *in* E^*.

The reader should see Exercise 3 for a strengthened form of Proposition E.11.

Proof. It is easy to check (do so) that the integrability of $\varphi \circ f$ follows from that of f. If f is a simple integrable function, attaining the nonzero values $a_1, \ldots, a_k$ on the sets $A_1, \ldots, A_k$, then each side of (10) is equal to $\sum_{i=1}^{k} \varphi(a_i)\mu(A_i)$; hence (10) holds for simple integrable functions. Next suppose that f is an arbitrary integrable function and that $\{f_n\}$ is a sequence of simple integrable functions such that $f(x) = \lim_n f_n(x)$ and $\sup_n |f_n(x)| \leq |f(x)|$ hold at each x in X (Proposition E.2). Then Theorems E.6 and 2.4.5 enable us to take limits in the relation $\int \varphi \circ f_n \, d\mu = \varphi(\int f_n \, d\mu)$, and (10) follows for arbitrary integrable functions.   □

The reader should note Exercises 5 and 7, which show some difficulties that arise in the extension of integration theory to vector-valued functions. The issues hinted at in these exercises have been the subject of much research over the years; see Diestel and Uhl [37] for a summary and for further references.

Exercises

1. Show that a simpler proof of Proposition E.2 could be given if the f_n's were not required to satisfy the inequality $|f_n(x)| \leq |f(x)|$.
2. Suppose that $(X, \mathscr{A})$ is a measurable space and that E is a Banach space. Show by example that the set of Borel measurable functions from X to E can fail to be a vector space. (Hint: Let E be a Banach space with cardinality greater than that of the continuum, and let $(X, \mathscr{A})$ be $(E \times E, \mathscr{B}(E) \times \mathscr{B}(E))$. See Exercise 5.1.8.)
3. Let $(X, \mathscr{A}, \mu)$ be a measure space, let E be a Banach space, and let $f: X \to E$ be Bochner integrable. Show that $\int f \, d\mu$ is the *only* element x_0 of E that satisfies $\varphi(x_0) = \int \varphi \circ f \, d\mu$ for each φ in E^*. (Hint: Use Corollary E.8.)
4. (This exercise hints at another, rather common, way to define strong measurability and Bochner measurability.) Suppose that $(X, \mathscr{A}, \mu)$ is a measure space and that E is a Banach space. Let $f: X \to E$ be a function for which there is a sequence $\{f_n\}$ of strongly measurable simple functions such that $f(x) = \lim_n f_n(x)$ holds at μ-almost every x in X.

   (a) Show by example that f need not have a separable range.
   (b) Show that there is a strongly measurable function $g: X \to E$ that agrees with f μ-almost everywhere.

(c) Show that $x \mapsto |f(x)|$ is measurable with respect to the completion $\mathscr{A}_\mu$ of $\mathscr{A}$ under μ.

(d) How should $\int f \, d\mu$ be defined if $\int |f| \, d\overline{\mu}$ is finite? (Of course $\overline{\mu}$ is the completion of μ.)

5. Let $(X, \mathscr{A})$ be a measurable space, and let E be a Banach space. An *E-valued measure* on $(X, \mathscr{A})$ is a function $v \colon \mathscr{A} \to E$ such that $v(\varnothing) = 0$ and such that $v(\cup_{i=1}^\infty A_i) = \sum_{i=1}^\infty v(A_i)$ holds for each infinite sequence $\{A_i\}$ of disjoint sets in $\mathscr{A}$. The *variation* $|v| \colon \mathscr{A} \to [0, +\infty]$ of the E-valued measure v is defined by letting $|v|(A)$ be the supremum of the sums $\sum_{i=1}^n |v(A_i)|$, where $\{A_i\}_{i=1}^n$ ranges over all finite partitions of A into $\mathscr{A}$-measurable sets.

(a) Show that the variation of an E-valued measure on $(X, \mathscr{A})$ is a positive measure on $(X, \mathscr{A})$.

(b) Show by example that the variation of an E-valued measure may not be finite. (Hint: Let X be $\mathbb{N}$, let $\mathscr{A}$ be $\mathscr{P}(\mathbb{N})$, let E be ℓ^2, and define $v \colon \mathscr{A} \to E$ by letting $v(A)$ be the sequence

$$n \mapsto \begin{cases} \frac{1}{n} & \text{if } n \in A, \\ 0 & \text{if } n \notin A.) \end{cases}$$

6. Let $(X, \mathscr{A}, \mu)$ be a measure space, let E be a Banach space, and let $f \colon X \to E$ be Bochner integrable. Define $v \colon \mathscr{A} \to E$ by $v(A) = \int \chi_A f \, d\mu$.

(a) Show that v is an E-valued measure on $(X, \mathscr{A})$.

(b) Show that the variation $|v|$ of v is finite.

7. Let λ be Lebesgue measure on $([0,1], \mathscr{B}([0,1]))$, and let E be the Banach space $L^1([0,1], \mathscr{B}([0,1]), \lambda, \mathbb{R})$. Define $v \colon \mathscr{B}([0,1]) \to E$ by letting $v(A)$ be the element of E determined by the characteristic function χ_A of A.

(a) Show that v is an E-valued measure on $([0,1], \mathscr{B}([0,1]))$.

(b) Show that $|v|$ is finite.

(c) Show that v is absolutely continuous with respect to λ (in other words, show that $v(A) = 0$ holds whenever A satisfies $\lambda(A) = 0$).

(d) Show that there is no Bochner integrable function $f \colon [0,1] \to E$ that satisfies $v(A) = \int \chi_A f \, d\lambda$ for each A in $\mathscr{B}([0,1])$. Thus the Radon–Nikodym theorem fails for the Bochner integral. (Hint: Use Proposition E.11.)

Appendix F
Liftings

Let $(X, \mathscr{A}, \mu)$ be a measure space. Throughout this appendix we will assume that the measure μ is finite but not the zero measure (see Exercise 2). Recall that $\mathscr{L}^\infty(X, \mathscr{A}, \mu, \mathbb{R})$ is the vector space of all bounded real-valued $\mathscr{A}$-measurable functions on X and that $L^\infty(X, \mathscr{A}, \mu, \mathbb{R})$ is the vector space of equivalence classes of functions in $\mathscr{L}^\infty(X, \mathscr{A}, \mu, \mathbb{R})$, where two functions are considered equivalent if they are equal μ-almost everywhere.[1] For simplicity, we will generally write $\mathscr{L}^\infty(X, \mathscr{A}, \mu)$, instead of $\mathscr{L}^\infty(X, \mathscr{A}, \mu, \mathbb{R})$. We will occasionally use the norm $\| \cdot \|_\infty$ on $\mathscr{L}^\infty(X, \mathscr{A}, \mu)$ defined by

$$\|f\|_\infty = \sup\{|f(x)| : x \in X\}.$$

Note that for this version of the norm $\| \cdot \|_\infty$ a function f satisfies $\|f\|_\infty = 0$ only if f vanishes *everywhere* on X; it is not enough for it to vanish almost everywhere.

It is natural to ask whether a function in $\mathscr{L}^\infty(X, \mathscr{A}, \mu)$ can be chosen from each equivalence class in $L^\infty(X, \mathscr{A}, \mu)$ in such a way the choice is linear and multiplicative. Since notation involving functions is simpler than notation involving equivalence classes, one generally deals with functions and makes the following definitions. A *lifting* of $\mathscr{L}^\infty(X, \mathscr{A}, \mu)$ is a function $\rho : \mathscr{L}^\infty(X, \mathscr{A}, \mu) \to \mathscr{L}^\infty(X, \mathscr{A}, \mu)$ such that for all f, g in $\mathscr{L}^\infty(X, \mathscr{A}, \mu)$ and all real numbers a and b we have

---

[1] In the present context (i.e., in cases where the measure μ is finite), it is the same to say that two functions agree almost everywhere as to say that they agree locally almost everywhere. Thus, for our current discussion the definition of $\mathscr{L}^\infty(X, \mathscr{A}, \mu, \mathbb{R})$ given here is consistent with the one in Chap. 4. We will use the current definition since it makes the exposition that follows simpler. If we were looking at liftings on very large measure spaces, we would speak of locally null sets and of equality locally almost everywhere; see [65].

D.L. Cohn, *Measure Theory: Second Edition*, Birkhäuser Advanced Texts Basler Lehrbücher, DOI 10.1007/978-1-4614-6956-8,
© Springer Science+Business Media, LLC 2013

(a) if $f = g$ almost everywhere, then[2] $\rho(f) = \rho(g)$,
(b) $\rho(f) = f$ almost everywhere,
(c) $\rho(af + bg) = a\rho(f) + b\rho(g)$,
(d) $\rho(fg) = \rho(f)\rho(g)$, and
(e) $\rho(1) = 1$.

Conditions (a) and (b) say that ρ can be interpreted as providing a choice of a function in $\mathscr{L}^\infty(X, \mathscr{A}, \mu)$ from each equivalence class in $L^\infty(X, \mathscr{A}, \mu)$.

The main theorem of this appendix (Theorem F.5, the lifting theorem) says that liftings of $\mathscr{L}^\infty(X, \mathscr{A}, \mu)$ exist, if the measure μ is complete.

If f is a nonnegative function in $\mathscr{L}^\infty(X, \mathscr{A}, \mu)$, then $\sqrt{f}$ also belongs to $\mathscr{L}^\infty(X, \mathscr{A}, \mu)$, and so $\rho(f) = \rho(\sqrt{f})\rho(\sqrt{f})$. It follows that

(f) if $f \geq 0$, then $\rho(f) \geq 0$.

A function $\rho : \mathscr{L}^\infty(X, \mathscr{A}, \mu) \to \mathscr{L}^\infty(X, \mathscr{A}, \mu)$ is called a *linear lifting* if it satisfies conditions (a), (b), (c), (e), and (f). We will encounter linear liftings while constructing liftings.

Recall that ℓ^∞ is the vector space of all bounded sequences of real numbers, with norm given by $\|\{x_n\}\|_\infty = \sup_n |x_n|$. Let c be the subspace of ℓ^∞ consisting of the sequences $\{x_n\}$ for which $\lim_n x_n$ exists; give c the norm it inherits from ℓ^∞.

F.1. (Lemma) *There is a linear functional $\Lambda : \ell^\infty \to \mathbb{R}$ such that*

(a) $\Lambda(\{x_n\}) = \lim_n x_n$ *for all $\{x_n\}$ in c,*
(b) $|\Lambda(\{x_n\})| \leq \|\{x_n\}\|_\infty$ *for all $\{x_n\}$ in ℓ^∞, and*
(c) $\Lambda(\{x_n\})$ *is positive, in the sense that $\Lambda(\{x_n\}) \geq 0$ whenever $\{x_n\}$ is a sequence in ℓ^∞ whose terms are nonnegative.*

In other words, if L is the linear functional defined on the subspace c of ℓ^∞ by $L(\{x_n\}) = \lim_n x_n$, then L can be extended to a linear functional on all of ℓ^∞ that has norm 1 and is positive.

Proof. As in the previous paragraph, define a linear functional L on c by $L(\{x_n\}) = \lim_n x_n$. Then L satisfies $|L(\{x_n\})| \leq \|\{x_n\}\|_\infty$ for all $\{x_n\}$ in c, and so the Hahn–Banach theorem (Theorem E.7 in Appendix E) gives a linear functional Λ on ℓ^∞ that satisfies conditions (a) and (b). If $\{x_n\}$ is a sequence in ℓ^∞ whose terms are nonnegative, and if $s = \sup_n x_n$, then

$$|\Lambda(\{x_n\}) - s/2| = |\Lambda(\{x_n - s/2\})| \leq \|\{x_n - s/2\}\|_\infty = s/2,$$

from which it follows that $\Lambda(\{x_n\}) \geq 0$.   □

---

[2]Note that when we say that two functions are equal, but don't give a qualification with the words "almost everywhere," then we are saying that the functions are identical. For example, condition (a) says that if $f(x) = g(x)$ for almost every x, then $\rho(f)(x) = \rho(g)(x)$ for *every* x.

Let C be a convex subset of a vector space E. An *extreme point* of C is a point x in C that cannot be written as a convex combination of points of C different from x. In other words, we are requiring that if $x = ty + (1-t)z$, where y and z belong to C and $0 < t < 1$, then $y = z = x$. More generally, an *extremal subset* of C is a nonempty subset C_0 of C such that if $x \in C_0$ and $x = ty + (1-t)z$, where y and z belong to C and $0 < t < 1$, then y and z belong to C_0. Thus a point x in C is an extreme point of C if and only if $\{x\}$ is an extremal subset of C.

As examples let us consider some subsets of $\mathbb{R}^2$. If C_1 is the disk defined by

$$C_1 = \{(x_1, x_2) \in \mathbb{R}^2 : x_1^2 + x_2^2 \le 1\},$$

then C_1 has infinitely many extreme points, namely the points on the circle that forms the boundary of C_1. On the other hand, if C_2 is the square defined by

$$C_2 = \{(x_1, x_2) \in \mathbb{R}^2 : -1 \le x_1 \le 1 \text{ and } -1 \le x_2 \le 1\},$$

then C_2 has only four extreme points, namely its corner points $(1,1)$, $(1,-1)$, $(-1,1)$, and $(-1,-1)$. The remaining boundary points of C_2 are not extreme points. The four line segments that make up the boundary of C_2 (that is, the line segments that join adjacent corners of C_2) are extremal subsets of C_2, as is the set that consists of all the boundary points of C_2. Finally, the open disk C_3 defined by

$$C_3 = \{(x_1, x_2) \in \mathbb{R}^2 : x_1^2 + x_2^2 < 1\}$$

is convex, but it has no extreme points.

We will need to know that certain sets have extreme points. If we assumed a substantial amount of functional analysis in the reader's background, we would simply appeal to the Krein–Milman theorem, which says that if K is a nonempty compact convex subset of a locally convex Hausdorff topological vector space, then K has extreme points and is in fact the smallest closed convex set that contains all the extreme points of K. However, all we need is given by the following lemma, which we can prove without too much work.

F.2. (Lemma) *Let S be a nonempty set and let E be the product space $\mathbb{R}^S$, considered as a vector space and as a topological space with the product topology. Then each nonempty compact convex subset of E has at least one extreme point.*

Proof. Let K be a nonempty compact convex subset of $\mathbb{R}^S$, and let $\mathscr{E}$ be the collection of all nonempty closed extremal subsets of K. Then $\mathscr{E}$ contains K, and so is nonempty. Let us view $\mathscr{E}$ as a partially ordered set, with $E_1 \le E_2$ holding if $E_2 \subseteq E_1$. (Be careful: sets that are larger with respect to the partial order $\le$ are smaller with respect to set inclusion.) We will use Zorn's lemma (see A.13) to get an element of $\mathscr{E}$ that is maximal with respect to $\le$ and hence minimal with respect to $\subseteq$. So suppose that $\mathscr{C}$ is a chain of elements of $\mathscr{E}$. The intersection of any finite subcollection of $\mathscr{C}$ belongs to $\mathscr{C}$ (it is a member of the subcollection), and so is nonempty. This, together with the compactness of K, implies that the intersection of all the members of $\mathscr{C}$ is nonempty (and, of course, closed). Furthermore, since each element of $\mathscr{C}$ is extremal, so is the intersection of the elements of $\mathscr{C}$. Thus $\mathscr{C}$, which was an arbitrary chain in $\mathscr{E}$, has an upper bound in $\mathscr{E}$. So we can apply Zorn's lemma, which gives a maximal element of $\mathscr{E}$, say E_0.

The maximality of E_0 says that E_0 has no subsets that belong to $\mathscr{E}$. What if E_0 contains more than one point? Each point in E_0 is a member of the product space $\mathbb{R}^S$ and so is a function from S to $\mathbb{R}$. If there are two members of E_0, say e_1 and e_2, then there must be a point s in S such that $e_1(s) \neq e_2(s)$. Let $m = \inf\{e(s) : e \in E_0\}$. Then, since E_0 is compact and the function $e \mapsto e(s)$ is continuous, the set

$$\{e \in E_0 : e(s) = m\}$$

is a proper subset of E_0 that is nonempty, closed, and extremal. This contradicts the maximality of E_0, and we conclude that E_0 can contain only one element, say e_0. It follows that e_0 is an extreme point of K.   □

The following two lemmas contain most of the technical details needed to prove the existence of liftings.

F.3. (Lemma) *Suppose that $(X, \mathscr{A}, \mu)$ is a probability[3] space, $\mathscr{A}_0$ is a sub-σ-algebra of $\mathscr{A}$, and ρ is a lifting of $\mathscr{L}^\infty(X, \mathscr{A}_0, \mu)$. If E_0 is a member of $\mathscr{A}$ that does not belong to $\mathscr{A}_0$, then ρ can be extended to a lifting of $\mathscr{L}^\infty(X, \sigma(\mathscr{A}_0 \cup \{E_0\}), \mu)$.*

Note the abuse of notation in the statement of Lemma F.3: μ first represents a measure on $\mathscr{A}$, then the restriction of that measure to the sub-σ-algebra $\mathscr{A}_0$, and finally the restriction of it to $\sigma(\mathscr{A}_0 \cup \{E_0\})$.

Proof. Recall that $\sigma(\mathscr{A}_0 \cup \{E_0\})$ consists of the sets of the form $(A \cap E_0) \cup (B \cap E_0^c)$, where A and B belong to $\mathscr{A}_0$ (see part (a) of Exercise 1.5.12), and that a function $f \colon X \to \mathbb{R}$ is $\sigma(\mathscr{A}_0 \cup \{E_0\})$-measurable if and only if there are $\mathscr{A}_0$-measurable real-valued functions f_0 and f_1 such that $f = f_0 \chi_{E_0} + f_1 \chi_{E_0^c}$ (see Exercise 2.1.9). It follows that the functions in $\mathscr{L}^\infty(X, \sigma(\mathscr{A}_0 \cup \{E_0\}), \mu)$ are those that have the form $f_0 \chi_{E_0} + f_1 \chi_{E_0^c}$ for some f_0, f_1 in $\mathscr{L}^\infty(X, \mathscr{A}_0, \mu)$.

Suppose that ρ_1 is a lifting of $\mathscr{L}^\infty(X, \sigma(\mathscr{A}_0 \cup \{E_0\}), \mu)$ that is an extension of ρ. Then there is a set E_1 in $\sigma(\mathscr{A}_0 \cup \{E_0\})$ such that $\rho_1(\chi_{E_0}) = \chi_{E_1}$ (see Exercise 3), and for each function of the form $f_0 \chi_{E_0} + f_1 \chi_{E_0^c}$, where f_0 and f_1 belong to $\mathscr{L}^\infty(X, \mathscr{A}_0, \mu)$, we have

$$\rho_1(f_0 \chi_{E_0} + f_1 \chi_{E_0^c}) = \rho_1(f_0)\rho_1(\chi_{E_0}) + \rho_1(f_1)\rho_1(\chi_{E_0^c})$$

$$= \rho(f_0)\chi_{E_1} + \rho(f_1)\chi_{E_1^c}.$$

---

[3] What we really want is for μ to be a finite measure such that $\mu(X) \neq 0$. It's easier, however, to say that we assume μ to be a probability measure, and if we prove our results for probability measures, we will also have proved them for all nonzero finite measures.

We need to construct such a lifting ρ_1; we'll do that by choosing a set E_1 in such a way that

$$\rho_1(f_0\chi_{E_0} + f_1\chi_{E_0^c}) = \rho(f_0)\chi_{E_1} + \rho(f_1)\chi_{E_1^c} \tag{1}$$

defines a lifting ρ_1 that is an extension of ρ.

Suppose that we produce a set E_1 in $\sigma(\mathscr{A}_0 \cup \{E_0\})$ such that

(a) $\chi_{E_0} = \chi_{E_1}$ a.e.,
(b) if functions f and f' in $\mathscr{L}^\infty(X, \mathscr{A}_0, \mu)$ agree almost everywhere on E_0, then $\rho(f)$ and $\rho(f')$ agree everywhere on E_1 (that is, if $(f - f')\chi_{E_0} = 0$ a.e., then $(\rho(f) - \rho(f'))\chi_{E_1} = 0$), and
(c) if functions f and f' in $\mathscr{L}^\infty(X, \mathscr{A}_0, \mu)$ agree almost everywhere on E_0^c, then $\rho(f)$ and $\rho(f')$ agree everywhere on E_1^c.

Then it follows that

$$\text{if } f_0\chi_{E_0} + f_1\chi_{E_0^c} = f_0'\chi_{E_0} + f_1'\chi_{E_0^c} \text{ a.e., then}$$
$$\rho(f_0)\chi_{E_1} + \rho(f_1)\chi_{E_1^c} = \rho(f_0')\chi_{E_1} + \rho(f_1')\chi_{E_1^c}$$

and

$$f_0\chi_{E_0} + f_1\chi_{E_0^c} = \rho(f_0)\chi_{E_1} + \rho(f_1)\chi_{E_1^c} \text{ a.e.}$$

This implies that Eq. (1) gives a well-defined function ρ_1 that is an extension of ρ and satisfies the first two conditions in the definition of a lifting. The remaining conditions (that ρ_1 is linear and multiplicative and that it satisfies $\rho_1(1) = 1$) are easy to check.

We turn to the construction of the set E_1. Choose a sequence $\{C_n\}$ of sets that belong to $\mathscr{A}_0$, satisfy $\chi_{C_n} \leq \chi_{E_0}$ a.e. for each n, and are such that

$$\sup_n \mu(C_n) = \sup\{\mu(C) : C \in \mathscr{A}_0 \text{ and } \chi_C \leq \chi_{E_0} \text{ a.e.}\};$$

then define a set F_1 by $F_1 = \cup_n C_n$. Then F_1 has maximal measure among the sets in $\mathscr{A}$ that are included (except perhaps for a null set) in E_0, and each $\mathscr{A}$-measurable set that is included (up to a null set) in E_0 is also included (up to a null set) in F_1. A similar construction produces an analogous set F_2 that is included (up to a null set) in E_0^c. Now let $G_1 = \rho(F_1)$ and $G_2 = \rho(F_2)$.

Claim. The sets G_1 and G_2 satisfy

$$G_1 \cap G_2 = \varnothing, \tag{2}$$

$$\mu(G_1 - E_0) = 0 \quad \text{(that is, } G_1 \subseteq E_0 \text{ to within a null set), and} \tag{3}$$

$$\mu(G_2 - E_0^c) = 0 \quad \text{(that is, } G_2 \subseteq E_0^c \text{ to within a null set).} \tag{4}$$

For (2), note that $\chi_{F_1}\chi_{F_2} = 0$ a.e., which implies that $\chi_{G_1}\chi_{G_2} = \rho(\chi_{F_1}\chi_{F_2}) = 0$. Relation (3) follows from the fact that $\chi_{G_1} = \chi_{F_1}$ a.e. and $\chi_{F_1} \leq \chi_{E_0}$ a.e., and (4) has a similar proof.

Now define E_1 by $E_1 = (E_0 \cup G_1) \cap G_2^c$. Then condition (a) above follows from (2)–(4). We turn to condition (b). Suppose that f and f' belong to $\mathscr{L}^\infty(X, \mathscr{A}_0, \mu)$ and agree almost everywhere on E_0. We need to show that $\rho(f) = \rho(f')$ on E_1. Let $D = \{x \in X : f(x) \neq f'(x)\}$. Then $\chi_D \leq \chi_{F_2}$ a.e., and so $\rho(\chi_D) \leq \rho(\chi_{F_2}) = \chi_{G_2}$. Since D was defined so that $(f - f')\chi_{D^c} = 0$, we have $(\rho(f) - \rho(f'))\rho(\chi_{D^c}) = 0$. It follows that $(\rho(f) - \rho(f'))\chi_{G_2^c} = 0$, and so $\rho(f)$ and $\rho(f')$ agree everywhere outside G_2 and hence on E_1. This completes the proof of (b). The proof of (c) is similar, and with that the lemma is proved.   □

F.4. (Lemma) *Suppose that $(X, \mathscr{A}, \mu)$ is a complete probability space and that L_0 is a linear lifting of $\mathscr{L}^\infty(X, \mathscr{A}, \mu)$. Then there is a lifting ρ of $\mathscr{L}^\infty(X, \mathscr{A}, \mu)$ such that*

$$\chi_{\{L_0(\chi_A)=1\}} \leq \rho(\chi_A) \leq \chi_{\{L_0(\chi_A)>0\}} \tag{5}$$

holds for each A in $\mathscr{A}$.

The significance of (5) will become clear when we use Lemma F.4 to prove Theorem F.5.

Proof. Let S be the Cartesian product $\mathscr{L}^\infty(X, \mathscr{A}, \mu) \times X$. We will identify linear liftings of $\mathscr{L}^\infty(X, \mathscr{A}, \mu)$ with functions from S to $\mathbb{R}$, that is, we will identify a linear lifting L of $\mathscr{L}^\infty(X, \mathscr{A}, \mu)$ with the function $L' : S \to \mathbb{R}$ defined by $L'(f, x) = L(f)(x)$. Thus we will view linear liftings as members of the product space $\mathbb{R}^S$. The plan for the current proof is to define a certain subset C of $\mathbb{R}^S$, to show that C is nonempty, compact, and convex, and then to show that the extreme points of C (which exist, according to Lemma F.2) are liftings that satisfy (5). That will complete the proof of the lemma.

Let us look at how the conditions defining liftings and linear liftings translate into conditions on elements of $\mathbb{R}^S$. For example, the condition that L satisfies $L(af + bg) = aL(f) + bL(g)$ for all a, b, f, and g becomes the condition that the corresponding function L' satisfies

$$L'(af + bg, x) = aL'(f, x) + bL'(g, x) \text{ for all } a, b, f, g, \text{ and } x. \tag{6}$$

Note also that, since all the coordinate projections $L' \mapsto L'(f, x)$ of $\mathbb{R}^S$ are continuous, those elements of $\mathbb{R}^S$ that satisfy (6) form a closed subset of $\mathbb{R}^S$.

We now define the set C to be the collection of all L' in $\mathbb{R}^S$ that satisfy the translations into conditions on L' of conditions (a), (c), (e), and (f) in the definition of a linear lifting, plus the translation of the relation

$$\chi_{\{L_0(\chi_A)=1\}} \leq L(\chi_A) \leq \chi_{\{L_0(\chi_A)>0\}}, \tag{7}$$

which is to hold for all A in $\mathcal{A}$. It is easy to check that C is closed and convex. Furthermore, conditions (c), (e), and (f) imply that

$$|L'(f,x)| \leq \|f\|_\infty \tag{8}$$

holds for all L' in C and all f and x; hence we can use Tychonoff's theorem to conclude that C is compact. Finally, the function in $\mathbb{R}^S$ that corresponds to L_0 belongs to C, and so C is nonempty.

It now follows from Lemma F.2 that C has at least one extreme point, say L_1'. Let us reverse our translation from linear liftings to elements of $\mathbb{R}^S$, and let L_1 be the function from $\mathscr{L}^\infty(X,\mathcal{A},\mu)$ to functions[4] on X that corresponds to the extreme point L_1'. We need to show that $L_1(f)$ is measurable and bounded, that $f = L_1(f)$ a.e., and that $L_1(fg) = L_1(f)L_1(g)$; the other conditions that L_1 must satisfy to be a lifting come from the conditions we placed on C.

It follows from (8) that $L_1(f)$ is bounded, and in fact that $\|L_1(f)\|_\infty \leq \|f\|_\infty$. We turn to the measurability of $L_1(f)$ and the requirement that $f = L_1(f)$ a.e. If we use (7), plus the fact that $f = L_0(f)$ a.e. (recall that L_0 is a linear lifting), we find that each f of the form χ_A satisfies $f = L_1(f)$ a.e. Since μ is complete, the measurability of $L_1(f)$ follows for such f. The measurability of $L_1(f)$ and the almost everywhere validity of $f = L_1(f)$ now follow first for simple $\mathcal{A}$-measurable functions and then for arbitrary $\mathcal{A}$-measurable functions (approximate an arbitrary function with simple functions, and use (8)).

We still need to show that L_1 is multiplicative,[5] in the sense that $L_1(fg) = L_1(f)L_1(g)$ holds for all f and g. It is easy to see that we only need to check the identity $L_1(fg) = L_1(f)L_1(g)$ in the case where $0 \leq g \leq 1$ (use the linearity of L_1 and the fact that $L_1(1) = 1$). So assume that g belongs to $\mathscr{L}^\infty(X,\mathcal{A},\mu)$ and satisfies $0 \leq g \leq 1$, and define functions $L_{1+}, L_{1-} : \mathscr{L}^\infty(X,\mathcal{A},\mu) \to \mathscr{L}^\infty(X,\mathcal{A},\mu)$ by

$$L_{1+}(f) = L_1(f) + (L_1(fg) - L_1(f)L_1(g)) \text{ and}$$

$$L_{1-}(f) = L_1(f) - (L_1(fg) - L_1(f)L_1(g)).$$

It is easy to check that L_{1+} and L_{1-} are linear liftings. We want to verify that they correspond to members of C, and for this we need to check that they satisfy (7). The keys to this will be the fact that $L_1 = \frac{1}{2}L_{1+} + \frac{1}{2}L_{1-}$, together with the fact that if $A \in \mathcal{A}$, then (since L_{1+} and L_{1-} are linear liftings) the values of the functions $L_{1+}(\chi_A)$ and $L_{1-}(\chi_A)$ belong to the interval $[0,1]$. Since L_1 corresponds to an element of C, it satisfies (7); thus if $L_0(\chi_A)(x) = 1$, then we can conclude that $L_1(\chi_A)(x) = 1$ (use (7)) and then that $L_{1+}(\chi_A)(x) = L_{1-}(\chi_A)(x) = 1$ (use the fact that $L_1 = \frac{1}{2}L_{1+} + \frac{1}{2}L_{1-}$, plus the fact that the values of $L_{1+}(\chi_A)$ and

---

[4]We cannot yet say "from $\mathscr{L}^\infty(X,\mathcal{A},\mu)$ to $\mathscr{L}^\infty(X,\mathcal{A},\mu)$," because we still need to verify that the functions $x \mapsto L_1(f,x)$ are measurable and bounded.

[5]Here is where we use the fact that L_1 corresponds to an extreme point in C.

$L_{1-}(\chi_A)$ belong to $[0,1]$). Likewise, if $L_0(\chi_A)(x) = 0$, then $L_1(\chi_A)(x) = 0$ and so $L_{1+}(\chi_A)(x) = L_{1-}(\chi_A)(x) = 0$. It follows that L_{1+} and L_{1-} satisfy (7) and so correspond to elements of C. However, L_1 corresponds to an extreme point of C and satisfies $L_1 = \frac{1}{2}L_{1+} + \frac{1}{2}L_{1-}$, and so we have $L_1 = L_{1+} = L_{1-}$. But this implies that $L_1(fg) - L_1(f)L_1(g) = 0$, and the multiplicativity of L_1 follows. Thus L_1 is a lifting that satisfies (5), and the proof is complete. $\square$

F.5. (Theorem) *If $(X, \mathscr{A}, \mu)$ is a complete probability space, then there is a lifting of $\mathscr{L}^\infty(X, \mathscr{A}, \mu)$.*

Proof. Let $\mathscr{T}$ be the collection of all pairs $(\mathscr{B}, \rho)$, where $\mathscr{B}$ is a sub-σ-algebra of $\mathscr{A}$ that contains all the μ-null sets in $\mathscr{A}$ and where ρ is a lifting of $\mathscr{L}^\infty(X, \mathscr{B}, \mu)$. (Of course, by $\mathscr{L}^\infty(X, \mathscr{B}, \mu)$ we really mean $\mathscr{L}^\infty(X, \mathscr{B}, \mu_{\mathscr{B}})$, where $\mu_{\mathscr{B}}$ is the restriction of μ to the sub-σ-algebra $\mathscr{B}$ of $\mathscr{A}$. Such abuse of notation will occur often in this proof.) Let us define a relation $\leq$ on $\mathscr{T}$ by defining $(\mathscr{B}_1, \rho_1) \leq (\mathscr{B}_2, \rho_2)$ to mean that $\mathscr{B}_1 \subseteq \mathscr{B}_2$ and ρ_1 is the restriction of ρ_2 to $\mathscr{L}^\infty(X, \mathscr{B}_1, \mu)$. Then $\leq$ is a partial order on $\mathscr{T}$.

We'll check that $\mathscr{T}$ is nonempty and that each chain in $\mathscr{T}$ has an upper bound in $\mathscr{T}$, and so Zorn's lemma will provide a maximal element $(\mathscr{B}', \rho')$ of $\mathscr{T}$. Then $\mathscr{B}'$ must be equal to $\mathscr{A}$ (and the proof will be complete), since otherwise Lemma F.3 would provide an extension of ρ' to $\mathscr{L}^\infty(X, \mathscr{B}'', \mu)$ for some still larger sub-σ-algebra $\mathscr{B}''$ of $\mathscr{A}$, and $(\mathscr{B}', \rho')$ would not be maximal.

We turn to the details. First let us check that $\mathscr{T}$ is nonempty. Let $\mathscr{B}_0$ be the collection of μ-null sets in $\mathscr{A}$, together with their complements. Then $\mathscr{B}_0$ is a σ-algebra, $\mathscr{L}^\infty(X, \mathscr{B}_0, \mu)$ consists of the bounded measurable functions that are almost everywhere constant, and the operator that assigns to each such function f the constant function that is almost everywhere equal to f is a lifting.

Next suppose that $\mathscr{C}$ is a chain in $\mathscr{T}$; we will produce an upper bound for $\mathscr{C}$. Let us consider two cases.

In the first case there is an increasing sequence $\{(B_n, \rho_n)\}_{n=1}^\infty$ in $\mathscr{C}$ that is *cofinal*, in the sense that for every $(\mathscr{B}, \rho)$ in $\mathscr{C}$ there is an n such that $(\mathscr{B}, \rho) \leq (\mathscr{B}_n, \rho_n)$. Let us construct an upper bound $(\mathscr{B}_\infty, \rho_\infty)$ of $\mathscr{C}$. We'll use conditional expectations and the martingale convergence theorem (see Sect. 10.4) to do so. Define $\mathscr{B}_\infty$ by $\mathscr{B}_\infty = \sigma(\cup_n \mathscr{B}_n)$. Choose a linear functional Λ on ℓ^∞ as given by Lemma F.1, and note that for each x in X the sequence $\{\rho_n(E(f|\mathscr{B}_n))(x)\}$ belongs to ℓ^∞ (of course, $E(f|\mathscr{B}_n)$ is only determined up to a null set, but then ρ_n, as a lifting, gives the same result whatever version of $E(f|\mathscr{B}_n)$ is used). Thus we can define an operator L on $\mathscr{L}^\infty(X, \mathscr{B}_\infty, \mu)$ by $L(f)(x) = \Lambda(\{\rho_n(E(f|\mathscr{B}_n))(x)\})$. It follows from Proposition 10.4.12 that if $f \in \mathscr{L}^\infty(X, \mathscr{B}_\infty, \mu)$, then the sequence $\{E(f|\mathscr{B}_n)\}$ converges almost everywhere to f. Hence $\{\rho_n(E(f|\mathscr{B}_n))\}$ also converges almost everywhere to f and so $L(f) = f$ a.e. In particular, since μ is complete, $L(f)$ is measurable. It is easy to check that L is a linear lifting that extends each lifting ρ_n. Now use Lemma F.4 to get a lifting ρ_∞ that satisfies

$$\chi_{\{L(\chi_A)=1\}} \le \rho_\infty(\chi_A) \le \chi_{\{L(\chi_A)>0\}} \tag{9}$$

for every A in $\mathscr{B}_\infty$. If $A \in \mathscr{B}_n$, then, since L is an extension of ρ_n and since 0 and 1 are the only possible values for the function $\rho_n(\chi_A)$, we have $\{L(\chi_A) = 1\} = \{L(\chi_A) > 0\}$. It now follows from (9) that $\rho_\infty(\chi_A) = L(\chi_A) = \rho_n(\chi_A)$, and so ρ_∞ is an extension of ρ_n (to check this, approximate functions in $\mathscr{L}^\infty(X, \mathscr{B}_\infty, \mu)$ with simple functions—see the proof of Lemma F.4). Thus we have an upper bound $(\mathscr{B}_\infty, \rho_\infty)$ for the chain $\mathscr{C}$.

Finally, we need to produce an upper bound for the chain $\mathscr{C}$ in the case where $\mathscr{C}$ has no cofinal sequences. Suppose that $\mathscr{C}$ is the family $\{(\mathscr{B}_\alpha, \rho_\alpha)\}_\alpha$, where α ranges over some index set. Then $\cup_\alpha \mathscr{B}_\alpha$ is a σ-algebra and $\mathscr{L}^\infty(X, \cup_\alpha \mathscr{B}_\alpha, \mu) = \cup_\alpha \mathscr{L}^\infty(X, \mathscr{B}_\alpha, \mu)$ (see Exercise 5). We can define a lifting ρ on $\mathscr{L}^\infty(X, \cup_\alpha \mathscr{B}_\alpha, \mu)$ by letting $\rho(f)$ be $\rho_\alpha(f)$, where α is an index such that $f \in \mathscr{L}^\infty(X, \mathscr{B}_\alpha, \mu)$ (the index α depends, of course, on f). With this we have an upper bound for the chain $\mathscr{C}$, and the proof is complete.   □

Exercises

1. Let $X = \{1, 2, 3\}$, let $\mathscr{A}$ be the set of all subsets of X, and let μ be the measure on $(X, \mathscr{A})$ defined by $\mu = \frac{1}{3}\delta_1 + \frac{2}{3}\delta_2$.
   (a) Find a lifting of $\mathscr{L}^\infty(X, \mathscr{A}, \mu)$.
   (b) Find all liftings of $\mathscr{L}^\infty(X, \mathscr{A}, \mu)$.
2. Suppose that $(X, \mathscr{A}, \mu)$ is a measure space such that X is nonempty but $\mu(X) = 0$. Show that there are no liftings of $\mathscr{L}^\infty(X, \mathscr{A}, \mu)$.
3. Suppose that ρ is a lifting of $\mathscr{L}^\infty(X, \mathscr{A}, \mu)$. Show that if $E \in \mathscr{A}$, then there is a set E' in $\mathscr{A}$ such that $\rho(\chi_E) = \chi_{E'}$ and $\mu(E \triangle E') = 0$.
4. Let $(X, \mathscr{A}, \mu)$ be a measure space. A function $\rho' : \mathscr{A} \to \mathscr{A}$ is a *lifting* of $\mathscr{A}$ if

   (i) $\rho'(A) = \rho'(B)$ whenever $\mu(A \triangle B) = 0$,
   (ii) $\mu(A \triangle \rho'(A)) = 0$ for all A in $\mathscr{A}$,
   (iii) $\rho'(\varnothing) = \varnothing$ and $\rho'(X) = X$,
   (iv) $\rho'(A \cup B) = \rho'(A) \cup \rho'(B)$ for all A and B in $\mathscr{A}$, and
   (v) $\rho'(A \cap B) = \rho'(A) \cap \rho'(B)$ for all A and B in $\mathscr{A}$.

   Suppose that for each lifting ρ of $\mathscr{L}^\infty(X, \mathscr{A}, \mu)$ we define a function $\rho' : \mathscr{A} \to \mathscr{A}$ by $\chi_{\rho'(A)} = \rho(\chi_A)$. Show that $\rho \mapsto \rho'$ is a bijection of the set of all liftings of $\mathscr{L}^\infty(X, \mathscr{A}, \mu)$ onto the set of all liftings of $\mathscr{A}$.
5. Let $(X, \mathscr{A}, \mu)$ be a measure space and let $\{\mathscr{B}_\alpha\}_\alpha$ be a linearly ordered family of sub-σ-algebras of $\mathscr{A}$. Suppose that for each countable subfamily $\{\mathscr{B}_{\alpha_n}\}_n$ of $\{\mathscr{B}_\alpha\}_\alpha$ there is an element $\mathscr{B}_{\alpha'}$ of $\{\mathscr{B}_\alpha\}_\alpha$ such that $\mathscr{B}_{\alpha_n} \subseteq \mathscr{B}_{\alpha'}$ holds for every n. Show that

(a) $\sigma(\cup_\alpha \mathscr{B}_\alpha) = \cup_\alpha \mathscr{B}_\alpha$, and
(b) $\mathscr{L}^\infty(X, \sigma(\cup_\alpha \mathscr{B}_\alpha), \mu) = \cup_\alpha \mathscr{L}^\infty(X, \mathscr{B}_\alpha, \mu)$.

6. In this exercise we look at a proof of the existence of liftings in the particular case of $\mathscr{L}^\infty([0,1], \mathscr{A}, \lambda)$, where $\mathscr{A}$ is the σ-algebra of Lebesgue measurable subsets of $[0,1]$. The proof outlined here has the advantage that it is simpler than the one given above and relates liftings to differentiation theory. However, it depends on a basic but nontrivial result about Banach algebras that is quoted below but not proved, and it only gives liftings in the case of certain measure spaces.

Let A be a commutative Banach algebra (see Sect. 9.4). We assume that A has a multiplicative identity element 1 that satisfies $\|1\| = 1$. Recall that an *ideal* in A is a subset I of A that is a vector subspace of A, is a proper subset of A, and is such that $xy \in I$ whenever $x \in A$ and $y \in I$. A *maximal ideal* is an ideal that is included in no larger ideal. We will be looking at Banach algebras over the field $\mathbb{C}$, because complex-variable techniques are used in the proof of the result we quote below. We will assume that the Banach algebras that we consider have an *involution* $x \mapsto x^*$ that satisfies

(i) $(x+y)^* = x^* + y^*$,
(ii) $(xy)^* = x^* y^*$,
(iii) $(\alpha x)^* = \overline{\alpha} x^*$ (where $\overline{\alpha}$ is the complex conjugate of α), and
(iv) $x^{**} = x$

for all x and y in A and all α in $\mathbb{C}$. In the case where $A = \mathscr{L}^\infty([0,1], \mathscr{A}, \lambda, \mathbb{C})$, the operator that takes a function f to the complex conjugate of f is an involution (in fact, it is the only involution we will need to consider).

The result we need to quote says that if A is a Banach algebra over $\mathbb{C}$ that has an involution, and if M is a maximal ideal in A, then there is a linear functional ϕ on A such that

(i) $\|\phi\| \leq 1$,
(ii) $\phi(xy) = \phi(x)\phi(y)$ holds for all x, y in A,
(iii) $\phi(1) = 1$,
(iv) $\phi(x^*) = \overline{\phi(x)}$ holds for all x in A, and
(v) $M = \{x \in A : \phi(x) = 0\}$.

(see Simmons [109, Chapters 12 and 13], Hewitt and Ross [58, Appendix C], or Lax [82, Chapters 18 and 19]).

(a) Let A be the Banach algebra $\mathscr{L}^\infty([0,1], \mathscr{A}, \lambda, \mathbb{C})$. For each t in $[0,1]$ let I_t be the subset of A consisting of those functions f such that $F'(t)$ exists and is equal to 0, where F is the function defined by $F(u) = \int_0^u |f(s)|\, ds$ (note the absolute value signs around $f(s)$). Show that I_t is an ideal in A.

(b) Show that for each t there is a maximal ideal M_t in A that includes I_t. (Hint: Use Zorn's lemma.)

(c) Suppose that for each t we apply the result quoted above to the maximal ideal M_t, thereby producing a family of function $\{\phi_t\}_t$. Show that if f is a real-valued function in $\mathscr{L}^\infty([0,1], \mathscr{A}, \lambda, \mathbb{C})$, then for each t the value $\phi_t(f)$ is a real number.

(d) Define an operator ρ on $\mathscr{L}^\infty([0,1],\mathscr{A},\lambda,\mathbb{R})$ by $\rho(f)(t) = \phi_t(f)$. Show that for each f the function $\rho(f)$ is bounded and measurable, and moreover that ρ is a lifting of $\mathscr{L}^\infty([0,1],\mathscr{A},\lambda,\mathbb{R})$.

Notes

The existence of liftings was first proved by von Neumann [119] and by Maharam [87]. In the 1960s A. and C. Ionescu Tulcea were very active in studying liftings; see [64, 65]. The paper by Strauss et al. [115] surveys much more recent work.

Appendix G
The Banach–Tarski Paradox

The usual informal statement of the Banach–Tarski paradox is as follows:

> A pea can be divided into a finite number of pieces, and these pieces, after being moved by rigid motions, can be reassembled in such a way as to produce the sun.

For a more precise statement, let us replace the pea and the sun with subsets P and S of $\mathbb{R}^3$ that are bounded and have nonempty interiors. Then the Banach–Tarski paradox says that there exist a positive integer n, disjoint subsets $A_1, A_2, \ldots, A_n$ of P, and disjoint subsets $B_1, B_2, \ldots, B_n$ of S such that

(a) $P = A_1 \cup A_2 \cup \cdots \cup A_n$,
(b) $S = B_1 \cup B_2 \cup \cdots \cup B_n$, and
(c) for each i there is a rigid motion of $\mathbb{R}^3$ that maps A_i onto B_i.

There are a couple of things to note here. First, this paradox depends on the axiom of choice, and so the sets $A_1, \ldots$ and $B_1, \ldots$ are produced in a very nonconstructive way. Second, the Banach–Tarski paradox implies that there is no way to extend Lebesgue measure to the collection of all subsets of $\mathbb{R}^3$ in such a way that the extension is invariant under rigid motions and is at least finitely additive.

Let us turn to the mathematical concepts that we need for a proof of the Banach–Tarski paradox. Let G be a group and let X be a nonempty set. Suppose (for definiteness) that the group operation on G is written multiplicatively and that e is the identity element of G. An *action* of G on X is a mapping $(g, x) \mapsto g \cdot x$ of $G \times X$ to X that satisfies

(a) $g_1 \cdot (g_2 \cdot x) = (g_1 g_2) \cdot x$ and
(b) $e \cdot x = x$

for all g_1, g_2 in G and all x in X. We often abbreviate $g \cdot x$ with gx. One sometimes says that G *acts* on X when we are dealing with an action of G on X.

If G acts on X, if $g \in G$, and if A is a subset of X, then gA or $g \cdot A$ is the set $\{y \in X : y = g \cdot a \text{ for some } a \text{ in } A\}$. Likewise, if H is a subset of G and A is a subset of X, then $H \cdot A$ is the set $\{y \in X : y = h \cdot a \text{ for some } h \text{ in } H \text{ and some } a \text{ in } A\}$.

D.L. Cohn, *Measure Theory: Second Edition*, Birkhäuser Advanced
Texts Basler Lehrbücher, DOI 10.1007/978-1-4614-6956-8,
© Springer Science+Business Media, LLC 2013

G.1. (Examples)

(a) Let d be a positive integer and let G be a subgroup of the group of all invertible d by d matrices. For S in G and x in $\mathbb{R}^d$ let Sx be the usual product of the matrix S and the vector x, where x is regarded as a column vector. Then $(S,x) \mapsto Sx$ gives an action of G on $\mathbb{R}^d$.

(b) Recall that a d by d matrix $S = (s_{ij})$ is *orthogonal* if its columns are orthogonal to one another and have norm 1 (with respect to the usual Euclidean norm $\|\cdot\|_2$). In other words, S is orthogonal if $\sum_i s_{ij} s_{ik}$ is 1 if $j = k$ and is 0 if $j \neq k$. The set of all d by d orthogonal matrices with determinant 1 is a group, which is called the *special orthogonal group* and is denoted by $SO(d)$. Such groups are, of course, groups of the sort described in the previous example.

(c) Now let G_3 be the set of all rigid motions $T \colon \mathbb{R}^3 \to \mathbb{R}^3$ of the form $T(x) = Sx + b$, where $S \in SO(3)$ and $b \in \mathbb{R}^3$. Thus G_3 is a group; it acts on $\mathbb{R}^3$ by $(T,x) \mapsto T(x)$.

(d) Let G be an arbitrary group. Then $(g,g') \mapsto g \cdot g'$, where $\cdot$ is the group operation of G, gives an action of G on G.

Equidecomposability

Now suppose that G acts on the set X and that A and B are subsets of X. Then A and B are called *G-equidecomposable* (or simply *equidecomposable*), or A is said to be *G-equidecomposable with* B if there exist a positive integer n, disjoint subsets A_1, $\ldots$, A_n of A, disjoint subsets B_1, $\ldots$, B_n of B, and elements g_1, $\ldots$, g_n of G such that

(a) $A = A_1 \cup A_2 \cup \cdots \cup A_n$,

(b) $B = B_1 \cup B_2 \cup \cdots \cup B_n$, and

(c) $B_i = g_i \cdot A_i$ holds for each i.

Thus A and B are G-equidecomposable if and only if there is a bijection $f \colon A \to B$ that is *defined piecewise*[1] by the action of G on X—that is, for which there are disjoint subsets A_1, $\ldots$, A_n of A that satisfy $A = A_1 \cup A_2 \cup \cdots \cup A_n$ and elements g_1, $\ldots$, g_n of G such that f is given by $f(x) = g_i \cdot x$ if $x \in A_i$, for $i = 1, \ldots, n$.

It is easy to check that if $g \colon A \to B$ and $f \colon B \to C$ are bijections that are defined piecewise by the action of G on X (see the preceding paragraph), then $f \circ g \colon A \to C$ is also a piecewise defined bijection. Since the identity map (from a subset A of X to itself) is such a piecewise defined bijection, as are the inverses of such bijections, it follows that the relation of G-equidecomposability is an equivalence relation.

Recall the Schröder–Bernstein theorem from set theory: if the set A has the same cardinality as some subset of the set B, and if B has the same cardinality as some subset of A, then A and B have the same cardinality. In other words, if there is a

---

[1] This is perhaps not entirely standard terminology.

bijection from A onto a subset of B and a bijection from B onto a subset of A, then there is a bijection from A onto B (see A.7 in Appendix A).

The following proposition gives an analogous result for G-equidecomposability.

G.2. (Proposition) *Suppose that the group G acts on the set X and that A and B are subsets of X. If A is G-equidecomposable with a subset of B and if B is G-equidecomposable with a subset of A, then A and B are G-equidecomposable with one another.*

Proof. Suppose that A and B are as in the statement of the proposition. Then there are injections $f \colon A \to B$ and $g \colon B \to A$ that are defined piecewise by the action of G on X. Let us look at how elements of A and B arise as images of elements of B and A under the functions g and f. As is rather standard in proving versions of the Schröder–Bernstein theorem, we express this in terms of ancestors. Consider an element a of A. We call an element b of B a *parent* of a if $a = g(b)$, and an element a' of A a *grandparent* of a if $a = g(f(a'))$. We continue in this way, considering great-grandparents, ... We view the parents, grandparents, ..., as *ancestors*. In a similar way, we define the ancestors of the elements of B. For example, the ancestors of b are the elements of the sequence $f^{-1}(b)$, $g^{-1}(f^{-1}(b))$, $f^{-1}(g^{-1}(f^{-1}(b)))$, ... Since f and g are injective but not necessarily surjective, these sequences may be of any length, containing $0, 1, 2, \ldots$, or even infinitely many terms. Let us define subsets A_e, A_o, and A_∞ of A to be the sets of elements of A for which the corresponding sequence is of even length, of odd length, or infinitely long. We define subsets B_e, B_o, and B_∞ of B similarly. It is not difficult to check that f maps A_e onto B_o and A_∞ onto B_∞, and that g maps B_e onto A_o. It follows that we can define a bijection $h \colon A \to B$ by

$$h(x) = \begin{cases} f(x) & \text{if } x \in A_e \text{ or } x \in A_\infty, \text{ and} \\ g^{-1}(x) & \text{if } x \in A_o. \end{cases}$$

Since f and g are injective and defined piecewise by the action of G, h is also defined piecewise by the action of G, and the proof is complete.   □

Finally, here is a precise version of the Banach–Tarski paradox; we prove it below.

G.3. (Theorem—the Banach–Tarski paradox) *Let A and B be subsets of $\mathbb{R}^3$ that are bounded and have nonempty interiors, and let G_3 be the group of rigid motions discussed in Example G.1(c). Then A and B are G_3-equidecomposable.*

Note that the Banach–Tarski paradox says that if $\{A_i\}$ and $\{B_i\}$ are the sets into which A and B are decomposed, then each A_i can be mapped onto the corresponding set B_i using a rigid motion from G_3. It does not say that the pieces A_i into which A is decomposed can be moved along continuous paths, eventually becoming the corresponding pieces B_i and never colliding with the other pieces. It was long an open problem whether such a continuous decomposition is possible. However, Wilson [129] has recently proved that such decompositions are possible.

In particular, he proves that there are continuous maps $t \mapsto g_t^i$ from $[0, 1]$ to G_3 such that

(a) $g_0^i \cdot A_i = A_i$ for all i,
(b) $g_1^i \cdot A_i = B_i$ for all i, and
(c) $g_t^i \cdot A_i \cap g_t^j \cdot A_j = \varnothing$ for all t in $[0, 1]$ and all i and j for which $i \neq j$.

Paradoxical Sets

Suppose that the group G acts on the set X. A subset A of X is G-*paradoxical*, or simply *paradoxical*, if it is equal to $A_1 \cup A_2$ for some pair A_1, A_2 of disjoint subsets of A, each of which is G-equidecomposable with A.

The following consequence of the Schröder–Bernstein-like theorem above makes it slightly easier to prove that a set is paradoxical: we can show that a set A is paradoxical by producing disjoint subsets A_1 and A_2 of A that are equidecomposable with A; we do not need to check that $A = A_1 \cup A_2$.

G.4. (Corollary) *Suppose that the group G acts on the set X. A subset A of X is G-paradoxical if it includes disjoint subsets A_1 and A_2, each of which is G-equidecomposable with A.*

Proof. Suppose that A, A_1, and A_2 are as in the statement of the corollary. Then $A - A_1$ is equidecomposable with a subset of A (it *is* a subset of A), and A is equidecomposable with a subset of $A - A_1$, namely with A_2. Thus Proposition G.2 implies that A and $A - A_1$ are equidecomposable, and so A_1 and $A - A_1$ form the required partition of A. $\square$

It is a consequence of the Banach–Tarski paradox that

$$\text{the ball } \{x \in \mathbb{R}^3 : \|x\| \leq 1\} \text{ is } G_3\text{-paradoxical} \tag{1}$$

(if we divide the ball into two pieces by cutting it with a plane through the origin, then the Banach–Tarski paradox says that the ball is equidecomposable with each of the two pieces).

Let us check that we can also derive the Banach–Tarski paradox from (1). So suppose that (1) holds. Certainly if some closed ball is G_3-paradoxical, then so are all closed balls (two sets that are equidecomposable are still equidecomposable if they are translated or if both are scaled by the same constant). Let A and B be the sets in the statement of the Banach–Tarski paradox, let B_0 be a closed ball included in A, and let r be the radius of B_0. Let $B_1, B_2, \ldots$ be disjoint closed balls, each with radius r. Since B_0 is the union of a pair of disjoint sets, each of which is equidecomposable with B_0, it follows that B_0 is equidecomposable with $B_1 \cup B_2$. By repeating that argument we can conclude that B_0 is equidecomposable with $B_1 \cup B_2 \cup B_3$, and eventually that it is equidecomposable with $B_1 \cup B_2 \cup \cdots \cup B_n$ for an arbitrary n. Since the set B in the statement of the Banach–Tarski paradox is bounded, we can choose n large enough that B can be covered with n closed balls of radius r. This implies that B is equidecomposable with a subset of $B_1 \cup B_2 \cup \cdots \cup B_n$, and hence with a subset of B_0, which is itself a subset of A. A similar argument tells us that A is equidecomposable with a subset of B, and then Proposition G.2 implies that A and B are equidecomposable. Thus the Banach–Tarski paradox follows from (1).

We will prove the Banach–Tarski paradox by proving (1). We need to gather some more tools.

Generators and Free Groups

Let G be a group, let S be a set of elements of G, and let $S^{-1} = \{u \in G : u = v^{-1} \text{ for some } v \text{ in } S\}$. The smallest subgroup of G that includes S is called the subgroup *generated by* S. The subgroup of G generated by S has a more constructive description; namely it consists of the elements of G that are represented[2] by a *word* of the form

$$s_1 s_2 \cdots s_n,$$

where n is a nonnegative integer and $s_1, \ldots, s_n$ are elements of $S \cup S^{-1}$.

Now suppose that S generates G and that $S \cap S^{-1} = \varnothing$. Note that if $s \in S$, then the words ss^{-1}, $ss^{-1}ss^{-1}$, $ss^{-1}ss^{-1}ss^{-1}$, ... all represent the same element of G, namely e. Furthermore, a word can be modified by repeatedly removing substrings of the form ss^{-1} or $s^{-1}s$, where $s \in S$, without changing the element of G represented by the word. We can continue this process until we reach a word in which no element of S appears adjacent to its inverse. A word in which no element of S appears adjacent to its inverse is called a *reduced* word.

Let us continue to assume that $S \cap S^{-1} = \varnothing$. The group G is said to be *free* on S, or to be *freely generated* by S, if S generates G and each element of G can be represented in only one way by a reduced word over S. If G is free on S and if S has n elements, then one sometimes says that G is free on n generators.

G.5. (Proposition) *Let F be a free group on two generators. Then the set F is paradoxical under the action of the group F on it.*

Proof. Suppose that F is freely generated by σ and τ and that e is the identity element of F. Let F_σ be the set of all elements of F that can be represented with reduced words that begin with σ, and define $F_{\sigma^{-1}}$, F_τ, and $F_{\tau^{-1}}$ analogously. The sets $\{e\}$, F_σ, $F_{\sigma^{-1}}$, F_τ, and $F_{\tau^{-1}}$ then form a partition of the set F. We can check that F and $F_\sigma \cup F_{\sigma^{-1}}$ are F-equidecomposable by writing $F = F_\sigma \cup (\{e\} \cup F_{\sigma^{-1}} \cup F_\tau \cup F_{\tau^{-1}})$ and noting that $F_\sigma = e \cdot F_\sigma$ and $F_{\sigma^{-1}} = \sigma^{-1} \cdot (\{e\} \cup F_{\sigma^{-1}} \cup F_\tau \cup F_{\tau^{-1}})$. A similar argument shows that F is also F-equidecomposable with $F_\tau \cup F_{\tau^{-1}}$. Since F is F-

---

[2]The word $s_1 s_2 \cdots s_n$ is the sequence $\{s_i\}_{i=1}^n$, and the element of G represented by the word is the group-theoretic product of $s_1, s_2, \ldots, s_n$. The empty word, where $n = 0$, gives the identity element of G.

equidecomposable with $F_\sigma \cup F_{\sigma^{-1}}$ and with $F_\tau \cup F_{\tau^{-1}}$, it follows from Corollary G.4 that F is F-paradoxical. □

G.6. (Proposition) *The special orthogonal group $SO(3)$ has a subgroup that is free on two generators.*

Proof. Let us begin with the question of how we might check that suitably chosen elements σ and τ of $SO(3)$ freely generate a subgroup of $SO(3)$. We need to show that distinct reduced words w_1 and w_2 in σ, σ^{-1}, τ, and τ^{-1} represent distinct elements of $SO(3)$. So assume that w_1 and w_2 are distinct reduced words that represent the same element of $SO(3)$. We can assume that they do not begin (on the left) with the same element, since otherwise we can remove elements from the left until w_1 and w_2 no longer begin with equal elements (this does not change whether the elements of $SO(3)$ represented by w_1 and w_2 are equal or different). So we can assume that either w_1 and w_2 begin with different ones of σ, σ^{-1}, τ, and τ^{-1}, or else one of w_1 and w_2 is the empty word and the other is not. Our job is to choose σ and τ in such a way that we can conclude that the elements of G represented by such w_1 and w_2 are necessarily distinct.

Suppose that we can find an element u of $\mathbb{R}^3$, plus disjoint subsets S_+, S_-, T_+, and T_- of $\mathbb{R}^3$ (none of which contains u), such that operating on u by the element of G represented by a non-null reduced word w gives an element of S_+, S_-, T_+, or T_-, according as the left-hand element of w is σ, σ^{-1}, τ, or τ^{-1}. If we can find such an element u and sets S_+, S_-, T_+, and T_-, and if w_1 and w_2 are distinct reduced words as described in the preceding paragraph, then operating on u by the group elements represented by w_1 and w_2 will give different elements of $\mathbb{R}^3$, and we will have a proof that w_1 and w_2 represent different elements of $SO(3)$.

The argument just outlined will work if we can verify that our choices of σ, τ, u, S_+, S_-, T_+, and T_- (with the choices still to be made) satisfy

$$\sigma(S_+ \cup T_+ \cup T_- \cup \{u\}) \subseteq S_+,$$

$$\sigma^{-1}(S_- \cup T_+ \cup T_- \cup \{u\}) \subseteq S_-,$$

$$\tau(S_+ \cup S_- \cup T_+ \cup \{u\}) \subseteq T_+, \text{ and}$$

$$\tau^{-1}(S_+ \cup S_- \cup T_- \cup \{u\}) \subseteq T_-.$$

Now let us define elements σ and τ of $SO(3)$ by

$$\sigma = \begin{pmatrix} 3/5 & 4/5 & 0 \\ -4/5 & 3/5 & 0 \\ 0 & 0 & 1 \end{pmatrix} \text{ and } \tau = \begin{pmatrix} 1 & 0 & 0 \\ 0 & 3/5 & -4/5 \\ 0 & 4/5 & 3/5 \end{pmatrix},$$

an element u of $\mathbb{R}^3$ by $u = (0,1,0)^t$, and subsets S_+, S_-, T_+, and T_- of $\mathbb{R}^3$ by

$$S_+ = \{\frac{1}{5^k}(x,y,z)^t : k \geq 1, x = 3y \bmod 5, x \neq 0 \bmod 5, \text{ and } z = 0 \bmod 5\},$$

$$S_- = \{\tfrac{1}{5^k}(x,y,z)^t \; : \; k \geq 1, x = -3y \bmod 5, x \neq 0 \bmod 5, \text{ and } z = 0 \bmod 5\},$$

$$T_+ = \{\tfrac{1}{5^k}(x,y,z)^t \; : \; k \geq 1, z = 3y \bmod 5, z \neq 0 \bmod 5, \text{ and } x = 0 \bmod 5\}, \text{ and}$$

$$T_- = \{\tfrac{1}{5^k}(x,y,z)^t \; : \; k \geq 1, z = -3y \bmod 5, z \neq 0 \bmod 5, \text{ and } x = 0 \bmod 5\}$$

(in these definitions k, x, y, and z are integers; furthermore, the t's on the vectors here indicate transposes, and so we are dealing with column vectors, rather than with the row vectors that are listed). It is now a routine calculation, which is left to the reader, to show that the sets S_+, S_-, T_+, and T_- are disjoint, that they do not contain u, and that the inclusions specified above indeed hold. With that we have shown that σ and τ freely generate a subgroup of $SO(3)$, and the proof of the proposition is complete.

$\square$

Details for the Banach–Tarski Paradox

The following proposition will let us use the free group on two generators that we just constructed to get some paradoxical subsets of $\mathbb{R}^3$. It is here that the axiom of choice is used.

We will be using the fact that every element of $SO(3)$, when interpreted as an action on $\mathbb{R}^3$, is a rotation about a line through the origin,[3] and the fact that each such rotation is given by an element of $SO(3)$. For proofs of these results, see the exercises at the end of this appendix.

G.7. (Proposition) *Let G be a group for which the action of G on G is paradoxical, let $(g,x) \mapsto g \cdot x$ be an action of G on a set X, and suppose that this action has no nontrivial fixed points (in other words, suppose that if $g \cdot x = x$ holds for some g and x, then $g = e$). Then the action of G on X is paradoxical.*

Proof. Let x be an element of X, and let $o(x)$ be the orbit of x under the action of G. That is, $o(x) = \{g \cdot x : g \in G\}$. Define a relation $\sim$ on X by letting $x \sim y$ hold if and only if $y = g \cdot x$ for some g in G. It is easy to check that $\sim$ is an equivalence relation and that the equivalence classes of $\sim$ are the orbits of the action of G on X. Use the axiom of choice to create a set C that contains one point from each orbit. We'll use the set C to show that X is G-paradoxical.

Since G is G-paradoxical, there is a partition $A \cup B$ of G such that G is G-equidecomposable with A and with B. Then $X = G \cdot C$, and the sets $A \cdot C$ and $B \cdot C$ form a partition of X (to check the disjointness of $A \cdot C$ and $B \cdot C$, use the assumption that the action of G on X has no fixed points, together with the fact that C contains exactly one element from each equivalence class under $\sim$). Since G is

---

[3]The identity element of $SO(3)$ may seem to be an exception. However, its action on $\mathbb{R}^3$ can be viewed as a rotation through the angle 0 about an arbitrary line through the origin.

equidecomposable with A, we can choose a partition $G_1, G_2, \ldots, G_n$ of G, a partition $A_1, A_2, \ldots, A_n$ of A, and elements $g_1, g_2, \ldots, g_n$ of G such that $A_i = g_i \cdot G_i$ for each i. Then the sets $G_1 \cdot C, G_2 \cdot C, \ldots, G_n \cdot C$ form a partition of X, the sets $A_1 \cdot C, A_2 \cdot C,$ $\ldots, A_n \cdot C$ form a partition of $A \cdot C$, and $A_i \cdot C = g_i \cdot (G_i \cdot C)$ holds for each i. In other words, X and $A \cdot C$ are equidecomposable. A similar argument shows that X and $B \cdot C$ are equidecomposable, and so X is G-paradoxical. $\qquad\square$

Let S be the unit sphere $\{x \in \mathbb{R}^3 : \|x\| = 1\}$, and let B be the unit ball $\{x \in \mathbb{R}^3 : \|x\| \le 1\}$.

G.8. (Proposition) *Let F be a subgroup of $SO(3)$ that is free on two generators. Then there is a countable subset D of the sphere S such that $S - D$ is F-paradoxical and hence $SO(3)$-paradoxical.*

Proof. The elements of F, since they belong to $SO(3)$, are distance-preserving as operators on $\mathbb{R}^3$; hence we can view them as acting on the sphere S. Each element of F (other than the identity element) is a nontrivial rotation about a line through the origin (see the remarks just before the statement of Proposition G.7) and so has exactly two fixed points on S. Let D be the collection of all fixed points on S of elements of F other than e. Since the group F is countable, D is also countable.

The elements of F have no fixed points in $S - D$, and $S - D$ is closed under the action of elements of F (for if $x \in S - D$, $f \in F$, and $fx \in D$, then fx would be a fixed point of some nontrivial element f' of F, from which it would follow that $f^{-1}f'fx = x$ and hence that $f^{-1}f'f = e$, which contradicts the assumption that $f' \ne e$). It now follows from Proposition G.7 that $S - D$ is F-paradoxical. Since F is a subgroup of $SO(3)$, $S - D$ is also $SO(3)$-paradoxical. $\qquad\square$

G.9. (Proposition) *The sphere S is $SO(3)$-paradoxical.*

Proof. Let F be a subgroup of $SO(3)$ that is free on two generators, and let D be a countable subset of S such that $S - D$ is F-paradoxical (see Proposition G.8). We begin the proof by constructing an element ρ_0 of $SO(3)$ such that the sets $D, \rho_0(D),$ $\rho_0^2(D), \ldots$ are disjoint. First we choose as axis for ρ_0 a line L that passes through the origin but through none of the points in D. We can describe the nontrivial rotations with axis L in terms of values (i.e., angles) in the interval $(0, 2\pi)$. For each pair of points x, y in $S - D$ there is at most one rotation about L that takes x to y. Thus there are only countably many rotations ρ about L for which $D \cap \rho(D)$ is nonempty. A similar argument shows that for each n there are at most countably many rotations ρ for which $D \cap \rho^n(D)$ is nonempty. Since there are uncountably many rotations about L, we can choose a rotation ρ_0 such that for every n the sets D and $\rho_0^n(D)$ are disjoint. It follows that for all k and n the sets $\rho_0^k(D)$ and $\rho_0^{k+n}(D)$ are disjoint, and hence that the sequence $D, \rho_0(D), \rho_0^2(D), \ldots$ consists of disjoint sets.

Claim. The sets S and $S - D$ are $SO(3)$-equidecomposable.

Let $D^{1,\infty} = \cup_{i=1}^{\infty} \rho_0^i(D)$ and let $D^{0,\infty} = \cup_{i=0}^{\infty} \rho_0^i(D) = D \cup D^{1,\infty}$. Then $S = (S - D^{0,\infty}) \cup D^{0,\infty}$ and $S - D = (S - D^{0,\infty}) \cup D^{1,\infty}$. Since $D^{1,\infty} = \rho_0 \cdot D^{0,\infty}$, it follows that S and $S - D$ are $SO(3)$-equidecomposable, and the claim is established.

Since S and $S - D$ are equidecomposable, while $S - D$ is paradoxical, it follows from Corollary G.4 that S is paradoxical.   □

G.10. (Proposition) *The ball B with its center removed, $\{x \in \mathbb{R}^3 : 0 < \|x\| \le 1\}$, is $SO(3)$-paradoxical.*

Proof. For each subset E of S let $c(E)$ be the conical piece of the ball B defined by

$$c(E) = \{x \in \mathbb{R}^3 : x = ts \text{ for some } t \text{ in } (0, 1] \text{ and some } s \text{ in } E\}.$$

Thus, for example, $c(S)$ is the ball B with its center removed. We know from Proposition G.9 that the sphere S is $SO(3)$-paradoxical. If $S = C \cup D$ is a partition of S into sets that are $SO(3)$-equidecomposable with S, then $c(S) = c(C) \cup c(D)$ is a partition of $c(S)$ into sets that are $SO(3)$-equidecomposable with $c(S)$; to see this, for instance, in the case of $c(S)$ and $c(C)$, take a bijection $f : S \to C$ that is piecewise defined by the group action, and note that $tx \mapsto tf(x)$ gives a bijection from $c(S)$ to $c(C)$ that is piecewise defined by the group action. Since $c(S)$ is the ball with its center removed, the proof is complete.   □

Now we can complete the proof of (1) and hence of the Banach–Tarski paradox:

G.11. (Theorem) *The ball B is G_3-paradoxical, where G_3 is the group of isometries defined in Example G.1(c).*

Proof. Let L be a line in $\mathbb{R}^3$ that does not pass through the origin 0 but lies close enough to it that none of the rotations about L map 0 to a point outside the ball B (note that the rotations about L belong to G_3 but not to $SO(3)$). Let ρ_0 be a rotation about L through an angle θ, where $\theta/2\pi$ is irrational, in which case the points 0, $\rho_0(0)$, $\rho_0^2(0)$, ... are distinct. Let $D^0 = \{0\} \cup \{\rho_0^n(0) : n \ge 1\}$ and $D^1 = \{\rho_0^n(0) : n \ge 1\}$. Then $B = (B - D^0) \cup D^0$ and $B - \{0\} = (B - D^0) \cup D^1$, and we can modify the last part of the proof of Proposition G.9 to conclude first that B is G_3-equidecomposable with $B - \{0\}$ and then, since $B - \{0\}$ is $SO(3)$-paradoxical (Proposition G.10), that B is G_3-paradoxical.   □

Exercises

Some of the linear algebra needed for this section is developed in the following exercises. In particular, these exercises give a proof that the rotations of $\mathbb{R}^3$ about lines through the origin are exactly the actions on $\mathbb{R}^3$ induced by the elements of $SO(3)$.

1. Let V be a subspace of $\mathbb{R}^d$ (possibly equal to $\mathbb{R}^d$), let $\{e_i\}$ be an orthonormal basis[4] of V, and let A be the matrix of T with respect to $\{e_i\}$. Show that the conditions

   (i) $(Tx, Ty) = (x, y)$ holds for all x, y in V,
   (ii) A is an orthogonal matrix, and
   (iii) $A^t A = I$

   are equivalent. Thus we can call the operator T *orthogonal* if its matrix with respect to some (and also every) orthonormal basis of V is an orthogonal matrix.

2. Suppose that T is an orthogonal operator on $\mathbb{R}^3$.

   (a) Show that $\det(T)$ is 1 or -1.
   (b) Show that T has at least one real eigenvalue. (Hint: The characteristic polynomial of T is a cubic polynomial.)
   (c) Show that every real eigenvalue of T has absolute value 1.

3. Let T be an orthogonal operator on $\mathbb{R}^3$, let λ be a real eigenvalue of T, and let x be an eigenvector of T that corresponds to the eigenvalue λ.

   (a) Let $x^\perp$ be the set of all vectors y in $\mathbb{R}^3$ that are *orthogonal* to x (i.e., the set of all y such that $(x, y) = 0$). Show that $x^\perp$ is a linear subspace of $\mathbb{R}^3$ that is invariant under T, in the sense that $T(y) \in x^\perp$ whenever $y \in x^\perp$.
   (b) Let $T_{x^\perp}$ be the restriction of T to $x^\perp$. Show that the determinants of T and $T_{x^\perp}$ are related by $\det(T) = \lambda \det(T_{x^\perp})$.

4. (a) Let S be an orthogonal operator on $\mathbb{R}^2$, or on a two-dimensional subspace of $\mathbb{R}^3$, and suppose that $\det(S) = -1$. Show that 1 and -1 are both eigenvalues of S. (Hint: This can be proved using elementary calculations involving the matrix of S; no big theorems are needed.)
   (b) Use part (a) to show that if T is an orthogonal operator on $\mathbb{R}^3$ that has determinant 1 and has -1 among its eigenvalues, then the eigenvalues of T are -1 (with multiplicity 2) and 1 (with multiplicity 1).
   (c) Conclude that if T is an orthogonal operator on $\mathbb{R}^3$ that has determinant 1 and has -1 among its eigenvalues, then T is a rotation through an angle of π about some line through the origin.

5. (a) Let S be an orthogonal operator on $\mathbb{R}^2$, or on a two-dimensional subspace of $\mathbb{R}^3$, and suppose that $\det(S) = 1$. Show that for any orthonormal basis of the two-dimensional space, there are real numbers a and b such that $a^2 + b^2 = 1$ and such that the matrix of S with respect to that basis is $\begin{pmatrix} a & -b \\ b & a \end{pmatrix}$ and hence has the form $\begin{pmatrix} \cos\theta & -\sin\theta \\ \sin\theta & \cos\theta \end{pmatrix}$ for some real number θ.

---

[4] An *orthonormal basis* for a finite-dimensional inner product space V is a basis $\{e_i\}$ of V such that $(e_i, e_j) = 0$ if $i \neq j$ and $(e_i, e_j) = 1$ if $i = j$.

(b) Use part (a) to show that if T is an orthogonal operator on $\mathbb{R}^3$ that has determinant 1 and has 1 among its eigenvalues, then there is an orthonormal basis of $\mathbb{R}^3$ with respect to which T has matrix

$$
\begin{pmatrix}
1 & 0 & 0 \\
0 & \cos\theta & -\sin\theta \\
0 & \sin\theta & \cos\theta
\end{pmatrix},
$$

where θ is a real number. Conclude that T is a rotation through an angle of θ about some line through the origin.

6. The preceding exercises outline a proof that every matrix in $SO(3)$ gives a rotation of $\mathbb{R}^3$ about some line through the origin. Prove the converse: every rotation of $\mathbb{R}^3$ about a line through the origin corresponds to a matrix in $SO(3)$.

Notes

The fundamental paper by Banach and Tarski is [2]. The book by Wagon [122] is very thorough and rather up-to-date.

Appendix H
The Henstock–Kurzweil and McShane Integrals

In this appendix we look at the consequences of making what may seem to be a small change to the definition of the Riemann integral. The modified definition gives what is often called the Henstock–Kurzweil integral or the generalized Riemann integral. It will be easy to see that the Henstock–Kurzweil integral is an extension of the Riemann integral; we will see later that it is in fact also an extension of the Lebesgue integral.

Near the end of this appendix we look at another modification of the definition of the Riemann integral; this modification gives the McShane integral. We will see that the McShane integral turns out to be equivalent to the Lebesgue integral.

Most of the results in this appendix are presented as exercises, often with hints.

Let $[a,b]$ be a closed bounded interval. Recall (see Sect. 2.5) that a *partition* of $[a,b]$ is a finite sequence $\{a_i\}_{i=0}^k$ of real numbers such that

$$a = a_0 < a_1 < \cdots < a_k = b,$$

and that a *tagged partition* of $[a,b]$ is a partition of $[a,b]$, together with a sequence $\{x_i\}_{i=1}^k$ of real numbers (called *tags*) such that $a_{i-1} \leq x_i \leq a_i$ holds for each i (in other words, such that for each i the value x_i belongs to the interval $[a_{i-1}, a_i]$). We will often denote a partition or a tagged partition by a letter such as $\mathscr{P}$. Recall also that the *norm* or *mesh* of a partition or tagged partition $\mathscr{P}$, written $\|\mathscr{P}\|$, is defined by $\|\mathscr{P}\| = \max_i(a_i - a_{i-1})$.

Let f be a real-valued function on an interval $[a,b]$, and let $\mathscr{P}$ be a tagged partition of $[a,b]$. Recall that the *Riemann sum* $\mathscr{R}(f,\mathscr{P})$ corresponding to f and $\mathscr{P}$ is the weighted sum of values of f given by

$$\mathscr{R}(f,\mathscr{P}) = \sum_{i=1}^k f(x_i)(a_i - a_{i-1}).$$

We saw in Proposition 2.5.7 that the Riemann integral of f over the interval $[a,b]$ is the limit of Riemann sums $\mathscr{R}(f,\mathscr{P})$, where the limit is taken as the mesh of $\mathscr{P}$

D.L. Cohn, *Measure Theory: Second Edition*, Birkhäuser Advanced Texts Basler Lehrbücher, DOI 10.1007/978-1-4614-6956-8, © Springer Science+Business Media, LLC 2013

approaches 0. More precisely, f is *Riemann integrable*, with *integral L*, if and only if for every positive number ε there is a positive number δ such that

$$|\mathscr{R}(f,\mathscr{P}) - L| < \varepsilon \text{ holds for every } \mathscr{P} \text{ that satisfies } \|\mathscr{P}\| < \delta.$$

It seems plausible that it might be worthwhile to require some of the subintervals in a tagged partition $\mathscr{P}$ to be rather narrow (perhaps in regions where the function f is varying rapidly), while allowing other subintervals to be wider. This is what the Henstock–Kurzweil integral does; we turn to the details.

A real-valued function δ whose domain includes the interval $[a,b]$ is said to be a *gauge* on $[a,b]$ if it satisfies $\delta(x) > 0$ at each x in $[a,b]$. Given a gauge δ, a tagged partition $\mathscr{P}$ of $[a,b]$ is said to be δ-*fine*, or *subordinate to δ*, if

$$[a_{i-1},a_i] \subseteq (x_i - \delta(x_i), x_i + \delta(x_i))$$

holds for each i. So the subintervals in a δ-fine tagged partition $\mathscr{P}$ must be very short in the parts of $[a,b]$ where all the values of δ are close to 0, while the subintervals in other parts of $[a,b]$ can be longer.

Now consider a function $f\colon [a,b] \to \mathbb{R}$. Note that, in contrast to our discussion of the Riemann integral, we are *not* assuming that f is bounded, although we are for now still assuming that it is real-valued (and not $[-\infty, +\infty]$-valued). Then f is *Henstock–Kurzweil integrable* on $[a,b]$ if there is a number L such that for every positive number ε there is a gauge δ on $[a,b]$ such that

$$|\mathscr{R}(f,\mathscr{P}) - L| < \varepsilon \text{ holds for every } \delta\text{-fine tagged partition } \mathscr{P} \text{ of } [a,b].$$

The number L is called the *Henstock–Kurzweil integral* of f over the interval $[a,b]$ and is denoted by $(H)\int_a^b f$ or by $(H)\int_a^b f(x)\,dx$. In cases where there does not seem to be a significant chance of confusion, we may simply write $\int_a^b f$ or $\int_a^b f(x)\,dx$.

See Exercises 11 and 12 for some nontrivial examples of Henstock–Kurzweil integrable functions.

The preceding definition would not make sense if for some function f there were two values of L, each satisfying the definition of the integral of f. The following exercise gives the tool needed to check (in Exercise 2) that such pathology does not occur.

Exercises

1. Cousin's lemma says that if δ is a gauge on an interval $[a,b]$, then there is a δ-fine partition of $[a,b]$. Prove Cousin's lemma

   (a) with a bisection argument (if $[a,b]$ fails to have a δ-fine partition, then so does either its left half or its right half, ...), and

(b) by analyzing

$$\sup\{t \in [a,b] : \text{there is a } \delta\text{-fine partition of } [a,t]\}.$$

2. Show that the value of the Henstock–Kurzweil integral is well defined. That is, show that if f is Henstock–Kurzweil integrable and if L_1 and L_2 are real numbers, each of which satisfies the definition of the Henstock–Kurzweil integral of f, then $L_1 = L_2$. (Hint: Use Exercise 1.)

3. Show that if $f: [a,b] \to \mathbb{R}$ is Riemann integrable, then f is Henstock–Kurzweil integrable and $(H) \int_a^b f = (R) \int_a^b f$. (The proof can be very short.)

4. Show that the set of Henstock–Kurzweil integrable functions on $[a,b]$ is a vector space and that the Henstock–Kurzweil integral is a positive linear functional on it.

5. (Cauchy criterion for Henstock–Kurzweil integrability) Show that a function $f: [a,b] \to \mathbb{R}$ is Henstock–Kurzweil integrable if and only if for every positive number ε there is a gauge δ such that $|\mathscr{R}(f,\mathscr{P}_1) - \mathscr{R}(f,\mathscr{P}_2)| < \varepsilon$ holds whenever $\mathscr{P}_1$ and $\mathscr{P}_2$ are δ-fine tagged partitions of $[a,b]$.

6. Suppose that δ is a gauge on $[a,b]$ and that x is a point in $[a,b]$. Then there is a gauge δ' that satisfies $\delta' \leq \delta$ and is such that each δ'-fine tagged partition contains x as one of its tags. In many situations this allows us to force specified points to be tags in the partitions under consideration. (Hint: Use

$$\delta'(t) = \begin{cases} \min(\delta(t), |t-x|/2) & \text{if } t \neq x, \text{ and} \\ \delta(x) & \text{if } t = x \end{cases}$$

to define δ'.)

7. Suppose that δ is a gauge on $[a,b]$ and that $\mathscr{P}$ is a δ-fine tagged partition of $[a,b]$ that contains x among its tags. If x belongs to the interior of one of the subintervals of $\mathscr{P}$, say, $a_{i-1} < x < a_i$, and if we define a partition $\mathscr{P}'$ to contain the same intervals and tags as $\mathscr{P}$, except that the interval $[a_{i-1}, a_i]$ is replaced with the two intervals $[a_{i-1}, x]$ and $[x, a_i]$, with x serving as tag in each of these new intervals, then $\mathscr{P}'$ is also a δ-fine partition of $[a,b]$ and $\mathscr{R}(f, \mathscr{P}') = \mathscr{R}(f, \mathscr{P})$ holds for each function f on $[a,b]$.

8. Show that if $f: [a,b] \to \mathbb{R}$ is Henstock–Kurzweil integrable on $[a,b]$ and if $g: [a,b] \to \mathbb{R}$ agrees with f everywhere in $[a,b]$ except perhaps at a finite number of points, then g is Henstock–Kurzweil integrable on $[a,b]$ and $\int_a^b g = \int_a^b f$.

9. Show that if $a < c < b$ and if f is Henstock–Kurzweil integrable on $[a,c]$ and on $[c,b]$, then f is Henstock–Kurzweil integrable on $[a,b]$ and $\int_a^b f = \int_a^c f + \int_c^b f$. (Hint: Use Exercises 6 and 7.)

10. Show that if f is Henstock–Kurzweil integrable on $[a,b]$ and if $[c,d]$ is a subinterval of $[a,b]$, then f is Henstock–Kurzweil integrable on $[c,d]$.

11. Let $f\colon [0,1] \to \mathbb{R}$ be defined by

$$f(x) = \begin{cases} n & \text{if } x \in [1 - \frac{1}{2^{n-1}}, 1 - \frac{1}{2^n}), \ n = 1, 2, \ldots, \text{ and} \\ 0 & \text{if } x = 1. \end{cases}$$

Using only the definition and basic properties of the Henstock–Kurzweil integral (that is, without using deeper results, such as those given in Exercises 14 and 17), verify that f is Henstock–Kurzweil integrable on $[0,1]$ and that

$$\int_0^1 f\,dx = \sum_{n=1}^{\infty} \frac{n}{2^n}.$$

12. Let $f\colon [0,1] \to \mathbb{R}$ be the characteristic function of the set of rational numbers in $[0,1]$. Show that f is Henstock–Kurzweil integrable, with $\int_0^1 f$ equal to 0. (Hint: Let $\{r_n\}_1^{\infty}$ be an enumeration of the rationals in $[0,1]$. Given a positive value ε, define a gauge δ by letting $\delta(r_n) = \varepsilon/2^{n+1}$ for each n, while letting $\delta(x) = 1$ for all other values of x. Check that each δ-fine partition $\mathscr{P}$ satisfies $|\mathscr{R}(f, \mathscr{P})| < \varepsilon$.)

13. (a) Let $f\colon [a,b] \to \mathbb{R}$ be a function that vanishes almost everywhere. Show that f is Henstock–Kurzweil integrable, with $\int_a^b f$ equal to 0. (Hint: Suppose that $\varepsilon > 0$. For each positive integer n first define A_n by $A_n = \{x \in [a,b] : n-1 < |f(x)| \le n\}$ and then choose an open set U_n such that $A_n \subseteq U_n$ and $\lambda(U_n) < \varepsilon/n2^n$. Define a gauge δ by letting $\delta(x)$ be the distance from x to the complement of U_n if $x \in A_n$ and letting $\delta(x) = 1$ if $x \notin \cup_n A_n$. Find an upper bound for $|\mathscr{R}(f, \mathscr{P})|$ that is valid for all δ-fine partitions $\mathscr{P}$ of $[a,b]$.)

(b) Suppose that the functions $f,g\colon [a,b] \to \mathbb{R}$ agree almost everywhere and that f is Henstock–Kurzweil integrable. Show that g is Henstock–Kurzweil integrable and that $\int_a^b g = \int_a^b f$.

We can now define the Henstock–Kurzweil integral for $[-\infty, +\infty]$-valued functions: one calls a function $f\colon [a,b] \to [-\infty, +\infty]$ *Henstock–Kurzweil integrable* if there is a function $g\colon [a,b] \to \mathbb{R}$ that is Henstock–Kurzweil integrable and agrees with f almost everywhere. The *Henstock–Kurzweil integral* of f is then defined to be that of g. Exercise 13(b) implies that the resulting concepts of integrability and integral are well defined. One can deal in a similar way with the Henstock–Kurzweil integral for functions that are defined only almost everywhere.

A *tagged subpartition* of an interval $[a,b]$ is a finite indexed collection $\{[c_i,d_i]\}_{i=1}^{k}$ of nonoverlapping[1] subintervals of $[a,b]$, together with tags $\{x_i\}_{i=1}^{k}$ such that $x_i \in [c_i,d_i]$ holds for each i. So a tagged subpartition is like a tagged partition, except that the intervals involved may not cover the entire interval $[a,b]$. Note that with subpartitions we cannot do as we did with partitions and use a

---

[1] Let $\{I_i\}$ be an indexed collection of intervals. These intervals are *nonoverlapping* if for all i and j, the intersection $I_i \cap I_j$ contains at most one point.

sequence of division points $\{a_i\}$ to specify the subintervals, since now there may be gaps between the subintervals.

Let δ be a gauge on $[a,b]$. A tagged subpartition is said to be δ-*fine*, or *subordinate to* δ, if $[c_i,d_i] \subseteq (x_i - \delta(x_i), x_i + \delta(x_i))$ holds for each i. The *Riemann sum* associated to a function f and tagged subpartition $\mathscr{P}$ is, of course, defined by $\mathscr{R}(f,\mathscr{P}) = \sum_i f(x_i)(d_i - c_i)$.

The following result gives some useful estimates involving Riemann sums over subpartitions.

14. (Saks–Henstock lemma) Suppose that $f\colon [a,b] \to \mathbb{R}$ is Henstock–Kurzweil integrable, that ε is a positive number, and that δ is a gauge on $[a,b]$ such that every δ-fine tagged partition $\mathscr{P}$ of $[a,b]$ satisfies $|\mathscr{R}(f,\mathscr{P}) - (H)\int_a^b f| < \varepsilon$. Show that if $\mathscr{P}'$ is a δ-fine tagged subpartition of $[a,b]$, with subintervals $\{[c_i,d_i]\}$ and tags $\{x_i\}$, then

$$\left| \sum_i f(x_i)(d_i - c_i) - \sum_i (H)\int_{c_i}^{d_i} f \right| \le \varepsilon \qquad (1)$$

and

$$\sum_i \left| f(x_i)(d_i - c_i) - (H)\int_{c_i}^{d_i} f \right| \le 2\varepsilon. \qquad (2)$$

(Hint: Suppose that f, ε, δ, and $\mathscr{P}'$ are as specified above. Let $\{[g_j,h_j]\}$ be the closures of the maximal subintervals of $[a,b]$ that are disjoint from all the subintervals of $\mathscr{P}'$, and for each j choose a partition $\mathscr{P}_j$ of $[g_j,h_j]$ that is subordinate to δ and moreover is such that $\mathscr{R}(f,\mathscr{P}_j)$ is extremely close to $(H)\int_{g_j}^{h_j} f$. To prove (1), consider the partition of $[a,b]$ formed by combining $\mathscr{P}'$ and all the $\mathscr{P}_j$. What happens when the partitions $\mathscr{P}_j$ are made finer and finer? In order to derive (2) from (1), look at two subpartitions, one where the differences $f(x_i)(d_i - c_i) - (H)\int_{c_i}^{d_i} f$ are all positive, and one where they are all negative.)

15. Suppose that $f\colon [a,b] \to \mathbb{R}$ is Henstock–Kurzweil integrable and that $F\colon [a,b] \to \mathbb{R}$ is defined by $F(x) = \int_a^x f$. Show that F is continuous. (Hint: Use the Saks–Henstock lemma (Exercise 14) to show that given a positive ε and an element x_0 of $[a,b]$, we have $|F(x) - F(x_0) - f(x_0)(x - x_0)| < \varepsilon$ for all x sufficiently close to x_0.)

16.(a) Suppose that $f\colon [a,b) \to \mathbb{R}$ is Henstock–Kurzweil integrable on $[a,c]$ for each c in (a,b). Show that for each positive ε there is a positive function δ on $[a,b)$ such that for each c in (a,b) and each δ-fine partition $\mathscr{P}$ of $[a,c]$ we have $|\mathscr{R}(f,\mathscr{P}) - \int_a^c f| < \varepsilon$. (Hint: Let $\{a_n\}_1^\infty$ be a strictly increasing sequence such that $a_1 = a$ and $\lim_n a_n = b$. For each n choose a gauge δ_n on $[a_n,a_{n+1}]$ such that each δ_n-fine partition $\mathscr{P}$ of $[a_n,a_{n+1}]$ satisfies $|\mathscr{R}(f,\mathscr{P}) - \int_{a_n}^{a_{n+1}} f| < \varepsilon/2^n$. Form δ by combining the gauges δ_n, $n = 1$, $2, \ldots$, suitably. See Exercises 6, 7, and 14.)

(b) Show that if $f\colon [a,b] \to \mathbb{R}$ is Henstock–Kurzweil integrable on $[a,c]$ for each c in (a,b) and if $\lim_{c\to b} \int_a^c f$ exists, then f is Henstock–Kurzweil integrable on $[a,b]$ and $\int_a^b f = \lim_{c \to b} \int_a^c f$. Thus the improper Henstock–Kurzweil integral is no more general than the Henstock–Kurzweil integral. (Hint: By modifying f, if necessary, we can assume that $f(b) = 0$. Use the function δ from part (a) of this exercise in your proof.)

17. (The monotone convergence theorem) This exercise is devoted to a proof of the monotone convergence theorem for the Henstock–Kurzweil integral, which can be stated as follows: Suppose that f and $f_1, f_2, \ldots$ are $[-\infty, +\infty]$-valued functions on $[a,b]$ that are finite almost everywhere and satisfy

$$f_1(x) \le f_2(x) \le \cdots \tag{3}$$

and

$$f(x) = \lim_n f_n(x) \tag{4}$$

at almost every x in $[a,b]$. If each f_n is Henstock–Kurzweil integrable and if the sequence $\{(H)\int_a^b f_n\}$ is bounded above, then f is Henstock–Kurzweil integrable and $(H)\int_a^b f = \lim_n (H)\int_a^b f_n$.

(a) Check that for proving the monotone convergence theorem it is enough to consider the case where all the functions involved are $[0, +\infty)$-valued and relations (3) and (4) hold at *every* x in $[a,b]$.

(b) Prove the monotone convergence theorem. (Hint: Let L be the limit of the sequence $\{(H)\int_a^b f_n\}$. Here is a strategy for showing that f is integrable, with integral L: Let ε be a positive number, and for each n let δ_n be a gauge such that each δ_n-fine partition $\mathscr{P}$ satisfies $|\mathscr{R}(f_n, \mathscr{P}) - \int_a^b f_n| < \varepsilon/2^n$. For each x in $[a,b]$ let $n(x)$ be the smallest of those positive integers n that satisfy $\int f_n > L - \varepsilon$ and $f_n(x) > f(x) - \varepsilon$. Use the δ_n's to create a gauge δ by letting $\delta(x) = \delta_{n(x)}(x)$ for each x. Let $\mathscr{P}$ be a δ-fine partition, with division points $\{a_i\}$ and tags $\{x_i\}$. To bound $|\mathscr{R}(f, \mathscr{P}) - L|$, let m and M be the smallest and largest values of $n(x_i)$ as x_i ranges over the set of tags of $\mathscr{P}$, note that

$$\left| \sum f(x_i)(a_i - a_{i-1}) - \sum \int_{a_{i-1}}^{a_i} f_{n(x_i)} \right|$$
$$\le \left| \sum f(x_i)(a_i - a_{i-1}) - \sum f_{n(x_i)}(x_i)(a_i - a_{i-1}) \right|$$
$$+ \left| \sum f_{n(x_i)}(x_i)(a_i - a_{i-1}) - \sum \int_{a_{i-1}}^{a_i} f_{n(x_i)} \right|,$$

use the definition of δ and the Saks–Henstock lemma to verify that the right side of the formula displayed above is at most $(b-a)\varepsilon + \varepsilon$, and then note that $\sum \int_{a_{i-1}}^{a_i} f_{n(x_i)}$ lies between $\int_a^b f_m$ and $\int_a^b f_M$, both of which are close to L.)

18. The goal of this exercise is to prove that the Henstock–Kurzweil integral is an extension of the Lebesgue integral—that is, that

$$f \text{ is Henstock–Kurzweil integrable and } (H)\int f = (L)\int f \tag{5}$$

holds for each Lebesgue integrable function $f: [a,b] \to \mathbb{R}$.

(a) Show that (5) holds if f is the characteristic function of a Borel subset of $[a,b]$. (Hint: Use Theorem 1.6.2.)

(b) Show that (5) also holds if f is the characteristic function of a Lebesgue measurable subset of $[a,b]$.

(c) Show that (5) holds if f is a nonnegative Lebesgue integrable function on $[a,b]$. (Hint: Use the monotone convergence theorems for the Lebesgue and Henstock–Kurzweil integrals.)

(d) Finally, show that (5) holds if f is an arbitrary Lebesgue integrable function.

19. Suppose that $f: [a,b] \to \mathbb{R}$ is Henstock–Kurzweil integrable, and let $F: [a,b] \mapsto \mathbb{R}$ be its indefinite integral—that is, the function defined by $F(x) = \int_a^x f$ for each x in $[a,b]$. Then F is differentiable, with derivative given by $F'(x) = f(x)$, at almost every x in $[a,b]$. (Hint: Define D^+ by

$$D^+(x) = \limsup_{t \to x^+} \frac{F(t) - F(x)}{t - x}.$$

Let α and ε be positive numbers, and use the Vitali covering theorem and the Saks–Henstock lemma to show that if the set $\{x : D^+(x) > f(x) + \alpha\}$ is nonempty, then we can choose a sequence $\{[a_i, b_i]\}$ of disjoint intervals that cover it up to a Lebesgue null set and satisfy

$$\varepsilon > \sum_i (F(b_i) - F(a_i) - f(a_i)(b_i - a_i)) > \alpha \lambda^*(\{x : D^+(x) > f(x) + \alpha\}).$$

Conclude that $D^+ \leq f$ almost everywhere. Prove analogous results for lower limits and for limits from the left.)

20. Show that each Henstock–Kurzweil integrable function is Lebesgue measurable. (Hint: Use Exercises 15 and 19.)

21. Is every Henstock–Kurzweil integrable function Borel measurable?

22.(a) Show that the converse to part (c) of Exercise 18 also holds. Thus a nonnegative function is Henstock–Kurzweil integrable if and only if it is Lebesgue integrable. (Hint: Why was this not included as a part of Exercise 18, but delayed to this point?)

(b) Show that part (a) fails if the non-negativity condition is omitted. (Hint: Take a function on $[a,b]$ that has an improper Riemann integral but is not Lebesgue integrable.)

23. (A version of Theorem 6.3.11 for the Henstock–Kurzweil integral) Suppose that the function $F: [a,b] \to \mathbb{R}$ is continuous on $[a,b]$ and is differentiable at all but a countable collection of points in $[a,b]$. Then its derivative F' is Henstock–Kurzweil integrable on $[a,b]$, and

$$(H) \int_a^b F'(x)\,dx = F(b) - F(a).$$

(Hint: It is enough to deal with the function f that agrees with F' where F is differentiable and that vanishes elsewhere. Let $\{t_i\}$ be a sequence consisting of the points at which F is not differentiable. Suppose that $\varepsilon > 0$, and define δ on the points t_i by choosing positive values $\delta(t_i)$ that are so small that $\sum_i |(F(b_i) - F(a_i)| < \varepsilon$ whenever $\{[a_i, b_i]\}$ is a finite sequence of intervals such that $t_i \in [a_i, b_i]$ and $[a_i, b_i] \subseteq (t_i - \delta(t_i), t_i + \delta(t_i))$ hold for each i. Check that δ can be extended to a gauge (also called δ) on $[a, b]$ such that each δ-fine partition $\mathscr{P}$ of $[a, b]$ satisfies $|\mathscr{R}(f, \mathscr{P}) - (F(b) - F(a))| < 2\varepsilon$.)

The McShane integral is another generalization of the Riemann integral; its definition is given by a slight modification of the definition of the Henstock–Kurzweil integral.

Let us consider a generalization of the concept of a tagged partition in which the tags x_i are no longer required to belong to the corresponding intervals $[a_{i-1}, a_i]$. More precisely, a *freely tagged*[2] partition of $[a, b]$ is a partition $\{a_i\}_{i=0}^k$ of $[a, b]$, together with a sequence $\{x_i\}_{i=1}^k$ of real numbers (tags) such that $x_i \in [a, b]$ for each i; it is not required that $x_i \in [a_{i-1}, a_i]$. If δ is a gauge on $[a, b]$, then a δ-*fine* freely tagged partition is a freely tagged partition such that

$$[a_{i-1}, a_i] \subseteq (x_i - \delta(x_i), x_i + \delta(x_i))$$

holds for each i. Thus the subintervals in a δ-fine freely tagged partition are required to lie close to the corresponding tags, but are not required to contain the tags.

Note that every δ-fine tagged partition of $[a, b]$ is a δ-fine freely tagged partition of $[a, b]$, but that the converse does not hold. Note also that the δ-fine tagged partitions of $[a, b]$ are exactly the δ-fine freely tagged partitions of $[a, b]$ that are in fact tagged partitions.

Riemann sums are defined for freely tagged partitions just as they are for tagged partitions: if the freely tagged partition $\mathscr{P}$ has division points $\{a_i\}_{i=0}^k$ and tags $\{x_i\}_{i=1}^k$, then for a function $f : [a, b] \to \mathbb{R}$ we have $\mathscr{R}(f, \mathscr{P}) = \sum_{i=1}^k f(x_i)(a_i - a_{i-1})$.

A function $f : [a, b] \to \mathbb{R}$ is *McShane integrable* on $[a, b]$ if there is a number L such that for every positive number ε there is a gauge δ on $[a, b]$ such that

$$|\mathscr{R}(f, \mathscr{P}) - L| < \varepsilon \text{ holds for every } \delta\text{-fine freely tagged partition } \mathscr{P} \text{ of } [a, b];$$

the number L is then called the *McShane integral* of f over $[a, b]$. We will denote the McShane integral of f over the interval $[a, b]$ by $(M) \int_a^b f$ or $(M) \int_a^b f(x)\, dx$; in cases where there does not seem to be a significant chance of confusion, we may write simply $\int_a^b f$ or $\int_a^b f(x)\, dx$.

Arguments that show that the Henstock–Kurzweil integral is well defined (see Exercise 2) can also be used to show that the McShane integral is well defined.

---

[2] Another term for a freely tagged partition is a *free tagged* partition.

Furthermore, it is easy to see that the Henstock–Kurzweil integral is an extension of the McShane integral: since every δ-fine tagged partition is a δ-fine freely tagged partition, it follows that if L is a value such that $|\mathscr{R}(f, \mathscr{P}) - L| < \varepsilon$ holds for every δ-fine freely tagged partition, then this same inequality holds for every δ-fine tagged partition. We will soon see that the McShane integral is equivalent to the Lebesgue integral.

24. Show that the McShane integral is an extension of the Riemann integral: if $f: [a,b] \to \mathbb{R}$ is Riemann integrable, then f is McShane integrable and the McShane and Riemann integrals of f are equal. (Hint: Modify the proof of Proposition 2.5.7.)

25. Show that the set of McShane integrable functions on $[a,b]$ is a vector space and that the McShane integral is a positive linear functional on it (see Exercise 4).

26. Formulate and prove a Cauchy criterion for McShane integrability (see Exercise 5).

27. Show that Exercises 9 and 10, which relate integrals on an interval to integrals on its subintervals, also hold for the McShane integral.

28. Prove a version of Exercise 13 for the McShane integral. That is, prove that sets of Lebesgue measure zero behave as might be expected.

29. Formulate and prove the Saks–Henstock lemma (see Exercise 14) for the McShane integral (your new version should involve freely tagged partitions and subpartitions, and not just tagged ones).

30. Formulate and prove the monotone convergence theorem (see Exercise 17) for the McShane integral.

31. Show that a nonnegative function $f: [a,b] \to \mathbb{R}$ is McShane integrable if and only if it is Lebesgue integrable, and that in that case $(M) \int_a^b f = (L) \int_a^b f$. (Hint: Use ideas from Exercises 18 and 22.)

32. In this exercise, we prove that the McShane and Lebesgue integrals (for functions on $[a,b]$) are equivalent.

(a) Show that if $f: [a,b] \to \mathbb{R}$ is McShane integrable, then $|f|$ is also McShane integrable. (Hint: Use the Cauchy criterion for McShane integrability. Suppose that $\mathscr{P}_1$ and $\mathscr{P}_2$ are δ-fine freely tagged partitions of $[a,b]$, where $\mathscr{P}_1$ has subintervals[3] $\{I_i\}$ and tags $\{x_i\}$ and $\mathscr{P}_2$ has subintervals $\{J_j\}$ and tags $\{y_j\}$. We will consider freely tagged partitions $\mathscr{P}_3$ and $\mathscr{P}_4$ of $[a,b]$ whose subintervals are the nondegenerate intervals of the form $I_i \cap J_j$ and whose tags (where $\mathscr{P}_3$ has tags $\{u_{i,j}\}$ and $\mathscr{P}_4$ has tags $\{v_{i,j}\}$) are such that both $u_{i,j}$ and $v_{i,j}$ belong to the set $\{x_i, y_j\}$. Check that in such cases $\mathscr{P}_3$ and $\mathscr{P}_4$ are both δ-fine. Check also that for each i and j we can choose $u_{i,j}$ and $v_{i,j}$ such that

$$\bigl|\,|f(x_i)| - |f(y_j)|\,\bigr| \leq f(u_{i,j}) - f(v_{i,j}),$$

---

[3]Here we name the subintervals, rather than the division points, since we will also be considering partitions consisting of subintervals of the form $I_i \cap J_j$; we will need to relate $I_i \cap J_j$ to I_i and J_j, and this is awkward to do in terms of division points.

and that with this choice of $\mathscr{P}_3$ and $\mathscr{P}_4$ we have

$$|\mathscr{R}(|f|,\mathscr{P}_1) - \mathscr{R}(|f|,\mathscr{P}_2)| \leq \mathscr{R}(f,\mathscr{P}_3) - \mathscr{R}(f,\mathscr{P}_4).$$

Use this inequality to derive the Cauchy condition for $|f|$ from the Cauchy condition for f.)

(b) Show that if $f\colon [a,b] \to \mathbb{R}$ is McShane integrable, then f^+ and f^-, the positive and negative parts of f, are McShane integrable. (Hint: Express f^+ and f^- as simple algebraic expressions involving $|f|$ and f.)

(c) Conclude that the McShane integral is equivalent to the Lebesgue integral. In other words, an arbitrary function $f\colon [a,b] \to \mathbb{R}$ is McShane integrable if and only if it is Lebesgue integrable, and in that case $(M)\int_a^b f = (L)\int_a^b f$. (See Exercise 31.)

(d) Show that part (a) of this exercise cannot be extended to the Henstock–Kurzweil integral. That is, show by example that the Henstock–Kurzweil integrability of a function $f\colon [a,b] \to \mathbb{R}$ does not imply the Henstock–Kurzweil integrability of $|f|$. (Hint: Once again, consider a function on $[a,b]$ that has an improper Riemann integral but is not Lebesgue integrable.)

Notes

There are many books and papers on the Henstock–Kurzweil integral. Two standard and thorough ones are by Bartle [5] and Gordon [52]. See also the paper by Bongiorno [16] in the handbook edited by Pap [95].

References

1. Artin, M.: Algebra, 2nd edn. Addison-Wesley, Reading (2011)
2. Banach, S., Tarski, A.: Sur la decomposition des ensembles de points en parties respective-ment congruents. Fund. Math. **6**, 244–277 (1924)
3. Bartle, R.G.: The Elements of Integration. Wiley, New York (1966)
4. Bartle, R.G.: The Elements of Real Analysis, 2nd edn. Wiley, New York (1976)
5. Bartle, R.G.: A Modern Theory of Integration. American Mathematical Society, Providence (2001)
6. Benedetto, J.J., Czaja, W.: Integration and Modern Analysis. Birkhäuser, Boston (2012)
7. Berberian, S.K.: Measure and Integration. Macmillan, New York (1965). Reprinted by AMS Chelsea Publishing, 2011
8. Billingsley, P.: Probability and Measure. Wiley, New York (1979)
9. Birkhoff, G., MacLane, S.: A Survey of Modern Algebra, 4th edn. Macmillan, New York (1977). Reprinted by A.K. Peters, 1998
10. Blackwell, D.: A Borel set not containing a graph. Ann. Math. Statist. **39**, 1345–1347 (1968)
11. Blackwell, D.: On a class of probability spaces. In: Proceedings of the 3rd Berkeley Symposium on Mathematical Statistics and Probability, vol. II, pp. 1–6. University of California Press, Berkeley (1956)
12. Bledsoe, W.W., Morse, A.P.: Product measures. Trans. Amer. Math. Soc. **79**, 173–215 (1955)
13. Bledsoe, W.W., Wilks, C.E.: On Borel product measures. Pacific J. Math. **42**, 569–579 (1972)
14. Blumenthal, R.M., Getoor, R.K.: Markov Processes and Potential Theory. Pure and Applied Mathematics, vol. 29. Academic, New York (1968). Reprinted by Dover, 2007
15. Bogachev, V.I.: Measure Theory, 2 vols. Springer, Berlin (2007)
16. Bongiorno, B.: The Henstock–Kurzweil integral. In: Pap, E. (ed.) Handbook of Measure Theory, 2 vols, pp. 587–615. North Holland (Elsevier), Amsterdam (2002)
17. Bourbaki, N.: General Topology, Part 2. Addison-Wesley, Reading (1966)
18. Bourbaki, N.: Intégration, Chaps. 1–4, 2nd edn. Hermann, Paris (1965); Bourbaki, N.: Intégration, Chap. 5, 2nd edn. Hermann, Paris (1967); Bourbaki, N.: Intégration, Chap. 6, Hermann, Paris (1959); Bourbaki, N.: Intégration, Chaps. 7–8. Hermann, Paris (1963); Bourbaki, N.: Intégration, Chap. 9. Hermann, Paris (1969)
19. Bredon, G.E.: A new treatment of the Haar integral. Michigan Math. J. **10**, 365–373 (1963)
20. Breiman, L.: Probability. Addison-Wesley, Reading (1968). Reprinted by SIAM, 1992
21. Bruckner, A.M.: Differentiation of integrals. Herbert Ellsworth Slaught Memorial Papers. Amer. Math. Monthly **78** (Suppl.) (1971)
22. Bruckner, A.M.: Differentiation of Real Functions. Lecture Notes in Mathematics, vol. 659. Springer, Berlin (1978)

D.L. Cohn, *Measure Theory: Second Edition*, Birkhäuser Advanced
Texts Basler Lehrbücher, DOI 10.1007/978-1-4614-6956-8,
© Springer Science+Business Media, LLC 2013

23. Bruckner, A.M., Bruckner, J.B., Thomson, B.S.: Real Analysis, 2nd edn. ClassicalRealAnalysis.com (2008)
24. Cartan, H.: Sur la mesure de Haar. C. R. Acad. Sci. Paris **211**, 759–762 (1940)
25. Cartier, P.: Processus aléatoires généralisés. In: Séminaire Bourbaki, 1963–1964, exposé 272. Benjamin, New York (1966)
26. Castaing, C., Valadier, M.: Convex Analysis and Measurable Multifunctions. Lecture Notes in Mathematics, vol. 580. Springer, Berlin (1977)
27. Chatterji, S.D.: Differentiation along algebras. Manuscripta Math. **4**, 213–224 (1971)
28. Choquet, G.: Theory of capacities. Ann. Inst. Fourier (Grenoble) **5**, 131–295 (1953–1954)
29. Christensen, J.P.R.: Topology and Borel Structure. North-Holland Mathematics Studies, vol. 10. North-Holland, Amsterdam (1974)
30. Cohen, P.J.: Set Theory and the Continuum Hypothesis. Benjamin, New York (1966)
31. Conway, J.B.: A Course in Functional Analysis, 2nd edn. Springer, New York (1990)
32. Daniell, P.J.: A general form of integral. Ann. of Math. (2) **19**, 279–294 (1917–1918)
33. de Guzmán, M.: Differentiation of Integrals in $\mathbb{R}^n$. Lecture Notes in Mathematics, vol. 481. Springer, Berlin (1975)
34. de Leeuw, K.: The Fubini theorem and convolution formula for regular measures. Math. Scand. **11**, 117–122 (1962)
35. Dellacherie, C.: Quelques exemples familiers, en probabilités, d'ensembles analytiques non boréliens. In: Séminaire de Probabilités XII. Lecture Notes in Mathematics, vol. 649, pp. 746–756. Springer, Berlin (1978)
36. Dellacherie, C.: Une démonstration du théorème de Souslin-Lusin. In: Séminaire de Probabilités VII. Lecture Notes in Mathematics, vol. 321, pp. 48–50. Springer, Berlin (1973)
37. Diestel, J., Uhl Jr., J.J.: Vector Measures. Mathematical Surveys, Number 15. American Mathematical Society, Providence (1977)
38. Doob, J.L.: Stochastic Processes. Wiley, New York (1953)
39. Dudley, R.M.: On measurability over product spaces. Bull. Amer. Math. Soc. **77**, 271–274 (1971)
40. Dudley, R.M.: Real Analysis and Probability, 2nd edn. Cambridge University Press, Cambridge (2002)
41. Dudley, R.M.: Uniform Central Limit Theorems. Cambridge University Press, Cambridge (1999)
42. Dunford, N., Schwartz, J.T.: Linear Operators. Part I: General Theory. Pure and Applied Mathematics, vol. VII. Interscience, New York (1958)
43. Dynkin, E.B.: Die Grundlagen der Theorie der Markoffschen Prozesse. Die Grundlehren der mathematischen Wissenschaften, Band 108. Springer, Berlin (1961)
44. Federer, H.: Geometric Measure Theory. Die Grundlehren der mathematischen Wissenschaften, Band 153. Springer, New York (1969)
45. Folland, G.B.: Real Analysis: Modern Techniques and Their Applications, 2nd edn. Wiley, New York (1999)
46. Fremlin, D.H.: Measure Theory, 5 vols. www.essex.ac.uk/maths/people/fremlin/mt.htm
47. Fremlin, D.H.: Topological measure spaces: two counter-examples. Math. Proc. Cambridge Philos. Soc. **78**, 95–106 (1975)
48. Gelbaum, B.R., Olmsted, J.M.H.: Counterexamples in Analysis. Holden-Day, San Francisco (1964). Reprinted by Dover, 2003
49. Gleason, A.M.: Fundamentals of Abstract Analysis. Addison-Wesley, Reading (1966). Reprinted by A.K. Peters, 1992
50. Gödel, K.: The Consistency of the Axiom of Choice and of the Generalized Continuum-Hypothesis with the Axioms of Set Theory. Annals of Mathematics Studies, vol. 3. Princeton University Press, Princeton (1940)
51. Godfrey, M.C., Sion, M.: On products of Radón measures. Canad. Math. Bull. **12**, 427–444 (1969)
52. Gordon, R.A.: The Integrals of Lebesgue, Denjoy, Perron, and Henstock. American Mathematical Society, Providence (1994)

53. Halmos, P.R.: Finite-Dimensional Vector Spaces, 2nd edn. Van Nostrand, Princeton (1958). Reprinted by Springer, 1974
54. Halmos, P.R.: Measure Theory. Van Nostrand, Princeton (1950). Reprinted by Springer, 1974
55. Halmos, P.R.: Naive Set Theory. Van Nostrand, Princeton (1960). Reprinted by Springer, 1974
56. Hayes, C.A., Pauc, C.Y.: Derivation and Martingales. Ergebnisse der Mathematik und ihrer Grenzgebiete, Band 49. Springer, Berlin (1970)
57. Herstein, I.N.: Topics in Algebra. Blaisdell, New York (1964)
58. Hewitt, E., Ross, K.A.: Abstract Harmonic Analysis I. Die Grundlehren der Mathematischen Wissenschaften, Band 115. Springer, Berlin (1963)
59. Hewitt, E., Stromberg, K.: Real and Abstract Analysis. Springer, New York (1965)
60. Hoffman, K.: Analysis in Euclidean Space. Prentice-Hall, Englewood Cliffs (1975). Reprinted by Dover, 2007
61. Hoffman, K.M., Kunze, R.: Linear Algebra. Prentice-Hall, Englewood Cliffs (1971)
62. Hoffmann-Jørgensen, J.: The Theory of Analytic Spaces. Various Publications Series, No. 10. Aarhus Universitet, Matematisk Institut, Aarhus (1970)
63. Hrbacek, K., Jech, T.: Introduction to Set Theory. Monographs and Textbooks in Pure and Applied Mathematics, vol. 45. Marcel Dekker, New York (1978)
64. Ionescu Tulcea, A., Ionescu Tulcea, C.: On the lifting property (I). J. Math. Anal. Appl. **3**, 537–546 (1961)
65. Ionescu Tulcea, A., Ionescu Tulcea, C.: Topics in the Theory of Lifting. Ergebnisse der Mathematik und ihrer Grenzgebiete, Band 48. Springer, Berlin (1969)
66. Jacobs, K.: Measure and Integral. Academic, New York (1978)
67. Kakutani, S.: Concrete representation of abstract (M)-spaces (a characterization of the space of continuous functions). Ann. of Math. (2) **42**, 994–1024 (1941)
68. Kechris, A.S.: Classical Descriptive Set Theory. Springer, New York (1995)
69. Kelley, J.L.: General Topology. Van Nostrand, Princeton (1955). Reprinted by Springer, 1975
70. Kindler, J.: A simple proof of the Daniell–Stone representation theorem. Amer. Math. Monthly **90**, 396–397 (1983)
71. Klenke, A.: Probability Theory. Springer, London (2008)
72. Kolmogorov, A.N.: Grundbegriffe der Wahrscheinlichkeitsrechnung. Springer, Berlin (1933); Kolmogorov, A.N.: Foundations of the Theory of Probability. Chelsea, New York (1956)
73. Kolmogorov, A.N., Fomin, S.V.: Introductory Real Analysis. Prentice-Hall, Englewood Cliffs (1970). Reprinted by Dover, 1975
74. Kölzow, D.: Differentiation von Massen. Lecture Notes in Mathematics, vol. 65. Springer, Berlin (1968)
75. Krantz, S.G., Parks, H.R.: Geometric Integration Theory. Birkhäuser, Boston (2008)
76. Krickeberg, K.: Probability Theory. Addison-Wesley, Reading (1965)
77. Kuratowski, K.: Topology, vol. 1. Academic, New York (1966)
78. Kuratowski, K., Mostowski, A.: Set Theory. Studies in Logic and the Foundations of Mathematics, vol. 86. North-Holland, Amsterdam (1976)
79. Lamperti, J.: Probability. W.A. Benjamin, New York (1966)
80. Lang, S.: Algebra. Addison-Wesley, Reading (1965)
81. Lang, S.: Analysis I. Addison-Wesley, Reading (1968)
82. Lax, P.: Functional Analysis, Wiley-Interscience, New York (2002)
83. Ljapunow, A.A., Stschegolkow, E.A., Arsenin, W.J.: Arbeiten zur deskriptiven Mengenlehre. VEB Deutscher Verlag der Wissenschaften, Berlin (1955)
84. Loomis, L.H.: An Introduction to Abstract Harmonic Analysis. Van Nostrand, Princeton (1953). Reprinted by Dover, 2011
85. Loomis, L.H., Sternberg, S.: Advanced Calculus. Addison-Wesley, Reading (1968)
86. Mackey, G.W.: Borel structure in groups and their duals. Trans. Amer. Math. Soc. **85**, 134–165 (1957)
87. Maharam, D.: On a theorem of von Neumann. Proc. Amer. Math. Soc. **9**, 978–994 (1958)

88. Mazurkiewicz, S.: Über die Menge der differenzierbaren Funktionen. Fund. Math. **27**, 244–249 (1936)
89. Morgan, F.: Geometric Measure Theory: A Beginner's Guide. Academic, San Diego (2000)
90. Moschovakis, Y.N.: Notes on Set Theory, 2nd edn. Springer, New York (2006)
91. Munkres, J.R.: Topology: A First Course. Prentice-Hall, Englewood Cliffs (1975)
92. Munroe, M.E.: Measure and Integration, 2nd edn. Addison-Wesley, Reading (1971)
93. Nachbin, L.: The Haar Integral. Van Nostrand, Princeton (1965)
94. Novikoff, P.: Sur les fonctions implicites mesurables B. Fund. Math. **17**, 8–25 (1931)
95. Pap, E. (ed.): Handbook of Measure Theory, 2 vols. North Holland (Elsevier), Amsterdam (2002)
96. Parthasarathy, K.R.: Probability Measures on Metric Spaces. Probability and Mathematical Statistics, vol. 3. Academic, New York (1967). Reprinted by AMS Chelsea Publishing, 2005
97. Pollard, D.: A User's Guide to Measure Theoretic Probability. Cambridge University Press, Cambridge (2002)
98. Pontryagin, L.S.: Topological Groups, 2nd edn. Gordon and Breach, New York (1966)
99. Riesz, F., Sz-Nagy, B.: Functional Analysis. Ungar, New York (1955). Reprinted by Dover, 1990
100. Rogers, C.A.: Hausdorff Measures. Cambridge University Press, Cambridge (1970)
101. Rogers, C.A. (ed.): Analytic Sets. Academic, London (1980)
102. Royden, H.L.: Real Analysis, 2nd edn. Macmillan, New York (1968)
103. Royden, H.L., Fitzpatrick, P.: Real Analysis, 4th edn. Pearson, Upper Saddle River (2010)
104. Rudin, W.: Principles of Mathematical Analysis, 3rd edn. McGraw-Hill, New York (1976)
105. Rudin, W.: Real and Complex Analysis, 2nd edn. McGraw-Hill, New York (1974)
106. Saks, S.: Theory of the Integral, 2nd revised edn. Dover, New York (1964). Reprint of 2nd revised edition, 1937
107. Schwartz, L.: Radon Measures on Arbitrary Topological Spaces and Cylindrical Measures. Oxford University Press, London (1973)
108. Segal, I.E., Kunze, R.A.: Integrals and Operators, 2nd edn. Die Grundlehren der mathematischen Wissenschaften, Band 228. Springer, Berlin (1978). First edition published by McGraw-Hill in 1968
109. Simmons, G.F.: Introduction to Topology and Modern Analysis. McGraw-Hill, New York (1963)
110. Solovay, R.M.: A model of set-theory in which every set of reals is Lebesgue measurable. Ann. of Math. (2) **92**, 1–56 (1970)
111. Spivak, M.: Calculus. Benjamin, New York (1967)
112. Srivastava, S.M.: A Course on Borel Sets. Springer, New York (1998)
113. Stone, A.H.: Analytic sets in non-separable metric spaces. In: Rogers, C.A. (ed.) Analytic Sets, pp. 471–480. Academic, London (1980)
114. Stone, M.H.: Notes on integration. Proc. Nat. Acad. Sci. U.S.A. **34**, 336–342, 447–455, 483–490 (1948); Stone, M.H.: Notes on integration. Proc. Nat. Acad. Sci. U.S.A. **35**, 50–58 (1949)
115. Strauss, W., Macheras, N.D., Musiał, K.: Liftings. In: Pap, E. (ed.) Handbook of Measure Theory, 2 vols, pp. 1131–1183. North Holland (Elsevier), Amsterdam (2002)
116. Taylor, A.E.: General Theory of Functions and Integration. Blaisdell, Waltham (1965). Reprinted by Dover, 1985
117. Thomson, B.S., Bruckner, J.B., Bruckner, A.M.: Elementary Real Analysis, 2nd edn. ClassicalRealAnalysis.com (2008)
118. van Dalen, D., Doets, H.C., de Swart, H.: Sets: Naive, Axiomatic and Applied. International Series in Pure and Applied Mathematics, vol. 106. Pergamon, Oxford (1978)
119. von Neumann, J.: Algebraische Repräsentanten der Funktionen bis auf eine Menge von Masse Null. J. Reine Angew. Math. **165**, 109–115 (1931)
120. von Neumann, J.: On rings of operators III. Ann. of Math. (2) **41**, 94–161 (1940)
121. Wagner, D.H.: Survey of measurable selection theorems. SIAM. J. Control Optim. **15**, 859–903 (1977)

122. Wagon, S.: The Banach–Tarski Paradox. Cambridge University Press, Cambridge (1984)
123. Walker, P.L.: On Lebesgue integrable derivatives. Amer. Math. Monthly **84**, 287–288 (1977)
124. Walsh, J.B.: Knowing the Odds: An Introduction to Probability. American Mathematical Society, Providence (2012)
125. Walter, W.: A counterexample in connection with Egorov's theorem. Amer. Math. Monthly **84**, 118–119 (1977)
126. Weil, A.: L'intégration dans les groupes topologiques et ses applications, 2nd edn. Hermann, Paris (1965)
127. Wheeden, R.L., Zygmund, A.: Measure and Integral. Monographs and Textbooks in Pure and Applied Mathematics, vol. 43. Marcel Dekker, New York (1977)
128. Williams, D.: Probability with Martingales. Cambridge University Press, Cambridge (1991)
129. Wilson, T.M.: A continuous movement version of the Banach–Tarski paradox: a solution to De Groot's problem. J. Symbolic Logic **70**, 946–952 (2005)
130. Zaanen, A.C.: An Introduction to the Theory of Integration. North-Holland, Amsterdam (1958)

Index of notation

D.L. Cohn, *Measure Theory: Second Edition*, Birkhäuser Advanced
Texts Basler Lehrbücher, DOI 10.1007/978-1-4614-6956-8,
© Springer Science+Business Media, LLC 2013

Index

A

$\mathscr{A}$-measurable
  function, 42
  set, 2
a.e., 50
a.e.$[\mu]$, 50
a.s., 319
absolute continuity, 122
  for functions from $\mathbb{R}$ to $\mathbb{R}$, 135
  for signed and complex measures, 125
  uniform, 129
absolutely continuous part of a measure, 130
absolutely convergent series, 88
act, 417
action, 417
adapted, 345
algebra, 300
  Banach, 300
  of functions, 392
  of sets, 1
algebraic dual space, 106
almost everywhere, 50
almost everywhere differentiability
  of finite Borel measures, 167
  of functions of finite variation, 171
  of monotone functions, 171
almost surely, 319
analytic
  measurable space, 270
  set, 248
    measurability, 262
    that is not a Borel set, 254
ancestor, 419
approximate identity, 305
atom of a σ-algebra, 272
axiom of choice, 27, 377

B

Baire
  category theorem, 395
  measure, 197
  set, 197
  σ-algebra, 197, 226
Banach
  algebra, 300
  space, 87
Banach–Tarski paradox, 417, 419
base
  for a family of neighborhoods, 280
  for a topological space, 390
basis
  Hamel, 30
Beppo Levi's theorem, 62
Bernoulli distribution, 315
bijection, bijective function, 375
binary expansion, 315–316, 382
binomial distribution, 318
Blackwell's theorem on analytic measurable
     spaces, 272
Bochner
  integrable function, 399
  integral, 399
Borel
  function, 42
  isomorphism, 259
  measurability of the image of a Borel set
     under an injective Borel function, 260
  measurable function, 42, 189, 397
  measure, 11, 189
  product, regular, 222
  σ-algebra, 4, 189
  subsets, 4, 189
Borel–Cantelli lemmas, 320

D.L. Cohn, *Measure Theory: Second Edition*, Birkhäuser Advanced
Texts Basler Lehrbücher, DOI 10.1007/978-1-4614-6956-8,
© Springer Science+Business Media, LLC 2013

Printed in the United States
By Bookmasters